Heat Transfer in Single and Multiphase Systems

Mechanical Engineering Series

Frank Kreith - Series Editor

Published Titles

Distributed Generation: The Power Paradigm for the New Millennium
Anne-Marie Borbely & Jan F. Kreider
Elastoplasticity Theory
Vlado A. Lubarda
Energy Audit of Building Systems: An Engineering Approach
Moncef Krarti
Entropy Generation Minimization
Adrian Bejan
Finite Element Method Using MATLAB, 2nd Edition
Young W. Kwon & Hyochoong Bang
Fluid Power Circuits and Controls: Fundamentals and Applications
John S. Cundiff
Fundamentals of Environmental Discharge Modeling
Lorin R. Davis
Introductory Finite Element Method
Chandrakant S. Desai & Tribikram Kundu
Intelligent Transportation Systems: New Principles and Architectures
Sumit Ghosh & Tony Lee
Mathematical & Physical Modeling of Materials Processing Operations
Olusegun Johnson Ileghus, Manabu Iguchi & Walter E. Wahnsiedler
Mechanics of Composite Materials
Autar K. Kaw
Mechanics of Fatigue
Vladimir V. Bolotin
Mechanism Design: Enumeration of Kinematic Structures According to Function
Lung-Wen Tsai
Nonlinear Analysis of Structures
M. Sathyamoorthy
Practical Inverse Analysis in Engineering
David M. Trujillo & Henry R. Busby
Principles of Solid Mechanics
Rowland Richards, Jr.
Thermodynamics for Engineers
Kau-Fui Wong
Viscoelastic Solids
Roderic S. Lakes
Heat Transfer in Single and Multiphase Systems
Greg F. Naterer

Forthcoming Titles

Engineering Experimentation
Euan Somerscales
Mechanics of Solids & Shells
Gerald Wempner & Demosthenes Talaslidis

Heat Transfer in Single and Multiphase Systems

Greg F. Naterer

CRC Press is an imprint of the
Taylor & Francis Group, an **informa** business

CRC Press
Taylor & Francis Group
6000 Broken Sound Parkway NW, Suite 300
Boca Raton, FL 33487-2742

First issued in paperback 2019

ISBN-13: 978-0-8493-1032-4 (hbk)
ISBN-13: 978-0-367-39586-5 (pbk)
Library of Congress Card Number 2002025922

Library of Congress Cataloging-in-Publication Data

Greg F. Naterer. Heat transfer in single and multiphase systems
p. cm.
Includes bibliographical references and index.
ISBN 0-8493-1032-6
1. Heat--Transmission. 2. Multiphase flow.

TJ260 .H39454 2002
621.402′2--dc21 2002025922

Visit the Taylor & Francis Web site at
http://www.taylorandfrancis.com

and the CRC Press Web site at
http://www.crcpress.com

To my wife Josie,

our children Jordan and Julia,

and my Mother and Father

Preface

Heat transfer is a major part of many engineering technologies. From power generation and energy storage to materials processing, aircraft de-icing, and many other applications, the mechanisms of heat transfer contribute substantially to various technological advances. A primary motivation for writing this book comes from discovering the need for a single source of material to cover each mode of multiphase heat transfer, as well as the fundamentals of heat transfer. Traditionally, these topics have usually been segregated into various sources or focused on a specific part of multiphase systems. As a result, analogies and a unified framework with a common nomenclature could not be established in a single source, despite the similarities between each mode of multiphase heat transfer.

This book has been developed based on course notes from both undergraduate and graduate-level courses. It includes material from various engineering projects at both levels. Thus, this book can serve both introductory and follow-up courses in heat transfer (such as advanced topic courses, i.e., multiphase flows with heat transfer), as well as graduate-level heat transfer. It should properly follow a first course in fluid mechanics. The student is expected to have knowledge of vector calculus and differential equations.

The text is organized into five main parts: (i) introduction (Chapter 1), (ii) single phase heat transfer (Chapters 2 to 4), (iii) multiphase heat transfer (Chapters 5 to 9), (iv) heat exchangers (Chapter 10), and (v) computational heat transfer (Chapter 11). The introduction provides the reader with fundamentals of heat transfer. The modes of single phase heat transfer, including conduction, convection, and radiation, are covered in the second part. Then, the reader may focus on all multiphase systems (Chapters 5 to 9) or on any particular system, such as liquid–solid systems (Chapter 8), without a loss of continuity. Finally, heat exchangers and numerical heat transfer are presented in the fourth and fifth parts, respectively.

During the past several years, numerous colleagues and students have contributed in significant ways to the development and preparation of the material in this book. I would like to express my sincere gratitude for this valuable input, particularly to the following reviewers: Dr. D.W. Fraser (University of Manitoba), Dr. J.A. Camberos (U.S. Air Force Research Laboratory), Dr. J.A. Esfahani (Ferdowsi University of Mashhad), Dr. D.C. Roach (University of New Brunswick), R. Xu, M. Milanez, P.S. Glockner, O.B. Adeyinka, and Dr. J.T. Bartley (University of Manitoba).

Author

G.F. Naterer is an associate professor in the Department of Mechanical and Industrial Engineering at the University of Manitoba, Winnipeg, Canada. He has written numerous publications on heat transfer, multiphase flows, Second Law Analysis, and fluid dynamics. He is a member of the AIAA, CSME, ASME, and APEGM, where he has been involved with international conferences as a session chairman, as a committee member, as an editorial board member (CSME Transactions) and in other roles. He is currently serving on the Thermophysics Technical Committee of AIAA, as well as on the education and international activities committees therein. Dr. Naterer received his Ph.D. in mechanical engineering from the University of Waterloo, Ontario, Canada, in 1995.

Author

G.F. Naterer is an associate professor in the Department of Mechanical and Industrial Engineering at the University of Manitoba, Winnipeg, Canada. He has written numerous publications on heat transfer, multiphase flows, second law analysis, and fluid dynamics. He is a member of the AIAA, CSME, ASME, and AFEOM, where he has been involved with international conferences as a session chairman, as a conference member, as an editorial board member (CSME Transactions), and in other roles. He is currently serving on the Thermophysics Technical Committee of AIAA, as well as on the education and international affairs committees therein. Dr. Naterer received his PhD in mechanical engineering from the University of Waterloo, Ontario, Canada in 1995.

Contents

List of Symbols

A	Area, m^2
AF	Air–fuel ratio
B_0	Boiling number
Bi	Biot number
c	Speed of light, m/sec; capacitance matrix
c_d	Drag coefficient
c_f	Friction coefficient
c_v, c_p	Specific heats, kJ/kgK
C	Concentration
C_k	Volume fraction (phase k)
D	Mass diffusivity, m^2/sec; diameter, m
e	Total specific energy, kJ/kg
$\hat{e}$	Internal energy, kJ/kg
E	Total energy, kJ; emissive power, W/m^2
Ec	Eckert number
f	Friction factor
f_b	Bubble volume fraction
f_k	Fraction of phase k
f_v	Void fraction
F	Blackbody function; radiation view factor
F'	Collector efficiency factor
F_r	Heat removal factor
Fr	Froude number
g	Gravitational acceleration, m^2/sec
G	Gibbs free energy, kJ/kg; irradiation, W/m^2
Gr	Grashof number
h	Convection coefficient, W/m^2K
h_{fg}	Latent heat of vaporization, kJ/kg
H	Bed height, m
i, j	Unit vectors
I	Intensity of radiation, W
j	Colburn factor
J	Mass diffusive flux, kg/m^2sec
k	Turbulent kinetic energy, m^2/sec^2
Le	Lewis number
$\dot{m}$	Mass flow rate
mf	Mass fraction
M	Figure of Merit
n	Surface normal
N	Shape function; number of tubes
$\dot{N}''_i$	Molar flux of constituent i, kmol/sec m^2
nel	Number of elements
NTU	Number of transfer units
Nu	Nusselt number
p	Pressure, Pa
P	Production; perimeter, m^2
Pr	Prandtl number
q	Heat flow rate
q''	Heat flux, W/m^2
r	Radial position, m
R	Thermal resistance, K/W; residual
Ra	Rayleigh number
Re	Reynolds number
s	Entropy, kJ/kgK
S	Shape factor, m; source; surface area, m^2
Sc	Schmidt number
Sh	Sherwood number
St	Stanton number
Ste	Stefan number
STP	Temperature of 25°C, pressure of 1 atm
t	Time, sec
T	Characteristic time period, sec
u, v	x and y direction velocities, m/sec
U	Freestream or reference velocity, m/sec
v'''	Specific volume, m^3/kg
v	Fluid velocity vector, m/sec
w	Complex coordinate $(u + iv)$
W	Work, kJ; width, m; weight function

K	Permeability, m^2
Kn	Knudsen number
l	Length, m
L	Latent heat of fusion, kJ/kg; length, m
$\mathscr{L}$	Operator
We	Weber number
x, y	Cartesian coordinates
X	Phase interface position
X_{tt}	Martinelli parameter
y_i	Mole fraction

Greek letters

α	Thermal diffusivity, m^2/sec; absorptivity
β	Thermal expansion coefficient, 1/K
γ	Surface tension, N/m
Γ	Mass flow rate per unit width, kg/msec
δ	Thickness, m; distance, m
ε	Emissivity; heat exchanger effectiveness; fluidized bed porosity; perturbation parameter
η	Efficiency; similarity variable
θ	Angle, rads; dimensionless temperature
κ	Boltzmann's constant (1.38×10^{-23} J/K)
λ	Wavelength, m; mean free path, m
μ	Dynamic viscosity, kg/msec
μ_b	Bingham viscosity, kg/msec
v	Kinematic viscosity, m^2/sec; frequency, 1/sec
ρ	Fluid density, kg/m^3; reflectivity
σ	Normal stress, N/m^2; Stefan–Boltzmann constant (5.67×10^{-8} W/m^2K); yield stress, N/m^2
ϕ	Angle, rads; relative humidity; velocity potential
Φ	Dissipation function, $1/sec^2$
ψ	Stream function
τ	Shear stress, N/m^2; transmissivity
ω	Solid angle, sr
χ	Flow weighting factor for permeability
ζ	Local coordinate

Subscripts

a	Air; ambient
A, B	Constituents A and B in a binary mixture
b	Base
bndry	Boundary
c	Cross section; collector; cold
civ	Civic
conv	Convection
crit	Critical
D	Diameter
e	East; mean beam length; eutectic
eff	Effective
f	Fluid; fin; fusion (phase change point)
g	Gas; ground; glass
gen	Generation
h	Hot; high; horizontal
i	Inner; interface; initial; surface index
j	Surface index
k	Phase number
ki	Kinematic
l, L	Low; laminar; liquid
loc	Local
m	Mixing length; mean; melting point
mf	Minimum fluidization
n	North
nb	Neighboring
o	Outer
p	Point p; particle
r	Reference; relative
rad	Radiation
s	Surface; entropy; solar; solid; south
sat	Saturated
sg	Superficial gas
sl	Superficial liquid; solid–liquid (together)
sub	Subcooled
t	Thermal; turbulent
tp	Two-phase
v	Vapor
w	Wall; wick; water; water vapor; west
x, y, z	Cartesian coordinates
1, 2	Node numbers

1

Introduction

For many centuries, heat was interpreted as an invisible form of matter called caloric. This caloric view persisted until about 1840, when the British physicist James Joule showed that heat was not a material substance, but rather a form of energy. Heat is not a property; instead, it is a mechanism to transfer energy across the boundaries of a system. The fundamental modes of heat transfer are conduction, convection, and radiation. Conduction heat transfer occurs from one part of a body to another, or between bodies in contact, without any movement on a macroscopic level. On the other hand, convection occurs when heat is transferred between a solid surface and a fluid region, or between different fluid regions, due to bulk fluid movement. In the case of forced convection, external processes (such as pressure-induced forces) drive the fluid motion; these external effects may result from devices such as pumps, fans, or atmospheric winds. In free convection, buoyancy (rather than external forces) drives the fluid motion. Finally, radiative heat transfer results from the emission of electromagnetic waves (or photons, i.e., packets of energy) by all surfaces above absolute zero. These processes will be described in detail in individual chapters devoted to each mode of heat transfer.

Furthermore, multiphase systems, such as gas–liquid and liquid–solid flows, arise in many engineering and scientific applications. For example, predicting and controlling the operation of condensers in an effective manner requires an understanding of transport phenomena in gas–liquid flows with phase change. In material processing, such as extrusion or casting, the liquid metal flow during solidification plays a significant role in final material properties (i.e., tensile strength), due to the resulting alignment of grain boundaries, shrinkage flows, etc. In this book, various solution techniques and engineering applications dealing with multiphase systems will be considered.

A common challenge in thermal engineering is finding ways to reduce (or increase) the heat transfer to a minimum (or maximum) value. For example, consider liquid metal solidification in a turbine blade casting (see Figure 1.1). Several complex processes arise during the solidification process, i.e., shrinkage flow at the phase interface, thermal–solutal convection with turbulence in the bulk liquid, and radiative heat transfer. An important aspect of this example is proper thermal control so that grain boundaries are aligned parallel to the blade axes during solidification. In this way, the material can effectively resist conditions of maximum stress during turbine operation, thereby improving system efficiency. Another common challenge facing the thermal engineer is how to achieve a specified heat transfer rate as efficiently and economically as possible. For example, this may include a method of cooling electronic circuits that is more effective than conventional cooling with a fan.

In many practical situations, heat transfer is complicated due to the interaction of various nonlinear and coupled processes. For example, consider ice formation on an aircraft. This ice formation increases drag and weight, and it presents serious danger to air safety. It can damage downstream structures whenever the ice breaks off, and ingested ice can damage the jet engine. In this instance, several heating and cooling modes arise, such as dissipative heating, convection, internal heating, latent heat of fusion, etc. In this example and others, the achievement of various technological goals is highly dependent on a full understanding of the heat transfer processes.

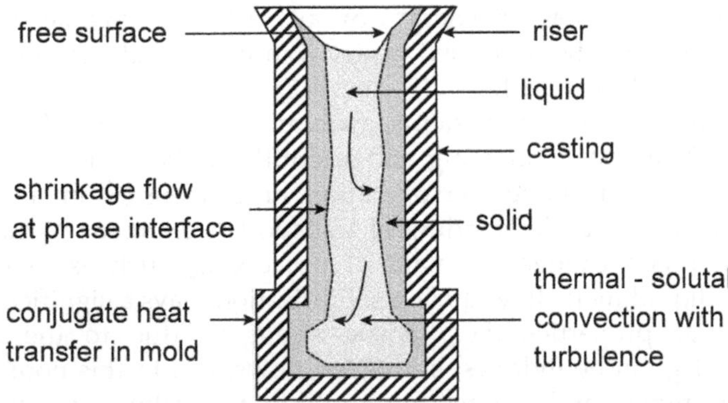

FIGURE 1.1
Example of solid–liquid phase change in turbine blade casting.

1.1 Vector and Tensor Notations

A brief overview of vectors and tensors is given here. A vector is a directed line segment with a specified magnitude and direction. It will usually be denoted by boldface letters in this book, i.e., $\boldsymbol{v}$ refers to the fluid velocity vector. A unit vector is a vector of unit magnitude. For example, $\boldsymbol{i}$ and $\boldsymbol{j}$ refer to the unit vectors in the x and y directions, i.e., (1, 0) and (0, 1), respectively. The symbol $|\boldsymbol{v}|$ designates the magnitude of the indicated vector.

The dot product between two vectors can be represented by a summation of their respective individual components. For example, if $\mathbf{u} = u_x\mathbf{i} + u_y\mathbf{j}$ and $\mathbf{v} = v_x\mathbf{i} + v_y\mathbf{j}$, then

$$\mathbf{u} \cdot \mathbf{v} = u_x v_x + u_y v_y \tag{1.1}$$

In an analogous way, matrices are contracted when their individual entries are multiplied with each other and summed. For example, if

$$\mathbf{A} = \begin{bmatrix} a_{11} & a_{12} \\ a_{21} & a_{22} \end{bmatrix}; \quad \mathbf{B} = \begin{bmatrix} b_{11} & b_{12} \\ b_{21} & b_{22} \end{bmatrix} \tag{1.2}$$

then

$$\mathbf{A} : \mathbf{B} = a_{11}b_{11} + a_{12}b_{12} + a_{21}b_{21} + a_{22}b_{22} \tag{1.3}$$

Tensors represent generalized notations for scalars (rank of zero), vectors (rank of 1), matrices (rank of 2), and so on. A tensor is denoted by a variable with appropriate subscripts. For example, a_{ij} represents the previously described matrix, where the range of subscripts is $i = 1, 2$ and $j = 1, 2$. When tensors use indices in this way, the notation is called *indicial notation*. The *summation convention* of tensors requires that repetition of an index in a term denotes a summation with respect to that index over its range. For example, in the previously cited case (dot product) involving two vectors,

$$u_i v_i = u_1 v_1 + u_2 v_2 \tag{1.4}$$

The range of the index is a set of specified integer values, such as $i = 1, 2$ in the previous equation. A *dummy index* refers to an index that is summed, whereas a *free index* is not summed. The rank of the tensor is increased for each index that is not repeated. For example, a_{ij} contains two nonrepeating indices, thereby indicating a tensor of rank 2 (i.e., matrix). Further details describing the operations of vectors and tensors can be found in standard undergraduate calculus or continuum mechanics textbooks.

1.2 Fundamental Concepts and Definitions

Both microscopic and macroscopic processes affect the transfer of heat. Consider the structural differences between solids, liquids, and gases, where the latter two states may be considered fluids. In solids, atomic bonds exist to create a compact structure in the material. There are large intermolecular forces binding the material together. Solids typically resist shear and compression forces, and they are self-supporting. Solid materials can be broadly characterized as metals, ceramics, or polymers. An important part of understanding various heat transfer processes, particularly during phase change, involves the atomic structures of solids and fluids. The main characteristics of these atomic structures will be described briefly.

Ceramics are compounds based predominantly on ionic bonding. Some common examples of ceramics are brick and porcelain. Ceramic phase diagrams have similar configurations as metal–metal systems. Metals usually exhibit less complex crystal structures than ceramics, and less energy is required to dislocate atoms in their atomic structure. Metals typically have lower yield stresses and lower hardness than ceramics. Ceramics are harder, but usually more brittle and more difficult to plastically deform than metals.

On the other hand, polymers are organic in nature and their atomic structure involves covalent bonding. Common examples of polymers are hydrocarbons, such as C_2H_4 (ethylene), plastics, rubbers, and CH_4 (methane). Polymers are utilized in coatings, adhesives, films, foam, and many other applications. Polymers are neither as strong nor as stiff as metals and ceramics. They form as molecular chains. Properties such as phase change temperature and material strength depend on their degree of crystallinity and the ability of the molecules to resist molecular chain motion. Unlike discrete phase change in metals, a continuous phase change between liquid and solid phases is observed in polymers.

The crystal structure of polymers usually involves *spherulites*. Spherulites are analogous to grains in metal formation. The extremities of spherulites impinge on one another to form linear boundaries in polymer materials. A region of high crystallinity is formed by thin layers called *lamellae* (typically of the order of 10 μm in length). *Amorphous* regions refer to regions within the polymer that are noncrystalline. In terms of thermophysical properties, the densities of ceramic materials are typically larger than those of polymers, but less than those of metals. Metals usually have melting temperatures higher than those of polymers but less than those of ceramics.

Also, the thermal conductivity of polymers is usually about two orders of magnitude less than that in metals and ceramics.

Unlike solids, there is a molecular freedom of movement with no particular structure in liquids. Also, common experience suggests that liquids need a container for storage purposes. Liquids cannot resist imposed shear stresses. However, they can resist compression. These characteristics show some clear differences between solids and liquids from a microscopic point of view. Some materials, such as slurries, tar, toothpaste, and snow, exhibit multiple characters. For example, tar resists shear at small stresses, but it flows at high stresses. The study of these hybrid materials is the subject of *rheology*.

In order to determine macroscopic properties, such as density and thermal conductivity, instead of microscopic properties (i.e., spatial distribution of molecules), it is often assumed that the fluid is a continuous medium. This assumption is called the *continuum assumption*. The following example describes how the continuum assumption applies to the definition of density.

Example 1.2.1
Density.
Consider the definition of density: mass divided by volume. In this definition, we need to choose a volume large enough such that the density is properly defined. In particular, the mass of molecules is considered to be distributed uniformly across the volume. The number of molecules within the volume can vary whenever the volume size approaches the scale of the mean free path. If the volume size is less than the mean free path, then significant variations in density can arise due to molecular fluctuations. In other words, molecules can fluctuate randomly in and out of a selected control volume. On the other hand, if the volume is large on a macroscopic scale, then variations associated with the spatial density distribution would be observed.

As a result, there is a specific range to be defined as the proper volume size for the continuum assumption to be applied. The control volume size is greater than the mean free path dimensions, but less than the characteristic macroscopic dimensions, to properly define the local fluid density. Continuum assumptions apply well to fluids beyond a minimum of 10^{12} molecules/mm^3. The continuum assumption becomes invalid in certain circumstances, such as atmospheric reentry of a spacecraft at high altitudes or other examples involving extreme gradients in velocity or temperature.

In this book, we will apply other fundamental concepts. In particular, we will refer to terms such as control volumes, control surfaces, and steady state. A control volume will refer to an open system consisting of a fixed region in space, and its boundary surfaces will be called control surfaces. Also, steady state will refer to conditions independent of time, i.e., negligible changes in the problem variables over time.

1.3 Eulerian and Lagrangian Descriptions

Different techniques (called *Eulerian* and *Lagrangian techniques*) may be used for descriptions of a fluid flow. In the Eulerian approach, various flow quantities are observed from a fixed location in space, whereas individual fluid particles are tracked in a Lagrangian method. For example, consider a gas particle trajectory in a heated duct. If a thermocouple is placed in the duct, then the temperature varies according to the selected position, as well as time. We have selected a fixed location, corresponding to the Eulerian approach.

On the other hand, if individual gas particles are tracked, then a Lagrangian approach is adopted. In this case, the temperature of a specific particle is a function of time alone along its particular trajectory. Since it is often impractical to try to trace all particle trajectories within a flow, the Eulerian approach will generally be adopted throughout this text. However, it should be noted that in some applications, such as free surface studies (i.e., tagged particles following the ocean currents in climate studies), a Lagrangian description may be more useful.

Mathematical expressions are considered separately in both approaches. For example, consider the definition of fluid acceleration. In the Lagrangian method, a specific fluid particle is selected and tracked over a time interval. In this case, the particle velocity has a functional dependence on both time and trajectory (i.e., spatial coordinates of the path that also vary with time). Now we can find the acceleration by taking the rate of change of velocity with respect to time. The results would show that the acceleration consists of a local part, due to changes in time, as well as a convective part, arising from changes in the particle's location. In other words, we can write the total rate of change of a scalar quantity, B, using

$$\frac{DB}{Dt} = \frac{\partial B}{\partial t} + \mathbf{v} \cdot \nabla B \tag{1.5}$$

where B refers to velocity in our example of the acceleration field. In general terms, Equation (1.5) uses the notation of DB/Dt to refer to the total (or material, or substantial) derivative of B. The latter term in Equation (1.5) gives the component of ∇B tangent to the trajectory.

If we use an Eulerian approach instead, we would then consider a stationary control volume and examine the change of velocity across the edges of this volume, as well as the changes in time. If we construct these differences and take appropriate limits as the grid spacing and time step become small, we obtain the same result as in Equation (1.5) for the total derivative and acceleration field. In other words, both Eulerian and Lagrangian descriptions provide the same mathematical expressions for the flow quantity in this case. In general, transport processes would have to be considered on an individual basis when converting between Lagrangian and Eulerian formulations.

1.4 Properties of a System

There are essentially four different types of properties of a system: thermodynamic, kinematic, transport, and other. In this section, we will discuss some primary fundamentals of these properties.

1.4.1 Thermodynamic Properties

Firstly, we have thermodynamic properties, such as pressure, density, temperature, and entropy. Density (mass per unit volume), ρ, is the reciprocal of specific volume, v'''. The fluid enthalpy, h, is defined by

$$h = \hat{e} + pv''' \tag{1.6}$$

where $\hat{e}$ and p refer to internal energy and pressure, respectively. Intensive properties are independent of size, whereas extensive properties (such as the total energy) are dependent on the system size. From the state postulate of thermodynamics for a simple compressible substance, we have the result that the number of intensive independent properties of a system must equal the number of relevant reversible work modes plus 1. We add 1 because even if we hold all properties constant within a system, we can still change one property, such as temperature, through heat transfer.

An important property of a system is pressure. Pressure, p, is the normal force per unit area acting on a fluid particle. Pressure arises from a momentum flux of fluid particles across a surface due to the motion of molecules. Consider the force applied to a surface on a wall as a result of molecular motion and momentum exchange on the other side of the wall.

For example, the impulse of a specific fluid particle is the change in momentum between some initial point and the final state (zero velocity at the wall) as the molecule contacts the side of the wall. Summing over many molecules and taking the average normal velocity of all molecules, we can obtain an expression for the average force per unit area exerted by the molecules on the wall. In performing these calculations, we obtain the ideal gas equation of state relating pressure to density and temperature. Thus, we can interpret pressure as an average force divided by area exerted by fluid particles in a flow at a particular point in space.

The pressure field acts normal to a surface. Also, pressure is a scalar field whose magnitude adjusts to conserve mass. For example, consider an air gap in a window cell with a buoyant internal flow arising from differential heating of both sides in the cavity. Due to buoyancy, warm air ascends near the hot boundary until it reaches the top corner. All fluid cannot only ascend, since mass would not be conserved. As a result, an adverse pressure gradient (i.e., increasing pressure in the flow direction) occurs as the fluid ascends, thereby causing the airflow to change directions and properly conserve mass.

Another important thermodynamic property is energy. For our purposes, energy refers to the sum of internal, kinetic, and potential energy parts. From the first law of thermodynamics, we know that total energy is a conserved quantity. Temperature is a scalar function that characterizes the internal energy of the system. Energy is often converted from one form to another form in engineering systems.

Example 1.4.1

Energy Exchange in a Falling Object.

As an example, consider the exchange of energy as we lift an object and then drop it. Initially, we must perform work in lifting the object from one height to another higher elevation. The work performed in this exchange is merely the change in potential energy between the two elevations. In other words, work has been mainly transferred into potential energy of the block at the higher elevation.

Now we drop the object. Since total energy is conserved, the sum of potential and kinetic energies must remain the same as the object hits the floor. We find that potential energy has now been transferred to kinetic energy as the block or object hits the floor. At this point, the total energy is still conserved, yet the potential energy and kinetic energy have been diminished. From experience, we know that the kinetic energy has been transferred into internal energy (i.e., small temperature rise), and possibly structural vibrations of the surface and noise. In practice, these energy exchanges occur regularly in the transport of fluid particles throughout a system. We can interpret the motion of a fluid particle during fluid flow analogously to the discrete falling object.

From the first law of thermodynamics, there are two ways to change energy: through work and heat transfer. Work and heat are the forms that energy takes in order to cross the boundaries of a system. A force alone does not change energy. From a visible or macroscopic scale, we can interpret work to be a mechanism to change the potential and kinetic energy of a system. On the other hand, internal energy changes occur at a microscopic, or subvisible, scale, due to heat transfer. In a certain way, heat transfer corresponds to work at a microscopic, or subvisible, scale.

In a manner similar to that in mass and energy, every system above a temperature of absolute zero has microscopic disorder, or entropy. Entropy represents an uncertainty about the system's microscopic state. It gives a quantitative value for randomization or disorder at the molecular level. Unlike mass and energy, entropy can be produced, but it can never be destroyed for an isolated system. The Second Law of Thermodynamics requires that the entropy of a system, including its surroundings (i.e., isolated system), never decrease. In other words, conservation of mass and energy requires that mass and total energy are not produced, whereas the second law states that entropy production is greater than or equal to zero (note: zero entropy production for reversible processes).

Example 1.4.2

Entropy Change in a Freezing Process.

It appears that less microscopic disorder arises in a liquid when it freezes. Does entropy decrease during a freezing process when a container with water is placed in a refrigerator?

When the substance freezes, its entropy decreases. However, the combined entropy of the system and its surroundings (room air) is not reduced. Thus, the Second Law of Thermodynamics is not violated.

In a perfect crystal of a pure substance at absolute zero temperature, the molecules are completely motionless and they are stacked precisely in accordance with their crystal structure. Since entropy is a measure of microscopic disorder, the entropy at zero absolute temperature equals zero, i.e., there is no uncertainty about the crystal's microscopic state (called the third law of thermodynamics). Entropy per unit mass, s, or total (extensive) entropy, S, can be interpreted from a statistical viewpoint, i.e.,

$$S = -\kappa \sum_i \Omega_i ln \Omega_i \tag{1.7}$$

where Ω_i refers to the probability of quantum state i and κ is Boltzmann's constant (1.38×10^{-23} J/K). For a given system, an increase in microscopic disorder (i.e., entropy increase) results in a loss of ability to perform useful work. Consider an example with a restrained expansion of an ideal gas from a half-cavity into an adjacent evacuated side of the remaining half-cavity. A partition initially divides the two sides of the entire cavity. When the partition is removed, the total energy remains constant, as well as temperature, since one side was initially evacuated (i.e., vacuum). During this process, there is a loss of ability to perform useful work. If the first side had a piston instead of a partition, the initial state could perform work, whereas now the final state cannot perform work since the gas has expanded into both sections. In the final state, there is also less certainty about a particle's location because it moves randomly within the entire volume, rather than only within the partial volume during the initial state. Thus, the entropy has increased, while a loss of ability to perform useful work occurred.

Total entropy is an extensive property with dimensions of energy divided by temperature. In other words, the total entropy of a combined system containing parts A and B is given by the sum of entropies of parts A and B. Although entropy, s, cannot be measured directly, it can be evaluated through the following Gibbs equation for a simple compressible substance:

$$Tds = d\hat{e} + pdv''' \tag{1.8}$$

where $\hat{e}$ and v''' refer to the internal energy and specific volume, respectively.

Entropy is produced during any irreversible process. Examples of irreversible processes are friction and heat transfer over a finite temperature difference. We consider them irreversible because it is highly unlikely (i.e., very low probability from a statistical viewpoint) that random molecular motion in the form of heat can order itself by its own means (without external effects) and convert itself into organized, macroscopic energy, such as kinetic energy. In other words, heat created during friction is unlikely to reorganize itself through its own means and convert to a form of organized work, such as a moving block.

Another thermodynamic property of importance, particularly in the equilibrium of multiphase systems, is the *Gibbs free energy*, G, defined by

$$G = H - TS \tag{1.9}$$

where H, T, and S refer to enthalpy (extensive), temperature, and entropy (extensive), respectively. Also, the chemical potential, μ, is defined as the rate of change of the Gibbs free energy with respect to a change in the number of moles in the mixture, n, i.e., for component A in a multicomponent mixture:

$$\mu_A = \frac{\partial G_A}{\partial n_A}\bigg|_{T,p} \tag{1.10}$$

where the derivative is evaluated at constant pressure and temperature. Thermodynamic equilibrium is identified by the condition that $dG = 0$, i.e., spatial uniformity of the Gibbs free energy.

1.4.2 Kinematic Properties

The second type of property of a system refers to a kinematic property. Examples of kinematic properties are fluid velocity and acceleration, which are governed by the conservation principles of mass and momentum. The detailed analysis of fluid motion is essential to the mechanisms of convective heat transfer (described in Chapter 3), as well as multiphase flows (Chapters 5 to 9).

1.4.3 Transport Properties

Thirdly, we often require transport properties in the system, such as thermal conductivity, diffusion coefficient, fluid viscosity, and specific heat. Examples of thermal conductivities of various solids and gases are shown in Figure 1.2. Also, specific heats and other thermophysical properties are given in the appendices. For example, specific heat is a thermophysical property defined as $c_v = d\hat{e}/dT$, which can be interpreted as the amount of heat required to increase the temperature of 1 kg of material by 1°C. It can be observed in the appendices that thermophysical properties can vary with temperature or pressure.

In convective heat transfer problems, fluid viscosity is an important transport property. The fluid viscosity is a measure of the frictional resistance of a fluid element when applying a shear stress. It is the ratio between the shear stress and the local strain rate. As an example, consider a Newtonian liquid contained between a lower, fixed plate and an upper, moving plate. For common fluids, such as oil and water, experimental measurements of the applied force show that the shear stress in the liquid is directly proportional to the strain rate in the liquid. Fluids are classified as *Newtonian fluids* when the shear stress varies linearly with strain rate. For non-Newtonian fluids, such as paint films, water–sand mixtures, or liquid polymers, the applied shear stress and resulting fluid strain rates are related in a nonlinear fashion.

An understanding of microscopic phenomena is needed to explain the trends and patterns observed in thermophysical property variations. For example, it can be observed from Figure 1.2 that metals have relatively high

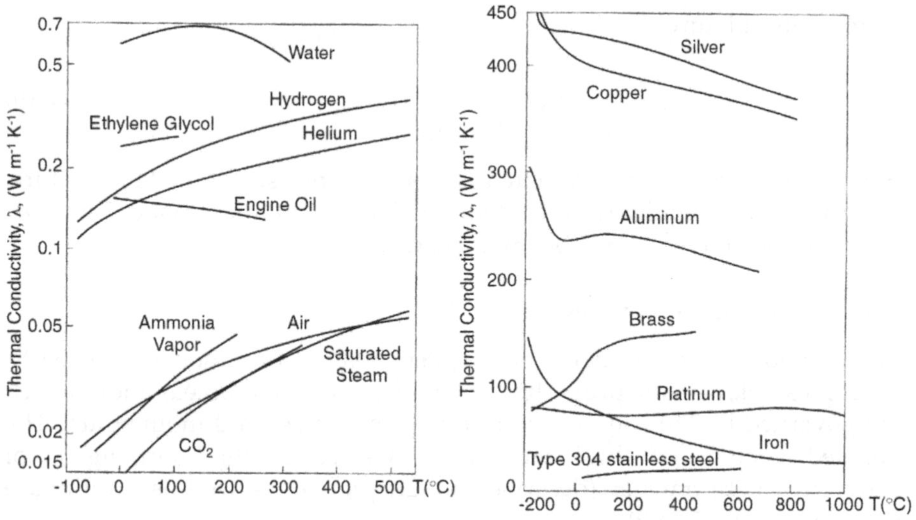

FIGURE 1.2
Thermal conductivities of solids and fluids. (From Hewitt, G.F., Shires, G.L., and Polezhaev, Y.V., Eds., *International Encyclopedia of Heat and Mass Transfer*, Copyright 1997, CRC Press, Boca Raton, FL. Reprinted with permission.)

thermal conductivities. In metallic bonds, metal atoms give up their free outer-shell electrons (one, two, or three valence electrons) to an electron gas and take up a regular arrangement. For example, Mg^+ ions are attracted to a negative electron cloud. These loosely held electrons in an electron cloud lead to high thermal conductivities since the electrons move readily through the solid.

In ionic bonds, an atomic bond is formed between metal and nonmetal ions. The metal gives up its valence electrons to the outer shell of the nonmetal. The positive metal ions and negative nonmetal ions are attracted to one another. For example, in sodium chloride, Na (one valence electron) reacts with Cl (seven valence electrons), with a Cl stable outer shell of eight valence electrons. The electrons are tightly held, thereby yielding a lower thermal conductivity than metallically bonded materials. Also, the electric fields of opposing ions on different planes repel each other, which leads to brittle crystal fractures.

In covalent bonds, electrons are shared by participating atoms and held tightly together. For example, in diamonds, a carbon atom is located at the center of a tetrahedron structure with a total of four valence electrons shared by adjoining atoms. Finally, in van der Waals bonds, secondary bonding between molecules is due to a charge attraction. This results from an unsymmetrical charge distribution in the material. For example, water molecules are attracted to each other by negatively and positively charged sides of adjacent molecules. All of these microscopic phenomena lead to the observed variations in transport properties.

In the evaluation of transport properties, macroscopic approximations are usually required to quantify microscopic transport phenomena. For example, recall our earlier Eulerian description of transport processes, such as conduction heat transfer, involving the continuum assumption. In order to estimate the macroscopic heat transfer rate in terms of random molecular fluctuations in or out of the selected differential control volume, we adopt a phenomenological law called *Fourier's law*. The term *phenomenological* refers to the notion that it cannot be proven rigorously from mathematical or first principles, but rather, it is an experimentally observable phenomenon. In Fourier's law, the conductivity is used as a macroscopic approximation of a microscopic transport phenomenon. In this way, we overcome the difficulty of tracking individual molecules, through their rotational, translational, and vibrational motions, and assessing their energy exchange during each interaction with each other.

1.4.4 Other Properties

Finally, we have other properties, such as surface tension or saturation pressure. These additional properties are often required whenever additional physical processes, such as a phase change, occur in a system. Vapor pressure is the pressure at which liquid–vapor phase transition occurs in a pure substance. This vapor pressure increases with saturation temperature, since more energy is required to break molecular bonds in the phase transition at higher temperatures. A variety of other properties is shown in the appendices.

Based on these properties and concepts, the fundamental modes of heat transfer may now be examined, namely, conduction, convection, radiation, and phase change heat transfer.

1.5 Conduction Heat Transfer

Heat conduction occurs when energy is transported by microscopic carriers (i.e., molecular motion) whose energy levels are modified during particle-to-particle interactions. These carrier motions refer to the translational, rotational, and vibrational movements of molecules. In this way, heat is transferred even though no net bulk movement of material occurs. For example, consider a motionless layer of liquid contained between a heated, upper plate and a cooled, lower plate. In this case, energy is transferred between the hotter side of the fluid and the colder side even though no bulk motion of the fluid occurs. Furthermore, consider an imaginary plane dividing the upper and lower regions of the fluid into equal portions. On average, an equal number of molecules crosses the imaginary plane in both directions; however, the molecules coming from the hotter section possess more energy. As a result, there is a net transfer of energy from the hotter side to the colder side (without any net transport of mass across the dividing plane).

Since it is impractical and difficult to track individual molecules during this microscopic transport, conduction heat transfer is typically approximated through macroscopic average quantities, such as temperature and an average diffusion coefficient (called thermal conductivity, k). In our previous one-dimensional example, the conduction process is governed by the following one-dimensional form of Fourier's law:

$$q = -kA\frac{\partial T}{\partial x} \tag{1.11}$$

where A and q refer to the heat flow area and heat transfer rate, respectively. Under steady-state, one-dimensional conditions without any heat sources or sinks, the heat transfer across a layer of thickness Δx becomes

$$q = -kA\frac{\Delta T}{\Delta x} \tag{1.12}$$

In multidimensional conditions, the partial derivative in Equation (1.11) becomes a gradient of temperature, so that q becomes a vector of heat flow components in the x, y, and z directions.

In Equation (1.12), the negative sign indicates that a negative temperature gradient (i.e., decreasing temperature in the positive x direction) multiplied by a minus sign yields a positive heat flow in the positive x direction. This situation is illustrated in Figure 1.3, where it can be observed that heat flows in the direction of decreasing temperature (as expected). The thermal conductivity varies largely, depending on the type of material, i.e., from 0.01 W/mK for many insulators up to 10 to 400 W/mK for most metals.

The result of a linear temperature profile in Equation (1.12) may be inferred from an energy balance, together with Fourier's law. In particular, a one-dimensional energy balance under steady-state conditions for a discrete

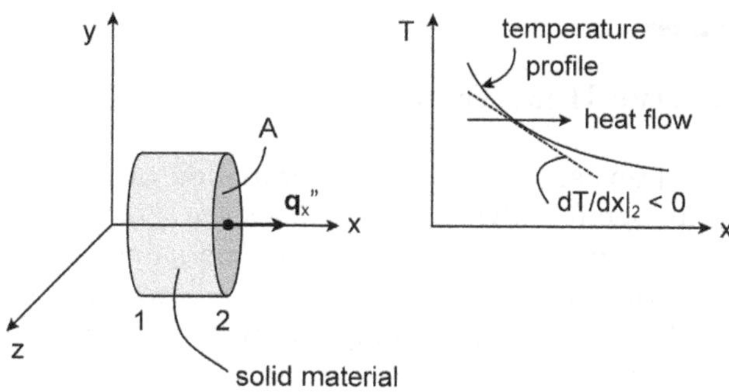

FIGURE 1.3
Heat flux by conduction.

control volume suggests that the rate of heat inflow balances the rate of heat outflow from the volume. From Fourier's law, this implies that the slope of the temperature profile at the inflow edge of the control volume matches the temperature slope (or gradient) at the outflow side. Furthermore, both of these temperature gradients are constant and equal to one another as a result of the steady-state energy balance (assuming zero sources and sinks of energy). Integrating this result reveals that the temperature profile is indeed linear across the entire volume to meet these heat balance requirements. If the thermal conductivity varies with temperature, then the temperature profile may become nonlinear.

As described earlier, the thermal conductivity is a transport property of the material that indicates the rate at which it can conduct heat. For example, a poor conductor like an insulating material has a low thermal conductivity. For solids, such as many pure metals, the thermal conductivity is often higher at low temperatures (i.e., rising above 4000 W/mK at temperatures below 80 K), but then falls to around 100 to 400 W/mK at room temperatures. On the other hand, metal alloy conductivities rise from low to higher conductivities as the temperature rises to room temperature. These trends indicate the close dependence on molecular structure; in particular, less uniform molecular structures in alloys often lead to lower thermal conductivities.

The thermal conductivity of water rises to about 0.7 W/mK at 400 K, but then gradually falls afterwards. Other liquids have similar trends in terms of thermal conductivity. In gases, an opposite type of trend is often observed, whereby the thermal conductivity rises with temperature above room temperature.

1.6 Convective Heat Transfer

Convective heat transfer refers to the combination of molecular diffusion and bulk fluid motion (or advection). *Newton's law of cooling* is widely used in the analysis of heat convection problems. The name is attributed to Sir Isaac Newton (1642–1727), largely due to the associated importance of *Newton's laws of motion* in governing the fluid motion as it occurs in convective heat transfer.

Consider a hot surface maintained at a constant temperature, T_s, and placed in contact with a uniform fluid stream flowing at T_f above the surface (see Figure 1.4). In this case, heat transfer to the fluid stream, q, from the plate, is given by

$$q = hA(T_s - T_f) \tag{1.13}$$

The convective heat transfer coefficient is denoted by h in the above equation. This coefficient depends on many factors, including the fluid properties, geometrical configuration, fluid velocity, surface roughness, and other factors.

Forced convection is induced by external means (i.e., fan, pump, atmospheric winds), whereas free convection involves buoyancy-induced fluid motion. Mixed convection includes both forced and free convection. The magnitude of heat transfer coefficient often indicates the general type of problem. In particular, the following values are usually encountered in engineering applications:

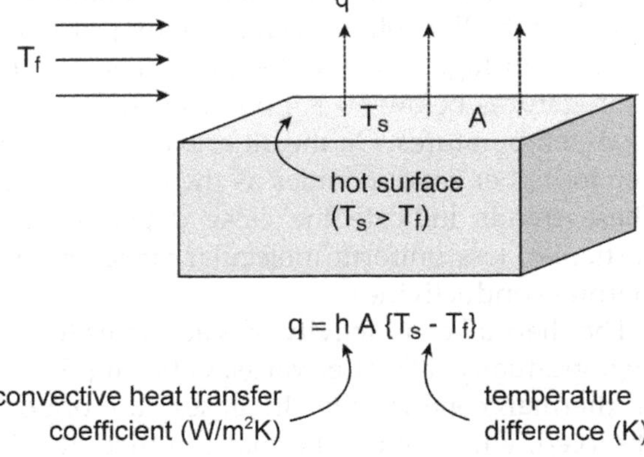

FIGURE 1.4
Convective heat transfer from a hot surface.

- 5–30 W/m^{2}K: free convection in air
- 100–500 W/m^{2}K: forced convection in air
- 100–15,000 W/m^{2}K: forced convection in water
- 5000–10,000 W/m^{2}K: water condensation

It can be observed that phase change and liquids with a higher heat capacity (i.e., specific heat of water is higher than that of air) are two factors leading to higher convective heat transfer coefficients.

In convection problems, the fluid temperature must be carefully specified. In particular, the fluid temperature in Newton's law of cooling depends on the type of problem under consideration. In the case of external flow, such as external flow past a circular cylinder, the fluid temperature is given by the freestream temperature. However, for internal flow, such as liquid flow in a pipe, the fluid temperature in Newton's law of cooling becomes the mean temperature of the fluid, since no freestream temperature is apparent in that situation. The mean temperature is obtained through spatial averaging of the velocity multiplied by temperature over the cross-sectional area of the pipe. In other words, the mean temperature is the mass flow weighted average temperature of the fluid in the pipe at the particular axial location within the pipe.

Another important feature of the heat transfer coefficient specification is distinguishing local and average convection coefficients. The local convection coefficient refers to the value of h at a particular point along the surface, whereas the average or total heat transfer coefficient refers to an integrated value around the surface.

A useful example indicating the differences between local and average convection coefficients is external flow past a circular cylinder. If we define θ as the angle from the upstream stagnation point around the circumference of the conductor, then a clear angular variation of h may be observed with θ. In particular, h decreases with θ on the upstream side as the boundary layer thickness grows, thereby reducing the gradient of temperature and heat transfer through the boundary layer. At low Reynolds numbers (below about 2.2×10^5), the boundary layer separation point is on the upstream side of the conductor. The convection coefficient decreases up to this separation angle (about 80°) and increases thereafter, since separation causes local flow reversal, enhanced mixing, and thus more heat transfer.

Variations in these angular trends due to boundary layer transition to turbulence on the upstream side of the cylinder are observed at higher Reynolds numbers (above about 2.2×10^5). The convection coefficient, h, initially decreases with θ, but then rises between 80° and 90°, decreases past 90° up to the separation point (around 140°), and increases thereafter. Transition to turbulence is responsible for the increase of h with θ on the upstream side, but this transition also delays boundary layer separation, thereby affecting the convection coefficient on the downstream side of the cylinder. The main observation in this discussion is to recognize that the heat transfer coefficient varies with the angle traversed around the conductor (i.e., local variations of h with θ). Also, the average, or total, coefficient can be obtained by spatial integration of this coefficient around the entire conductor.

1.7 Radiative Heat Transfer

Thermal radiation is a form of energy emitted by all matter above absolute zero temperature. The emissions are due to changes in the electron configurations of the constituent atoms or molecules, and they are transported through space as electromagnetic waves (or photons). Thermal radiation does not require the presence of a material medium between the objects exchanging energy; it occurs most efficiently in a vacuum.

The basic processes associated with radiative transport may be considered through two objects (numbered 1 and 2) exchanging heat by radiation. Energy in the form of heat is emitted by object 1 through space until it arrives at object 2. Once this incident radiation arrives at object 2, several possibilities occur. Some of the incident radiation is reflected from surface 2, while other incoming energy is absorbed into the surface of object 2. Following this absorption, some energy may be entirely transmitted through the object until it passes through its opposite side. Finally, some of the electromagnetic waves emitted from surface 1 may not be absorbed at all by surface 2 since they are not in the line of sight of that object.

In heat conduction problems, a main governing equation is Fourier's law, whereas Newton's law of cooling was adopted in heat convection problems. In radiative heat transfer, we can determine the emitted radiation by *Stefan–Boltzmann's law,* which is the analog of the aforementioned laws for the other modes of heat transfer. If a surface is maintained at a temperature of T_s (in absolute units, i.e., Kelvin or Rankine units) and its surface emissivity is ϵ, then the heat flux, q_s'', due to thermal radiation, is given by

$$q_s'' = \epsilon \sigma T_s^4 \tag{1.14}$$

where $\sigma = 5.67 \times 10^{-8}$ W/m^2 K^4 is the Stefan–Boltzmann constant. Also, the surface emissivity, ϵ, indicates the surface's ability to emit radiation in comparison to an ideal emitter (i.e., $\epsilon = 1$ for a *blackbody*).

Example 1.7.1

Small Object Enclosed by a Large Isothermal Cavity.

In this example, the net heat transfer between a small object and a surrounding cavity is required. Since all the radiation emitted from the small object (object 2) is absorbed by the large surrounding cavity (surface 1), the cavity behaves like a blackbody at T_1. Assuming that the object is hotter than its surroundings, i.e., $T_2 > T_1$, there is a net heat loss by radiation from the small object, given by

$$q_{rad} = \epsilon\sigma(T_2^4 - T_1^4)A \tag{1.15}$$

where A refers to the surface area of object 2. Also, ϵ denotes the emissivity of the object. The latter part of Equation (1.15) includes the temperature of the surroundings, T_1.

The previous expression and net radiation exchange between the object and surroundings, Equation (1.15), is nonlinear in the sense that the temperature difference contains an exponent of 4. In solutions of nonlinear problems, linearization is often required to obtain the final algebraic equations. As a result, we define an effective radiation heat transfer coefficient, h_{rad}, in the following manner,

$$q_{rad} = h_{rad}(T_2 - T_1)A \tag{1.16}$$

where

$$h_{rad} = \epsilon\sigma(T_2 + T_1)(T_2^2 + T_1^2) \tag{1.17}$$

In this way, the radiation coefficient can be combined with the convection coefficient to form a single net heat transfer coefficient accounting for combined modes of heat transfer.

1.8 Phase Change Heat Transfer

In addition to the previously described modes of heat transfer, phase change arises in many engineering applications. The variations of properties during phase change processes are often best understood with the aid of phase diagrams. These diagrams illustrate the various properties and phases that coexist in equilibrium.

For example, consider a process in which an initially solid substance is heated at a constant pressure until it eventually becomes entirely gas (see Figure 1.5). In the solid region (region 1), heat is added and the temperature increases until melting begins at the phase change temperature. In region 2, heat input is then transferred to a change of phase (i.e., solid–liquid phase change), without a change of temperature. Under further heating, the material is entirely melted in region 3, and the heat addition leads to sensible heating (i.e., change of temperature) until the liquid temperature reaches the saturation point (onset of boiling).

Now, in a similar way, heat is added to sustain a change of phase (region 4) until the saturated vapor point is reached, beyond which further heating is transferred as sensible heating of the gas (region 5). If these steps are repeated over several different pressure levels, then the resulting measurements from each experiment (i.e., temperature, specific volume, etc.) can be joined together. These resulting data points would provide the entire phase diagram for phase change of a pure material (depicted in Figure 1.5).

Additional phase diagrams, such as the phase diagram for solid–liquid mixtures at equilibrium, can show the effects of species concentration on equilibrium processes during phase change. Phase diagrams for common materials and liquids, such as water or molten steel, are readily available in both graphical and tabular forms. The following example outlines how this type of phase diagram can be constructed.

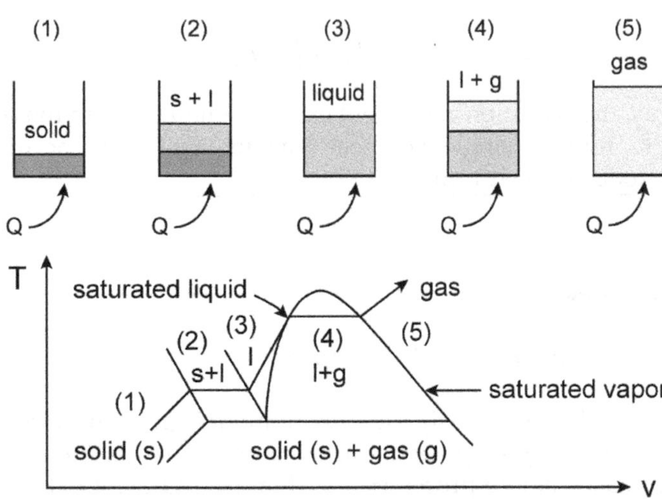

FIGURE 1.5
Phase diagram for solid–liquid–gas pure material.

Example 1.8.1

Constructing a Phase Equilibrium Diagram for a Binary Alloy.

A phase diagram can be constructed in a systematic fashion. Outline a procedure for a basic experiment that can be used to construct the solid–liquid phase equilibrium diagram for a mixture consisting of aluminum and silicon.

A graph is initially created with the vertical and horizontal axes representing temperature and solute (silicon) concentrations, respectively. An observation can be made that pure aluminum (i.e., 0% silicon composition) melts and freezes at 650°C, and so the intercept at the vertical axis is 650°C. Then, add a fixed mass of silicon to the bath containing both materials, such as 5% silicon (by mass). This mixture now begins freezing at 626°C (called the *liquidus temperature*) when cooled from the liquid phase, but the majority of the mixture still remains as a liquid at equilibrium.

If the mixture is further cooled, higher fractions of solid are formed at equilibrium, until additional cooling eventually completely freezes the entire mixture at 577°C (called the *solidus temperature*). In between these temperatures, varying fractions of solid and liquid coexist in equilibrium (called the *mushy* or *two-phase region*). The process is repeated at other concentrations of silicon, and liquidus and solidus points are connected together to form the phase equilibrium diagram for the binary mixture (see Figure 1.6).

The *eutectic temperature* refers to the minimum temperature of the two-phase region between the solidus and liquidus lines that contains unfrozen liquid. The *eutectic point* refers to the composition of the minimum freezing point, or the temperature at which a liquid of eutectic composition freezes. The eutectic temperature is the temperature at which a liquid of eutectic

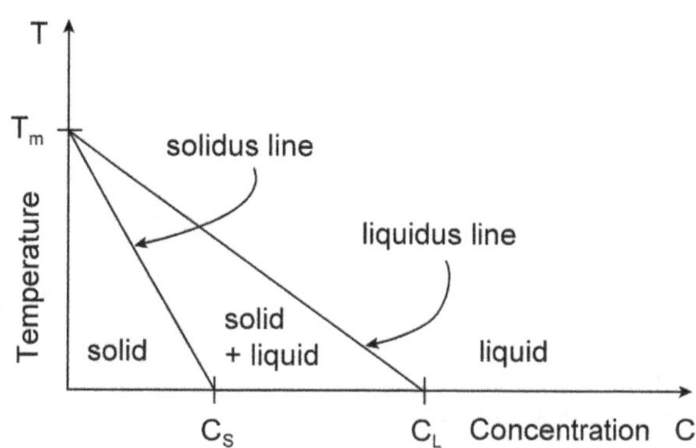

FIGURE 1.6
Phase diagram for solid–liquid binary mixture.

composition freezes to form two solids simultaneously under equilibrium conditions.

Below the eutectic temperature, simultaneous growth of two or more solid phases (i.e., different types of microstructures in the solid) from the liquid phase is observed. A material with a composition below the eutectic point is called a *hypoeutectoid*, whereas a material above the eutectic composition is called a *hypereutectoid*. This discussion was presented for a specific case of binary material solidification, but analogous concepts are used to establish the phase equilibrium diagrams for other multiphase systems.

1.9 Conservation of Energy

An energy balance is a fundamental part of a heat transfer analysis. Two overall types of energy balances may be adopted, namely, a control mass or a control volume approach. A closed system is identified as a system encompassing a fixed amount of mass. On the other hand, a fixed region in space characterizes a control volume (open system) analysis. The following energy balance gives a general balance of energy for a control volume:

$$\dot{E}_{cv} = \dot{E}_{in} - \dot{E}_{out} + \dot{E}_g \tag{1.18}$$

These various terms are illustrated in reference to an arbitrarily shaped control volume in Figure 1.7.

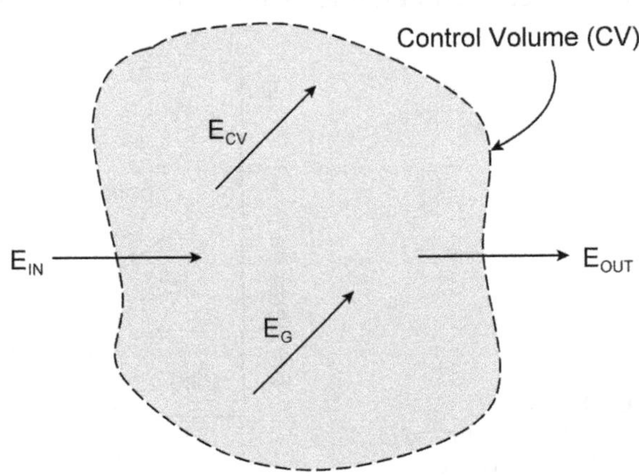

FIGURE 1.7
Schematic of energy balance for a control volume.

In Equation (1.18), the terms represent: (i) the rate of energy accumulation over time, (ii) the rate of energy inflow, (iii) the rate of energy outflow, and (iv) the rate of internal heat generation arising from processes such as chemical reactions. The energy equation essentially requires that the rate of the increase of energy within the control volume equals the net rate of energy inflow plus any heat generated. A dot notation above E (energy) refers to the rate of change with respect to time. On the left side of Equation (1.18), the rate of energy change may be written in terms of the mass multiplied by specific heat and rate of temperature change with respect to time. This change of variables comes as a result of the definition of the specific heat. An example that utilizes these various terms is given in the following problem.

Example 1.9.1

Electrically Heated Wire in a Large Enclosure.

An energy balance is required for a resistance heating element in a large enclosure, i.e., hot-wire anemometer in a wind tunnel or a power source in a furnace. The length, diameter, resistance, and current in the wire are represented by L, D, R, and I, respectively. It is assumed that temperature varies only with time, i.e., no temperature gradients in the wire and constant properties may be assumed.

Based on these assumptions and definitions, Equation (1.18) becomes

$$\rho ALc_p \frac{\partial T}{\partial t} = -q_{conv} - q_{rad} + I^2 R \tag{1.19}$$

This equation states that the rate of energy increase, given by the left side, is the difference between convective and radiative heat losses, and energy generated by the electrical resistance heating. The convection term may be written in terms of a temperature difference (between wire temperature and surrounding air temperature). Also, the radiative term may be written in terms of a temperature difference (involving fourth power) with the temperature of the wire, T, and the surface temperature of the furnace walls surrounding the wire. It is assumed that no energy is absorbed or scattered within the gas layer between the wire and the furnace walls.

In this way, the energy balance involves a first-order derivative of temperature with respect to time, linear temperature difference, and differences of temperature to their fourth powers, as well as heat generated electrically. The energy balance becomes a first-order nonlinear ordinary differential equation that may be solved by a Runge–Kutta numerical integration procedure. A direct, explicit solution is generally not available, but an iterative scheme may be adopted, whereby the radiation term is linearized through a linearized radiation coefficient (as discussed previously). In this way, an approximate solution may be obtained without requiring a full numerical integration.

Another generalized form of the energy balance in Equation (1.18), which includes both work and heat transfer, is given by the first law of thermodynamics. For a control mass, the first law can be obtained from Equation (1.18) for a finite period of time by integration over time, thereby removing the rates denoted by a dot above E. For work performed by the system (i.e., out of the control mass), denoted by W, the first law for the control mass becomes

$$E_i + Q - W = E_f \tag{1.20}$$

where the subscripts i and f refer to initial and final states, respectively, and Q refers to the net inflow of heat into the control mass (note: negative Q represents a heat outflow). An example of a typical work mode is expansion or compression of a gas.

For an open system (control volume), the energy balance in Equation (1.18) is used, where $\dot{Q}_{in}$ refers to the heat inflow across the solid part of the boundary, $\dot{W}_{in}$ is the flow work into the control volume arising from pressure, and $\dot{E}_{in}$ refers to the energy inflow carried by the mass inflow. Similar outflow terms are constructed, yielding the following form of the first law for a control volume:

$$\dot{E}_{cv} = \dot{Q}_{in} + \dot{W}_{shaft} + \dot{E}_{in} - \dot{E}_{out} + \dot{W}_{in} - \dot{W}_{out} \tag{1.21}$$

where $\dot{W}_{shaft}$ refers to the work or power input (or output) due to a protruding shaft across the boundary of the control volume. For example, a turbine shaft and blades would extract power from a control volume encompassing a steam turbine in a power plant. Although the form of the first law in Equation (1.21) refers to a single outlet and inlet, a similar expression can be obtained for multiple inlets and outlets by taking a summation over all inlets and outlets in Equation (1.21).

The flow work term in Equation (1.21) can be represented by a force (i.e., pressure, p_{in}, times flow area) multiplied by a characteristic distance (i.e., flow velocity multiplied by a specified time step). Writing the velocity and area in terms of the mass flow rate, $\dot{m}$, and expressing the flow work on a rate basis (i.e., per unit time),

$$\dot{W}_{in} = \dot{m}_{in} p_{in} v'''_{in} \tag{1.22}$$

where v'''_{in} is the specific volume of the fluid.

Substituting Equation (1.22) into Equation (1.21) and similarly for the outflow terms,

$$\dot{E}_{cv} = \dot{Q} + \dot{W}_{shaft} + [\dot{m}(e + pv''')]_{in} - [\dot{m}(e + pv''')]_{out} \tag{1.23}$$

where e refers to total specific energy (including internal, potential, and kinetic energy). The form of Equation (1.23) is based on heat and work inflow (i.e., positive Q and W) into the control volume. Assuming steady-state and steady-flow processes (i.e., $\dot{E}_{cv} = 0$ and $\dot{m}_{in} = \dot{m}_{out}$), as well as negligible changes in kinetic and potential energies across the control volume, Equation (1.23) becomes

$$\dot{Q} + \dot{W}_{shaft} = \dot{m}(h_{out} - h_{in}) \tag{1.24}$$

where h is the fluid enthalpy. These results correspond to conventional forms of the first law presented in most undergraduate thermodynamics textbooks. These overall forms of the total energy balance will be useful in various contexts throughout this book.

At the boundary of a system or a surface separating two distinct phases within the system, a control surface balance represents a special case of the energy balance. In particular, the control volume thickness shrinks to zero as it encompasses the region about the boundary or phase interface. In this case, the control volume energy balance term representing transient energy accumulation becomes zero since the mass of the volume approaches zero. Furthermore, the heat generated may be nonzero as a result of processes such as friction between two different phases or heat transfer due to latent heat evolved at a moving phase interface. The control surface energy balance may be regarded as a boundary condition applied on the system, but it still represents a balance of energy at that location.

Problems

1. Suppose that you touched and held two different materials, such as a piece of steel and a piece of wood, that were both initially at the same subzero temperature. Would either material "feel" colder than the other? Explain your response.

2. The heat loss through a common brick wall in a building is 1800 W. The height, width, and thickness of the wall are 3, 8, and 0.22 m, respectively. If the inside temperature of the wall is 22°C, find the outside surface temperature of the wall.
3. Heat flows in a one-dimensional direction through a layer of material of unknown thermal conductivity. Explain a method to estimate the material's conductivity based on temperature measurements recorded by thermocouples at various positions throughout the layer of material.
4. A heat loss of 1600 W is experienced through a glass window (1.4 m wide × 2.6 m high) at the outer wall of a building. The convection coefficient is 30 W/m^2K. Estimate the window temperature when the outside air temperature reaches −10 °C.
5. An aluminum sheet leaves the hot-roll section (called the entry point) of a mill at a temperature of 620°C. The surrounding air temperature is 20°C. For a desired uniform cooling rate of 38 kW/m^2 across the sheet, estimate the aluminum surface temperature at a position where the convection coefficient is 40% higher than the value at the entry point.
6. Air flows across a tube surface ($\epsilon = 0.9$) in a heat exchanger. The ambient air temperature and convection coefficient are 20°C and 120 W/m^2K, respectively. Compare the heat transfer by convection and radiation at surface temperatures of (a) 60°C, (b) 400°C, and (c) 1300°C.
7. A metal block (0.8 × 1.2 × 0.4 m) in a furnace is heated by radiation exchange with the walls at 1100°C. The block temperature is 420°C, and all surfaces may be approximated as blackbodies. Find the net rate of heat transfer to the block due to radiation exchange with the walls. How would the result be altered if the block was placed in a corner of the furnace?
8. Superheated steam at 240°C flows through an uninsulated pipe in a basement hallway at 20°C. The pipe emissivity is 0.7 and the coefficient of free convection is 20 W/m^2K. Find an expression for the total heat loss from the pipe (per unit length of pipe) in terms of the pipe diameter.
9. Water is heated and slowly boiled in a water distiller. The water boils, and superheated steam exits from the heating tank through a single tube. Once cooled, this steam is condensed and collected in a second storage tank of purified water. If the electrical element provides a net heat input of 400 W to the water, how many liters of distilled water are produced after 6 h of operation? Assume that external heat losses from the heating tank are negligible and the heat input is directed entirely into phase change (i.e., 6 h refers to time taken once the water reaches the saturation temperature).
10. An ice-making machine uses a refrigeration system to freeze water with a net heat removal rate of 11 kW. How much time is required to

produce 1200 kg of ice? It may be assumed that the water is cooled to slightly above 0°C before it enters the refrigeration unit.

11. The diameter of the sun is approximately 1.39×10^9 m. A heat flux of about 1353 W/m^2 from the sun's radiant heat reaches the outer atmosphere of the earth. Estimate the radius of the orbit of Earth's trajectory around the sun (i.e., Earth's distance from the sun). The sun can be approximated as a blackbody at 5800 K.
12. The wall of a walk-in freezer in a meat processing plant contains insulation with a thickness of 10 cm and a thermal conductivity of 0.02 W/mK. The inside wall temperature is −10°C. The outside air temperature and convection coefficient (outside) are 20°C and 5 W/m^2K, respectively. What additional thickness of insulation is required to reduce the heat gain (into the freezer) by 10%?
13. The temperature of the indoor side of a stone concrete wall in a building is 20°C. The wall thickness is 20 cm, while the convection coefficient and ambient air temperature for the outdoor air are 50 W/m^2K and −5°C, respectively. Determine the outer wall temperature of the concrete wall.
14. A thin plate is exposed to an incident radiation flux of 2 kW/m^2 on its top surface, and the bottom surface is well insulated. The exposed surface absorbs 80% of the incident radiation and exchanges heat by convection and radiation to the ambient air at 290 K. If the plate's emissivity is 0.8 and the measured wall temperature is 350 K, estimate the convection coefficient, h, under steady-state conditions.
15. The gas temperature in an autoclave is measured with a thermocouple (wire of 1-mm outer diameter). Composite material components for an aircraft are cured in the autoclave. The junction of the thermocouple ($\epsilon = 0.6$) is located at the end of the thermocouple wire, which protrudes into the autoclave and exchanges heat by radiation with the walls at 410°C. The freestream temperature and convection coefficient of gas flow past the thermocouple are 180°C and 30 W/m^2K, respectively. What error (i.e., difference between thermocouple reading and gas temperature) is expected in the measurement of the gas temperature? How can this temperature measurement error be reduced? Conduction losses through the thermocouple wire can be neglected.
16. Heat is generated electrically at a certain rate within an overhead power transmission cable. Air flows past the cable with a freestream temperature and convection coefficient of −6°C and 80 W/m^2K, respectively. For a specified copper cable (3-cm diameter), estimate the required heat generation rate (per unit length of cable) to maintain the surface temperature above at least 0°C.

17. The top cover of a solar collector is exposed to ambient air at 20°C with a convection coefficient of 6 W/m^2K. A surface coating is developed and applied to modify the radiative properties of this absorbing cover plate. The incident solar radiation is 860 W/m^2, and the outer surface temperature of the absorber plate is 70°C.
    (a) What plate emissivity is required to provide a conduction heat flux of 500 W/m^2 through the absorber plate?
    (b) What is the proportion of heat exchange by radiation between the cover plate and surroundings (ambient air) relative to the convective heat loss from the plate?
18. Heating or cooling loads must be supplied by heating, ventilating, and air conditioning (HVAC) equipment to maintain a room or space in a building at desired conditions. Identify the heat transfer processes and appropriate energy balance(s) to find the steady-state temperature of a room in a building.
19. Aircraft icing may occur when an aircraft passes through a cloud containing supercooled droplets. Describe the heat flows to and from an iced aircraft surface, and write an energy balance for the surface based on these heat flows.
20. Particles of pulverized coal are injected and burned in a boiler of a thermal power plant. Identify and briefly describe the relevant heat transfer modes that contribute to an energy balance for a pulverized coal particle.
21. Identify the relevant heat transfer processes that arise in a thermal balance of the human body. Explain how these processes are combined in an overall energy balance to find the total heat losses from the body.
22. Several technical problems are encountered by thermal engineers in the design of more efficient systems for heating and cooling of buildings. Perform a review of the recent technical literature (including journals and magazines) to describe the main trends and challenges facing thermal engineers in the development of new HVAC systems.

2

Conduction Heat Transfer

2.1 Introduction

In this chapter, the fundamental equations and analysis of problems involving conduction heat transfer will be considered. Heat conduction is a transport process that arises in both single and multiphase systems. For example, heat transfer by conduction occurs at the phase interface (solid–liquid or liquid–vapor) of two-phase flows. In this chapter, both one-dimensional and two-dimensional problems, including steady and transient conditions, will be examined.

Isotherms represent lines of constant temperature. Plotting these isotherms in a system graphically depicts the regions of high and low temperatures. Heat flows by conduction in a direction normal to the local isotherm. Thus, heat flow lines and isotherms are mutually perpendicular in heat conduction problems, so that the heat flow paths may be constructed by joining perpendicular crossing points between successive isotherms. Another way of interpreting this process is that heat flows by conduction in the direction of steepest temperature descent.

The magnitude of the heat flow is directly proportional to the temperature gradient. For example, a high rate of temperature descent suggests that more heat is transferred than does a shallow rate of temperature descent. The proportionality constant between this heat flow and temperature gradient, outlined in Equation (1.11), is represented by the thermal conductivity. A negative sign is placed in front of the temperature gradient in Fourier's law, Equation (1.11), to show that heat flows in the direction of decreasing temperature. In other words, a positive temperature gradient suggests that temperature is increasing, and heat must then flow in the opposite direction.

Heat transfer by conduction occurs through an exchange of rotational, vibrational, and translational energies at a molecular level. As briefly outlined in the previous chapter, heat conduction is governed by Fourier's law, which relates the magnitude and direction of heat flow with temperature. Since heat flows in an opposite direction to increasing temperature, a negative sign is used to accommodate this relationship. The temperature gradient, $\nabla\mathbf{T}$, and heat flux, $\mathbf{q}$, are vector quantities (designated by bold font). Fourier's law in Equation (1.11) can be generalized to multidimensional heat conduction in the following manner:

$$\mathbf{q}'' = -k\nabla T \tag{2.1}$$

This vector notation includes all coordinate systems. The heat flux, $\mathbf{q}''$, represents the heat flow per unit area, whereas $\mathbf{q}$ will refer to the total heat flow rate (units of W).

The heat flux can be represented in various coordinate directions by writing the gradient operator in the appropriate coordinate system. For example, in *Cartesian coordinates*, x, y, and z are the three coordinate directions, and the heat flux becomes

$$\mathbf{q}'' = -k\nabla T = -k\left(\frac{\partial T}{\partial x}, \frac{\partial T}{\partial y}, \frac{\partial T}{\partial y}\right) \tag{2.2}$$

where

$$-k\left(\frac{\partial T}{\partial x}, \frac{\partial T}{\partial y}, \frac{\partial T}{\partial y}\right) = \left(-k\frac{\partial T}{\partial x}\hat{\mathbf{i}}\right) + \left(-k\frac{\partial T}{\partial y}\hat{\mathbf{j}}\right) + \left(-k\frac{\partial T}{\partial z}\hat{\mathbf{k}}\right) \tag{2.3}$$

Other coordinate systems, including Cartesian, cylindrical, and spherical coordinates, are illustrated in Figure 2.1. In *cylindrical coordinates* (r, θ, and z),

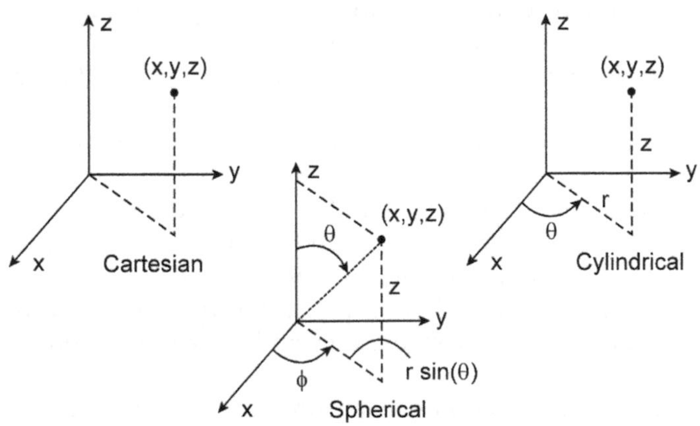

FIGURE 2.1
Cartesian, cylindrical, and spherical coordinates.

$$\mathbf{q}'' = -k\left(\frac{\partial T}{\partial r}, \frac{1}{r}\frac{\partial T}{\partial \theta}, \frac{\partial T}{\partial z}\right) \tag{2.4}$$

In *spherical coordinates* (r, θ, and ϕ),

$$\mathbf{q}'' = -k\left(\frac{\partial T}{\partial r}, \frac{1}{r}\frac{\partial T}{\partial \theta}, -\frac{1}{rsin\theta}\frac{\partial T}{\partial \phi}\right) \tag{2.5}$$

The use of coordinate systems should be selected to closely approximate the problem domain such that boundary conditions may be most readily specified. For example, in fluid flow through a pipe, it is more convenient to identify the pipe boundary through cylindrical coordinates, rather than Cartesian coordinates. In numerical methods for heat transfer and fluid flow, Cartesian coordinates are typically adopted since a variety of complex geometries are generally encountered and Cartesian coordinates generally contain the most flexibility in accounting for these geometries.

2.2 One-Dimensional Heat Conduction

The heat conduction equation may be derived from the general form of an energy balance over a control volume, i.e., from Equation (1.18):

$$\dot{E}_{cv} = \dot{E}_{in} - \dot{E}_{out} + \dot{E}_g \tag{2.6}$$

Control volumes for one-dimensional and two-dimensional heat balances are illustrated in Figure 2.2. Expressing the transient energy change, $\dot{E}_{cv}$, in terms of a temperature change (using the definition of the specific heat), and writing the heat outflux term based on a Taylor series expansion about the influx value,

$$\rho c_p A dx \frac{\partial T}{\partial t} = q''_x - \left[q''_x A + \frac{\partial}{\partial x}(q''_x A)dx + \ldots\right] + \dot{q}''' A dx \tag{2.7}$$

where $\dot{q}'''$ refers to the volumetric heat generation rate.

Furthermore, Fourier's law may be used, where

$$q''_x = -k\frac{\partial T}{\partial x} \tag{2.8}$$

Substituting Equation (2.8) into Equation (2.7), neglecting higher order

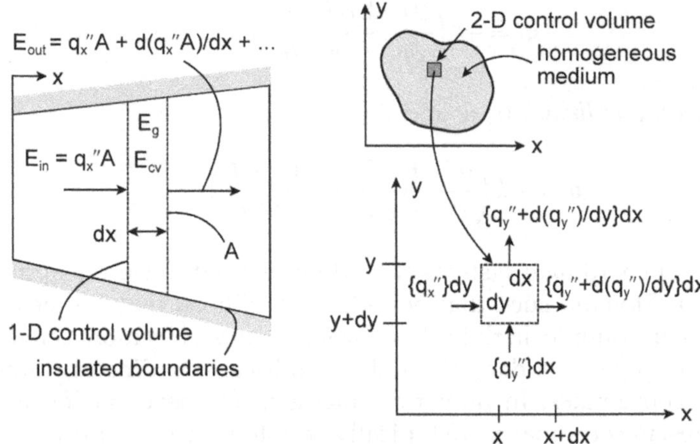

FIGURE 2.2
Energy balance for one- and two-dimensional control volumes.

terms, and rearranging, we obtain the following one-dimensional heat conduction equation:

$$\frac{1}{\alpha}\frac{\partial T}{\partial x}=\frac{\partial}{\partial x}\left(\frac{\partial T}{\partial x}\right)+\frac{\dot{q}'''}{k} \tag{2.9}$$

where $\alpha = k/(\rho c_p)$ is the thermal diffusivity of the material. An analogous form of the heat conduction equation may be readily obtained for cylindrical coordinates by writing the gradient operator in Fourier's law in cylindrical coordinates, rather than Cartesian coordinates. Alternatively, the heat balances outlined in Figure 2.2 could be performed over a cylindrical control volume. For a constant thermal conductivity, the value of k may be taken out of the bracketed term in Equation (2.9). In this case, a second-order spatial derivative of temperature governs the rate of heat conduction in the first term on the right side of Equation (2.9). In multidimensional problems, this term becomes the Laplacian operator acting on temperature (i.e., $\nabla^2 T$).

In the heat conduction equation, α may be interpreted in terms of a characteristic time, τ_c, required to change the temperature of an object. As an example, consider dropping a small sphere into a container of liquid. In this example, the sphere diameter (considered to be a characteristic length, L, of the problem) is 1 cm. The sphere is initially dropped from the air into a container of water where the water temperature is higher than the air temperature. Heat is transferred by conduction through the sphere and convection from the sphere, based on a characteristic time of L^2/α. At early periods of time ($t \ll \tau_c$), when the time elapsed is much less than the characteristic time indicated, the temperature of the sphere remains approximately the same as its initial air temperature. On the other hand, after a long period of time, after the sphere has settled out in the water ($t \gg \tau_c$), the sphere temperature approximately reaches the water temperature. During the intermediate times (approximately of the order of τ_c), a more detailed solution is required since the sphere temperature lies somewhere between the air and water temperatures.

Thus, the characteristic time for conduction is indicated by a length scale (squared) for the problem divided by the thermal diffusivity of the material. If the previous example was modified such that the material properties changed, while retaining the same water depth, then the ability of the material to conduct heat could be well interpreted through the characteristic time. For example, the characteristic time for a certain copper sphere dropped into water is 1 sec, whereas the characteristic time for a steel sphere of the same size would be 10 sec. Also, the characteristic time for a cork sphere dropped into the water is 10 min, thereby confirming our expected trend that cork would take much longer to reach the water temperature, since it is poor conductor in comparison with metals.

In addition to the previously mentioned governing equation for heat conduction, we also need boundary conditions for the completion of the problem specification. In heat conduction problems, we have three main types of boundary conditions: Dirichlet, Neumann, and Robin boundary conditions (see Figure 2.3).

The Dirichlet condition refers to a condition whereby a given surface temperature is specified at the boundary of the domain. For example, this condition may arise along the walls of a pipe where an internal condensing flow maintains the surface temperature at the phase change temperature.

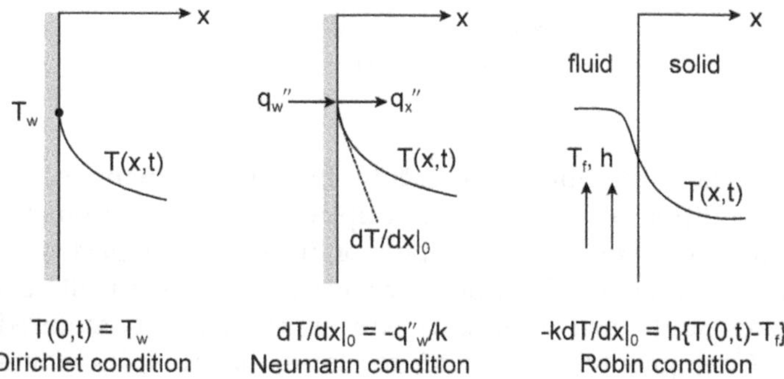

FIGURE 2.3
Types of boundary conditions.

For a Dirichlet condition at a wall ($x = 0$),

$$T(0, t) = T_w \tag{2.10}$$

where T_w is the known or specified temperature.

The second type of condition is a heat flux or temperature gradient specified condition (called Neumann condition). For example, at a solid boundary ($x = 0$), the specified wall heat flux balances the heat flow into the solid. Applying Fourier's law, this boundary condition becomes

$$\left.\frac{\partial T}{\partial x}\right|_0 = -\frac{q''_w}{k} \tag{2.11}$$

where q''_w refers to the known or specified boundary heat flux. If the boundary temperature gradient is zero, this represents an adiabatic (or insulated) boundary, since there is zero heat flow across the boundary. In other words, an insulated boundary is a special case of a Neumann boundary condition.

The third type of boundary condition is a convection condition (called the Robin condition). This type is a combination of the previous two types, that is, the temperature and temperature gradient appear in the boundary condition. It is commonly encountered at a convective boundary. Conduction through the solid involves a temperature gradient, whereas a temperature value is obtained on the fluid side of the boundary due to Newton's law of cooling. For example, at a boundary ($x = 0$), heat conduction into the boundary balances convective heat removal from the other side as required by an energy balance at the boundary. As a result,

$$-k\left.\frac{\partial T}{\partial x}\right|_0 = h[T(0, t) - T_f] \tag{2.12}$$

where T_f refers to the fluid temperature. In a similar manner, a linearized radiation condition at a boundary, coupled with heat conduction in the solid, would lead to a Robin-type condition. As expected, problems involving this type of condition are often more difficult, since both temperatures and their gradients are unknown yet required at the boundary. Conjugate problems refer to classes of problems where convection in the fluid and conduction in the adjoining solid are coupled and must be solved simultaneously (typically through an iterative procedure) at the boundary identified by a Robin condition.

Example 2.2.1

Electrically Heated Metal Slab Exposed to Convective Cooling.

In this example, a metal slab is insulated along the left boundary ($x = 0$) and a surrounding fluid is initially at a temperature of T_f at $x = L$ (right boundary of slab). Then, a resistance heating element is turned on and heat is generated electrically within the slab. Give the one-dimensional governing and boundary conditions for this problem and find the steady-state temperature of the fluid–solid boundary. Explain how the heat flux at this boundary and the other insulated boundary varies with time.

In this problem, the governing and boundary conditions are given by

$$\frac{1}{\alpha}\frac{\partial T}{\partial t} = \frac{\partial^2 T}{\partial x^2} + \frac{\dot{q}'''}{k} \tag{2.13}$$

$$\left.\frac{\partial T}{\partial x}\right|_0 = 0 \tag{2.14}$$

$$-k\left.\frac{\partial T}{\partial x}\right|_L = h[T_s(L, t) - T_f] \tag{2.15}$$

Also, the initial condition for this problem is that the temperature is T_f at $t = 0$. Equation (2.13) can be solved subject to the boundary conditions to find the steady-state temperature at $x = L$.

Alternatively, an overall heat balance can be applied at steady state. At steady state, the only remaining heat balance terms are the terms involving the heat generated and convective heat removal from the metal, so that

$$h[T(L, t \to \infty) - T_f] = \dot{q}'''L \tag{2.16}$$

From this equation, the steady-state temperature at the solid–fluid boundary can be obtained. Furthermore, the heat flux is zero at the left boundary ($x = 0$) since that boundary is insulated. Before steady state is reached at the solid–fluid boundary, the initial heat flux is zero, but then rises monotonically toward its steady-state value as time proceeds. Based on an energy balance for the metal slab itself, this steady-state heat flux must balance the rate of internal heat generation within the slab.

2.3 Thermal and Contact Resistances

In electrical systems, a widely used equation is *Ohm's law*, i.e.,

$$I = \frac{V_h - V_l}{R_e} \tag{2.17}$$

where I, $V_h - V_l$, and R_e refer to the electric current, electric (potential) difference, and electric resistance, respectively. This law essentially states that the current flow is proportional to the driving potential (difference between the high and low applied voltages) and inversely proportional to the resistance through the electrically conducting medium.

A thermal analogy of Ohm's law may be written as follows:

$$q = \frac{T_h - T_l}{R_t} \tag{2.18}$$

In this case, the thermal current is represented by the heat transfer rate, q. Also, rather than a voltage difference, the temperature difference, $T_h - T_l$, gives the driving potential for the flow of heat. The value R_t now refers to *thermal resistance*, instead of electrical resistance. Thus, in problems involving steady-state conditions without internal heat generation, we can use a thermal analogy to electrical circuits. In particular, we can construct a thermal circuit (like electrical circuits) with each stage of heat transfer involving a thermal resistance. Alternatively, the reciprocal of the thermal resistance multiplied by thermal conductivity is called the conduction *shape factor* (i.e., $S = 1/(R_t k)$).

The benefit of using an electrical analogy is that thermal circuit diagrams involving several heat transfer paths may be constructed to analyze thermal problems in a way similar to how electrical circuits are used in electrical problems. For one-dimensional conduction in Cartesian coordinates, Equations (1.12) and (2.18) suggest that $R_t = \Delta x / k$, where Δx refers to the thickness of the plane layer. For convective heat transfer, the thermal resistance of a boundary layer may be inferred from Newton's law of cooling. In this case, the convective heat transfer rate is written as a temperature difference (between surface and fluid) divided by the thermal resistance, $R_{t,conv}$, where

$$R_{t,conv} = \frac{1}{h_{conv} A} \tag{2.19}$$

Similarly, for radiation problems where the heat transfer coefficient is linearized (as described in Chapter 1), the radiative resistance becomes

$$R_{t,rad} = \frac{1}{h_{rad}A} \tag{2.20}$$

The following example demonstrates how a thermal resistance can be calculated for a cylindrical geometry.

Example 2.3.1

Radial Conduction in a Circular Pipe.

The objective of this example is to find the total thermal resistance between a cold fluid flowing inside a pipe and airflow around and outside the tube. Since the convection resistances (inside and outside of the pipe) are obtained by the reciprocals of the convection coefficient, it remains that the conduction resistance must be obtained for a cylindrical wall of the pipe. Then, once all three resistances are combined in series, the total resistance becomes the sum of two convection resistances and one conduction resistance. Finally, the temperature difference (between inner fluid mean temperature and the outside ambient air temperature) divided by the total resistance gives the total rate of heat transfer through the tube.

Performing a steady-state heat balance for a cylindrical control volume similar to that outlined in Equation (2.7),

$$\frac{d}{dr}(q_r''2\pi rL) = 0 \tag{2.21}$$

where r and L refer to radial position and length of pipe, respectively. Substituting Fourier's law and integrating the previous expression twice,

$$T(r) = -\frac{C_1 L}{k} ln(r) + C_2 \tag{2.22}$$

The coefficients of integration may be obtained by imposing specified temperature values at the inner and outer radii, r_1 and r_2, respectively, thereby yielding

$$T(r) = \left[\frac{T_h - T_l}{Lln(r_1/r_2)}\right] ln\left(\frac{r}{r_2}\right) + T_l \tag{2.23}$$

where the subscripts h and l refer to high and low, respectively. Differentiating Equation (2.23) with respect to r and using Fourier's law,

$$q_r = -kA\frac{dT}{dr} = \frac{T_h - T_l}{ln(r_2/r_1)/(2\pi Lk)} \tag{2.24}$$

From this result, it can be observed that the heat flux is written in terms of the temperature difference divided by an appropriate resistance, $R_{t,cond}$, where

$$R_{t,cond} = \frac{ln(r_2/r_1)}{2\pi Lk} \tag{2.25}$$

This result represents the thermal resistance to conduction for heat flow through a cylindrical wall with inner and outer radii of r_1 and r_2, respectively.

The previous example outlines a method to find the thermal resistance in a cylindrical geometry. This approach solved the heat equation and obtained the resulting heat flux by Fourier's law. Then this heat flux was written as a ratio involving temperature difference such that the denominator of the expression is identified as the thermal resistance. Following this procedure, thermal resistances for other geometrical configurations may be readily obtained, i.e.:

- Plane wall: $R_{cond} = L/(kA)$, where L and A refer to the width and cross-sectional area of the wall, respectively
- Spherical shell: $R_{cond} = (1/r_i - 1/r_o)/(4\pi k)$, where r_i and r_o refer to the inner and outer radii, respectively

It can be observed that these thermal resistances are essentially geometric factors describing the resistance to heat flow in a particular geometry. Thermal resistances (or their reciprocals, called shape factors) may be readily calculated for other basic geometries, such as hollow cylinders, hollow spheres, or a horizontal cylinder within a plate. For more complicated regions, solutions by other means may be required, i.e., numerical or graphical techniques.

Once individual thermal resistances are known, heat transfer through a composite system can be determined by a suitable thermal circuit. A sample circuit for heat transfer through a composite wall is described in the following example.

Example 2.3.2

Heat Conduction through a Composite Wall.

A composite wall in a house consists of drywall exterior sections and an inner section equally divided between wood studs and insulation

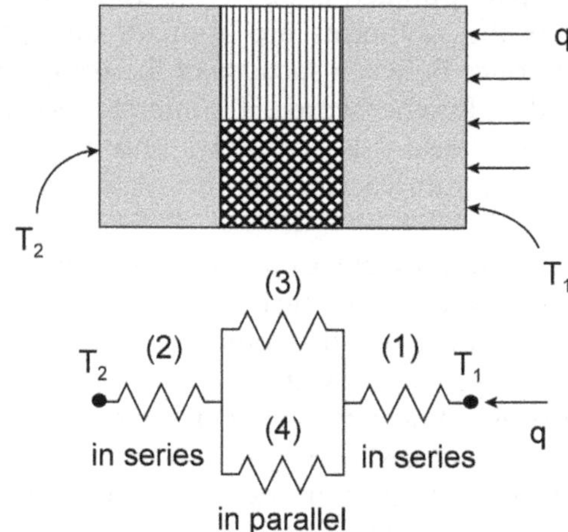

FIGURE 2.4
Thermal circuit for heat transfer in composite wall.

(see Figure 2.4). Construct a thermal circuit to depict the heat conduction through this composite wall. The outer boundaries are isothermal. Is your analysis a one-dimensional solution?

Analogous to Ohm's law, the heat flow through the wall can be written as the specified temperature difference divided by the total thermal resistance, R. This problem can be considered to consist of a thermal circuit in parallel (interior wood and insulation) plus remaining sections in series. Thus, the total resistance consists of sections in series and parallel (see Figure 2.4). Each individual conduction resistance is given by $R = \Delta x/k$, where Δx and k refer to the wall thickness and appropriate thermal conductivity, respectively.

The total resistance to heat flow through the composite wall, R_{tot}, may be expressed by a combination of resistances in series and parallel as follows:

$$R_{tot} = R_1 + \left(\frac{1}{R_3} + \frac{1}{R_4}\right)^{-1} + R_2 \tag{2.26}$$

such that

$$q = \frac{T_1 - T_2}{R_{tot}} \tag{2.27}$$

If the convection coefficients and fluid temperatures surrounding the walls are known, rather than the wall temperatures, then additional convection resistances could be added in series with other resistances in Equation (2.26).

This approach is quasi-one-dimensional in the sense that it is assumed that the isotherms are approximately perpendicular to the wall at the interface between sections of the composite wall in parallel (i.e., sections 3 and 4). In other words, this interface is considered to be essentially adiabatic in the sense that heat flows only in the x direction, thereby leading to a temperature variation in the x direction only.

In the previous example, it is interesting to consider why the front and back sections (i.e., sections 1 and 2) are not also approximated in parallel. This approximation would effectively provide adiabatic separation between the upper and lower parts of these sections, thereby distorting the actual process of heat transfer. Placing two sections in parallel within the thermal circuit suggests that heat flows concurrently through both sections in proportion to their respective thermal resistances; it occurs in the flow direction with no flow across the sections in parallel. The following example further illustrates the usefulness of thermal circuits with convection.

Example 2.3.3
Laboratory Testing of Aircraft De-Icing.
A thin film-type resistance heater is embedded on the inside surface of a wind tunnel observation window for de-fogging during testing of aircraft de-icing systems. The outside and inside air temperatures are 23 and -12°C, respectively, and the outside and inside convective heat transfer coefficients are 12 and 70 W/m^{2}K, respectively. The thermal conductivity of the observation window is 1.2 W/mK, and the window is 5 mm thick. It is desired to find the outside window surface temperatures for the following cases: (i) without a heater, and (ii) with a heater turned on with a heat input of 1.2 kW/m^2. In particular, it is required to find whether the film-type heating is sufficient to prevent frost formation on the window.

At these temperatures, it is anticipated that radiation effects are negligible in comparison to convection and conduction. Also, steady-state conditions and constant thermophysical properties are assumed in this problem. A thermal circuit in series can be constructed from the inside air, through the window, and to the outside ambient air. Based on these assumptions, all thermal resistances can be determined, and the heat flow is given, so that the temperature difference across the windows may be found.

In the first case (heater turned off), the thermal circuit yields

$$q_x = \frac{T_{h,\infty} - T_{l,\infty}}{1/(h_h A) + L/(kA) + 1/(h_l A)} = \frac{T_{h,\infty} - T_h}{1/(h_h A)} \tag{2.28}$$

where the subscripts h and l refer to values evaluated outside and inside the wind tunnel, respectively. Substituting the given values and solving for the window temperature, we obtain $T_h = -5.7°C$, which indicates that frost would form on that side of the window. As a result, heating is required to prevent the frost formation.

In the second case (heater turned on), a similar thermal circuit is constructed, but with a heat source (i.e., film-type heater) located on the outside surface of the window. In this case, the thermal circuit is expressed mathematically as

$$\frac{T_{h,\infty} - T_h}{1/(h_h A)} + q_h'' A = \frac{T_h - T_{l,\infty}}{L/(kA) + 1/(h_l A)} \quad (2.29)$$

Solving this equation in terms of the window surface temperature, we obtain $T_h = 12.5°C$. The temperature is above 0°C with the heater on. As a result, the film-type heater may be used to prevent frost formation on the observation windows during wind tunnel testing of the aircraft de-icing system.

In many practical applications, thermal contact between adjoining material layers is not perfect. In other words, some contact resistance is typically encountered at the interface between adjoining materials. This contact resistance is typically due to machined surfaces in contact that are not perfectly smooth. Some roughness elements are formed at each interface, and when rough surfaces come into contact, gaps between the surfaces and roughness elements lead to poorer convective and radiative transport between the surfaces (as compared with pure conduction alone).

For example, consider a typical contact point with a roughness gap of 50 μm across the contact interface. There may be a temperature drop across this thickness that differs from the anticipated temperature drop across that thickness without the presence of the rough interface. In particular, the thermal circuit across the contact interface now involves conduction together with possibly convection and radiation, rather than conduction alone, if a smooth interface is assumed. These three resistances arise in parallel since the conduction occurs across the points of contact of different roughness elements, whereas convection and radiation across the gaps may occur in between the points of contact. It is anticipated that in many cases these additional thermal resistances due to convection and radiation are small in comparison to conduction. However, in certain circumstances, involving high temperatures or high sensitivity to any temperature change, the contact resistance should be included in the overall thermal analysis.

In some cases, the surface roughness at the contact interface may be somewhat well known. For example, the mean roughness height and the waviness may be quantified, and so some estimate of the surface profile may be adopted. In this way, some analysis of two wavy surfaces in contact over a small distance may give an estimate of the thermal contact resistance. However, in many cases it is difficult to reliably and fully assess the microscopic behavior of roughness elements in contact with each other. As a result, in these cases, the thermal contact resistance is best determined by experimental measurements. Once an estimate of the thermal contact resistance is obtained, it may be used in the thermal circuit as a component in series at the location of the interface between the two materials.

2.4 Fins and Extended Surfaces

In many engineering technologies, such as electronics cooling and compact heat exchangers, enhancing the heat transfer (cooling or heating) from the object is an important design consideration. Fins, or extended surfaces from an object, are common techniques for heat transfer enhancement. Examples of finned surfaces on a tube are shown in Figure 2.5. Along a fin the solid experiences heat transfer by conduction within its edges, as well as heat transfer by convection (or radiation) between its edges and the surrounding fluid. In this section, we will investigate heat transfer through fins.

In the fin analysis, we will consider the case of a thin fin so that a one-dimensional idealization for heat transfer within the fin can be used. In actual cases, the isotherms within a fin may exhibit a two-dimensional character near the edges of the fin. However, in the thin fin approximation, we will use one-dimensional isotherms along the axial direction of the fin. The fin equation may be obtained in a manner similar to that of the procedure carried out previously for the general heat conduction equation. Consider a variable cross-section fin with the base at a temperature of T_b (see Figure 2.6).

Performing a heat balance over the control volume illustrated in Figure 2.6,

$$q_x''A_c = q_x''A_c + \frac{d}{dx}(q_x''A_c)dx + (dA_s)h(T - T_\infty) \tag{2.30}$$

where A_c and A_s refer to the cross-sectional and outer surface areas, respectively, for the differential control volume of thickness dx in Figure 2.6. Using Fourier's law, Equation (1.11), and rewriting it,

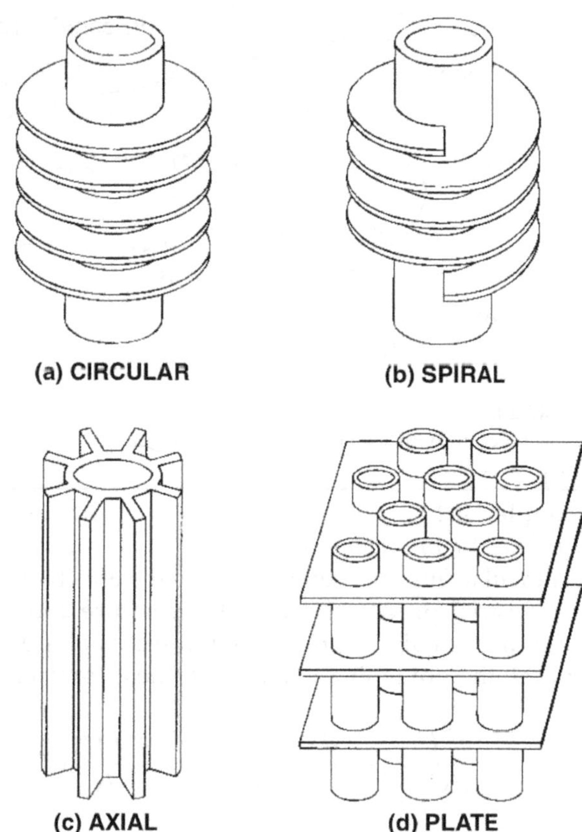

FIGURE 2.5
Finned surfaces. (From Hewitt, G.F., Shires, G.L., and Polezhaev, Y.V., Eds., *International Encyclopedia of Heat and Mass Transfer*, Copyright 1997, CRC Press, Boca Raton, FL. Reprinted with permission.)

$$\frac{d^2T}{dx^2} + \left(\frac{1}{A_c}\frac{dA_c}{dx}\right)\frac{dT}{dx} - \left(\frac{1}{A_c}\frac{h}{k}\frac{dA_s}{dx}\right)(T - T_\infty) = 0 \tag{2.31}$$

In one-dimensional problems, exact solutions of this fin equation are often available. Problems involving this type of linear system, particularly homogeneous equations with constant coefficients, can be solved with methods such as Laplace transforms or integrating factors. A description of these techniques is available in most undergraduate calculus textbooks involving differential equations.

In this section, we will consider the one-dimensional fin as an example where an exact solution is available by direct means. Analytical solutions are generally limited to idealized problems in simplified geometries. However, they provide useful reference solutions for examining particular trends in more complex systems or a source for comparisons with other numerical and experimental data.

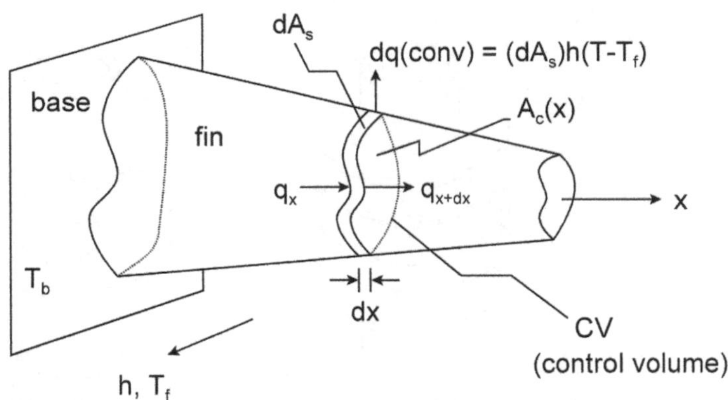

FIGURE 2.6
Schematic of fin with variable cross-sectional area.

Example 2.4.1
Heat Transfer in a Uniform Fin.
Rectangular fins are used to enhance the effectiveness of a heat exchanger in an industrial process. Find the temperature and heat flux distribution in a one-dimensional fin (uniform cross-sectional area; length of L).

Recall the governing equation for heat transfer in fins:

$$\frac{d^2T}{dx^2} + \left(\frac{1}{A_c}\frac{dA_c}{dx}\right)\frac{dT}{dx} - \left(\frac{1}{A_c}\frac{h}{k}\frac{dA_s}{dx}\right)(T - T_\infty) = 0 \tag{2.32}$$

In the case of fins with a uniform cross-sectional area,

$$\frac{d^2\theta}{dx^2} - m^2\theta = 0 \tag{2.33}$$

where $\theta = T(x) - T_\infty$, $m^2 = hP/(kA_c)$, and P, T_∞, and A_c refer to the perimeter, fluid temperature, and cross-sectional area of the fin, respectively. The solution of Equation (2.33) is

$$\theta(x) = D_1 sinh(mx) + D_2 cosh(mx) \tag{2.34}$$

where the coefficients D_1 and D_2 are obtained through specification of the boundary conditions.

For the case of an adiabatic fin tip with a specified base temperature, the following results are obtained for the temperature excess, θ, and heat flux, q_f, respectively:

$$\frac{\theta}{\theta_b} = \frac{cosh[m(L-x)]}{cosh(mL)} \tag{2.35}$$

$$q_f = -kA_c \left.\frac{d\theta}{dx}\right|_0 = Mtanh(mL) \tag{2.36}$$

where $M = \theta_b \sqrt{hPkA_c}$. It can be shown that as the fin length becomes very long (i.e., $L \to \infty$), the following results are obtained:

$$\frac{\theta}{\theta_b} = e^{-mx} \tag{2.37}$$

$$q_f = M \tag{2.38}$$

Another type of boundary condition is a specified temperature condition at the tip of the fin, i.e., $\theta(L) = \theta_L$. In this case, the results for temperature and heat flux are obtained as

$$\frac{\theta}{\theta_b} = \frac{(\theta_L/\theta_b)sinh(mx) + sinh[m(L-x)]}{sinh(mL)} \tag{2.39}$$

$$q_f = M\frac{(cosh(mL) - \theta_L/\theta_b)}{sinh(mL)} \tag{2.40}$$

These results were obtained for the case of fins with a constant cross-sectional area; nonuniform fin analysis must retain the terms involving A_c in Equation (2.32). The one-dimensional approximation allowed us to reduce the general heat conduction equation from a partial differential equation to an ordinary differential equation. It will be observed in subsequent sections that this type of problem reduction can also be extended to other types of boundary value problems, such as separable solutions of two-dimensional heat conduction problems in other geometries.

The previous fin results can be used to assess the thermal effectiveness or performance of the fin. The fin efficiency is defined as the ratio of the actual fin heat transfer, q_f, to the maximum heat transfer, $q_{f,max}$, which occurs if the base temperature is maintained throughout the entire fin. In that case, the maximum temperature difference for convection would be realized. The fin efficiency becomes

$$\eta_f = \frac{q_f}{q_{f,max}} = \frac{q_f}{hA_s\theta_b} \tag{2.41}$$

For example, in a fin of uniform cross-sectional area with an insulated tip, Equations (2.36) and (2.41) yield

$$\eta_f = \frac{tanh(mL)}{mL} \tag{2.42}$$

It is often observed that the fin efficiency decreases with longer fins, since minimal heat transfer occurs whenever the temperature along the fin approaches the fluid temperature (in long fins).

The fin efficiency is frequently depicted in graphical form in terms of an abscissa, X_c, where

$$X_c = L_c^{3/2}\left(\frac{h}{kA_p}\right)^{1/2} \tag{2.43}$$

which represents a dimensionless characteristic length of the fin. In this way, η_f typically decreases with X_c and various geometrical fin configurations can be compared, based on this abscissa, in order to evaluate the fin effectiveness under various conditions. Results for a rectangular fin, based on Equation (2.42), are shown in Figure 2.7. All design parameters that foster a temperature distribution closer to the base temperature (i.e., higher

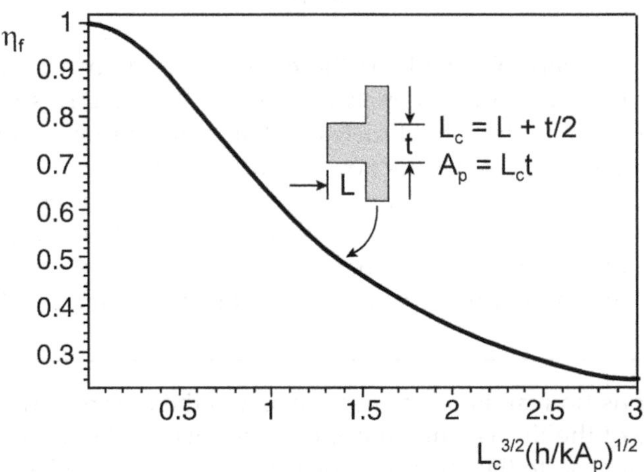

FIGURE 2.7
Fin performance curves.

fin conductivity, lower convection coefficient) lead to a higher overall efficiency of the fin.

Geometrical configuration can have a significant role in fin performance. For example, a triangular fin cross-sectional area often shows better performance (for a given surface length) since this geometry permits a surface temperature closer to the base temperature. In the case of a rectangular fin, the fin profile area, A_p, is $L_c t$, and the corrected fin length (or characteristic length), L_c, is $L + t/2$, where L and t refer to the protruding fin width and base thickness, respectively (see Figure 2.7). On the other hand, $L_c = L$ and $A_p = Lt/2$ for a triangular fin. For purposes of comparing different types of fins, the fin areas are typically matched.

It is useful to examine the trends of fin efficiency in terms of the dimensionless characteristic length, X_c, as shown in Figure 2.7. For shorter fins, L_c decreases, so the temperature profile remains closer to the base temperature, thereby raising η_f. However, the heat transfer is reduced (since the exposed area of the fin is reduced). As a result, a common design objective becomes the minimization of L_c to provide the required rate of heat transfer from the fin. In terms of conductivity trends in the abscissa, X_c, a higher value of k (i.e., copper rather than aluminum fin) allows the fin temperature to become closer to the base temperature, thereby raising the fin efficiency. As expected, the proper trend of an increasing η_f with k is observed in Figure 2.7.

Furthermore, as h (convection coefficient) increases, $T(x)$ along the fin deviates farther from the base temperature, and so η_f decreases. However, the fin heat transfer increases with h (from Newton's law of cooling). Although more heat is transferred, its efficiency of transfer is reduced, due to various possible factors, e.g., boundary layer separation from the fin or an increased pressure drop across the fin gaps. Alternatively, additional entropy is produced at higher values of h, which simultaneously reduces the system efficiency even though the heat transfer is increased. This trade-off between heat transfer and efficiency commonly arises in many thermodynamic systems. For example, increasing the number of baffles inside a heat exchanger typically increases the heat transfer to the fluid, but at the expense of increased pressure drops through the system, thereby requiring additional pumping power with higher system entropy production.

At a given A_p and L_c, while maintaining other parameters (such as convection coefficient) constant, the fin efficiency for a triangular fin is higher than the value for a rectangular fin. However, manufacturing of a triangular fin is generally more difficult and expensive, and rectangular fins can often be packed together more tightly than triangular fins. As a result, a trade-off arises. In the limiting case of $X_c \to 0$, the fin efficiency reaches its maximum value of 1, but this situation has limited practical value since it suggests that the convection coefficient approaches zero (i.e., no heat transfer). In practice, the objective of selecting a suitable fin configuration involves finding the highest efficiency to maintain a required heat transfer rate, while reducing costs and time of manufacturing the fin assembly.

Since heat transfer occurs through the fin and across the base surface, individual resistances are constructed for both parts. In particular, the fin resistance, R_f, and resistance of the base surface, R_b, are given as follows:

$$R_f = \frac{1}{\eta_f h_f A_f} \tag{2.44}$$

$$R_b = \frac{1}{h_b A_b} \tag{2.45}$$

where the subscripts f and b refer to fin and base, respectively.

Since the heat transfer occurs in parallel through the fin and base, the equivalent resistance is given by

$$R_e = \left(\frac{1}{R_f} + \frac{1}{R_b}\right)^{-1} \tag{2.46}$$

Substituting Equations (2.44) and (2.45) into Equation (2.46),

$$R_e = \frac{1}{h_b A_b + \eta_f h_f A_f} \tag{2.47}$$

The equivalent resistance on the left side of Equation (2.47) can be written in terms of a single overall surface efficiency, η_o, and an average convection coefficient, h, as follows:

$$R_e = \frac{1}{\eta_o h A} \tag{2.48}$$

where A refers to the total surface area (including fins and base).

Equating the expressions in Equations (2.47) and (2.48), it can be shown that

$$\eta_o = \frac{h_b A_b + \eta_f h_f A_f}{hA} \tag{2.49}$$

Also, substituting $A = A_f + A_b$ in Equation (2.49) and further assuming that the convection coefficients are uniform and constant, i.e., $h \approx h_b \approx h_f$,

$$\eta_o = 1 - \frac{A_f}{A}(1 - \eta_f) \tag{2.50}$$

This overall surface efficiency can then be used in Equation (2.48) to give the equivalent resistance. The heat transfer from a finned surface to the surrounding fluid can then be obtained by the temperature difference (between surface and fluid) divided by this resulting thermal resistance. This approach is particularly useful in the analysis of finned heat exchangers (Chapter 10) and other extended surface configurations (Kern, Kraus, 1972).

2.5 Multidimensional Heat Conduction

The previous sections have mainly examined heat transfer in one-dimensional systems, where temperature variations occur predominantly in one spatial direction only. However, in many other applications, heat transfer occurs across more than one spatial direction. For example, in three-dimensional heat conduction, Fourier's law becomes

$$\mathbf{q}'' = -k\left(\mathbf{i}\frac{\partial T}{\partial x} + \mathbf{j}\frac{\partial T}{\partial y} + \mathbf{k}\frac{\partial T}{\partial z}\right) \tag{2.51}$$

where **i**, **j**, and **k** (note: bold font; not thermal conductivity) refer to the unit vectors in the x, y, and z directions, respectively.

The minus sign in Equation (2.51) shows that heat flows in the direction of decreasing temperature. In a similar manner, diffusive mass transfer may be expressed in terms of a mass diffusion coefficient and concentration gradient (called *Fick's law*). In Equation (2.51), k refers to thermal conductivity. For example, typical values are $k \approx 10^{-2}$ W/mK for gases. Typical values of k in metal processing applications are $k \approx 35$ W/mK (solid lead) and $k \approx 15$ W/mK (liquid lead).

The governing equation for two-dimensional heat conduction can be obtained by a heat balance over a differential control volume. In particular, consider a homogeneous medium and select a differential control volume, as illustrated in Figure 2.2. From an energy balance for this control volume, the transient increase in energy within the volume must balance the net energy inflow minus the energy outflow, plus any energy generated within the control volume, i.e.:

$$\dot{E}_{cv} = \dot{E}_{in} - \dot{E}_{out} + \dot{E}_g \tag{2.52}$$

Inflow and outflow terms may be evaluated as the heat flux per unit area multiplied by the area of the face (unit depth) on the edge of the differential control volume. Across the upper and right edges of the control volume, the heat fluxes may be expanded with a Taylor series in terms of the corresponding heat fluxes at the inflow faces of the control volume (see Figure 2.2).

After substituting the individual heat flux terms into Equation (2.52) and extending the result to three-dimensional coordinates, we obtain the following heat conduction equation:

$$\rho c_p \frac{\partial T}{\partial t} = \frac{\partial}{\partial x}\left(k\frac{\partial T}{\partial x}\right) + \frac{\partial}{\partial y}\left(k\frac{\partial T}{\partial y}\right) + \frac{\partial}{\partial x}\left(k\frac{\partial T}{\partial z}\right) + \dot{q} \tag{2.53}$$

where $\dot{q}$ refers to a volumetric heat source. In cylindrical coordinates (see Figure 2.1),

$$\rho c_p \frac{\partial T}{\partial t} = \frac{k}{r}\frac{\partial}{\partial r}\left(r\frac{\partial T}{\partial r}\right) + \frac{k}{r^2}\frac{\partial^2 T}{\partial \phi^2} + k\frac{\partial^2 T}{\partial z^2} + \dot{q} \tag{2.54}$$

where a constant thermal conductivity, k, has been further assumed. Alternatively, in spherical coordinates,

$$\rho c_p \frac{\partial T}{\partial t} = \frac{k}{r^2}\left(\frac{1}{r^2}\frac{\partial T}{\partial r}\right) + \frac{k}{r^2 \sin\theta}\frac{\partial^2 T}{\partial \phi^2} + \frac{k}{r^2 \sin\theta}\frac{\partial}{\partial \theta}\left(\sin\theta\frac{\partial T}{\partial \theta}\right) + \dot{q} \tag{2.55}$$

Under steady-state conditions with no heat generation or source terms, the heat equation in Cartesian coordinates, Equation (2.53), is reduced to the following two-dimensional *Laplace's equation*:

$$\nabla^2 T \equiv \frac{\partial^2 T}{\partial x^2} + \frac{\partial^2 T}{\partial y^2} = 0 \tag{2.56}$$

Techniques for solving this heat equation will be described in subsequent sections. The following example shows how boundary conditions must also be outlined for fully describing a heat transfer problem.

Example 2.5.1

Heat Transfer in an Underground Pipeline.

An underground pipeline carrying warm oil is cooled when heat is lost to the surrounding soil. The pipe and ground temperature (above the pipe) are T_1 and T_2, respectively. The pipe radius is R, and its centerline is located at a distance of L below the ground level. The purpose of this example is to set up the governing equations and boundary conditions that fully describe the steady-state heat transfer through the ground.

If the origin of the Cartesian axes is located on the ground above the pipeline, the following governing equations and boundary conditions are obtained:

$$\frac{\partial^2 T}{\partial x^2}+\frac{\partial^2 T}{\partial y^2}=0 \tag{2.57}$$

subject to

$$T(x,\ 0)=T_2=T(x,\ y\rightarrow\infty) \tag{2.58}$$

$$T=T_1 \quad \text{on } x^2+(y-L)^2=R^2 \tag{2.59}$$

Also, based on the problem symmetry about $x=0$,

$$\left.\frac{\partial T}{\partial x}\right|_0=0 \tag{2.60}$$

These equations may be solved by analytical, numerical, or graphical techniques (discussed in subsequent sections and chapters) to find the temperature distribution. Then additional results, such as the rate of heat transfer from the pipe, can be obtained from these temperature results, based on Fourier's law.

Steady-state heat conduction with constant thermophysical properties and no internal heat generation can be classified as a *linear* problem. In linear systems, if several solutions satisfy the governing equation, then a linear combination of these solutions is also a solution of the governing equation. Since steady-state heat conduction is a linear problem, we can use the *superposition principle* by combining individual solutions of simplified problems to obtain the solution of more complicated problems. For example, consider a square plate with each of the four sides maintained at four separate temperatures, T_1, T_2, T_3, and T_4. Furthermore, consider four individual problems with the same square plate, but where one side is maintained at T_i (where $i=1, 2, 3$, or 4), and the remaining sides are held at 0°C. The side maintained at T_i is the same side corresponding to the former problem involving sides at four different temperatures.

Denoting the solution of each individual problem, in order, by T_a, T_b, T_c, and T_d, then each solution obeys Laplaces's equation for steady-state heat conduction, respectively, so that

$$\nabla^2 T_a + \nabla^2 T_b + \nabla^2 T_c + \nabla^2 T_d = 0 \tag{2.61}$$

Combining individual solutions together,

$$\frac{\partial^2}{\partial x^2}(T_a + T_b + T_c + T_d) + \frac{\partial^2}{\partial y^2}(T_a + T_b + T_c + T_d) = 0 \tag{2.62}$$

Thus, the following sum is a solution of the combined problem:

$$T(x, y) \equiv T_a(x, y) + T_b(x, y) + T_c(x, y) + T_d(x, y) \tag{2.63}$$

which indicates that the sum of solutions of each individual problem comprises the solution of the original overall problem. This feature is an important characteristic of linear problems.

The previous example showed that linear problems, particularly steady-state heat conduction, can be subdivided into a sum of individual subproblems involving Dirichlet (specified temperature) boundary conditions. In addition to Dirichlet conditions, the principle of superposition of individual solutions can be applied to linear problems involving mixed Dirichlet–Neumann (flux specified) boundary conditions. For example, consider the same plate described earlier, but now subjected to specified temperatures of T_c and T_h on the left and right boundaries, respectively, as well as an insulated top boundary, and a specified heat flux of q_w along the bottom boundary.

The two subproblems (both with an insulated top boundary) are (i) both sides maintained at T_c, q_w specified along bottom boundary, and (ii) left side held at 0°C, right side at $T_h - T_c$, and insulated bottom boundary. In this case with mixed Dirichlet–Neumann conditions, it can be readily verified that the solutions of individual problems, T_a and T_b, obey

$$T(x, y) = T_a(x, y) + T_b(x, y) \tag{2.64}$$

where

$$T_b(x, y) = \left(\frac{T_h - T_c}{L}\right)x \tag{2.65}$$

and L and x refer to the width of the plate and the Cartesian coordinate, respectively. The main result indicates that the sum of individual solutions of the subproblems is also the solution of the original, combined problem. This principle of superposition is a useful characteristic when analyzing linear problems such as steady-state heat conduction.

2.6 Graphical Solution Methods

In many cases, a geometric transformation between the domain of interest and a simpler solution domain (such as a circular cylinder) may lead to an approximation of the thermal resistance. In particular, if we (i) set equal inner areas between the domains and (ii) set equal volumes between the domains, then the heat flow lanes approach an analogous character. These steps can be used to find graphically based estimates of the thermal resistance.

Example 2.6.1

Thermal Resistance of a Concentric Square Region.

Find the thermal resistance of a region bounded between concentric squares (i.e., two different-sized squares having a common center point). If the ratio between the external side lengths becomes large, then heat flow through the region preserves an isotherm pattern similar to a region bounded between concentric circles. When the previous two-step procedure is applied for concentric squares with outer and inner side lengths of $L = 2\beta$ and $L = 2\alpha$, respectively, and concentric circles with inner and outer radii of $r = a$ and $r = b$, respectively, we obtain

$$kR = \frac{1}{S} \approx \left(\frac{1}{2\pi L}\right) \ln \sqrt{\frac{\pi}{4}\left[\left(\frac{\beta}{\alpha}\right)^2 - 1\right] + 1} \tag{2.66}$$

The thermal resistance of the concentric circle region has been used with a geometric transformation to estimate the resulting resistance between concentric squares.

It may be shown that the error inherent in this solution for steady-state heat conduction between concentric squares is less than 1% (in comparison with the analytic solution). The maximum error occurs for $\beta/\alpha \approx 2$, where the solution possesses its highest degree of two-dimensionality.

In a similar way, approximations of heat transfer rates can be achieved by graphical means whereby an orthogonal network of isotherms and heat flow lines is constructed throughout the domain. This method may be applied to two-dimensional conduction problems with isothermal or adiabatic boundaries. Some examples of commonly encountered configurations are listed as follows, where S denotes the conduction shape factor:

- Plane wall of thickness L: $S = 1/L$

- Isothermal sphere (diameter of D) buried at a depth of z in a semi-infinite medium: $S = 2\pi D/(1 - D/4z)$
- Horizontal isothermal cylinder (length of L, diameter of D) buried in a semi-infinite medium: $S = 2\pi L/\ln(4z/D)$

In general, geometries involving regions not bounded by some type of curvilinear coordinates typically require full numerical solutions for calculating the thermal resistances. Computational methods in heat transfer are discussed in Chapter 11.

In addition to calculating the thermal resistance, discrete graphical techniques can also be used to find upper and lower bound estimates of the thermal resistance (or shape factor). If the actual conduction resistance of a material (i.e., $R = \Delta T_{ref}/q$) cannot be determined by analytical means, approximations for the upper and lower bounds on the thermal resistance may be determined from the *parallel adiabat* and *parallel isotherm* methods, respectively (see the following example).

Example 2.6.2

Thermal Resistance of a Composite Material.

Consider a two-dimensional region in a composite material formed by filaments of conductivity k_2 inside a matrix of conductivity k_1 (where $k_1 \neq k_2$). Two surfaces are at constant temperatures (i.e., $T = T_1$ and $T = T_2$), while the other two boundaries are considered to be adiabatic since they represent lines of symmetry within the material. Explain how the upper and lower bounds of the thermal resistance can be calculated for the two-dimensional region.

In the two-dimensional region, heat flows from the hotter isothermal boundary to the other isothermal boundary. If the heat transfer were unrestrained (i.e., a uniform conductivity throughout the element), so that no cross-diffusion occurred against the principal heat flow direction, the thermal circuit would represent a series of one-dimensional conductances. In other words, if we construct a series of adiabats parallel to the adiabatic boundaries with zero thickness to approximate the heat flow lines, this condition would yield the maximum thermal resistance for the composite material. Thus, the parallel adiabats construction would yield an upper bound on the thermal resistance.

On the other hand, if the resistances between the different material sections were aligned in parallel, then this circuit would yield the minimum thermal resistance. For a parallel isotherm construction through the material, the resistance represents a lower bound on the material's thermal resistance.

Since the upper and lower bounds on the thermal resistance, R_{UB} and R_{LB}, represent asymptotes, the actual thermal resistance lies between these two limit cases. An arithmetic mean, $R = (R_{UB} + R_{LB})/2$, is an approximation of the actual thermal resistance. If the upper and lower bounds are far apart, then a geometric mean would be a more accurate representation for the actual resistance. An effective thermal conductivity of the material may then be evaluated based on this thermal resistance and Fourier's law.

This approach can be useful, particularly for problems involving complicated geometries.

2.7 Analytical Methods

If the governing differential equation can be solved directly, the resulting analytical solution often represents an effective and highly accurate way of analyzing a heat transfer problem (Myers, 1971; Carslaw, Jaeger, 1959). In addition to heat conduction problems, the analytical techniques described in this section can be extended to predict individual aspects of multiphase problems. For example, heat transfer between a nucleating bubble and a heated surface in boiling flows may be analyzed by analytical solutions of the heat conduction equation (Naterer et al., 1998). In this section, analytical techniques based on the *method of separation of variables* and *conformal mapping* will be described.

Boundary value problems refer to problems that seek to determine the solution of a differential equation subject to boundary conditions for the unknown function, such as temperature, at two or more values of the independent variable. Conduction heat transfer problems may be classified as boundary value problems because temperature (or its derivative, in the case of the heat flux) is specified in some part of the overall boundary conditions for the problem. The method of separation of variables may be used to find the exact solution of boundary value problems. In this approach, the solution (such as $T(x, y)$) is assumed to be separable, i.e., $T(x, y) = X(x)Y(y)$. Under this assumption, the governing equation can be separated into a set of ordinary differential equations whose solutions may be more straightforward than the original partial differential equation.

Example 2.7.1
Heat Conduction in a Two-Dimensional Domain.
Consider a heating process in which a rectangular metal ingot is heated by combustion gases in a furnace (see Figure 2.8). The bottom boundary is insulated, while the remaining boundaries are convectively heated. Find the steady-state temperature distribution in the solid and the rate of heat flow across the left boundary. Express your answer in terms of the heat transfer coefficients for each surface, as well as the ambient gas temperature, T_f.

For steady conduction heat transfer in Cartesian coordinates,

$$\frac{\partial^2 T}{\partial x^2} + \frac{\partial^2 T}{\partial y^2} = 0 \tag{2.67}$$

Neumann and Robin boundary conditions exist along the ingot surfaces, i.e.,

$$\left.\frac{\partial T}{\partial y}\right|_{y=0} = 0 \tag{2.68}$$

at $y = 0$. Also,

$$\frac{\partial T}{\partial y} = -\frac{h}{k}(T(x, b) - T_f) \tag{2.69}$$

at $y = b$ and

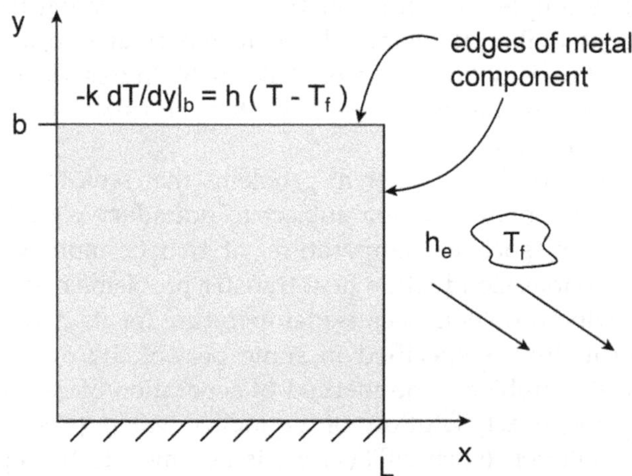

FIGURE 2.8
Schematic of two-dimensional heat conduction.

$$\frac{\partial T}{\partial x} = -\frac{h_e}{k}(T(L, y) - T_f) \tag{2.70}$$

at $x = L$. Along the boundary at $x = 0$, we have

$$\frac{\partial T}{\partial x} = -\frac{h_c}{k}(T_b - T(0, y)) \tag{2.71}$$

Now assume that $T(x, y)$ is separable and let $T(x, y) = X(x)Y(y)$. From Laplace's equation, Equation (2.67),

$$X''Y + XY'' = 0 \tag{2.72}$$

or

$$\frac{X''}{X} = -\frac{Y''}{Y} \tag{2.73}$$

Since the left side of Equation (2.73) is a function of only x and the right side of Equation (2.73) is a function of y, both sides must be constant; let the constant equal $k = \lambda^2$. The constant should be positive rather than negative, $k = -\lambda^2$, because the function Y should contain a periodic solution to be able to satisfy the first two boundary conditions.

Laplace's equation may then be reduced to a set of ordinary differential equations with exponential and trigonometric solutions for $X(x)$ and $Y(y)$, respectively, i.e.,

$$X(x) = c_1 cosh(\lambda x) + c_2 sinh(\lambda x) \tag{2.74}$$

$$Y(y) = c_3 cos(\lambda y) + c_4 sin(\lambda y) \tag{2.75}$$

The boundary conditions along the $y = 0$ and $y = b$ boundaries become $Y'(0) = 0$ and $Y'(b) = -hY(b)/k$, respectively. Substituting these conditions into Equation (2.75) yields $c_4 = 0$ and an equation for the *eigenvalues* of this *Sturm–Liouville problem*,

$$\frac{\lambda b}{Bi} = cot(\lambda b) \tag{2.76}$$

where $Bi = hb/k$ represents the Biot number. The roots of Equation (2.76) represent the eigenvalue solutions or separation constants for Equation (2.75). These roots lie approximately π apart. In order to locate the exact root positions, a Newton–Raphson root-searching algorithm can be used.

In order to satisfy the remaining boundary conditions along the $x = 0$ and $x = L$ boundaries, superpose all possible solutions to meet those conditions, i.e.,

$$T(x, y) = \sum_{n=1}^{\infty} [A_n cosh(\lambda_n x) + B_n sinh(\lambda_n x)] cos(\lambda_n y) \tag{2.77}$$

where λ_n refers to the set of n eigenvalues. Applying the third boundary condition along the boundary $x = L$,

$$B_n = -A_n \psi_n \tag{2.78}$$

where

$$\psi_n = \frac{Bi_e cosh(\lambda_n L) + (\lambda_n L) sinh(\lambda_n L)}{(\lambda_n L) cosh(\lambda_n L) + Bi_e sinh(\lambda_n L)} \tag{2.79}$$

For the asymptotes, $Bi_e = 0$ and $Bi_e \to \infty$, the coefficients in Equation (2.78) approach $\psi_n = tanh(\lambda L)$ and $\psi_n = coth(\lambda_n L)$, respectively. If $\lambda_n L > 2.65$, then $tanh(\lambda L) \approx coth(\lambda L)$.

The solution may then be rewritten in terms of only one unknown coefficient:

$$T(x, y) = \sum_{i=1}^{\infty} A_n [cosh(\lambda_n x) - \psi_n sinh(\lambda_n x)] cosh(\lambda_n y) \tag{2.80}$$

Applying the final boundary condition,

$$\sum_{n=1}^{\infty} A_n \left(\psi_n \lambda_n + \frac{h_c}{k} \right) cos(\lambda_n y) = \frac{h_c}{k} (T_b - T_f) \tag{2.81}$$

We may now multiply this equation by orthogonal functions $cos(\lambda_m y)$ and integrate from $y = 0$ to $y = b$ to obtain the final unknown coefficient, A_n.

From orthogonality relations (i.e., the integrated left side vanishes for all series terms except for the case $m = n$), the unknown coefficient may be simplified as follows:

$$A_n = \frac{(T_b - T_f) sin(\lambda_n b) h_c}{k \psi_n \lambda_n (\lambda_n b + sin(\lambda_n b) cos(\lambda_n b))} \tag{2.82}$$

This equation now permits the final solution for the temperature field, Equation (2.80), to be written in a closed form.

In addition, the total heat flow into the solid (across the left boundary) may be obtained from differentiation of Equation (2.80) and integration of the resulting Fourier heat flux across the exterior surface, i.e.,

$$q = 2\int_0^b \left(-k\frac{\partial T}{\partial x}\right)_0 dy = 2k\sum_{i=1}^{\infty} A_n \psi_n sin(\lambda_n b) \tag{2.83}$$

where k refers to the thermal conductivity of the solid.

Although solutions obtained by separation of variables usually involve complicated expressions with an infinite series, the series terms generally converge quickly. As a result, in many cases only the first few terms are required in the series in order to achieve good accuracy.

Conformal mapping is another useful technique for solving problems governed by the following Laplace's equation:

$$\frac{\partial^2 T}{\partial x^2} + \frac{\partial^2 T}{\partial y^2} = 0 \tag{2.84}$$

where T is the temperature at position (x, y). A solution of this equation that satisfies the boundary conditions will determine the temperature distribution and pattern of heat flow. The method of conformal mapping will be briefly introduced in this section; the reader is referred to complex analysis books by Mathews (1982) and Henrici (1986) for more detailed treatments regarding this method.

The solution may be obtained by transforming a given complicated geometry in the complex z-plane ($z = x + iy$ where $i = \sqrt{-1}$) with a conformal transformation to the w-plane ($w = u + iv$). In other words, a point (x, y) in the z-plane may be mapped to a corresponding point (u, v) in the w-plane through a transformation,

$$z = g(w) \tag{2.85}$$

where $g'(w)$ is nonzero. The temperature that satisfies Laplace's equation in the z-plane also satisfies the equation in the w-plane, and there will exist corresponding points in both planes with identical temperatures.

Example 2.7.2

Heat Transfer in an Irregularly Shaped Region.

Consider an irregular region formed by the gap between concentric polygons. Use the method of conformal mapping to find the temperature distribution in this region when Dirichlet (fixed temperature) boundary conditions are applied along the surfaces of both polygons.

Perform a conformal mapping of $z = g(w)$ where $z = x + iy$ defines coordinates in the irregular region. In the w-plane, the basic shape with a well-known solution is a concentric cylinder. In that domain, purely radial heat flow (one-dimensional in the r direction) arises subject to isothermal boundary conditions, i.e.,

$$T(r_2) = T_2 \tag{2.86}$$

$$T(r_1) = T_1 \tag{2.87}$$

The solution of Laplace's equation in radial coordinates,

$$\frac{\partial}{\partial r}\left(r\frac{\partial T}{\partial r}\right) = 0 \tag{2.88}$$

subject to the boundary conditions, may be written as

$$\frac{T - T_2}{T_1 - T_2} = \frac{ln(\sqrt{u^2 + v^2}/r_2)}{ln(r_1/r_2)} \tag{2.89}$$

where $r^2 = u^2 + v^2$. The temperature at any point (u, v) corresponds to the temperature at (x, y) in the z-plane, where the function $z = g(w)$ performs the mapping between the z and w planes.

The heat flow may then be determined by integration of the Fourier heat flux ($q'' = -k(dT/dr)$) along the boundary. For concentric cylinders,

$$Q = \left[\frac{2\pi}{ln(r_2/r_1)}\right]k(T_1 - T_2) \equiv Sk(T_1 - T_2) \tag{2.90}$$

per unit length of cylinder. Once mapped back to the z-plane, the results represent the solution of heat transfer in the original irregular domain.

If the transformation of variables between z- and w-planes cannot be handled by analytical means alone, then a point-by-point numerical matching and integration procedure, together with Equation (2.85), can provide an alternative numerical solution. For example, the transformation between concentric polygonal regions and concentric cylinders (Naterer, 1996) has been considered in terms of a Schwarz–Christoffel transformation with applications to conduction heat transfer through shaped fibers in composite materials.

2.8 Transient Heat Conduction

In problems where temperature varies with time, an additional transient term (representing rate of energy increase/decrease with time) must be considered in the heat equation. Various analytical techniques, including separation of variables, may be used to analyze these transient problems. Alternatively, scaling analysis provides an approximate solution technique that can be applied in many types of problems (involving conduction and other modes of heat transfer).

In discrete scaling, an approximate solution is obtained by a discrete, rather than differential, analysis of the problem. For example, a global energy balance can provide a reasonable estimate of the total heat transfer from a system in some cases without examining the specific details of the differential analysis. An approximate solution of the governing equations may be obtained by discrete estimates of characteristic scales in the problem. In many applications, this approach provides reasonable estimates.

Example 2.8.1

Transient Conduction in a Long Bar.

Consider one-dimensional heat conduction in a bar, initially at $T = T_o$, with the left boundary ($x = 0$) heated and maintained at $T = T_s$. Apply discrete scaling to find the heat flow across this left boundary.

Recall the governing differential equation for transient heat conduction,

$$\rho c_p \frac{\partial T}{\partial t} = k \frac{\partial^2 T}{\partial x^2} \tag{2.91}$$

In this problem, the left boundary transfers heat into the domain and the thermal disturbance moves rightward as heat is transferred in this direction from the boundary. Consider the resulting characteristic discrete scales in this problem. For example, let δ refer to the distance of propagation of the thermal disturbance. If the temperature distribution is approximated by a linear profile across the depth δ while varying in slope at different time levels, then we have the following discrete estimate of the governing equation:

$$\rho c_p \left(\frac{T_s - T_o}{t} \right) \approx k \left(\frac{T_s - T_o}{\delta^2} \right) \tag{2.92}$$

Solving Equation (2.92) in terms of δ and then calculating the heat flux, q, through Fourier's law by forming the temperature difference divided by δ, we obtain

$$q = k\left(\frac{T_s - T_o}{\sqrt{\alpha t}}\right) \tag{2.93}$$

which agrees with the exact solution (presented in an upcoming section), except for an additional factor of $\sqrt{\pi}$ in the denominator. In other words, the correct functional trend of the solution has been obtained by discrete scaling. The approximate solution is in the same form as the exact solution with only a missing factor of $1/\sqrt{\pi}$, due to the approximation of the linearized temperature profiles in the solution. Thus, the method of discrete scaling can indicate correct trends quickly in the solution without providing the exact numerical values.

The analysis of transient heat conduction is largely dependent on the extent of spatial temperature gradients within the material. For example, recall our earlier results involving steady-state conduction through a plane wall. One side of the wall is maintained at T_h. The other side is at a temperature of T_s and is exposed to a fluid at T_∞ (see Figure 2.9). A thermal circuit of this heat flow problem would have a conduction resistance (internal resistance) in series with a convection resistance between the wall and fluid temperature (external resistance). Then, computing the ratio of the internal to external thermal resistances,

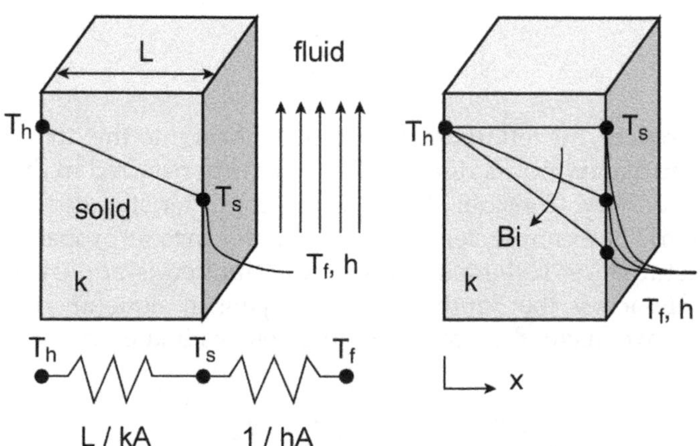

FIGURE 2.9
Schematic of transient heat conduction through a wall.

$$\frac{T_h - T_s}{T_s - T_f} = \frac{L/k}{1/h} = \frac{hL}{k} \equiv Bi \tag{2.94}$$

where Bi refers to the Biot number (dimensionless).

The Biot number is an important dimensionless parameter that identifies the relative significance of spatial temperature gradients within a material during heat conduction. Based on Equation (2.94) for Biot values much less than unity, the wall thickness, L, is usually small, or the thermal conductivity, k, is large. This condition occurs when the internal thermal resistance is much smaller than the external resistance or, alternatively, the difference between sidewall temperatures is much smaller than the difference between T_s and T_f. In other words, variations of temperature with positions inside the solid are so small that a *lumped analysis* may be used. In a lumped analysis, spatial gradients within the solid are neglected and the solid may be considered to be essentially isothermal (see Figure 2.9).

On the other hand, for very large values of the Biot number, the convective heat transfer coefficient is large, i.e., boiling heat transfer, turbulent flow, etc. As a result, the external resistance to heat transfer is small in comparison to conduction (internal resistance) within the solid. Spatial gradients of temperature become significant, since the temperature difference between the sidewall temperatures is large (relative to the difference between T_s and T_f). In this case, as well as in circumstances where the Biot number is approximately of order unity, a lumped analysis cannot be used; instead, the heat conduction equation must be solved with both transient and spatial derivatives present. The Biot number then becomes a main parameter in analyzing transient conduction problems. As a general rule, if $Bi \ll 1$ (i.e., less than about 0.1), then a lumped analysis may be used; otherwise, spatial temperature gradients should be retained in the analysis.

In many cases, such as a thin wall or a tiny droplet, the length scales alone can usually give an indication of the significance of spatial temperature variations within a solid, particle, or droplet. For example, in our earlier fin analysis, a one-dimensional temperature profile was assumed to be valid since only x direction temperature gradients (along the fin axis) were retained in the governing equation. We can now ask whether this assumption is justified based on the resulting Biot number. Based on the fin half-width, we find that the Biot number is typically much smaller than unity since the fin thickness is small and the thermal conductivity is high (i.e., copper fin to provide high fin efficiency). This computation suggests that spatial variations of temperature in the y direction are small, and so the temperature varies along the fin only, not across the fin. The y direction refers to the cross-flow direction corresponding to the direction of length used in the Biot number calculation. Similarly, in other applications such as multiphase flows with droplets, spatial gradients within the droplet may be neglected for droplets that are sufficiently small. Droplets with a characteristic length (such as diameter) yielding a Biot number of at least an order of magnitude smaller than unity may be considered small enough to be essentially isothermal.

Thus, lumped capacitance method may be used when $Bi \leq 0.1$ in heat conduction problems. In a lumped analysis, the temperature is assumed to be constant throughout the solid (or medium corresponding to the Biot number calculation). The following example outlines the procedure of the lumped capacitance method.

Example 2.8.2

Lumped Analysis of an Object Suddenly Immersed in a Fluid.

Consider an arbitrarily shaped solid, initially at a temperature of T_i. It is suddenly cooled once exposed to a convective environment at T_∞ with a convection coefficient of h. The volume and surface area of the object are V and A_s, respectively. The objective of the problem is to find the temperature variation of the solid object over time.

Based on the conservation of energy for a control volume defined by the solid object,

$$\dot{E}_{cv} = -\dot{E}_{out} \tag{2.95}$$

This energy balance states that the rate of decrease of energy of the object (positive $\dot{E}_{out}$ leaving the control volume) balances the rate of energy (heat) outflow from the object to the surrounding fluid. Writing the energy expressions in terms of temperature,

$$\rho c_p V \frac{\partial T}{\partial t} = -hA_s(T - T_\infty) \tag{2.96}$$

where $T > T_\infty$.

The factor $\rho c_p V$ refers to the lumped thermal capacitance. It essentially indicates the energy stored (or released) per degree of change of temperature for the entire mass of the object once it is immersed in the fluid at T_∞. The capacitance is lumped in the sense that the temperature change is considered to be uniform throughout the entire isothermal object. As a result, the method is called a lumped capacitance method.

The temperature excess, θ, is defined as the difference between the solid and fluid temperatures, so that integrating both sides of Equation (2.96), subject to the initial condition ($\theta(0) = \theta_i$), yields

$$\frac{\theta}{\theta_i} = \frac{T(t) - T_\infty}{T_i - T_\infty} = e^{-t/\tau} \tag{2.97}$$

where τ refers to the thermal time constant,

$$\tau = \frac{\rho c_p L_c}{h} \tag{2.98}$$

The characteristic length, L_c, may be regarded as the ratio between the object's volume and its surface area. In this lumped approach, the temperature is observed to vary with time, but not spatially within the object.

Alternatively, the solution may be expressed as

$$\frac{\theta}{\theta_i} = exp\left(-\frac{ht}{\rho c_p L_c}\right) = exp(-Bi \cdot Fo) \tag{2.99}$$

where Fo refers to the Fourier number ($Fo = \alpha t/L_c^2$), representing non-dimensional time. Bi is the Biot number.

In the previous example, τ characterizes the object's rate of temperature change. In particular, if τ is small, the temperature change is rapid, since it suggests that the convective heat transfer coefficient is high. Conversely, if the convection coefficient is small relative to the thermal capacitance, or numerator in Equation (2.98), then a resulting large τ suggests that the rate of temperature change with time is slow.

2.9 Combined Transient and Spatial Effects

In the previous section, a lumped capacitance method was used in problems where spatial gradients of temperature within the object were considered to be negligible (i.e., when $Bi < 0.1$). However, if $Bi \geq 0.1$, spatial variations of temperature are not negligible. For example, if the wall thickness is large, it is anticipated that some transient period is required before the thermal effects of convection near the wall are fully conducted throughout the wall, unlike a more rapid transfer of heat across a very thin wall. As a result, temperature varies with both position and time in transient problems where $Bi \geq 0.1$. In this section, solution methods for these types of problems will be presented. In the following example, spatial and transient effects are predicted for conduction heat transfer through a plane wall.

Example 2.9.1

Plane Wall with Convection.

Consider a plane wall, initially at T_i, suddenly exposed to a convective environment on both sides of the wall with a convection coefficient of h and an ambient temperature of T_∞. The wall half-thickness is L, and x is measured from the midpoint of the wall rightward. The purpose of this example is to find the temperature distribution within the wall, including the variations with time and position.

The governing equation for this problem is given by

$$\frac{1}{\alpha}\frac{\partial T}{\partial t} = \frac{\partial^2 T}{\partial x^2} \tag{2.100}$$

subject to the following initial and boundary conditions:

$$T(x,0) = T_i(x) \tag{2.101}$$

$$\left.\frac{\partial T}{\partial x}\right|_0 = 0 \tag{2.102}$$

where $x = 0$ is the midpoint and

$$-k\left.\frac{\partial T}{\partial x}\right|_L = h[T(L,t) - T_\infty] \tag{2.103}$$

The boundary condition at the midpoint position ($x = 0$) represents a symmetry condition in the temperature profile, since both sides of the wall are identically cooled by convection. The other boundary condition represents the balance between conduction on the solid side and convective heat transfer to the surrounding fluid at $x = L$.

The above equations may be solved by the method of separation of variables. Using this method,

$$\frac{\theta}{\theta_i} = \sum_{n=1}^{\infty} C_n exp(-\zeta_n^2 Fo)cos\left(\zeta_n \frac{x}{L}\right) \tag{2.104}$$

where

$$C_n = \frac{4sin\zeta_n}{2\zeta_n + sin(2\zeta_n)} \tag{2.105}$$

and ζ_n satisfies

$$\zeta_n tan(\zeta_n) = Bi \tag{2.106}$$

The above solution was obtained for a planar wall. A similar approach can be used to obtain solutions for other geometric configurations aligned with boundaries in curvilinear coordinates, such as a cylindrical or spherical domain. These solutions are typically illustrated in graphical form, i.e., *Heisler charts* (1947). For the previous example, the Heisler chart would graphically depict Equation (2.104) to show that the midplane temperature declines with Fourier number. At a given Fourier number (i.e., fixed point in time), a higher midplane temperature is realized at lower Biot numbers. This result occurs since the convective heat transfer coefficient decreases with the Biot number, thereby suggesting a lower rate of temperature change within the wall at low Biot numbers. These trends have been interpreted in the context of the previous example (plane wall), but similar charts and results are available for long cylinders (two-dimensional) and spherical configurations.

In the previous example, heat conduction occurred within a planar wall of finite thickness. Boundary conditions were applied at the midplane (symmetry conditions) and the convective boundary. Analytical solutions can be obtained for other transient conduction problems. In the following example, heat conduction is examined in a one-dimensional semi-infinite domain where end effects are not experienced. This configuration can also be used to represent heat conduction in a finite one-dimensional domain at early times before the effects of the opposite boundary are transmitted inward.

Example 2.9.2

Heat Conduction in a Semi-Infinite Solid.

The purpose of this example is to obtain the temperature and heat flux distributions in a semi-infinite solid subject to various types of boundary conditions. The governing equation is given by

$$\frac{1}{\alpha}\frac{\partial T}{\partial t} = \frac{\partial^2 T}{\partial x^2} \tag{2.107}$$

Find the temperature distributions for three types of boundary conditions: (i) constant surface temperature, (ii) constant surface heat flux, and (iii) surface convection condition.

Case (i): Constant Surface Temperature

In this case, the left boundary is maintained at T_s and heat is conducted into the solid from the boundary located at $x = 0$. The governing initial and boundary conditions are

$$T(x,0) = T_i \tag{2.108}$$

$$T(\infty, t) = T_i \tag{2.109}$$

$$T(0, t) = T_s \tag{2.110}$$

It is anticipated that the solution of this problem exhibits self-similarity in the sense that a thermal wave propagates into the material, and these thermal disturbances are self-similar with respect to each other over time. Based on this property, a similarity variable may be defined that allows all profiles to be collapsed into a single profile in terms of that similarity variable. In other words, the similarity variable is a type of stretching factor that is applied to the thermal profiles to allow them to be combined into a single profile. In this way, the governing equation can be separated into a set of ordinary differential equations that can be solved together to form the resulting similarity solution.

This previously described procedure (based on the similarity solution method described in Chapter 3) can be used as a possible way to obtain the following solution:

$$\frac{T(x,t) - T_s}{T_i - T_s} = \frac{2}{\pi}\int_0^w e^{-v^2} dv = erf\left(\frac{x}{2\sqrt{\alpha t}}\right) \tag{2.111}$$

Based on Fourier's law (i.e., differentiating the temperature distribution with respect to x and evaluating at the wall), we obtain the following heat flux:

$$q_s''(t) = -k\frac{\partial T}{\partial x}\bigg|_0 = \frac{k(T_i - T_s)}{\sqrt{\pi \alpha t}} \tag{2.112}$$

The result shows that the heat flux decreases with time since the slope of temperature falls with time as heat conducts further into the solid.

Case (ii): Constant Surface Heat Flux

In this case, the initial and boundary conditions are given by

$$T(x,0) = T_i \tag{2.113}$$

$$T(\infty, t) = T_i \tag{2.114}$$

$$-k\frac{\partial T}{\partial x}\bigg|_0 = q_o'' \tag{2.115}$$

For this problem, the exact solution is given by

$$T(x,t) = T_i + \frac{2q_o''}{k}\sqrt{\frac{\alpha t}{\pi}}exp\left(\frac{-x^2}{4\alpha t}\right) - \frac{q_o'' x}{k}erfc\left(\frac{x}{2\sqrt{\alpha t}}\right) \tag{2.116}$$

where $erfc(w) = 1 - erf(w)$ is the complementary error function. The heat flux can be determined from this result after differentiation of the temperature field with respect to x and evaluation at $x = 0$ (similarly as shown in case (i)).

Case (iii): Surface Convection

In this case, the initial and boundary conditions are given by

$$T(x,0) = T_i \tag{2.117}$$

$$T(\infty,t) = T_i \tag{2.118}$$

$$-k\frac{\partial T}{\partial x}\bigg|_0 = h[T_\infty - T(0,t)] \tag{2.119}$$

In this case, the analytic solution is

$$\frac{T(x,t) - T_i}{T_\infty - T_i} = 1 - erf\left(\frac{x}{2\sqrt{\alpha t}}\right) - \left[exp\left(\frac{hx}{k} + \frac{h^2\alpha t}{k^2}\right)\right] \times \left[1 - erf\left(\frac{x}{2\sqrt{\alpha t}} + \frac{h\sqrt{\alpha t}}{k}\right)\right] \tag{2.120}$$

Again, the heat flux can be computed from this temperature distribution, together with Fourier's law.

In addition to the methods described in the previous example problems, transient heat conduction problems are often separable, and as a result, they can be well analyzed by the method of separation of variables. Based on this approach, the temperature solution can be subdivided into a product of functions of spatial coordinate(s), as well as a function of time (representing the transient temperature variation). Further detailed treatments of more complex conduction problems are presented by Kakac and Yener (1988) and Schneider (1955).

References

H.S. Carslaw, and J.C. Jaeger. 1959. *Conduction of Heat in Solids*, Oxford: Oxford University Press.

M.P. Heisler. 1947. "Temperature charts for induction and constant temperature heating," *Trans. ASME* **69**: 227–236.

P. Henrici. 1986. *Applied and Computational Complex Analysis*, New York: John Wiley and Sons.

S. Kakac, and Y. Yener. 1988. *Heat Conduction*, Washington, D.C.: Hemisphere.

D.Q. Kern, and A.D. Kraus. 1972. *Extended Surface Heat Transfer*, New York: McGraw-Hill.

J.H. Mathews. 1982. *Basic Complex Variables*, Boston: Allyn and Bacon.

G.E. Myers. 1971. *Analytical Methods in Conduction Heat Transfer*, New York: McGraw-Hill.

G.F. Naterer. 1996. "Conduction shape factors of long polygonal fibres in a matrix," *Numerical Heat Transfer A* **30**: 721–738.

G.F. Naterer, et al. 1998. "Near-wall microlayer evaporation analysis and experimental study of nucleate pool boiling on inclined surfaces," *ASME J. Heat Transfer* **120**: 641–653.

P.J. Schneider. 1955. *Conduction Heat Transfer*, Cambridge, MA: Addison-Wesley.

Problems

1. A system of mass M is initially maintained at a temperature of T_0. Then the system is suddenly energized by an internal source of heat generation, $\dot{q}_o$, and the surface at $x = L$ is heated by a fluid at T_∞ with a convection coefficient of h. The boundary at $x = 0$ is well insulated and one-dimensional heat transfer is assumed to occur.
   (a) Write the mathematical form of the initial and boundary conditions.

   (b) Explain how the temperature varies with time and position within the system. Will a steady-state condition be reached? Explain your response.
   (c) Explain how the net heat flux varies with time along the planes of $x = 0$ and $x = L$.
2. A system of mass M involving one-dimensional heat transfer with constant properties is initially at a temperature of T_0. Suddenly the system is heated by an internal electrical source, q_o''', and the surface at $x = L$ is heated by a fluid at T_∞ with a convection coefficient of h. The boundary at $x = 0$ is maintained at $T(0, t) = T_0$.
   (a) Select an appropriate control volume and derive the differential heat equation for $T(x, t)$. Identify the initial and boundary conditions for this problem. Do not solve the differential equation.
   (b) How does the temperature distribution throughout the material vary with time and position? Will a steady-state temperature be reached? Explain your response.
   (c) Explain how the heat flux varies with time along the planes $x = 0$ and $x = L$.
   (d) Discuss how the relative magnitudes of the internal and external heat transfer processes affect the results in parts (b) and (c).
3. A rectangular system of mass M involving one-dimensional heat transfer with constant properties and no internal heat generation is initially at a temperature of T_i. Suddenly the surface at $x = L$ is heated by a fluid at T_∞ with a convection coefficient of h and a uniform radiative heat flux of q_o''. The boundaries at $x = 0$ and elsewhere are well insulated.
   (a) Select an appropriate control volume and derive the governing differential heat equation for $T(x, t)$. Identify the initial and boundary conditions for this problem. Do not solve the differential equation.
   (b) Explain how the temperature varies with x at the initial condition ($t = 0$) and for several subsequent times. Will a steady-state condition be reached? Explain your response.
   (c) Explain how the heat flux varies with time at $x = L$.
   (d) How would the results in parts (a) and (b) change if only a portion of the incoming radiation was absorbed by the surface at $x = L$ and the remainder was reflected from the surface?
4. Derive the three-dimensional heat conduction equation (including heat generation) in the spherical coordinate system.

5. An experimental apparatus for the measurement of thermal conductivity uses an electrically heated plate at a temperature of T_h between two identical samples that are pressed between cold plates at temperatures of T_c. It may be assumed that contact resistances between the surfaces in the apparatus are negligible. Thermocouples are embedded in the samples at a spacing l apart. The lateral sides of the apparatus are insulated (i.e., one-dimensional heat transfer), and the thermal conductivity varies with temperature approximately as $k(T) = k_0(1 + cT^2)$, where k_0 and c are constants and T refers to temperature.
   (a) The measurements indicate that a temperature difference of $T_{1,h} - T_{1,c}$ exists across the gap of width l. Determine an expression for the heat flow per unit area through the sample in terms of $T_{1,h}$, $T_{1,c}$, l, and the constants k_0 and c.
   (b) State an important advantage in constructing the apparatus with two identical samples surrounding the heater, rather than a single heater–sample combination.
6. A composite wall consists of a layer of brick (8 cm thick with $k =$ 0.6 W/mK) and a layer of fiber board (2 cm thick with $k = 0.1$ W/mK). Find the thickness that an additional insulation layer ($k = 0.06$ W/mK) should have in order to reduce the heat transfer through the wall by 50%. The convection coefficient is $h = 10$ W/m^{2}K along the brick side of the wall.
7. A composite wall inside a house consists of two external wood sections (thickness and height of 6 cm and 1.1 m, respectively) that surround an interior section of a width of 20 cm, consisting of insulation (height of 1 m) above wood (height of 0.1 m). The wood and insulation conductivities are 0.13 and 0.05 W/mK, respectively. Since the upper and lower portions of the wall are well insulated, quasi-one-dimensional heat conduction through the wall is assumed. Under steady-state conditions, a temperature difference of 46°C is recorded across the two end surfaces. Using a suitable thermal circuit for this system, find the total rate of heat transfer through the wall (per unit depth).
8. A window heater in a refrigerated storage chamber consists of resistance heating wires molded into the edge of the 4-mm-thick glass. The wires provide approximately uniform heating when power is supplied to them. On the warmer (interior) side of the window, the air temperature and convection coefficient are $T_{h,\infty} =$ 25°C and $h_h = 10$ W/m^{2}K, respectively, whereas on the exterior side they are $T_{l,\infty} = -10$°C and $h_l = 65$ W/m^{2}K.
   (a) Determine the steady-state glass surface temperatures, T_h and T_l, before the electrical heater has been turned on.

(b) Explain how the thermal circuit is constructed for this problem after the heater has been turned on for some time. Calculate the heater input, q''_{ht}, required to maintain an inner surface temperature of $T_h = 15°C$.

9. The cover of a press in an industrial process is heated by liquid flowing within uniformly spaced tubes beneath the surface of the top cover. Each tube is centrally embedded within a horizontal plate ($k = 18$ W/mK), which is covered by another top plate ($k = 80$ W/mK) and insulated from below. A heating rate of 1.5 kW/m to the top cover plate is required. Air flows at a temperature of 26°C with a convection coefficient of 180 W/m^{2}K above the top cover. The top plate thickness is 6 mm, and the inner tube diameter is 20 mm. The tube-to-tube spacing is 80 mm. What is the temperature of the top cover? The convection coefficient for the inner tube flow is 900 W/m^{2}K.
10. Show that the one-dimensional (radial) thermal resistance of a cylindrical wall approaches the correct limiting behavior as the cylinder radius becomes large relative to the wall thickness.
11. An oil and gas mixture at $T_i = 160°C$ is extracted from an offshore reservoir and transported through a pipeline with a radius of 3 cm. Ambient conditions surrounding the pipe are $T_\infty = 30°C$ with a convection coefficient of $h = 60$ W/m^{2}K. Determine the thickness of an asbestos insulation layer ($k = 0.1$ W/mK) such that the heat losses from the pipeline are reduced by 50%.
12. A spherical storage tank with an inner diameter of 120 cm is maintained at a temperature of 130°C. An insulation layer is required to cover the tank to reduce heat losses from the tank and ensure that the outer surface temperature of the tank does not exceed 50°C. The tank is surrounded by an airstream at 20°C with a convection coefficient of 20 W/m^{2}K. Determine the thermal conductivity and thickness of insulation required to reduce the heat losses from the tank by 72%. It may be assumed that the convection coefficient is unaltered by the insulation layer.
13. A cylindrical fuel rod within a nuclear reactor generates heat due to fission according to $q = q_o - q_o\ (r/R)$, where R and q_o refer to the radius of the fuel rod and an empirical coefficient, respectively. The boundary surface at $r = R$ is maintained at a uniform temperature, T_o. Find the coefficient, q_o, if the measured temperature difference between the centerline and the outer surface of the rod is 400 K. In this case, $R = 2.2$ cm and $k = 12$ W/mK.
14. A plane layer of a fuel element within a nuclear reactor generates heat by fission in the amount of q_o''' (W/m^3). The boundary surfaces at $x = 0$ and $x = L$ are maintained at temperatures of T_1

and T_2, respectively. The thermal conductivity of the element increases with temperature based on $k(T) = k_0 + k_1 T$, where k_0 and k_1 are constants. Assuming steady-state conditions, determine the heat flux across the boundary at the surface $x = L$.

15. Brazing is a metal-joining process that heats a base metal to a high temperature and applies a brazing material to the heated joint. The base metal melts the brazing alloy and fills the joint, and after it cools, the metal solidifies in the joint. Consider a copper alloy brazing rod with a thermal conductivity, diameter, and length of 70 Btu/h ft °F, 1/4 in., and 5 ft, respectively. The end of the rod is heated by a torch and reaches an average temperature of 1400°F. A heat transfer analysis is required to determine whether a machinist can grasp the rod. What is the rod temperature at a distance of 2 ft from the heated end? The ambient air temperature and convection coefficient are 70°F and 1 Btu/h ft °F, respectively.
16. A team member in a race car competition has proposed to air-cool the cylinder of the combustion chamber by joining an aluminum casing with annular fins ($k = 240$ W/mK) to the cylinder wall ($k = 50$ W/mK). This configuration involves an inner cylinder (radii of 50 and 55 mm) covered by an outer cylinder (radii of 55 and 60 mm) and annular fins. The outer radius, thickness, and spacing between the fins are 10 cm, 3 mm, and 2 mm, respectively. Air flows through the fins at a temperature of 308 K with a convection coefficient of $h = 100$ W/m^{2}K. The temporal average heat flux at the inner surface is 80 kW/m^2, and the wall casing contact resistance is negligible.
    (a) If the fin efficiency is 84%, then find the steady-state interior wall temperature (at $r = 50$ mm) for the cases of the cylinder with and without fins.
    (b) Briefly state what factors should be investigated prior to a final design with a specific fin configuration.
17. Pin fins are often used in electronic systems to cool internal components, as well as to support devices. Consider a pin fin that connects two identical devices of width $L_g = 1$ cm and surface area $A_g = 2$ cm^2. The devices are characterized by a uniform heat generation of $\dot{q} \approx 300$ mW. Assume that the back and top or bottom sides of the devices are well insulated. Also, assume that the exposed surfaces of the devices are at a uniform temperature, T_b, and heat transfer occurs by convection from the exposed surfaces to the surrounding fluid at $T_\infty = 20$°C with $h = 40$ W/m^{2}K. Select an appropriate control volume and determine the base (surface) temperature, T_b. The fin diameter and length are 1 and 25 mm, respectively, and the thermal conductivity of the fin is 400 W/mK.

18. Copper tubing is joined to a solar collector plate (thickness of t), and fluid flowing within each tube maintains the plate temperature above each tube at T_0. A net radiative heat flux of q_{rad} is incident on the top surface of the plate, and the bottom surface is well insulated. The top surface is also exposed to air at a temperature of T_∞ with a convection coefficient of h. Since t may be assumed to be small, the temperature variation within the collector plate in the y direction (perpendicular to the plate) is considered negligible.
    (a) Select an appropriate control volume and derive the governing differential equation that describes the heat flow and temperature variation in the x direction along the collector plate. Outline the boundary conditions.
    (b) Solve the governing differential equation to find the plate temperature halfway between the copper tubes. The convection coefficient, thermal conductivity, plate thickness, T_∞, T_0, q_{rad}, and tube-to-tube spacing are 20 W/m^{2}K, 200 W/mK, 8 mm, 20°C, 60°C, 700 W/m^2, and 0.1 m, respectively.
19. A cylindrical copper fin is heated by an airstream at 600°C with a convective heat transfer coefficient of 400 W/m^{2}K. The base temperature of the 20-cm-long fin is 320°C. What fin diameter is required to produce a fin tip temperature of 560°C? Also, find the rate of heat loss from the fin.
20. The temperature distribution within a solid material covering $-0.1 \le x \le 0.4$ and $-0.1 \le y \le 0.3$ (m) is given by $T = 20x^2 - 40y^2 + 10x + 200$ (K). The thermal conductivity of the solid is 20 W/mK. The purpose of this problem is to find the heat generated and heat flux at specified points and across various surfaces within the solid.
    (a) Find the rate of heat generation and the heat flux (magnitude and direction) at the point $(x, y) = (0.1, 0.1)$ in the solid.
    (b) Find the total heat flows (W/m; unit depth of solid) across each of the four surfaces of the rectangular region defined by $0 \le x \le 0.3$ and $0 \le y \le 0.2$ within the solid.
21. Find the upper and lower bounds on the thermal resistance of a composite cylindrical wall consisting of an inner layer (thickness of Δ_1, conductivity of k_1) and two adjoining outer half-cylinders (thickness of Δ_2, conductivities of k_2 and k_3). The contact resistances at the interfaces between the cylindrical layers may be neglected.
22. The top and side boundaries of a long metal bar are maintained at 0°C, and the bottom side is held at 100°C. If the square bar thickness is 1 cm, find the spatial temperature distribution inside the plate.

23. During laser heating of a metal rod of height L, a uniform heating rate of q_o is applied along half of the upper surface of the bar, whereas the remaining half is well insulated. The top end is also well insulated, and the other end is exposed to a fluid at T_∞ and a convection coefficient of h. Use the method of separation of variables to the find the steady-state temperature distribution within the rod.
24. The outer wall of a hollow half-cylinder ($r = r_o$, $-90° < \theta < 90°$) is subjected to a uniform heat flux between $-\beta \leq \theta \leq \beta$, and it is well insulated over the remaining angular range and the ends at $\theta = \pm 90°$. Along the inner boundary ($r = r_i$), the component is convectively cooled by a fluid with a temperature and convection coefficient of T_f and h, respectively. Use the method of separation of variables to find the steady-state temperature within the half-cylinder metal component.
25. After heat treatment in a furnace, a cylindrical metal ingot is removed and cooled until it reaches a steady-state temperature governed by two-dimensional heat conduction in the radial (r) and axial (z) directions. The top ($z = H$), bottom ($z = 0$), and outer surfaces ($r = r_o$) are maintained at T_w, T_w, and T_c, respectively. Use the method of separation of variables to find the temperature distribution and resulting total rate of heat transfer from the ingot.
26. Problems have been observed with the quenching of steel ingots under conditions when the surface cooling is not sufficiently fast to avoid the formation of soft pearlite and bainite microstructures in the steel. Consider a small steel sphere (diameter of 2 cm) that is initially uniformly heated to 900°C and then hardened by suddenly quenching it into an oil bath at 30°C. If the average convection coefficient is $h = 600$ W/m^{2}K, then what is the sphere surface temperature after 1 min of quenching time has elapsed? The properties of the steel alloy may be taken as $\rho = 7830$ kg/m^3, $c_p = 430$ J/kg K, and $k = 64$ W/mK.
27. Water is heated by hot combustion gases that flow through a cylindrical flue in a gas-fired water heater. In a conventional water heater, the gases flow upward through an unobstructed exit damper (i.e., open tube). A new energy storage scheme is proposed to increase the water heater performance by constructing a damper that consists of alternating flow spaces and rectangular plates (width of 0.01 m) aligned in the streamwise (longitudinal) direction. Each plate has an initial temperature of 25°C with $\rho = 2700$ kg/m^3, $k = 230$ W/mK, and $c_p = 1030$ J/kg K. Under typical operating conditions, the damper is thermally charged by passing

a hot gas through the hot spaces with a convection coefficient of $h = 100$ W/m^{2}K and $T_\infty = 600$°C.

(a) How long will it take to reach 90% of the maximum (steady state) energy storage in each central damper plate?

(b) What is the plate temperature at this time?

28. A spherical lead bullet with a diameter of $D = 5$ mm travels through the air at a supersonic velocity. A shock wave forms and heats the air to $T_\infty = 680$ K around the bullet. The convection coefficient between the air and the bullet is $h = 600$ W/m^{2}K, and the bullet leaves the barrel at $T_i = 295$ K. Determine the bullet's temperature upon impact if its time of flight is $\Delta t = 0.6$ sec. Assume constant lead properties with $\rho = 11{,}340$ kg/m^3, $c_p = 129$ J/kgK, and $k = 35$ W/mK.
29. A pulverized coal particle flows through a cylindrical tube prior to injection and combustion in a furnace. The inlet temperature and diameter of the coal particle are 22°C and 0.5 mm, respectively. The tubular surface is maintained at 900°C. What particle inlet velocity is required for it to reach an outlet temperature of 540°C over a tubular duct 5 m in length?
30. A cylinder wall of a metal combustion chamber ($\rho = 7850$ kg/m^3, $c_p = 430$ J/kgK) is held at an initially uniform temperature of 308 K. In a set of experimental tests, the outer surface is exposed to convection heat transfer with air at 308 K and a convection coefficient of $h = 80$ W/m^{2}K. An internal combustion process exchanges heat by radiation between the inner surface and the heat source at 2260 K (approximate flame temperature of combustion). Under these conditions, estimate the time required for melting of the cylinder to occur (note: melting temperature is about 1800 K). Use the lumped capacitance method. In practice, how would convective cooling be applied to prevent melting of the cylinder?

3

Convective Heat Transfer

3.1 Introduction

Convective heat transfer refers to the combination of molecular diffusion and bulk fluid motion (or advection). Since bulk motion of the fluid often contributes substantially to the transport of heat, knowledge of the fluid velocity distribution is essential in convection problems. As a result, both fluid flow and energy equations need to be considered during convective heat transfer. These governing equations and associated physical processes in convective heat transfer will be examined in this chapter.

As briefly described in Chapter 1, Newton's law of cooling may be used in convection problems. For example, consider a cold airflow above a hot plate, where the air temperature is T_∞ and the surface temperature of the hot plate is T_s. Based on Newton's law of cooling, the rate of heat transfer to the air may be written as

$$q = hA(T_s - T_\infty) \tag{3.1}$$

where h is the convective heat transfer coefficient. This coefficient depends on the fluid thermophysical properties, problem geometry, fluid velocity, and other factors. Rearranging Equation (3.1) to identify the convection thermal resistance, $R_{conv} = 1/(hA)$, we obtain

$$q = \frac{T_s - T_\infty}{R_{conv}} \tag{3.2}$$

We can often classify a problem by the magnitude of the convective heat transfer coefficient. For example, we have the following typical ranges for convective heat transfer: (i) free convection with air ($5 < h < 25$ [W/m^{2}K]), (ii) forced convection with air ($10 < h < 500$ [W/m^{2}K]), (iii) forced convection with water ($100 < h < 15{,}000$ [W/m^{2}K]), and (iv) condensation of water ($5{,}000 < h < 100{,}000$ [W/m^{2}K]).

Also, variations of fluid flow processes over a surface lead to a distinction between the local and total (or average) convection coefficients. In the following example, variations of the convection coefficient are described for external flow past a cylinder.

Example 3.1.1

Convective Heat Transfer from a Cylinder in a Cross-Flow.

Explain how local variations of the convective heat transfer coefficient are encountered for external flow past a cylinder.

An angular variation of convective heat transfer coefficient is observed for external flow past a cylinder. On the upstream side of the cylinder for laminar flow, the convective heat transfer coefficient decreases with angle (see Figure 3.1). This trend is a result of the growing boundary layer thickness with angle on the upstream side. However, on the back side of the cylinder, separation of the boundary layer from the surface, and the resulting increase of fluid mixing, causes a rise in the convection coefficient with the angle.

Similar trends can be observed in the case of turbulent flow. For example, the transition from a laminar to turbulent boundary layer on the front side of the cylinder causes a rise in the convective heat transfer coefficient. Boundary layer separation on the back side of the cylinder is delayed for turbulent flow. In all cases, we see that the local heat transfer coefficient varies with the angle along the surface of cylinder. If we consider the average (or total) heat transfer from the cylinder to the air, the average value of h would be used in Newton's law of cooling, rather than the local heat transfer coefficient shown in Figure 3.1.

The fluid temperature in Newton's law of cooling must be carefully specified. For external flows, such as the previous example, the fluid

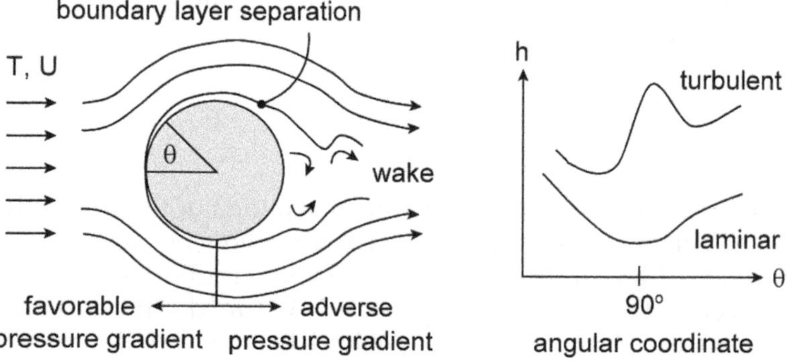

FIGURE 3.1
Heat transfer from a cylinder in a cross-flow.

temperature is usually selected to be the ambient fluid temperature. However, for internal flows (such as flow in a duct or a pipe), the fluid temperature in Newton's law of cooling is usually selected to be the mean temperature of the fluid, which typically varies with position. These selections are closely linked with the manner by which empirical correlations for the heat transfer coefficient are defined.

3.2 Convection Governing Equations

The convection equations include both the fluid flow equations and the First Law of Thermodynamics. The fluid flow equations are governed by mass conservation (or continuity) and the momentum (or Navier–Stokes) equations. The general balance equation for a scalar quantity, B, is given by

$$\dot{B}_{cv} = \dot{B}_{in} - \dot{B}_{out} + \dot{B}_g \tag{3.3}$$

where the subscripts cv, in, out, and g refer to control volume, inflow, outflow, and generation term, respectively. This equation gives the conservation (rather than transport) form of the governing equation, since it can be interpreted as a balance of B with respect to the selected control volume.

3.2.1 Mass and Momentum Equations

3.2.1.1 Conservation of Mass

A mass balance based on Equation (3.3), where B represents mass, is performed over a differential control volume (i.e., Figure 3.2). The rate of mass increase with time in a control volume may be represented by the partial derivative of density with respect to time, multiplied by the volume size (or area in two-dimensional flows). It will be assumed that no mass is generated in the control volume. Also, we have the following expressions for the mass flow at the inflow and outflow boundaries of the control volume (per unit width):

$$\dot{m}_x = \rho u dy \tag{3.4}$$

$$\dot{m}_{x+dx} = \dot{m}_x + \frac{\partial}{\partial x}(\dot{m}_x) \tag{3.5}$$

where a Taylor series expansion has been used (while neglecting higher order terms) at the position $x + dx$. Similar expressions can be obtained for the mass flow rates in the y direction.

After substituting and re-arranging the results in the general conservation equation, Equation (3.3), we obtain the following mass conservation equation for two-dimensional flow:

$$\frac{\partial \rho}{\partial t} + \frac{\partial(\rho u)}{\partial x} + \frac{\partial(\rho v)}{\partial y} = 0 \tag{3.6}$$

For incompressible flows ($\rho \approx constant$), Equation (3.6) may be simplified with the result that the divergence of the velocity field ($\nabla \cdot \mathbf{v}$) equals zero. The divergence of velocity may be interpreted as the net outflow from a control volume (fully occupied by fluid), which must equal zero at steady state, since any inflows are balanced by mass outflows.

3.2.1.2 Momentum Equations

The momentum equations represent a form of Newton's law. Forces on a fluid particle (i.e., pressure, shear forces, etc.) balance the particle's mass times its acceleration (i.e., total, or substantial, derivative of velocity). In one sense, it may be viewed that momentum is not strictly conserved since the action of forces on the fluid particles alters their momentum. However, since the form of the momentum equation closely resembles the standard form of a conservation equation, Equation (3.3), we often refer to the conservation of momentum equations. These equations are called the *Navier–Stokes equations*. In the upcoming mathematical derivations, we will consider a direct implementation of the relevant physical principles on an infinitesimally small fluid element fixed in space.

The conserved quantities are given by the x direction momentum (ρu) and the y direction momentum (ρv) per unit volume. With no momentum generation, the rate of increase of x direction momentum becomes

$$\dot{M}_{x,cv} = \left[\dot{M}_x + \sum F_x\right]_{in} - \left[\dot{M}_x + \sum F_x\right]_{out} \tag{3.7}$$

In this equation, we see that both the x momentum fluxes and the sum of the forces on the control volume can independently alter the rate of change of momentum in the control volume (see Figure 3.2). In calculating the x momentum flux terms at the inflow and outflow faces of the control volume, we can adopt the same procedure as our earlier mass conservation approach, whereby in this case the conserved quantity refers to the x direction momentum.

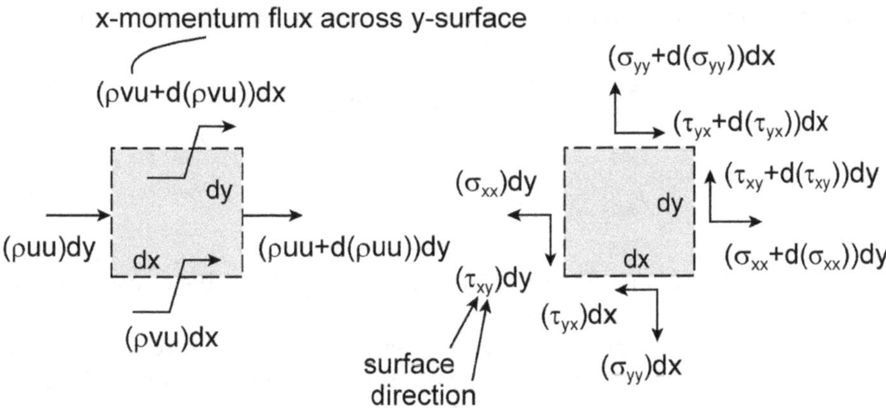

FIGURE 3.2
Momentum flux and force components for a control volume.

For example, the contributions of x momentum fluxes at the y faces (locations y and $y+dy$; see Figure 3.2) of the selected control volume are given by the following expressions:

$$\dot{M}_{x,in} = \rho vu \tag{3.8}$$

$$\dot{M}_{x,out} = \rho vu + \frac{\partial}{\partial y}(\rho vu)dy \tag{3.9}$$

where higher order terms have been neglected. These terms represent the x momentum flux terms due to mass flowing across the y surfaces. Also, a unit depth has been assumed in these two-dimensional calculations.

In order to obtain the remaining force terms for the momentum balance, we need to consider pressure, shear stress, and normal stress terms. Let the subscripts for stress terms refer to their respective surface and direction. The normal stress includes pressure normal to the face (edge of the control volume). Also, body forces may be required in the momentum balances. For example, F_{bx} refers to the body force in the x direction. In the case of natural convection along an inclined plate, this term would refer to a body force arising from fluid buoyancy in the tangential direction (x direction) along the plate (i.e., $F_{bx} = \rho g_x dx \cdot dy$). The body force acts at the center of the control volume.

The shear stress terms act in the tangential direction along the face of each control volume edge, whereas the normal stress terms act in the direction normal to the edge (see Figure 3.2). Let us consider the calculation of individual terms in the overall momentum balance, such as the normal stress and shear stress terms at locations $x + dx$ and $y + dy$, respectively.

$$F_{s,y+dy} = \left(\tau_{yx} + \frac{\partial \tau_{yx}}{\partial y} dy\right) dx \tag{3.10}$$

$$F_{n,x+dx} = \left(\sigma_{xx} + \frac{\partial \sigma_{xx}}{dx} dx\right) dy \tag{3.11}$$

In these expressions, Taylor series expansions have been used, while higher order terms were neglected.

Substituting all of the indicated momentum flux and force terms into Equation (3.7), which corresponds to Equation (3.3), with B representing the x direction momentum of the fluid, the following result is obtained:

$$\frac{\partial(\rho u)}{\partial t} + \frac{\partial(\rho uu)}{\partial x} + \frac{\partial(\rho vu)}{\partial y} = \frac{\partial \sigma_{xx}}{\partial x} + \frac{\partial \tau_{yx}}{\partial y} + F_{bx} \tag{3.12}$$

Similarly, the following result is obtained in the y direction:

$$\frac{\partial(\rho v)}{\partial t} + \frac{\partial(\rho vu)}{\partial x} + \frac{\partial(\rho vv)}{\partial y} = \frac{\partial \tau_{xy}}{\partial x} + \frac{\partial \sigma_{yy}}{\partial y} + F_{by} \tag{3.13}$$

These equations cannot be directly solved in this form since there are more unknowns (i.e., stresses, velocities, and pressure) than available equations. As a result, we need additional relations (called constitutive relations) between the stresses and velocities. In *Newtonian fluids*, the stresses are proportional to the rate of deformation (or strain rate). In contrast, examples of nonlinear relations (non-Newtonian fluids) include Bingham plastics, Ostwald–deWalls fluids (exponential relation) and Eyring fluids (hyperbolic relation), and others (see Figure 3.3).

For incompressible flows of Newtonian fluids, we have the following two-dimensional constitutive relations for stresses in terms of the other problem variables (pressure and velocity):

$$\sigma_{xx} = -p + 2\mu \frac{\partial u}{\partial x} \tag{3.14}$$

$$\sigma_{yy} = -p + 2\mu \frac{\partial v}{\partial y} \tag{3.15}$$

$$\tau_{yx} = \mu\left(\frac{\partial u}{\partial y} + \frac{\partial v}{\partial x}\right) = \tau_{xy} \tag{3.16}$$

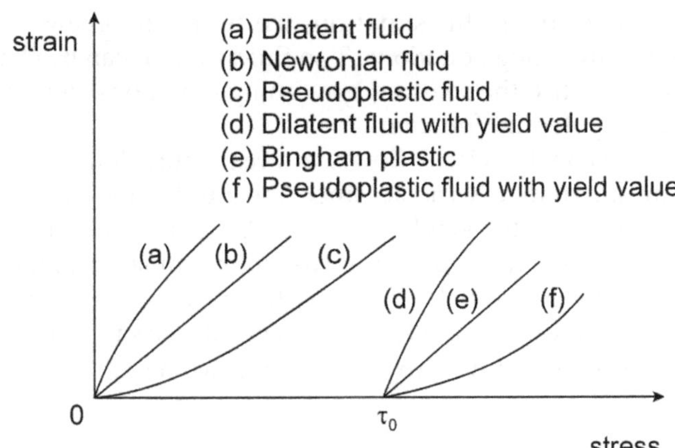

FIGURE 3.3
Newtonian and non-Newtonian fluids. (From Lane, G.A., *Solar Heat Storage: Latent Heat Materials*, Copyright 1996, CRC Press, Boca Raton, FL. Adapted and reprinted with permission.)

Substituting these constitutive relations into the previous momentum equations, we obtain

$$\rho\frac{\partial u}{\partial t}+\rho u\frac{\partial u}{\partial x}+\rho v\frac{\partial u}{\partial y}=-\frac{\partial p}{\partial x}+\mu\left(\frac{\partial^2 u}{\partial x^2}+\frac{\partial^2 u}{\partial y^2}\right)+F_{bx} \tag{3.17}$$

$$\rho\frac{\partial v}{\partial t}+\rho u\frac{\partial v}{\partial x}+\rho v\frac{\partial v}{\partial y}=-\frac{\partial p}{\partial x}+\mu\left(\frac{\partial^2 v}{\partial x^2}+\frac{\partial^2 v}{\partial y^2}\right)+F_{by} \tag{3.18}$$

which represent the two-dimensional Navier–Stokes equations. General solutions of these equations are limited because of the difficulties inherent in the nonlinear and coupled (with continuity) nature of the equations.

Fluid flow regions are often classified as either viscous or nearly inviscid regions. In a viscous region, such as a boundary layer, frictional forces are significant. A boundary layer refers to the thin diffusion layer near the surface of a solid body, where the fluid velocity decreases from its freestream value to zero at the wall over a short distance. The boundary layer equations (to be discussed in the following section) represent an approximation of the Navier–Stokes equations under conditions associated with boundary layer flows.

In contrast to viscous regions, frictional forces are often small in comparison to fluid inertia in regions far away from a surface or boundary layer. The *Euler equations* are a special form of the Navier–Stokes equations for frictionless (or inviscid) flow. An inviscid fluid refers to an idealized fluid with no viscosity. In this situation, the terms involving viscosity are absent from the governing equations. The fluid motion can be characterized as a *potential flow,* since the reduced governing equations can be written in terms of a scalar potential function.

For two-dimensional inviscid flows, a stream function, $\psi(x, y)$, is often adopted as an alternative variable since the fluid flow equations can be simplified in terms of this single variable. A streamline is tangent to the velocity field such that the slope of the streamline must equal the tangent of the angle that the velocity field makes with the x axis. We can label each streamline with a unique stream function. Since the stream function must be constant along each streamline (by definition), we have

$$u = \frac{\partial \psi}{\partial y} \tag{3.19}$$

$$v = -\frac{\partial \psi}{\partial x} \tag{3.20}$$

Based on an integration of the mass flow between two adjacent streamlines, we find that the change in streamline values between streamlines can be interpreted as the volume flow rate per unit depth between those streamlines.

Also, the concept of fluid rotation has an important connection with the governing equation for potential (inviscid, incompressible) flow. Vorticity is calculated by the vector curl of the velocity field. It represents the rotation of a fluid particle about the axes. For locally irrotational flow, the fluid vorticity is zero. However, irrotational flow does not require a straight flow path line. For example, a fluid particle could follow a straight path line while rotating about its own local axes due to frictional forces (i.e., rotational flow). Conversely, a fluid particle could flow in a circular path even though it is locally irrotational; there may be no rotation about its local axes. It may be shown that the irrotationality condition (i.e., zero vorticity) implies the existence of a scalar function, called the velocity potential, $\phi(x, y)$, such that the velocity field may be written as the gradient of this scalar potential function. Combining this result with conservation of mass for incompressible flow, we find that the governing equation for potential flow becomes Laplace's equation, Equation (2.56), in terms of the velocity potential (or stream function).

3.2.2 Mechanical and Internal Energy Equations

As we have considered earlier, energy is composed of both mechanical energy (i.e., kinetic, potential energy) and internal energy. In this section, an appropriate governing equation will be formulated for each type of energy. In particular, we will find the mechanical and total energy equation (i.e., First Law of Thermodynamics), and then by subtracting those results, we will establish the internal energy equation. Although the results are developed for two-dimensional flows, similar results can be derived for three-dimensional flows.

3.2.2.1 Mechanical Energy Equation

The mechanical energy equations can be obtained by multiplying each u_i momentum equation by u_i (where $i = 1, 2$ for two-dimensional flows) and adding together. Using the substantial derivative notation, we obtain

$$\frac{1}{2}\rho\frac{D}{Dt}(u^2+v^2) = -u\frac{\partial p}{\partial x} - v\frac{\partial p}{\partial y} + u\frac{\partial \tau_{xx}}{\partial x} + u\frac{\partial \tau_{yx}}{\partial y} + v\frac{\partial \tau_{xy}}{\partial x} + v\frac{\partial \tau_{yy}}{\partial y} + uF_x + vF_y \tag{3.21}$$

Using the product rule and generalizing to a vector notation, we obtain the following mechanical energy equation:

$$\frac{1}{2}\rho\frac{D}{Dt}(V^2) = -[\nabla\cdot(p\mathbf{v}) - p\nabla\cdot\mathbf{v}] + [\nabla\cdot(\tau\cdot\mathbf{v}) - \tau:\nabla\mathbf{v}] + \mathbf{v}\cdot\mathbf{F} \tag{3.22}$$

where $V = \sqrt{u^2+v^2}$ refers to the total resultant magnitude of the velocity. When Equation (3.22) is written in a general vector form, the result becomes valid for three-dimensional flows.

In Equation (3.22), the first term (left side) represents the rate of increase of kinetic energy of a fluid particle with respect to time. On the right side, the second and third terms give the flow work done by pressure on the differential control volume to increase its kinetic energy. The difference between the second and third terms gives the net flow work done by the surroundings on the fluid particle to increase its kinetic energy. This includes the total work done through pressure (second term) minus the work done in fluid compression (i.e., not increasing the kinetic energy of the fluid). The third term represents an energy sink in the mechanical energy equation and it becomes zero for incompressible flows. It is subtracted because the work is directed toward fluid compression or expansion, rather than a change in the fluid's kinetic energy.

The difference between the fourth and fifth terms on the right side of Equation (3.22) gives the net fluid work done by viscous stresses to increase the kinetic energy of the fluid within the control volume. In other words, it gives the total rate at which the surroundings do work on the fluid through viscous stresses (fourth term) minus the portion of work not going into kinetic energy (fifth term). The latter portion represents work lost through viscous dissipation, which does not change the fluid's kinetic energy. This viscous dissipation is a degradation of mechanical energy into internal energy through viscous effects, and it is often comparably small, except for very viscous flows or very high speed flows. In Equation (3.22), this viscous dissipation is represented by $\tau : \nabla \mathbf{v}$, which refers to the viscous stress tensor contracted with the velocity gradient. The matrices are contracted when corresponding entries in each matrix are multiplied together and all resulting products are summed.

For two-dimensional incompressible flows of a Newtonian fluid, it may be shown that the viscous dissipation term in Equation (3.22) can be written as

$$\tau : \nabla \mathbf{v} = 2\mu \left[\left(\frac{\partial u}{\partial x} \right)^2 + \left(\frac{\partial v}{\partial y} \right)^2 \right] + \mu \left(\frac{\partial u}{\partial y} + \frac{\partial v}{\partial x} \right)^2 \equiv \mu \Phi \tag{3.23}$$

where Φ refers to the positive-definite viscous dissipation function. This function is greater than or equal to zero. As a result, the conversion of mechanical energy into internal energy through viscous dissipation represents an energy sink in the mechanical energy equation. Thus, mechanical energy is not conserved, but instead, a portion of this energy is degraded and lost to internal energy through viscous dissipation. It is degraded in the sense that a certain quality of energy is lost in the irreversible transformation. We would anticipate that this energy sink corresponds to an energy source in the internal energy equation. Furthermore, both energy sinks and sources should cancel upon summation of mechanical and internal energy equations, thereby ensuring conservation of total energy (i.e., mechanical plus internal energy parts). It will be shown that these results do indeed occur from the energy equations.

The final remaining term in the mechanical energy equation is the sixth term on the right side of Equation (3.22). It refers to the net rate at which a fluid body force, **F**, performs work on the fluid, thereby increasing its kinetic energy within the control volume.

For a special case of zero heat and frictional losses (reversible, incompressible flow), it can be shown that Equation (3.22) can be integrated along a streamline to obtain the following Bernoulli equation for inviscid flow (end-of-chapter problem):

$$\frac{P_1}{\rho g} + \frac{V_1^2}{2g} + z_1 = \frac{P_2}{\rho g} + \frac{V_2^2}{2g} + z_2 \tag{3.24}$$

where both sides are constant along a given fluid streamline. Here z refers to surface elevation, and the subscripts 1 and 2 refer to two points along a given streamline where the integration is performed. In many applications, Bernoulli's equation can provide a useful interpretation of physical interactions between pressure, velocity, and gravity forces in fluid dynamics.

3.2.2.2 First Law of Thermodynamics (Total Energy)

The conservation of total energy (i.e., internal plus mechanical energy) is called the First Law of Thermodynamics. Referring back to the general conservation equation, Equation (3.3), and denoting B as the total energy, we obtain the following equation for the conservation of total energy:

$$\dot{E}_{cv} = (\dot{E}_{adv} + \dot{E}_{cond} + \dot{W})_{in} - (\dot{E}_{adv} + \dot{E}_{cond} + \dot{W})_{out} + \dot{E}_g \tag{3.25}$$

In order to obtain the inflow and outflow terms, consider advection (subscript *adv*), conduction (subscript *cond*), and the rate of work done on a differential control (see Figure 3.4). For example, at location $x + dx$ the conduction and advection flux terms at the edge of the control volume may be expressed as

$$q_{cond} = \left(q_x + \frac{\partial}{\partial x}(q_x)dx \right) dy \tag{3.26}$$

$$q_{adv} = \left(\rho ue + \frac{\partial}{\partial x}(\rho ue)dx \right) dy \tag{3.27}$$

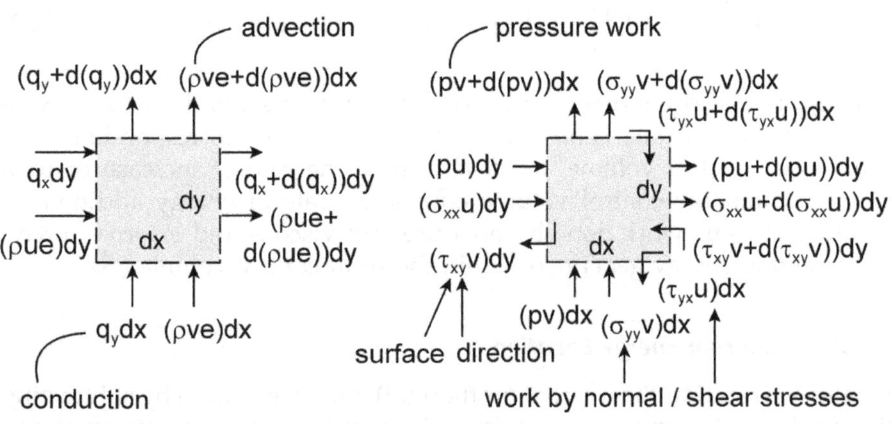

FIGURE 3.4
Heat flux and work terms for a control volume.

where a Taylor series expansion has been used and higher order terms have been neglected. Also, e refers to total energy per unit mass.

In terms of the rate of work done on the fluid control volume by various forces (i.e., pressure, shear, and normal stresses), consider the contributions from terms at the upper face of the control volume (position $y+dy$).

$$W_p = \left(pv + \frac{\partial p}{\partial y}dy\right)dx \tag{3.28}$$

$$W_n = \left(\tau_{yy}v + \frac{\partial(\tau_{yy}v)}{\partial y}dy\right)dx \tag{3.29}$$

$$W_s = \left(\tau_{yx}u + \frac{\partial(\tau_{yx}u)}{\partial y}dy\right)dx \tag{3.30}$$

After constructing similar terms at the other faces of the control volume, assembling all terms in Equation (3.25), and rearranging terms, we obtain

$$\rho\frac{\partial e}{\partial t} + \frac{\partial(\rho ue)}{\partial x} + \frac{\partial(\rho ve)}{\partial y}$$
$$= -\frac{\partial q_x}{\partial x} - \frac{\partial q_y}{\partial y} - \frac{\partial(pu)}{\partial x} - \frac{\partial(pv)}{\partial y} + \frac{\partial(\tau_{xx}u)}{\partial x} + \frac{\partial(\tau_{yx}u)}{\partial y} + \frac{\partial(\tau_{xy}v)}{\partial x} + \frac{\partial(\tau_{yy}v)}{\partial y} \tag{3.31}$$

Alternatively, by substituting continuity, generalizing to three-dimensional flows, and writing in a general vector form,

$$\rho\frac{D}{Dt}\left(\hat{e} + \frac{1}{2}V^2\right) = -\nabla\cdot\mathbf{q} - \nabla\cdot(p\mathbf{v} - \tau\cdot\mathbf{v}) + \mathbf{F}\cdot\mathbf{v} + \dot{S} \tag{3.32}$$

where $\hat{e}$ refers to internal energy. This result represents the First Law of Thermodynamics (i.e., conservation of total energy) corresponding to a differential control volume, $dx\cdot dy\cdot dz$. Thus, the rate of increase of total energy within the control volume equals the rate of energy addition by conduction, plus work done by pressure and viscous and external forces, plus internal energy generated within the fluid control volume ($\dot{S}$).

3.2.2.3 Internal Energy Equation

Finally, we can find the internal (or thermal) energy equation by subtracting the mechanical energy equation from the first law (total energy equation). Performing this subtraction and writing the results in a general vector form, we have

$$\rho \frac{D\hat{e}}{Dt} = -\nabla \cdot \mathbf{q} - p\nabla \cdot \mathbf{v} + \tau : \nabla \mathbf{v} + \dot{S} \tag{3.33}$$

where the fourth term (right side) refers to the viscous stress tensor contracted with the velocity gradient. It represents an energy source since it arises from the conversion of mechanical energy to internal energy through viscous dissipation. In the thermal energy equation, it provides an energy source, which corresponds to the energy sink previously observed in the mechanical energy equation. In other words, its magnitude is identical, but its sign changes in transposing from the mechanical to internal energy equations.

Thus, the thermal energy equation states that the rate of increase of thermal energy equals the heat addition by conduction, plus work done by fluid compression, plus viscous dissipation and internal heat generation. An application that outlines the conversion between various forms of energy is given in the following example.

Example 3.2.2.3.1

Energy Exchange in Couette Flows.

In this example, energy exchange mechanisms will be described for fluid flow beneath a moving plate (Couette-type flow). If the plate moves at a constant and steady velocity, u, in the positive x direction, then the velocity profile beneath the plate decreases linearly to the wall. This result may be obtained from a solution of the reduced Navier–Stokes equations subject to the appropriate boundary conditions.

As a result, taking the velocity gradient normal to the wall and multiplying by viscosity, we find that the sheer stress remains constant throughout the liquid beneath the moving plate. Under these circumstances, we find that the rate of increase of kinetic energy is zero as a result of the steady plate motion. On the other hand, multiplying the shear stress by velocity gradient, with the appropriate sign convention of positive velocity in the positive x direction, we find that the rate of increase of internal energy is positive. In other words, the total energy added to the plate, as it moves right, is transferred entirely into internal energy. This result is consistent with the result indicating that viscous forces are the predominant mechanisms in the momentum balance. As a result, work is entirely converted into internal energy in this viscous flow.

In a similar way, consider a deformable solid subjected to an applied force at its upper boundary, while maintaining contact with a solid surface such that $\tau_w \approx 0$ at the lower wall. We may imagine this example as a type of soap bar on a table. In this case, the solid accelerates from rest as a constant force is applied at its upper boundary. It moves as a solid object; thus the velocity remains constant through the object, while its magnitude varies with time. The shear stress decreases linearly from a specified positive value at its upper boundary to about zero at the wall (beneath the solid object).

If we now consider the kinetic energy and internal energy terms, we find that the velocity multiplied by the gradient of shear stress is a positive value, and the kinetic energy increases since the object is accelerating. However, the rate of increase of internal energy in this case is zero. Since it behaves as a single rigid body, no internal velocity gradient or friction arises, so there is no corresponding change of internal energy.

The previous thermal energy equations may be written in a variety of other forms, including temperature or enthalpy equations. Although temperature is not conserved (unlike total energy), we often use a temperature equation (instead of energy) because temperature can be measured directly, whereas internal energy cannot be measured directly. Using the definitions of specific heat, where

$$c_v = \frac{\partial \hat{e}}{\partial T}\bigg|_{v'''} \tag{3.34}$$

$$c_p = \frac{\partial h}{\partial T}\bigg|_{p} \tag{3.35}$$

it can be shown that the thermal energy equation with constant thermophysical properties can be rewritten as the following temperature equation for incompressible two-dimensional flows:

$$\rho c_v \frac{DT}{Dt} = k\frac{\partial^2 T}{\partial x^2} + k\frac{\partial^2 T}{\partial y^2} + \mu\Phi + \dot{S} \tag{3.36}$$

Alternatively,

$$\rho c_p \frac{DT}{Dt} = \beta T\frac{Dp}{Dt} + k\frac{\partial^2 T}{\partial x^2} + k\frac{\partial^2 T}{\partial y^2} + \mu\Phi + \dot{S} \tag{3.37}$$

where

$$\beta = -\frac{1}{\rho}\left(\frac{\partial \rho}{\partial T}\right)_p \tag{3.38}$$

is called the thermal expansion coefficient. For an ideal gas, this expansion coefficient can be evaluated as

$$\beta = -\frac{1}{\rho}\left(\frac{-p}{RT^2}\right) = \frac{1}{T} \tag{3.39}$$

3.3 Velocity and Thermal Boundary Layers

The concepts of velocity and thermal boundary layers will be explained in relation to fluid flow over a flat plate (see Figure 3.5). As the fluid flows over the plate, it is restrained by friction and the presence of the surface such that its velocity and temperature decrease from freestream values (just outside the friction layer) to zero and the wall temperature, respectively, at the wall. The boundary layer grows in thickness in the downstream direction as a result of continued cross-stream momentum and thermal diffusion (normal to the plate). The boundary layer development displaces fluid momentum in the y direction. In the region close to the leading edge of the plate, the boundary layer remains thin. For example, if we consider air flowing at 100 km/h over a surface with a length of 1.5 m, the thickness of the velocity boundary layer is about 3 cm at the end of the surface.

Within the boundary layer, frictional effects dampen the inertia of the fluid motion, thereby permitting a smooth, laminar flow parallel to the wall. However, further downstream fluid inertia may no longer be sufficiently restrained by friction in the outer portion of the boundary layer due to the

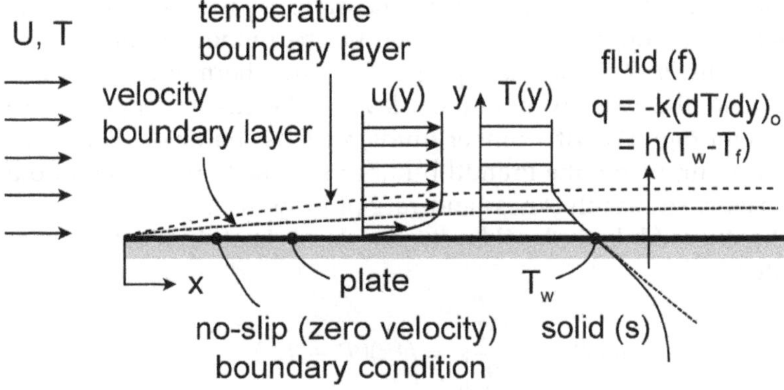

FIGURE 3.5
Velocity and thermal boundary layers.

growth of the boundary layer thickness. This may lead to turbulence, or intense mixing and rotational flow (not parallel to the wall), exhibited by a fluid element as it moves downstream. The boundary layer thickness is defined as the outer edge of the layer where friction effects remain important. Alternatively, the edge of the velocity boundary layer corresponds to the point where the fluid velocity reaches 99% of its freestream value. In practice, the boundary layer is a very thin region.

The slope of the velocity profile in the boundary layer at the wall is related to the shear stress acting on the wall. For flat plate boundary layer flow of a Newtonian fluid, the shear stress is represented by

$$\tau_w(x) = \mu \left. \frac{\partial u}{\partial y} \right|_0 \tag{3.40}$$

Once the shear stress is evaluated, the local skin friction coefficient, c_f, may be obtained by

$$c_f(x) = \frac{\tau_w(x)}{\rho U^2/2} \tag{3.41}$$

where U refers to the freestream velocity (outside the boundary layer). The drag force on an object due to friction may be obtained by integrating the local friction coefficient over the surface area of the object.

In an analogous manner, the temperature boundary layer thickness represents the edge of the diffusion layer where the fluid temperature reaches 99% of its freestream value. The thickness of the temperature boundary layer is generally different than the thickness of the velocity boundary layer since the respective roles of diffusion of heat and momentum are typically different. The ratio of the velocity to thermal boundary layer thicknesses is indicated by the Prandtl number ($Pr = \nu/\alpha$). In fluids with $Pr \ll 1$, such as liquid metals, the thermal boundary layer is thicker than the velocity boundary layer, whereas an opposite trend is observed for high Prandtl number fluids, such as engine oil. For fluids such as water or air, where the Prandtl number is close to unity, both boundary layers have a similar thickness and rate of growth.

In the thermal boundary layer, the local heat transfer rate can be computed by

$$q_s''(x) = -k \left. \frac{\partial T}{\partial y} \right|_w = h(x)(T_s - T_\infty) \tag{3.42}$$

Thus, the local convection coefficient may be expressed in terms of the wall temperature gradient, fluid conductivity, and the temperature difference (between the wall and the fluid). Integrating the heat transfer coefficient over the surface of the object yields the total heat transfer from the object. This total heat transfer can be written as the average heat transfer coefficient, $\bar{h}$, multiplied by the temperature difference, $T_s - T_\infty$, where

$$\bar{h} = \frac{1}{A_s}\int_{A_s} h dA_s \tag{3.43}$$

In a similar way to heat transfer, an analogous set of equations and coefficients can be obtained for mass transfer arising from species concentration differences in the flow field. For example, the convective mass transfer coefficient, h_m, mass diffusivity, D_{ab} for constituents a (such as water vapor) and b (such as air), concentration gradient, $\partial\rho_a/\partial y$, and mass transfer rate, $\dot{m}''$, are analogous to the convection coefficient, h, conductivity, k, temperature gradient, and heat transfer rate. The concentration of constituent a, given by ρ_a, is governed by diffusion due to spatial gradients of ρ_a and convection, in a way similar to heat transfer due to temperature differences in the flow field. For instance, a moist airflow over a liquid creates a concentration boundary layer of water vapor above the liquid, in a way similar to how temperature and velocity gradients occur there. As a result, boundary layer approximations of the species transport equation could be used so that the resulting boundary layer equations would give the spatial distribution of water vapor concentration. The following example clarifies how mass transfer can be examined in a way analogous to the principles developed for heat transfer analysis.

Example 3.3.1

Evaporative Cooling of Liquid-Filled Container.

A moist airstream with a velocity of U_∞ and a temperature of T_∞ flows over a container holding liquid at a temperature of T_s. The freestream concentration of water vapor (subscript a) is $\rho_{a,\infty}$, and the relative humidity is ϕ_∞. The purpose of this problem is to find an expression for the convective mass transfer coefficient, h_m, in terms of the measured recession rate of the liquid, dh/dt, in the container.

The relative humidity of a moist airstream may be defined in the following manner:

$$\phi_a(T) \equiv \frac{P_a(T)}{P_{a,sat}(T)} \tag{3.44}$$

where the numerator represents the partial pressure of the constituent and the denominator refers to the saturation pressure of the mixture. Also, from

Dalton's law for mixtures, the total pressure, P_{tot}, is given by the sum of constituent partial pressures, $P_a + P_b$.

Thus, the relative humidity may be written as the fraction of partial pressure of water vapor to the saturation pressure at the mixture temperature, T. Also, assuming that the air and water vapor components of the mixture each behave as ideal gases (individually),

$$\phi = \frac{RT_w/v_w'''}{RT_g/v_g'''} = \frac{\rho_w}{\rho_g} \tag{3.45}$$

where R and v''' refer to the gas constant and specific volume, respectively. The subscripts g and w refer to evaluation at the saturation pressure and partial pressure of the mixture.

The control volume is selected as the liquid in the container. Then the conservation of mass requires that the rate of mass loss balances the loss of mass by evaporation due to the concentration gradient between the liquid in the container and water vapor in the surrounding moist air. On a per unit area basis,

$$\frac{d}{dt}(\rho_l h) = -\dot{m}_{evap} = \overline{h}_m(\rho_{a,s} - \rho_{a,\infty}) \tag{3.46}$$

where the subscripts l, s, and ∞ refer to liquid, surface, and freestream, respectively. Also, h (without subscript) refers to the height of the liquid in the container.

Rearranging these expressions and rewriting the freestream vapor concentration in terms of the relative humidity, we obtain the following expression with respect to a constant liquid density:

$$\overline{h}_m = -\frac{\rho_l(dh/dt)}{\rho_g(T_s) - \phi_\infty \rho_g(T_\infty)} \tag{3.47}$$

It can be observed that the averaged convective mass transfer coefficient can be obtained from measurements of the rate of change of liquid height, dh/dt, densities of the saturated vapor (or reciprocals of the specific volumes) and relative humidity, respectively. The result confirms that a larger convective mass transfer coefficient is obtained when a faster recession rate is recorded (as expected). A steeper concentration gradient at higher flow velocities leads to increased convection and more rapid evaporation of liquid from the container.

The convection coefficients for mass, momentum, and heat exchange are implicitly related to the slope of the appropriate variable's profile (i.e., temperature for heat exchange) at the boundary. Within the boundary layer, it is usually assumed that the velocity component normal to the wall is much smaller than the velocity component in the flow (streamwise) direction. Also, since the boundary layer is typically very thin, it is generally assumed that spatial gradients of any flow quantity in the streamwise direction remain much less than the corresponding gradients in the cross-stream direction. For a flat plate boundary layer, the momentum equation in the cross-stream direction indicates that pressure varies hydrostatically, such that pressure outside the boundary layer prescribes the pressure inside the boundary layer at some position x.

If the y direction momentum equation yields an approximately zero pressure gradient in the cross-stream (y) direction, then the pressure inside the boundary layer at a given point, x, matches the pressure at the same position, x, outside the boundary layer. In this case, the streamwise pressure variation can be determined outside the boundary layer from a potential flow analysis (i.e., solution of the inviscid flow equations) or Bernoulli's equation. Using either approach and writing the resulting pressure gradient in terms of velocity, the following reduced two-dimensional boundary layer equations are obtained for mass, momentum, and energy transport, respectively:

$$\frac{\partial u}{\partial x} + \frac{\partial v}{\partial y} = 0 \tag{3.48}$$

$$\rho \frac{\partial (uu)}{\partial x} + \rho \frac{\partial (uv)}{\partial y} = \rho U \frac{dU}{dx} + \mu \frac{\partial^2 u}{\partial y^2} \tag{3.49}$$

$$\rho c_p u \frac{\partial T}{\partial x} + \rho c_p v \frac{\partial T}{\partial y} = k \frac{\partial^2 T}{\partial y^2} + \mu \left(\frac{\partial u}{\partial y} \right)^2 \tag{3.50}$$

where $U(x)$ refers to the freestream velocity distribution (assumed to be known for a specified geometrical configuration).

In the previous boundary layer equations, the following assumptions have been adopted: (i) steady-state conditions, (ii) constant thermophysical properties, (iii) nonreacting, two-dimensional incompressible flow of a Newtonian fluid, and (iv) negligible internal heat generation. Also, it has been assumed that the characteristic length scale in the y direction, i.e., boundary layer thickness of $\delta(x)$, is much smaller than that of the streamwise (x) direction, L (length of plate). Thus, spatial gradients of variables in the y direction are more significant. Furthermore, the cross-stream (y) direction velocity component, v, has been assumed to be much smaller than the x direction velocity, u, within the boundary layer. Then the two-dimensional boundary layer equations provide three equations to be solved for three unknowns (u, v, T). An exact similarity solution of these boundary layer equations was first obtained by Blasius in 1908 (to be described in an upcoming section).

In addition to the previous equations, boundary conditions are required for problem closure. In the fluid flow equations, the boundary conditions involve zero fluid velocity on the surface (i.e., no-slip condition) and freestream velocity conditions far from the wall. Heat transfer occurs by conduction through the wall and convective heat transfer to and from the fluid. In the temperature equation, a Robin condition is specified whereby the Fourier heat flux at the wall balances the convective heat transfer outlined by Newton's law of cooling.

3.4 Nondimensional Form of Governing Equations

A nondimensional form of the governing equations allows us to reduce the number of independent variables in a problem, thereby simplifying the overall analysis. Reducing the number of independent variables can also allow us to reduce associated experimental costs. For example, if we can summarize the heat transfer coefficient in terms of one dimensionless group of variables, rather than three separate dimensional variables, then we can perform fewer experiments in capturing the full problem behavior over a wide range of conditions. Also, dimensional analysis can provide the scaling parameters between model and prototype designs. For example, if we can match appropriate dimensionless geometric and flow properties, then the dimensional analysis could allow us to infer equivalence between heat transfer parameters between the model and the prototype.

Various techniques are available for the dimensional analysis of the heat transfer equations. In particular, the *Buckingham–Pi theorem* allows us to obtain the group of dimensionless products associated with a specific problem in an efficient manner. This approach allows us to reduce a dimensional homogeneous equation with k variables to a relationship involving $k-r$ dimensionless products, where r refers to the minimum number of reference dimensions of the variables.

In the Buckingham–Pi method, all problem parameters are listed and the k parameters are written in terms of r primary dimensions, such as mass, length, and time. We would then set the number of pi terms equal to $k-r$

terms, while selecting r independent and repeating problem variables. In the procedure, we would then form a pi term by multiplying a nonrepeating variable by the product of repeating variables, each raised to an exponent that would reduce the entire combination dimensionless. Then these steps would be repeated for each remaining repeating variable. Finally, we would have an expression of the final equation as a relationship among several dimensionless groups. As a result, we would obtain a functional relationship for the problem in terms of dimensionless variables only.

An alternative is a direct approach that takes a governing equation and transforms it into a dimensionless equation through the appropriate definition of dimensionless variables. In this way, we can view the governing equations in their nondimensional form and observe the equivalence of dimensionless parameters in the equations. The resulting equations would have all of the relevant dimensionless groups for the problem.

Example 3.4.1

Convective Heat Transfer from an Isothermal Object.

Find the dimensionless parameters governing the heat transfer from an arbitrarily shaped isothermal object to a surrounding freestream flow. It may be assumed that density differences are negligible, except where they drive the mixed convection flow. Also, assume constant thermophysical properties and steady-state conditions. The surface temperature of the object is T_w, and the freestream velocity and temperature are U_∞ and T_∞, respectively. The gravitational force acts in the negative y direction.

The two-dimensional governing equations for mass, momentum (x and y directions), and energy in this convection problem may be written as follows:

$$\frac{\partial u}{\partial x}+\frac{\partial v}{\partial y}=0 \tag{3.51}$$

$$\rho\frac{Du}{Dt}=-\frac{\partial p_{ki}}{\partial x}+\mu\left(\frac{\partial^2 u}{\partial x^2}+\frac{\partial^2 u}{\partial y^2}\right) \tag{3.52}$$

$$\rho\frac{Dv}{Dt}=-\left(\frac{\partial p_{ki}}{\partial x}-\rho_\infty g\right)+\mu\left(\frac{\partial^2 v}{\partial x^2}+\frac{\partial^2 v}{\partial y^2}\right)-\rho g \tag{3.53}$$

$$\rho c_p \frac{DT}{Dt} = \beta T\left(\frac{Dp_{ki}}{Dt} - v\rho_\infty g\right) + k\left(\frac{\partial^2 T}{\partial x^2} + \frac{\partial^2 T}{\partial y^2}\right) + \mu\Phi + S \tag{3.54}$$

The variable p_{ki} refers to the kinematic pressure, i.e., the deviation from the hydrostatic pressure, such that

$$p(x, y) = p_\infty(y) + p_{ki}(x, y) \tag{3.55}$$

where

$$p_\infty(y) = p_o - \int_{y_o}^{y} \rho_\infty(y) g dy \tag{3.56}$$

In this way, we can rewrite the buoyancy forces in terms of a temperature difference through a linearized equation of state involving the thermal expansion coefficient. Also, let L refer to a characteristic length of the object, such as height, and define the following set of dimensionless variables:

$$t^* = \frac{tU_\infty}{L} \tag{3.57}$$

$$x^* = \frac{x}{L}; \quad y^* = \frac{y}{L} \tag{3.58}$$

$$u^* = \frac{u}{U_\infty}; \quad v^* = \frac{v}{U_\infty} \tag{3.59}$$

$$p_{ki}^* = \frac{p_{ki}}{\rho U_\infty^2/2} \tag{3.60}$$

$$\theta = \frac{T - T_\infty}{T_w - T_\infty}; \quad \Delta T = T_w - T_\infty \tag{3.61}$$

We can now use the chain rule of calculus, together with these dimensionless variables, to rewrite the governing equations in a nondimensional form. For example, consider the first term in the continuity equation and apply the chain rule so that

$$\frac{\partial u}{\partial x} = \frac{\partial}{\partial x^*}\left[\frac{\partial x^*}{\partial x}(u^* U_\infty)\right] \tag{3.62}$$

Similar transformations can be applied to all other terms in the governing equations with the following result for mass, momentum, and energy equations:

$$\frac{\partial u^*}{\partial x^*}+\frac{\partial v^*}{\partial y^*}=0 \tag{3.63}$$

$$\frac{Du^*}{Dt^*}=-\frac{\partial p^*_{ki}}{\partial x^*}+\frac{1}{Re}\left(\frac{\partial^2 u^*}{\partial x^{*2}}+\frac{\partial^2 u^*}{\partial y^{*2}}\right) \tag{3.64}$$

$$\frac{Dv^*}{Dt^*}=-\frac{\partial p^*_{ki}}{\partial y^*}+\frac{1}{Re}\left(\frac{\partial^2 v^*}{\partial x^{*2}}+\frac{\partial^2 v^*}{\partial y^{*2}}\right)+\left(\frac{Gr}{Re^2}\right)\theta \tag{3.65}$$

$$\frac{D\theta}{Dt}=-(Ec\cdot\beta T)\frac{Dp^*_{ki}}{Dt^*}-\left(\frac{\beta TgL}{c_p\Delta T}\right)v^*+\frac{1}{Re\cdot Pr}\left(\frac{\partial^2\theta}{\partial x^{*2}}+\frac{\partial^2\theta}{\partial y^{*2}}\right)+\left(\frac{Ec}{Re}\right)\Phi^* \tag{3.66}$$

where Re, Gr, Ec, and Pr refer to the dimensionless parameters of the Reynolds number ($Re_L=\rho U_\infty L/\mu$), Grashof number ($Gr_L=\beta\Delta TgL^3\rho^2/\mu^2$), Eckert number ($Ec=U_\infty^2/(c_p\Delta T)$), and Prandtl number ($Pr=\mu c_p/k$), respectively.

The governing equations are subject to the following boundary conditions:

$$\mathbf{v}^*=0;\quad \theta=1 \tag{3.67}$$

on the boundary, and

$$u^*=1;\quad v^*=0;\quad \theta=1 \tag{3.68}$$

in the freestream.

From these results, it can be observed that the solution of the problem has the following functional form:

$$\mathbf{v}^* \quad and \quad \theta=f\left(x^*,y^*,t^*,Re,Gr,Ec,Pr,\frac{\beta TgL}{c_p(T_w-T_\infty)}\right) \tag{3.69}$$

The convective heat transfer is outlined in terms of the Nusselt number ($Nu_L=hL/k$), where

$$\overline{Nu}_L=\frac{hL}{k}=\frac{(q''/\Delta T)L}{k} \tag{3.70}$$

$$\overline{Nu_L} = \frac{-(\partial T/\partial n)_w L}{k(T_w - T_\infty)} = -\left.\frac{\partial \theta}{\partial n}\right|_w \tag{3.71}$$

Furthermore, we can find the average heat transfer by integrating the local Nusselt number over the object's surface, i.e.,

$$\overline{Nu_L} = \frac{\int_s Nu \cdot dS}{\int_s dS} = \overline{Nu}\left(Re, Pr, Gr, Ec, \frac{\beta TgL}{c_p \Delta T}\right) \tag{3.72}$$

In Equation (3.71), n and w refer to normal to the surface and wall, respectively. Spatial integration and steady-state conditions allowed us to remove the spatial and temporal dependencies in the previous functional relation. Certain parameters can be dropped from the functional relation depending on the type of flow; i.e., Gr can be dropped for forced convection in high-speed flows.

The widely used parameter of Pr is named in recognition of Ludwig Prandtl (see Figure 3.6). His discovery of the boundary layer concept has become fundamental material of fluid mechanics, convective heat transfer, and aeronautics.

FIGURE 3.6
Ludwig Prandtl (1875–1953). (From Hewitt, G.F., Shires, G.L., and Polezhaev, Y.V., Eds., *International Encyclopedia of Heat and Mass Transfer*, Copyright 1997, CRC Press, Boca Raton, FL. Reprinted with permission.)

Based on the previous example, the following functional relations can be summarized for special cases of convection problems at steady state (laminar and turbulent):

- forced convection: $\overline{Nu} = \overline{Nu}(Re, Pr)$
- high-speed flows: $\overline{Nu} = \overline{Nu}(Re, Pr, Ec)$
- free convection: $\overline{Nu} = \overline{Nu}(Re, Pr, Gr)$

Other dimensionless groups would arise when additional processes are present, i.e., phase change, chemical reactions, etc.

An important parameter in the earlier definition of the nondimensional velocity is the reference velocity, i.e., U_∞. In some problems, such as natural convection, this value may not be explicitly apparent. For example, the ambient conditions may be quiescent, and $U_\infty = 0$ for a free convection problem. As a result, another reference velocity may be required. In this case of free convection, we can use a scaling approximation from Equation (3.164) in order to obtain the following reference velocity:

$$U_{ref} = \sqrt{g\beta\Delta TL} \tag{3.73}$$

Alternatively, using a time scale of $t_{ref} = L^2/\alpha$ from conduction heat transfer, we can adopt

$$U_{ref} = \frac{k}{\rho L c_p} \tag{3.74}$$

Based on these parameters, the only dimensionless parameters appearing in free convection problems at steady state are $Ra = Gr \cdot Pr$ (Rayleigh number) and Pr (Prandtl number).

As discussed earlier, a useful benefit of dimensional analysis is that laboratory model experiments can be used to predict larger scale prototype behavior when geometric, kinematic, and dynamic similarity (i.e., equivalent Re, Pr, etc.) are maintained between the model and prototype. However, in practice, exact similarity is often not possible due to various practical limitations, such as limited fluid choices in the laboratory testing, roughness on the surface, and other factors. Nevertheless, based on the previously described procedures applied to the boundary layer equations, the following reduced dimensionless equations are obtained:

$$\frac{\partial u^*}{\partial x^*} + \frac{\partial v^*}{\partial y^*} = 0 \tag{3.75}$$

$$u^*\frac{\partial u^*}{\partial x^*}+v^*\frac{\partial u^*}{\partial y^*}=-\frac{\partial p^*}{\partial x^*}+\left(\frac{1}{Re}\right)\frac{\partial^2 u^*}{\partial y^{*2}} \tag{3.76}$$

$$u^*\frac{\partial T^*}{\partial x^*}+v^*\frac{\partial T^*}{\partial y^*}=\left(\frac{1}{RePr}\right)\frac{\partial^2 T^*}{\partial y^{*2}} \tag{3.77}$$

These governing equations are subject to appropriate boundary conditions, such as specified conditions at the wall, including no-slip ($\mathbf{v}=0$) and specified temperature values at the wall. The similarity parameters in the dimensionless equations are the Reynolds number ($Re = UL/\nu$) and the Prandtl number ($Pr = \nu/\alpha$). The Reynolds number may be interpreted as the ratio of characteristic inertial forces in the flow field to viscous forces, whereas the Prandtl number characterizes the ratio of momentum (viscous) to thermal diffusivities.

The functional form of the solution of the boundary layer equations becomes

$$u^* = u^*\left(x^*, y^*, Re, \frac{dp^*}{dx^*}\right) \tag{3.78}$$

$$T^* = T^*\left(x^*, y^*, Re, Pr, \frac{dp^*}{dx^*}\right) \tag{3.79}$$

where the dependence on pressure gradient in the temperature relationship arises due to the dependence on velocities. The pressure gradient is assumed to be known at the edge of the boundary layer for a given surface geometry, based on a potential flow solution outside the boundary layer. For example, the streamwise pressure gradient in the boundary layer will equal the freestream pressure gradient for boundary layer problems, and furthermore, it equals zero for a flat plate boundary layer. As a result, the functional dependence on pressure gradient can be effectively removed since it will be known for boundary layer problems.

Then the previous functional relations can be written in terms of the friction coefficient, c_f, and heat transfer coefficient, Nu, as follows:

$$c_f = \frac{\tau_w}{\rho u_\infty^2/2} = \frac{\mu(\partial u/\partial y)_0}{\rho u_\infty^2/2} \tag{3.80}$$

$$c_f = \frac{2}{Re}\left(\frac{\partial u^*}{\partial y^*}\right)_0 = \frac{2}{Re} f_1(x^*, Re) \tag{3.81}$$

$$Nu = \frac{hL}{k_f} = \left(\frac{\partial T^*}{\partial y^*}\right)_0 = f_2(x^*, Re, Pr) \quad (3.82)$$

where f_1, f_2, L, and k_f refer to functions 1 and 2, plate length, and fluid conductivity, respectively.

Similar results can be obtained for mass transfer, but the Prandtl and Nusselt numbers are replaced by the Schmidt number ($Sc = \nu/D_{ab}$) and the Sherwood number ($Sh = h_m \ L/D_{ab}$), respectively. The Sherwood number includes the convective mass transfer coefficient, h_m, analogous to the convection coefficient for heat transfer problems.

3.5 Heat and Momentum Analogies

From Equations (3.78) and (3.79), it can be observed that the functional form of both equations becomes identical when $Pr = 1$ with a zero pressure gradient. This observation suggests that $u^* = T^*$ under these conditions when the velocity and thermal boundary conditions (dimensionless) are identical. Then f_1 in Equation (3.81) and f_2 in Equation (3.82) would become identical, so we obtain the following *Reynolds analogy* between momentum and heat transfer (with extension by analogy to mass transfer):

$$Re\left(\frac{c_f}{2}\right) = Nu = Sh \quad (3.83)$$

Thus, when conditions permit this analogy, we find that any analytical, computational, or experimental results from the fluid mechanics problem can be related (through the analogy) to the heat transfer problem.

Alternatively, in terms of the Stanton numbers,

$$\frac{c_f}{2} = St = St_m \quad (3.84)$$

where

$$St = \frac{Nu}{RePr} \quad (3.85)$$

is the Stanton number for heat transfer and

$$St_m = \frac{Sh}{ReSc} \tag{3.86}$$

refers to the Stanton number for mass transfer.

Departures from the previous analogy due to a typical pressure gradient often remain within the limit of experimental errors associated with heat transfer measurements. Furthermore, corrections to account for cases where $Pr \neq 1$ and $Sc \neq 1$ include the following *Chilton–Colburn analogy*:

$$\frac{c_f}{2} = StPr^{1-n} = St_m Sc^{1-n} \tag{3.87}$$

which is applicable for $Pr \neq 1$ and $Sc \neq 1$. For many applications, it is often assumed that $n = 1/3$. The middle and last expressions in Equation (3.87) are called the *Colburn j factors*, j_H and j_m, for heat and mass transfer, respectively. In addition to the currently discussed analogies, j_H is widely used in the analysis and design of heat exchangers (see Chapter 10).

Equation (3.87) allows one convection coefficient, such as h_m, to be written in terms of the other coefficient, h, i.e.,

$$\frac{hL/k}{Pr^{1/3}} = \frac{h_m L/D_{ab}}{Sc^{1/3}} \tag{3.88}$$

which can be rearranged to give h/h_m in terms of the Lewis number, Le, where $Le = Sc/Pr$.

Once the form of the appropriate correlation is determined, experimental data can be gathered and summarized based on this form to reduce the number of independent parameters under investigation. Experimental measurements are usually required to establish or validate the correlations. In most cases, these correlations are presented in a nondimensional form, i.e., in terms of the appropriate dimensionless groups, such as Nu, Re, and Pr. A sample technique for acquiring the coefficients associated with these correlations is given in the following example.

Example 3.5.1

Measurement of Convection Coefficients.

Outline a procedure (or a basic experiment) that can be used to measure the convection coefficient and resulting convection correlation for heat transfer by forced convection over a flat, heated surface. A cold airstream flows at T_∞ across a heater plate, and heat is transferred by convection through the boundary layer to the air.

An experiment can be constructed by a heater embedded within the surface. A current is then supplied through the heater in order to maintain a constant surface temperature, T_S. Then, from an energy balance on the plate, the rate of heat transfer by convection to the air must balance the electrical heat supplied within the plate.

As a result, we can find the average convective heat transfer coefficient, $\overline{h}$, in terms of the applied electrical current, I, electrical resistance, R, and the temperature difference between the surface and the airstream, i.e.,

$$\overline{h} \approx \frac{I^2 R}{A_s(T_s - T_\infty)} \tag{3.89}$$

Then we can perform several experiments with different Reynolds and Prandtl numbers and determine a correlation based on experimental data in the following form:

$$\overline{Nu} = C Re_L^m Pr^n \tag{3.90}$$

For example, after performing the experiment with air, water, and oil and also changing the incoming velocities, we would likely obtain $n \approx 1/3$ and $m \approx 1/2$ for laminar forced convection, but $m \approx 4/3$ for turbulent forced convection.

Similar experiments could be performed for problems involving other geometries, processes, and flow conditions. The previous example represents a possible way of experimentally finding the heat transfer coefficient for convection problems.

3.6 External Forced Convection

In forced convection, the bulk fluid motion is induced by external means, such as pumps, blowers, or atmospheric winds. In external flows, the fluid motion is restricted only by the presence of a single boundary (i.e., airflow past an airplane wing), whereas internal flows are completely contained within solid boundaries (i.e., water flow in a pipe). Some problems involve a combination of external and internal flow, i.e., a fan delivering cool air over a circuit board (external flow) through a set of hot electrical components (internal flow). In this section, the main features and analysis of external flows will be examined.

3.6.1 External Flow Past Flat Plate

This section considers external flow over a horizontal plate. The reduced governing equations and boundary conditions for steady, incompressible, and laminar boundary layer flow past a flat plate (see Figure 3.4) were obtained as:

$$\frac{\partial u}{\partial x} + \frac{\partial v}{\partial y} = 0 \tag{3.91}$$

$$\rho \frac{Du}{Dt} \equiv \rho u \frac{\partial u}{\partial x} + \rho v \frac{\partial v}{\partial y} = -\frac{\partial p}{\partial x} + \mu \left(\frac{\partial^2 u}{\partial y^2} \right) \tag{3.92}$$

$$\rho c_p \frac{DT}{Dt} \equiv \rho c_p u \frac{\partial T}{\partial x} + \rho c_p v \frac{\partial T}{\partial y} = k \left(\frac{\partial^2 T}{\partial y^2} \right) \tag{3.93}$$

subject to $u = 0 = v$ and $T = T_w$ at $y = 0$ (on the wall) and $u = u_\infty$, $v = 0$, and $T = T_\infty$ (in the freestream). For flow past a flat plate, the streamwise pressure gradient, $\partial p/\partial x$, is zero since the streamwise velocity component is assumed to be uniform and constant outside the boundary layer.

The following methods will be presented for the analysis of laminar forced convection in this boundary layer flow: (i) similarity solution, (ii) discrete scaling, and (iii) integral solution. The detailed understanding of various solution techniques is important in view of their potential application to other similar types of parallel flows, including multiphase flows with heat transfer. In addition to laminar flows, this section will also present appropriate correlations for the corresponding turbulent flows.

3.6.1.1 Similarity Solutions

Similarity solutions may be available whenever one independent coordinate in the solution domain exists and physical influences are carried only in that one direction (see Figure 3.7). In the current boundary layer example, the term *similarity* means that if the y coordinate (in the cross-stream direction) is stretched at each location x by the factor $g(x)$, then all solution curves over the y direction collapse onto a single curve. In other words, the solution becomes a function of $y \cdot g(x)$ alone.

The existence of similarity solutions may be established by group theoretical methods. Group theory is based on invariance of the governing equations under continuous (Lie) groups of transformations. For example, if the governing equations for the heat transfer problem are invariant (i.e., retain same functional form) under the transformations $u \to u$, $v \to v/a$, $x \to$

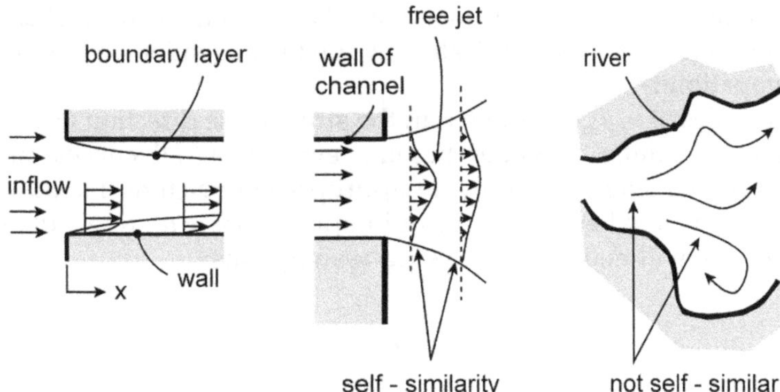

FIGURE 3.7
Similarity characteristics of various typical flows.

a^2x, and $y \to ay$, then $y/\sqrt{x}$, u and $v/\sqrt{x}$ would become the invariants of the group with the transformations of $u = f(\eta)$, $v = g(\eta)/\sqrt{x}$, and $\eta = y/\sqrt{x}$.

A group of transformations that leaves a given system invariant is a topic of mathematical group theory. Examples of these transformations include the rotation or translation of coordinates. Other miscellaneous methods exist for determining similarity solutions, i.e., deductive methods, the method of free parameters, and dimensional analysis. In this section, similarity solutions will be derived based on physical interpretations of the related problem variables. From a physical viewpoint, similarity may be present whenever one independent coordinate in the solution domain exists; as a result, the physical transport processes will follow this single direction. For example, if downstream disturbances do not appreciably affect the upstream flow structure, then the flow profiles in the flow direction would remain "self-similar."

Consider three problems in terms of flow similarity (see Figure 3.7). In the first case (boundary layer), self-similarity can be observed by the developing nature of the flow in the duct. In a similar way, a free jet exhibits self-similarity in terms of its flow structure. Finally, if we consider the diverse flow behavior in a river flow, then it appears evident that self-similarity cannot be sustained therein. In this case, the propagation of disturbances is not carried in a single flow direction.

Example 3.6.1.1.1

Boundary Layer Past a Flat Plate.

Use the similarity solution method to find the velocity and temperature profiles for boundary flow past an isothermal flat plate. The solution can proceed under the assumption that a similarity profile can be obtained; if the assumption is incorrect, indicate how the final equations would indicate the incompatibility.

If the profiles of u/u_∞ are plotted in the streamwise direction (Figures 3.5, 3.7), their self-similarity would become evident, as all curves could be collapsed onto a single curve by an appropriate stretching factor, $g(x)$, at each position x. In other words, it would appear that u/u_∞ depends only on $\eta = y \cdot g(x)$. We will make the following assumptions:

$$\frac{u}{u_\infty} = f'(\eta) \tag{3.94}$$

$$\eta = y \cdot g(x) \tag{3.95}$$

For a specific problem, these assumptions can be deduced in various ways (i.e., method of unknown coefficients; see discussion following this example problem). The derivative of the arbitrary function, $f'(\eta)$, rather than the function itself, is adopted to simplify the resulting mathematics involving integrations. If the assumption in Equation (3.94) is incorrect, the resulting equations will make it clearly evident. We will now transform coordinates from (x,y) to (x, η) coordinates.

In order to implicitly satisfy the continuity equation, a stream function, ψ, is defined such that

$$u = \frac{\partial \psi}{\partial x} \tag{3.96}$$

$$v = -\frac{\partial \psi}{\partial y} \tag{3.97}$$

Using Equations (3.95) and (3.97), together with boundary conditions at the surface of the plate, we obtain

$$\psi = u_\infty g_2(x) f(\eta) + C \tag{3.98}$$

where

$$\eta = y \cdot g_1(x) \tag{3.99}$$

and

$$g_2(x) = \frac{1}{g_1(x)} \tag{3.100}$$

is introduced for simplicity. Also, C represents an arbitrary constant of integration.

This stream function can now be differentiated to give both velocity components in Equations (3.96) and (3.97). Substituting these components and their derivatives (as required) into the momentum equation, Equation (3.92), and rearranging terms,

$$-\frac{f'''(\eta)}{f(\eta)\cdot f''(\eta)} = \frac{u_\infty g_2'(x)}{\nu g_1(x)} \equiv C_1 \tag{3.101}$$

where $dp/dx = 0$ has been assumed for flat plate boundary layer flow. The result in Equation (3.101) requires that both sides equal a constant, called C_1, since the left term is a function of η alone, whereas the middle term is a function of x alone. If our original similarity assumption in Equation (3.95) was incorrect, then a separable constraint, i.e., Equation (3.101), could not be achieved. Thus, the separability of Equation (3.101) indicates that flat plate boundary layer flows are self-similar with each other in the downstream direction (as expected).

Solving the result in Equation (3.101) and applying the boundary conditions, we obtain the following *Blasius equation*:

$$f'''(\eta) + \frac{1}{2} f(\eta) f''(\eta) = 0 \tag{3.102}$$

where

$$\eta = y\sqrt{\frac{u_\infty}{\nu x}} \tag{3.103}$$

subject to

$$f'(\eta \to \infty) \to 1 \tag{3.104}$$

and

$$f'(0) = 0; \quad f(0) = 0 \tag{3.105}$$

This system can be readily solved by numerical integration, such as the Runge–Kutta method.

The result will provide the velocity distribution throughout the boundary layer, which will be used to determine the heat transfer from the plate. Transforming Equation (3.93) to the similarity variables, (x, η), based on a nondimensional temperature using $\theta(\eta) = (T - T_w)/(T_\infty - T_w)$, and following a procedure similar to that carried out earlier in the momentum equation, we obtain

$$\theta''(\eta) + \frac{Pr}{2}\theta'(\eta)f(\eta) = 0 \tag{3.106}$$

subject to $\theta(0) = 0$ and $\theta(\eta \to \infty) = 1$. Since Equation (3.102) is de-coupled from Equation (3.106), the solutions from the Blasius equation for $f(\eta)$ can be used for the heat transfer in solving Equation (3.106).

We can obtain the result for $\theta(\eta)$ and u/u_∞ by isolating $\theta'(\eta)$ and integrating Equation (3.106) twice, and similarly for Equation (3.102), to obtain

$$\theta(\eta) = \frac{\int_0^\eta \left[exp(-Pr\int_0^\eta f(\eta)d\eta)/2 \right] d\eta}{\int_0^\infty \left[exp(-Pr\int_0^\eta f(\eta)d\eta)/2 \right] d\eta} \tag{3.107}$$

$$\frac{u}{u_\infty} = f'(\eta) = \frac{\int_0^\eta \left[exp(-\int_0^\eta f(\eta)d\eta)/2 \right] d\eta}{\int_0^\infty \left[exp(-\int_0^\eta f(\eta)d\eta)/2 \right] d\eta} \tag{3.108}$$

where Pr refers to the Prandtl number.

TABLE 3.1

Similarity Functions for Flat Plate Boundary Layer Flow

η	$f(\eta)$	$f'(\eta)$	$f''(\eta)$
0	0.000	0.000	0.332
0.6	0.060	0.199	0.330
1.2	0.238	0.394	0.317
1.8	0.530	0.575	0.283
2.4	0.922	0.729	0.228
3.0	1.397	0.846	0.161
3.6	1.930	0.923	0.098
4.2	2.498	0.967	0.051
4.8	3.085	0.988	0.021
5.4	3.681	0.996	0.008
6.0	4.280	0.999	0.002
6.6	4.879	1.000	0.001
7.2	5.479	1.000	0.000

As a result, we can now find $f(\eta)$ from Equation (3.102) by numerical integration (see Table 3.1 for sample results). Then we can use Equations (3.107) and (3.108) to find the temperature, θ, and velocity, u/u_{∞}, distributions. Upon investigating those results, we find that θ rises to θ_{∞} faster as $Pr \to \infty$ (compared with $Pr \ll 1$) since the thermal boundary thickness is small relative to the velocity boundary layer thickness for large Prandtl numbers.

In addition, we can now find the local Nusselt number and the wall heat flux by taking the temperature gradient normal to the wall at $\eta = 0$, i.e.,

$$q_w''(x) = -k\frac{\partial T}{\partial y}\bigg|_{y=0} = k\Delta T\theta'(0)\sqrt{\frac{u_{\infty}}{\nu x}} \tag{3.109}$$

$$Nu_x = \frac{hx}{k} = \frac{(q_w''/\Delta T)x}{k} = \theta'(0)Re_x^{1/2} \tag{3.110}$$

From the results of the temperatures in Equation (3.107), it can be shown that the slope of the dimensionless temperature profile is observed to vary with Pr according to $\theta'(0) \approx 0.332Pr^{1/3}$ so that

$$Nu_x = 0.332{Re_x}^{1/2}Pr^{1/3} \tag{3.111}$$

where x refers to the position along the plate. This result agrees closely with experimental data for heat transfer from a flat plate in steady, incompressible flow.

Finding a suitable similarity variable at the beginning of a problem is often difficult. In some cases, we may consider the *method of unknown coefficients*, where the definition of the similarity variable is based on a product of all independent problem variables, such as x and y, each raised to an exponent (unknown coefficient). The specific values of the unknown coefficients may then be determined by proceeding in a manner similar to that in the previous example, but then imposing constraints on the unknown coefficients in order to eliminate their dependence on the independent variables in the boundary conditions. In this way, we can have a standard procedure for finding the similarity variables. In most cases, the equations themselves will indicate whether the multivariable problem can be reduced to a separable system with a similarity solution.

3.6.1.2 Discrete Scaling Analysis

Another alternative approach for analyzing the forced convection equations is an approximate method based on discrete scaling of the governing equations. In the context of the boundary layer equations, the quantities δ, L, v_o, and u_∞ will be used to refer to the boundary layer thickness, plate length, cross-stream reference velocity, and freestream velocity, respectively. Each of these reference variables can be substituted into the appropriate places in the governing equations to represent their respective scales in the overall problem.

For example, from the continuity equation the approximate order of the cross-stream velocity component becomes $v_o \approx \delta u_\infty / L$. Furthermore, approximate order-of-magnitude scaling of the momentum equation yields

$$\rho u_\infty \frac{u_\infty}{L} + \rho \left(\frac{\delta}{L} u_\infty \right) \frac{u_\infty}{\delta} \approx \mu \frac{u_\infty}{\delta^2} \tag{3.112}$$

which can be rewritten in terms of the boundary layer thickness (note: order of magnitudes only, without leading numerical coefficients) as follows:

$$\delta \approx \frac{L}{\sqrt{Re}} \tag{3.113}$$

where

$$Re_L = \frac{\rho u_\infty L}{\mu} \tag{3.114}$$

Using this result and calculating the skin friction coefficient, we find

$$\bar{c}_f = \frac{\bar{\tau}_w}{\rho u_\infty^2 / 2} \approx \frac{\mu u_\infty / \delta}{\rho u_\infty^2 / 2} = \frac{2}{\sqrt{Re_L}} \tag{3.115}$$

The previous exact solution of the Blasius equation yields the same functional form, except that the coefficient in the numerator is 1.328 (rather than 2). As a result, the approximate scaling analysis agrees well with the exact result in terms of functional form and the approximate order of magnitude of the results.

The following general steps and guidelines can be used when performing a discrete scaling analysis:

- Identify the relevant length scale(s) in the problem.
- The order of magnitude of a sum (or difference) of terms is determined by the dominant term. For example, if $a = b + c$ and $\mathcal{O}(b) > \mathcal{O}(c)$, then $\mathcal{O}(a) = \mathcal{O}(b)$, where the notation of $\mathcal{O}(\,)$ refers to the order of magnitude.

- The order of magnitude of a sum of terms each having the same order of magnitude is the same as the order of magnitude of each individual term.
- The order of magnitude of a product (or quotient) of terms is the same as the product (or quotient) of orders of magnitude of the individual terms.

These guidelines become useful, particularly when many terms appear in the governing equations.

The discrete scaling technique can be used in various other problems, such as multiphase heat transfer. In addition to providing reasonable estimates of the dependent variables (such as temperature and heat flux), it can also be used for discretization involving certain independent variables, such as the grid spacing and time step size in numerical models. Discrete scaling represents an approximation of continuous quantities by a discrete distribution. In the following section, another type of approach for the discrete analysis of differential systems, called integral solution methods, will be described.

3.6.1.3 Integral Solution Method

In integral solution methods, the number of problem variables is reduced through integration of the differential equations over a specified independent variable(s). This procedure removes the variation of the dependent variables in the integrated coordinate direction (or time) since those variations are effectively averaged by the integration. However, this benefit comes at the expense of requiring additional information regarding the profile variations in the integrated direction. In other words, an approximation of the dependent variable is required in a certain direction (or time), where knowledge of its variation exists. Then spatial variations in the other directions are obtained from the integral solution.

The level of the integral solution may be classified according to the number of degrees of freedom. For example, if the profile of an integrated quantity is available from the exact solution of another governing equation, then zero degrees of freedom are used. However, if empirical estimates or other nonexact approximations are adopted in the assumed profiles prior to the integration, then multiple degrees of freedom are adopted, since certain freedom is adopted in the assumed profile (i.e., parabolic profile rather than a cubic profile of temperature in the boundary layer region).

In this analysis, we will use an integral method where the integrated quantities are not available from additional equations in the problem. As a result, some degrees of freedom, such as the profile shape in the cross-stream direction, will be applied in the solution procedure. In this case, the integrated unknown variable is not available by analytic means. Instead, an assumed profile is used for the unknown quantity. In the integration procedure, the following Leibnitz rule will be used:

$$\frac{\partial}{\partial x}\int_{a(x)}^{b(x)} f(x,y)dy = \int_{a(x)}^{b(x)} \frac{\partial f(x,y)}{\partial x}dy + f(a,y)\frac{db}{dx} - f(b,y)\frac{da}{dx} \tag{3.116}$$

In many problems, property changes in one direction may be well known, but changes in the other direction(s) (i.e., streamwise direction) are unknown. The integral method considers a control volume with a finite extent in all directions except the direction where the unknown variations arise. As a result, an ordinary differential equation arises in terms of that one coordinate direction. In the integral technique involving multiple degrees of freedom, the governing differential equations are directly integrated across the direction where property changes are known (such as the cross-stream direction in the following boundary layer example).

Example 3.6.1.3.1

Integral Solution of Boundary Layer Flow.

Perform an integration of the governing equations across the boundary layer thickness (δ) to find the spatial variations of velocity and temperature in boundary layer flow along a flat plate.

Integrating the continuity equation across the boundary layer from $y = 0 \rightarrow \delta$ and using Leibnitz's rule, Equation (3.116),

$$\frac{\partial}{\partial x}\left(\int_0^{\delta} \rho u dy\right) - \rho u_{\infty}\frac{\partial \delta}{\partial x} + \rho v_{\delta} - \rho v_o = 0 \tag{3.117}$$

where $v_o = 0$ (no-slip condition at the wall).

In a similar fashion, we have the following result after integration of the boundary layer momentum equation:

$$\frac{\partial}{\partial x}\left(\int_0^{\delta} \rho u u dy\right) - \frac{\partial \delta}{\partial x}\rho u_{\infty}^2 + \rho u_{\infty} v_{\delta} - \rho u_{\infty} v_o$$
$$= -\frac{\partial}{\partial x}\int_0^{\delta} p dy + p\frac{\partial \delta}{\partial x} + \mu\left(\frac{\partial u}{\partial y}\right)_y - \mu\left(\frac{\partial u}{\partial y}\right)_0 \tag{3.118}$$

where the velocity gradient at the edge of the boundary layer approaches zero, i.e., $\partial u/\partial y|_{\delta} = 0$. The additional velocity gradient term (i.e., velocity gradient at the wall) can be written in terms of the wall shear stress, τ_w. Furthermore, substituting the momentum equation at $y = \delta$ (inviscid freestream) into Equation (3.118) allows us to rewrite the pressure gradient in terms of the freestream velocity, u_∞, so that

$$\frac{d}{dx}\int_0^\delta \rho u(u_\infty - u)dy + \frac{du_\infty}{dx}\int_0^\delta \rho(u_\infty - u)dy = \tau_w \tag{3.119}$$

At this stage, we need to provide the assumed profile for the velocity across the boundary layer (in the cross-stream direction) so that the integration in Equation (3.119) can be completed. The resulting equation yields the governing ordinary differential equation. The velocity profile within the boundary layer depends on the pressure gradient and the surface geometry. In general, we need to specify a suitable profile based on these considerations.

In particular, let us specifically consider a boundary layer along a flat plate with a constant freestream velocity, u_∞. In this case, the following cubic profile is assumed to represent the variation of velocity across the boundary layer:

$$\frac{u}{u_\infty} = a_o + a_1\left(\frac{y}{\delta}\right) + a_2\left(\frac{y}{\delta}\right)^2 + a_3\left(\frac{y}{\delta}\right)^3 \tag{3.120}$$

In this way, we are using an integral scheme with one degree of freedom since one empirical profile (i.e., streamwise velocity) is specified in the solution. Upon substitution of four appropriate boundary conditions into Equation (3.120), such as $u(0) = 0$ and a zero gradient at the edge of the boundary layer, we find that $a_o = 0$, $a_1 = 3/2$, $a_2 = 0$, and $a_3 = -1/2$.

Also, after substituting this profile into Equation (3.119) and performing the integration, we obtain the following ordinary differential equation:

$$\delta\frac{d\delta}{dx} = \frac{140}{13}\frac{\mu}{\rho u_\infty^2} \tag{3.121}$$

This equation can be solved to give the boundary layer growth with position, i.e., $\delta(x)$.

Then the wall shear stress can be obtained from that result. Differentiating the velocity profile and evaluating at $y = 0$ to determine the wall shear stress, we obtain

$$\frac{\tau_w}{\rho u_\infty^2} = \frac{0.323}{\sqrt{Re_x}} \tag{3.122}$$

This result agrees very closely with the known exact (Blasius) solution, except that the coefficient 0.323 is replaced by 0.332. In other words, this integral analysis agrees within 3% of the exact solution, and the correct functional dependence on the Reynolds number is achieved. A similar procedure, involving the energy equation, can be followed for estimates of the temperature distribution and heat transfer along the plate.

Many problems of practical and industrial relevance can be analyzed by the integral technique. For example, a thermal plume leaving the top of an industrial stack can be examined in the streamwise direction (plume flow direction) with a control volume of finite cross-stream width (i.e., symmetrical profile assumed) and differential streamwise thickness for an analysis in the flow direction.

Furthermore, in subsequent chapters we will consider this technique in its application to a variety of multiphase problems. The previous examples have outlined the fundamental principles in the application of the technique. Extensions to multiphase problems will be based on these principles. For example, insights into two-phase (liquid–gas) flow in a pipe may be evident by an integral analysis involving a control volume of finite cross-stream width (i.e., diameter of the pipe) and differential thickness in the streamwise direction (flow direction). The integral analysis would require an assumed radial velocity distribution, in order to solve for and find the remaining profiles in the flow direction. In terms of heat transfer, the resulting temperature profiles in both cross-stream and streamwise directions can be used to find the wall heat flux.

3.6.1.4 Correlations for Laminar and Turbulent Flow

In the previous analysis, laminar external flow past a horizontal plate was examined. In practice, the structure of the flat plate boundary layer often involves a laminar part from the leading edge up to a transition point (where $Re_x = U_\infty x/\nu \approx 5 \times 10^5$) and then transition to turbulence thereafter. Near the leading edge of the plate (laminar region), a fluid particle flowing parallel to the wall within the boundary layer is sheared by viscous action. The resulting change in the streamwise velocity component, u, is accompanied by a change in the cross-stream velocity component, v. In this way, mass is conserved while the boundary layer thickness grows with x in the direction of the flow.

Once transition to turbulence is reached, the chaotic nature of turbulence leads to an increase of cross-stream velocity fluctuations, and thus the turbulent boundary layer grows more rapidly than the previous analysis involving laminar boundary layer flow. In turbulent boundary layer flow, the fluid particles are far enough from the wall that their chaotic motion is not restrained by the frictional restraining effects of the wall. In this section, the previous laminar flow correlations will be summarized, together with corresponding correlations for turbulent boundary layer flow.

The following laminar flow results for the thicknesses of the velocity boundary layer, $\delta(x)$, temperature boundary layer, $\delta_t(x)$, and concentration boundary layer, $\delta_m(x)$ were available from the exact similarity solution:

$$\delta(x) = \frac{5x}{\sqrt{Re_x}} \tag{3.123}$$

$$\frac{\delta_t(x)}{\delta(x)} = Pr^{-1/3} \tag{3.124}$$

$$\frac{\delta_m(x)}{\delta(x)} = Sc^{-1/3} \tag{3.125}$$

For turbulent flow past the transition point in the boundary layer (Schlichting, 1979), the following correlations are based on experimental data:

$$\delta(x) = \frac{0.37x}{Re_x^{1/5}} \approx \delta_t(x) \approx \delta_m(x) \tag{3.126}$$

It can be observed that the turbulent boundary layer grows faster than the laminar portion of the boundary layer, partly as a result of the turbulent eddy motion.

The results for wall shear stress and heat transfer in flat plate boundary layer flow are summarized in Table 3.2.

The results in Table 3.2 indicate that the wall shear stress and convective heat transfer coefficient, h, decrease as $x^{-1/2}$ in the positive x direction

TABLE 3.2

Correlations for Laminar and Turbulent Flow Past a Flat Plate

Conditions	Wall Shear	Heat Transfer
	$c_f = \frac{\tau_w}{\rho U_\infty^2/2}$	$Nu = \frac{hx}{k}$
Laminar ($Re_x < 5 \times 10^5$)	$c_f(x) = 0.664\ Re_x^{-1/2}$	$Nu(x) = 0.332\ Re_x^{1/2}\ Pr^{1/3}$ ($Pr > 0.6$)
Turbulent ($Re_x \geq 5 \times 10^5$)	$c_f(x) = 0.059\ Re_x^{-1/5}$	$Nu(x) = 0.0296\ Re_x^{4/5}\ Pr^{1/3}$ ($0.6 < Pr < 60$)

TABLE 3.3

Summary of Correlations for Average Heat Transfer

Conditions	Flow	Heat Transfer
$x_c/L > 0.95$	Laminar	$\overline{Nu}_L = 0.664 Re_L^{1/2} Pr^{1/3}$; $0.6 < Pr < 50$
$x_c/L < 0.1$	Turbulent	$\overline{Nu}_L = 0.037 Re_L^{4/5} Pr^{1/3}$; $0.6 < Pr < 60$
$0.1 \le x_c/L \le 0.95$	Mixed	$\overline{Nu}_L = (0.037 Re_L^{4/5} - 871) Pr^{1/3}$; $(0.6 < Pr < 60)$

(laminar regime). As expected, the growth of the boundary layer leads to a decreasing velocity and temperature gradient at the wall in the positive x direction, thereby reducing the wall shear stress and convection coefficient. In the turbulence transition region, τ_w and h increase abruptly, but then start decreasing as $x^{-1/5}$ in the turbulent regime for reasons similar to those cited earlier.

Since the boundary layer typically contains both laminar and turbulent parts, the average heat transfer coefficient, $\overline{h}_L$, over the entire plate length, L, is obtained by integration of both laminar and turbulent results over the appropriate range, i.e.,

$$\overline{h}_L = \frac{1}{L}\left[\int_0^{x_c} h(lam)dx + \int_{x_c}^{L} h(turb)dx\right] \tag{3.127}$$

where x_c refers to the critical point (transition to turbulence). Performing this integration using the appropriate correlations, the results in Table 3.3 are obtained.

In Table 3.3, the laminar flow results assume that the laminar flow correlation can be used across the entire integration range, i.e., laminar flow occupies the majority of the plate (more than 95%). Similarly, turbulent flow is assumed to arise predominantly over the entire plate in the latter turbulent flow result. The mixed-flow correlation uses a combination of laminar and turbulent flow correlations when the transition to turbulence occurs past the front 10% or before the last 5% of the plate.

3.6.2 External Flow Past a Circular Cylinder

Heat transfer with fluid flow past a circular cylinder is a commonly encountered configuration in various engineering applications. On the front side of the cylinder (facing the incoming flow), the flow experiences a favorable pressure gradient (i.e., decreasing pressure) as it accelerates toward the top or bottom surfaces of the cylinder (see Figure 3.1). On the other hand, an adverse pressure gradient (i.e., increasing pressure) occurs on the back side of the cylinder. A boundary layer is formed along the cylinder surface where viscous effects are significant. For certain flow velocities, the increasing pressure along the back side of the cylinder causes the boundary layer to separate from the wall and create a local flow reversal with vortices shed from the cylinder. The boundary layer is detached from the wall due to the adverse pressure gradient, which causes the pressure to act against the fluid motion in the streamwise direction.

At low Reynolds numbers ($Re_D < 100$, where $Re = U_\infty D/\nu$ and D refers to the cylinder diameter), *Stokes creeping flow* is observed. Separation of the boundary layer on the back side of the cylinder first occurs at $Re_D \approx 6$. Downstream vortices at the top and bottom of the cylinder are formed. Shedding of the vortices is first observed at $Re_D \approx 60$. The separation of the boundary layer from the cylinder surface first appears at 180° and rapidly moves back to 80° as the Reynolds number increases. The circumferential angle is measured around the cylinder starting from the front edge directly facing the incoming flow (see Figure 3.1). In this low-*Re* regime, the drag coefficient (i.e., drag force divided by $\rho U_\infty^2/2$) decreases with Re_D in view of the increasing freestream velocity. The flow remains laminar in this Stokes flow regime.

However, between Reynolds numbers of 100 and 2×10^5, the drag coefficient remains approximately constant ($c_d \approx 1.2$) since the boundary layer separation remains nearly stationary at approximately 80°. In terms of the heat transfer, the Nusselt number decreases with circumferential angle, θ, up to the separation point, as the growing boundary layer thickness reduces the near-wall temperature gradient. Beyond the separation point, Nu_D increases with θ since boundary layer separation increases the near-wall fluid and thermal mixing. Changes with varying Re_D are mainly observed in the structure of vortex shedding downstream of the cylinder. Although more turbulent mixing occurs in the downstream wake, its effects on the resulting pressure and friction drag forces are nominal. In this intermediate range of Reynolds numbers, the boundary layer remains laminar, but the downstream wake and mixing are inherently turbulent.

At $Re_D \approx 2 \times 10^5$ the boundary layer becomes turbulent, which causes it to remain attached to the wall longer, so that separation of the boundary layer is delayed to approximately 140° (i.e., moves to the downstream side of the cylinder). Turbulence causes the boundary layer to become more resistant to separation induced by the adverse pressure gradient. A narrower downstream wake and the resulting change in pressure therein (i.e., smaller recirculating low-pressure wake) cause an abrupt decrease of the drag coefficient ($c_d \approx 0.3$) following the transition to turbulence. Beyond this point, the drag coefficient rises with Re_D due to the increasing role of turbulent mixing on the flow structure.

Also, the Nusselt number decreases with circumferential angle, θ, up to the separation point on the front side for reasons similar to those cited earlier (i.e., growing boundary layer thickness, reduced near-wall temperature gradient). Beyond about 80° in Figure 3.1, Nu_D increases with θ (front side of the cylinder) due to transition to turbulence in the boundary layer. On the back side of the cylinder (between 90 and 140°), Nu_D decreases with θ since the increasing thickness of the attached boundary layer leads to a reduced near-wall temperature gradient and wall heat flux. Beyond the separation point of the turbulent boundary layer at about 140°, Nu_D increases as a result of the increased fluid and thermal mixing in the separated, recirculating flow.

Based on experimental data for flow past a cylinder over a wide range of Reynolds numbers, the following heat transfer correlations are summarized in a tabular form (Zhukauskas, 1972; see Table 3.4):

$$\overline{Nu}_D = CRe_D^m Pr^n \left(\frac{Pr}{Pr_s}\right)^{1/4}; \quad 0.7 < Pr < 500 \tag{3.128}$$

For Equation (3.128) and Table 3.4, all thermophysical properties are evaluated at the freestream temperature, except the Prandtl number, Pr_s, which is evaluated at the surface temperature, T_s. Due to the inherent instability and unsteadiness of vortex shedding from the cylinder, and other physical factors limiting the accuracy of measurements in these flows, Equation (3.128) is anticipated to have only $\pm 20\%$ accuracy. Other correlations are available for specific ranges of Re_D, but the Zhukauskas correlation is given here in view of its general applicability over a wide range of Reynolds and Prandtl numbers.

TABLE 3.4

Correlations for External Flow Past a Circular Cylinder

Re_D	C	m	$n(Pr \leq 10)$	$n(Pr > 10)$
$1-40$	0.75	0.4	0.37	0.36
$40-10^3$	0.51	0.5	0.37	0.36
$10^3-2\times10^5$	0.26	0.6	0.37	0.36
$2\times10^5-10^6$	0.076	0.7	0.37	0.36

3.6.3 External Flow Past Other Configurations

In this section, heat transfer correlations for external flows past other configurations (i.e., sphere, tube bundles) will be given.

3.6.3.1 Sphere

External flow past a sphere exhibits similar characteristics as the previously described flow past a circular cylinder. In particular, boundary layer separation and transition to turbulence involve similar processes. For external flow at a velocity of U_∞ and a temperature of T_∞ past an isothermal sphere at T_w, Whitaker (1972) recommends

$$\overline{Nu}_D = 2 + (0.4Re_D^{1/2} + 0.06Re_D^{2/3})Pr^{0.4}\left(\frac{\mu_\infty}{\mu_w}\right)^{1/4} \tag{3.129}$$

which is accurate within $\pm 30\%$ in the range of $0.71 < Pr < 380$, $3.5 < Re_D < 76{,}000$, and $1.0 < (\mu_\infty/\mu_w) < 3.2$. The subscript D refers to evaluation based on the sphere diameter, D. All properties should be evaluated at the temperature of T_∞, except μ_w, which is evaluated at the surface temperature, T_w. Typical trends of heat transfer from a sphere in an airstream are shown in Figure 3.8.

An alternative expression was recommended by Achenbach (1978), based on

$$\overline{Nu} = 2 + \left(\frac{Re_D}{4} + 3 \times 10^{-4} Re_D^{1.6}\right)^{1/2} \tag{3.130}$$

for $100 < Re_D < 2 \times 10^5$. This correlation was determined from measurements of the average heat transfer coefficient for convective heat transfer from a constant temperature sphere to air.

3.6.3.2 Tube Bundles

Another geometry of practical significance is a cross-flow past a large number of regularly spaced (parallel) cylinders. This geometry is commonly encountered in industrial applications such as tube bundles in heat exchangers, tubes inside condensers, and air conditioner coils. In these applications, an external flow past the tube bundles usually transfers heat to or from fluid moving inside the tubes. Zhukauskas (1972) has recommended the following correlation:

$$\overline{Nu}_D = CRe_{D,max}^m Pr^{0.36}\left(\frac{Pr_\infty}{Pr_w}\right)^{1/4} \tag{3.131}$$

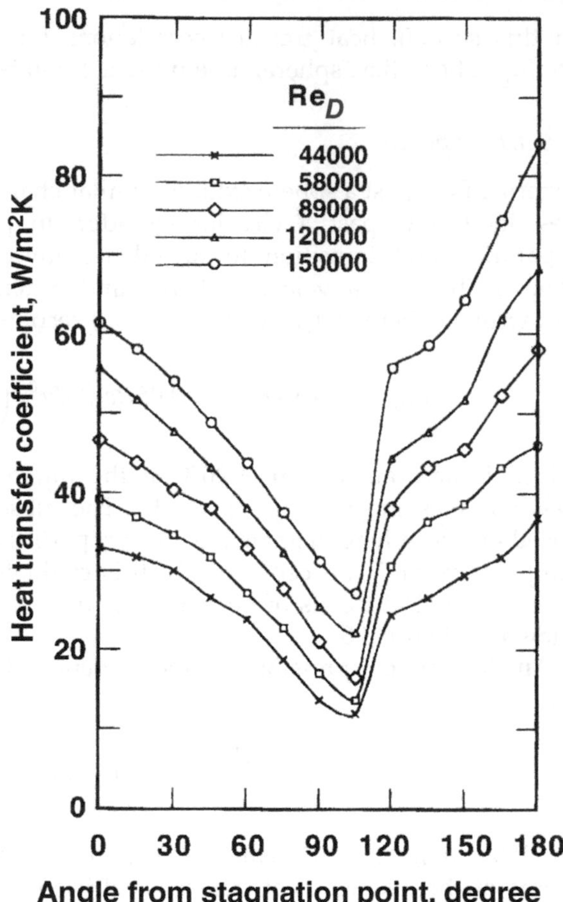

FIGURE 3.8
Local heat transfer coefficients for a sphere in an airstream. (From Hewitt, G.F., Shires, G.L., and Polezhaev, Y.V., Eds., *International Encyclopedia of Heat and Mass Transfer*, Copyright 1997, CRC Press, Boca Raton, FL. Reprinted with permission.)

where the subscripts ∞ and w refer to evaluation at the freestream and tube wall temperatures, respectively. This correlation is valid under the following conditions: 20 or more rows of tubes ($N_L \geq 20$), $0.7 < Pr < 500$, and $1{,}000 < Re_{D,max} < 2 \times 10^6$. The properties are evaluated at the arithmetic mean of the fluid inlet and outlet temperatures, except Pr_w, which is evaluated at the wall temperature.

The tube bundle configurations are characterized by the tube diameter, D, transverse pitch (measured between tube centers perpendicular to the flow direction), S_T, longitudinal pitch (parallel to the flow direction), S_L, and the ratio of pitches, $S_R = S_T/S_L$. Also, the constants C and m are summarized and listed below.

3.6.3.2.1 Aligned Tubes

- For $10 < Re_{D,max} < 100$, $C = 0.8$ and $m = 0.4$.
- Between $100 < Re_{D,max} < 1{,}000$, the flow can be approximated by correlations for a single (isolated) cylinder.
- For $1{,}000 < Re_{D,max} < 2 \times 10^5$ with $S_R < 0.7$ (higher S_R ratios yield inefficient heat transfer), $C = 0.27$ and $m = 0.63$.
- For $2 \times 10^5 < Re_{D,max} < 2 \times 10^6$, $C = 0.021$ and $m = 0.84$.

3.6.3.2.2 Staggered Tubes

- For $10 < Re_{D,max} < 100$, $C = 0.9$ and $m = 0.4$.
- For $100 < Re_{D,max} < 1{,}000$, the flow can be approximated by correlations for a single (isolated) cylinder.
- For $1{,}000 < Re_{D,max} < 2 \times 10^5$, $m = 0.6$ and $C = 0.35\, S_R^{0.2}$ when $S_R < 2$ and $C = 0.4$ when $S_R > 2$.
- For $2 \times 10^5 < Re_{D,max} < 2 \times 10^6$, $C = 0.022$ and $m = 0.84$.

For other flow conditions, such as $N_L < 20$, additional correction factors can be applied to these correlations (Zhukauskas, 1972) Other configurations of external flows, such as impinging jets, are documented by Incropera and DeWitt (1990).

3.7 Internal Forced Convection

Internal flows with heat transfer occur in many applications, such as water or oil flows in pipes, airflows in ventilating ducts, and the design of shell-and-tube and concentric tube heat exchangers (Chapter 10). In this section, the structure and analysis of internal flows will be considered, including the role of fluid mechanics in contributing to the convective heat transfer.

3.7.1 Internal Flow in Pipes

For uniform flow entering a pipe, an entry region is formed where a growing boundary layer develops along the walls of the pipe (i.e., Figure 3.9). Up to some critical distance called the entry length, x_c, the flow is considered to be developing since the velocity profile changes in the

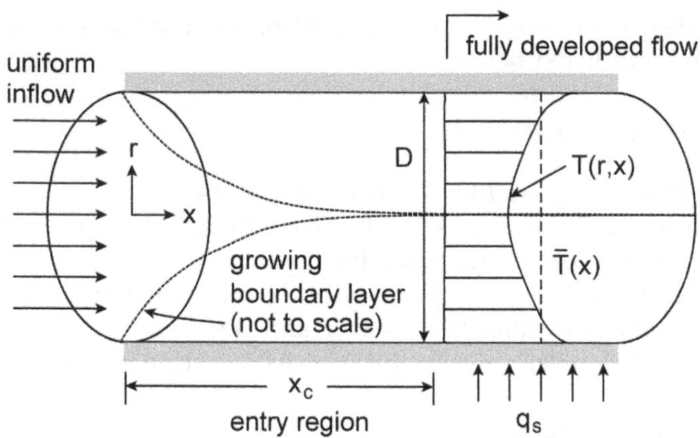

FIGURE 3.9
Schematic of developing pipe flow.

streamwise (x) direction as a result of the viscous shear action on the entering flow. This critical distance is given by $x_c/D \approx 0.05Re_D$ for laminar flow and $10 < x_c/D < 60$ for turbulent flow. At the entry distance, x_c, and beyond this position, the boundary layers growing inward from the walls have merged together and the subsequent downstream velocity profiles remain uniform. A zero velocity gradient in the streamwise direction occurs in the fully developed region (see Figure 3.9).

For pipe flows, it is convenient to write the governing equations in polar coordinates. The following mass and reduced momentum (x direction) equations govern the fluid flow in pipe flow:

$$\frac{\partial u}{\partial x}+\frac{1}{r}\frac{\partial (rv)}{\partial r}=0 \tag{3.132}$$

$$\rho u\frac{\partial u}{\partial x}+\rho v\frac{\partial u}{\partial r}=-\frac{\partial p}{\partial x}+\mu\frac{\partial}{\partial r}\left(r\frac{\partial u}{\partial r}\right) \tag{3.133}$$

The momentum equation in the radial direction leads to the result that pressure remains a function of x (axial coordinate in the flow direction) and not r (radial coordinate beginning at the center of pipe). Also, since the radial velocity component is zero, the continuity equation gives the result that the axial velocity is a function of radial position only (not axial position).

Heat transfer in the pipe flow is governed by the following thermal energy equation:

$$\rho c_p u \frac{\partial T}{\partial x} + \rho c_p v \frac{\partial T}{\partial r} = k \left[\frac{\partial^2 T}{\partial x^2} + \frac{1}{r} \left(r \frac{\partial T}{\partial r} \right) \right] \tag{3.134}$$

Once the velocity components are obtained from the solution of the mass and momentum equations, these components are used in Equation (3.134) to give a direct solution of the radial temperature distribution in the fluid. Then this expression can be differentiated with respect to r to yield the radial heat flux at the wall (based on Fourier's law). When this heat flux is equated with the convective heat transfer to or from the fluid (based on Newton's law of cooling), the convective heat transfer coefficient is obtained.

The solutions of the fluid flow equations will be obtained based on the assumptions of steady, laminar, fully developed, and incompressible flow in the pipe. Under these assumptions, the momentum equation in the axial direction becomes

$$0 = -\frac{\partial p}{\partial x} + \frac{\mu}{r} \frac{\partial}{\partial r} \left(r \frac{\partial u}{\partial r} \right) \tag{3.135}$$

The only way that this equation can be satisfied is when the pressure gradient in the axial direction ($\partial p / \partial x$) remains constant (since it is not a function of r like the velocity gradient term).

The mean velocity, $\overline{u}$, is defined as the average (integrated) velocity across the pipe cross-sectional area. Solving the pipe flow equations, subject to no-slip and symmetry boundary conditions at $r = r_o$ and $r = 0$, respectively,

$$\frac{u(r)}{\overline{u}} = 2 \left[1 - \left(\frac{r}{r_o} \right)^2 \right] \tag{3.136}$$

where

$$\overline{u} = -\frac{r_o^2}{8\mu} \left(\frac{dp}{dx} \right) \tag{3.137}$$

This result holds for laminar flow. For turbulent flows, the bracketed term in Equation (3.136) is replaced by $(r_o - r)^{1/n} / r_o$ where r_o refers to outer pipe radius, $n = n(Re)$ and $0.6 \leq n \leq 10$.

The pressure drop in the pipe is closely related to pipe roughness and the friction factor. For fluid flow in pipes, the Moody friction factor, f, is defined by

$$f = \frac{-D(dp/dx)}{\rho \overline{u}^2 / 2} \tag{3.138}$$

For laminar flow based on the analytical solution in Equation (3.136), this factor becomes

$$f = \frac{64}{Re_D} \tag{3.139}$$

For turbulent flow in a smooth pipe, the following correlations may be adopted:

$$f = 0.316 Re_D^{-1/4} \quad (Re_D < 2 \times 10^4) \tag{3.140}$$

$$f = 0.184 Re_D^{-1/5} \quad (Re_D \geq 2 \times 10^4) \tag{3.141}$$

With increasing wall roughness, the resulting trends in friction factor are illustrated in the Moody chart. From the above correlations and the Moody chart, it can be observed that the friction factor decreases with Re_D (except in the transition region), partly since this corresponds to an increasing velocity. In other words, the increased velocity in the denominator of the definition of friction factor overrides the corresponding change in wall shear stress.

In terms of the heat transfer, the fully developed thermal condition may be expressed in the following manner in terms of the nondimensional temperature, θ :

$$\theta_1 = \left(\frac{T(r,x) - T_s(x)}{\overline{T}(x) - T_s}\right)_1 = \left(\frac{T(r,x) - T_s(x)}{\overline{T}(x) - T_s}\right)_2 = \theta_2 \tag{3.142}$$

where the subscripts 1, 2, and s refer to an upstream location, downstream location, and surface, respectively. The mean temperature of the fluid is denoted by $\overline{T}(x)$, where

$$\overline{T}(x) = \frac{1}{\overline{u}A_c} \int_{A_c} u(r,x) T(r,x) dA_c \tag{3.143}$$

and A_c is the cross-sectional area of the pipe.

It still remains that the variation of the mean temperature with axial position be found. Unlike the fluid flow problem, where the mean velocity remains constant, the value of $\overline{T}$ changes in the downstream direction. Otherwise, if $\overline{T}$ was constant, then no heat transfer would occur. As a result, zero streamwise derivatives of θ are assumed for thermally fully developed conditions, rather than zero streamwise derivatives of $\overline{T}$. Also, the convective heat transfer coefficient, h, remains constant for fully developed thermal conditions.

Boundary conditions are required for the evaluation of $\overline{T}(x)$. In this section, we will consider two different types: constant heat flux and constant surface temperature.

3.7.1.1 Specified (Constant) Wall Heat Flux

Consider an energy balance over a control volume within the pipe (disk of thickness dx; perimeter of P around the pipe). The change of energy within the control volume balances the rate of heat addition or removal (q_s) from the boundary so that

$$q_s''P = \dot{m}c_p \frac{d\overline{T}}{dx} \tag{3.144}$$

where $\dot{m}$ refers to the mass flow rate. This energy balance suggests that $d\overline{T}/dx$ is constant, and so $\overline{T}$ varies linearly throughout the pipe. At the inlet section $\overline{T} = \overline{T}_{in}$, and so the following linear profile of mean temperature is obtained:

$$\overline{T}(x) = \overline{T}_{in} + \left(\frac{q_s''P}{\dot{m}c_p}\right)x \tag{3.145}$$

From this result and Newton's law of cooling, it can be shown that the difference between the surface (wall) temperature and the mean temperature remains constant, or alternatively, q_s''/h remains constant in the fully developed region.

3.7.1.2 Constant Surface (Wall) Temperature

In this case, when the pipe wall temperature remains constant, the energy conservation equation yields

$$-\frac{d(\Delta T)}{\Delta T} = \left(\frac{1}{\dot{m}c_p}\right)Ph(x)dx \tag{3.146}$$

where $\Delta T = T_s - \overline{T}(x)$. We will define the average convective heat transfer coefficient as

$$\overline{h} = \frac{1}{x}\int_0^x h(x)dx \tag{3.147}$$

Integrating the energy balance and exponentiating both sides of the resulting expression involving ΔT,

$$\frac{\Delta T}{\Delta T_{in}} = \frac{T_s - \overline{T}(x)}{T_s - \overline{T}_{in}} = exp\left(\frac{-Px\overline{h}}{\dot{m}c_p}\right) \tag{3.148}$$

where $\overline{h}$ refers to the spatially averaged convection coefficient along the length of pipe.

If we consider an overall energy balance along the entire pipe (i.e., not a differential segment, dx), then we obtain

$$q_{net} = \overline{h}A\left[\frac{\Delta T_{out} - \Delta T_{in}}{ln(\Delta T_{out}/\Delta T_{in})}\right] \tag{3.149}$$

where the expression inside the bracketed term (right side) is called the *log mean temperature difference*. Here ΔT refers to the difference between the surface and mean fluid temperatures. In Equation (3.149), $\overline{h}$ is obtained from a suitable convection correlation.

The previous results can be readily extended to other, more complicated flows, such as a combined internal flow with external flow past the outer walls of the pipe. In this case, the ambient fluid temperature (T_∞) may be known, instead of the surface (wall) temperature of the pipe. As a result, we can use the previous results, but we need to replace $T_s \rightarrow T_\infty$. Also, the convective heat transfer coefficient is replaced by the average (overall) heat transfer coefficient, $\overline{U}$, where

$$\frac{1}{\overline{U}} = \frac{1}{\overline{h}_i} + \frac{1}{\overline{h}_o} \tag{3.150}$$

which corresponds to thermal resistances in series, i.e., primarily convection resistances inside (subscript i) and outside (subscript o) the pipe.

The heat transfer results and correlations obtained from the previous analysis are summarized in Table 3.5. Important trends in the results, particularly involving temperature, are shown in Figure 3.10.

The names attributed to the turbulent flow correlations are shown in Table 3.5 in parentheses. These results apply to convective heat transfer in circular tubes. For analogous studies with noncircular tubes, the same correlations may be used, but the diameter adopted in the Nusselt number should be replaced with the equivalent hydraulic diameter, $D_h = 4A_c/P$, where A_c and P refer to the cross-sectional area and perimeter of the noncircular duct, respectively.

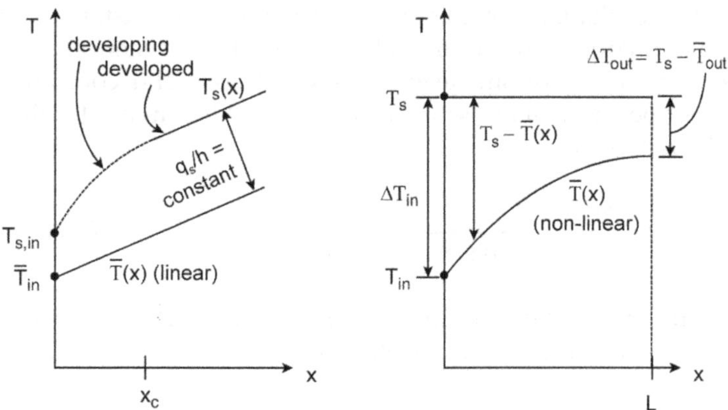

FIGURE 3.10
Temperature profiles for pipe flow with various boundary conditions.

TABLE 3.5

Heat Transfer Correlations for Internal Flow in a Pipe

Flow	Conditions	Heat Transfer ($Nu_D = \bar{h}D/k$)
Laminar	Constant wall heat flux, $Re_D < 10^4$	$Nu_D = 4.36$
Laminar	Constant wall temperature, $Re_D < 10^4$	$Nu_D = 3.66$
Turbulent	Constant wall heat flux or temperature, $Re_D \geq 10^4; L/D > 10$; $0.7 \leq Pr \leq 16,700$ (Seider–Tate)	$Nu_D = 0.027 Re^{4/5} Pr^{1/3} (\mu/\mu_s)^{0.14}$
Turbulent	Constant wall heat flux or temperature, $Re_D \geq 10^4$; $0.7 \leq Pr \leq 160 n =$ 0.4 (heating); $n = 0.3$ (cooling) (Dittus–Boelter correlation)	$Nu_D = 0.023\ Re^{4/5}\ Pr^n$

Example 3.7.1.2.1

Exhaust Gases from an Industrial Smokestack.

Exhaust gases leave the outlet of a smokestack (with a height of 8 m and diameter of 0.6 m) at a mean temperature of 600°C. The mass flow rate of air inside the cylindrical smokestack is 1 kg/sec. Ambient air flows past the outside surface of the smokestack with a velocity of 10 m/sec and a freestream temperature of 5°C. The purpose of this problem is to determine the mean temperature of the air at the base of the smokestack and the heat loss from the air flowing through the smokestack. It may be assumed that radiative heat exchange is negligible and the thermal resistance of the wall is negligible in comparison to the other convection resistances.

In the following solution, steady-state conditions and constant thermo-phyhsical properties will be assumed. The Reynolds number for the internal flow is calculated as

$$Re_D = \frac{\rho VD}{\mu} = \frac{4\dot{m}}{\pi D \mu} = \frac{4 \times 1}{\pi \times 0.6 \times 390 \times 10^{-7}} = 54,412 \quad (turbulent) \qquad (3.151)$$

As a result, from the definition of the Nusselt number, $Nu_D = \overline{h}_i D/k$, together with the Seider–Tate correlation,

$$\overline{h}_i = \frac{k}{D} \times 0.027 Re_D^{4/5} Pr^{1/3} \left(\frac{\mu}{\mu_s}\right)^{0.14} = 15 \quad [\mathrm{W/m^2K}] \qquad (3.152)$$

For the external cross-flow past the outside surface of the smokestack, $Re_D = 4.3 \times 10^5$ (i.e., turbulent flow). Thus, the convection coefficient is obtained by

$$\overline{h}_o = \frac{k}{D} \times 0.26 Re_D^{0.6} Pr^{0.37} \left(\frac{Pr}{Pr_s}\right)^{1/4} = 22.7 \quad [\mathrm{W/m^2K}] \qquad (3.153)$$

In this problem, the temperature of the external airflow, T_∞, rather than the tube surface temperature, T_s, is fixed. The analytic solution may be used to approximate the pipe flow with T_s replaced by T_∞, and the convection coefficient, h, replaced by the overall heat transfer coefficient, $\overline{U}$, i.e.,

$$\frac{T_\infty - \overline{T}_o}{T_\infty - \overline{T}_i} = exp\left[\frac{-PL\overline{U}}{\dot{m}c_p}\right] \qquad (3.154)$$

where

$$\frac{1}{\overline{U}} = \frac{1}{\overline{h}_i} + \frac{1}{\overline{h}_o} \qquad (3.155)$$

Substituting numerical values into these equations,

$$\overline{T}_i = 5 - (5 - 600) exp\left[\frac{\pi \times 0.6 \times 8}{1 \times 1104}\left(\frac{1}{1/15 + 1/22.7}\right)\right] \qquad (3.156)$$

which gives $\overline{T}_i = 678.1°\mathrm{C}$. Then, constructing a thermal circuit with internal and external convection resistances in series,

$$q_s'' = \frac{\overline{T}_i - \overline{T}_o}{R_{tot}} = \frac{\overline{T}_i - \overline{T}_o}{1/\overline{h}_i + 1/\overline{h}_o} = 5.37 \quad [\text{kW/m}^2] \tag{3.157}$$

In order to refine this estimate, iterations can be performed for the evaluation of thermophysical properties, based on the result obtained for the inlet temperature, $\overline{T}_i$. From a practical perspective, these operating conditions should be properly controlled so that the discharge gases do not condense within the smokestack. Also, gases discharged by thermal buoyancy and convection from the smokestack must be well understood, particularly in view of their potential contamination of nearby farmland and other human activities.

The previous analysis and problems have examined convective heat transfer under conditions where the fluid flow is induced by external means. In the following section, the fluid motion is induced by buoyancy in the analysis of free convection problems.

3.8 Free Convection

Free convection (or natural convection) arises from buoyancy forces due to density differences caused by temperature variations within the fluid. For example, warm air ascends above a hot wire placed in a cool room due to buoyancy forces arising from lighter, warmer air near the surface of the hot wire. The resulting flow entrains ambient air into the thermal plume rising above the hot wire. Mixed convection refers to problems involving aspects of both forced and free convection. In this section, the governing equations and processes of free convection will be examined.

3.8.1 Governing Equations and Correlations

The laminar boundary layer equations for free convection are similar to our earlier boundary layer equations with the exception of an additional body force (i.e., buoyancy force) term in the momentum equation. Consider free convection under steady-state conditions along a vertical plate in which gravity acts in the negative x direction. The coordinate x refers to the streamwise direction, whereas the coordinate y refers to the direction normal to the wall (i.e., the cross-stream direction). Assume that density differences remain negligible in all terms, except the buoyancy forces, where the density difference induces the fluid motion (*Boussinesq approximation*).

If the velocity component perpendicular to the plate is assumed to be negligibly small in comparison to flow velocities in the streamwise direction (x direction), then the momentum equation in the y direction is reduced to yield an approximately zero pressure gradient in the y direction. As a result, $p = p(x)$ only. In other words, the pressure within the boundary layer (at a given position x) is the same as the pressure value at the same x position outside the boundary layer where the velocity components are zero (quiescent freestream air).

Based on a static balance of forces on a stationary fluid particle outside the boundary layer,

$$pdy - \left(p + \frac{\partial p}{\partial x}dx\right)dy - \rho_{\infty}gdxdy = 0 \tag{3.158}$$

where the second term represents a Taylor Series expansion of pressure at the position $x + dx$ about the pressure value at point x. Higher order terms are neglected in this expansion. Rearranging the previous result indicates that the pressure gradient in the x direction is represented by a hydrostatic variation, i.e.,

$$\frac{\partial p}{\partial x} = -\rho_{\infty}g \tag{3.159}$$

Also, the steady-state mass and momentum (x direction) equations can be written as

$$\frac{\partial u}{\partial x} + \frac{\partial v}{\partial y} = 0 \tag{3.160}$$

$$\rho u\frac{\partial u}{\partial x} + \rho v\frac{\partial u}{\partial y} = -\frac{\partial p}{\partial x} + \mu\frac{\partial^2 u}{\partial y^2} - \rho g \tag{3.161}$$

where the streamwise diffusion term is neglected since flow changes in the x direction are anticipated to be much smaller than cross-stream variations in the near-wall region.

The hydrostatic term (arising from $\partial p/\partial x$) and fluid weight term, ρg, can now be combined in the momentum equation. The resulting density difference between local and freestream density values in the x direction momentum equation can be written in terms of a corresponding temperature difference by introducing the thermal expansion coefficient,

$$\beta = -\frac{1}{\rho}\left(\frac{\partial \rho}{\partial T}\right)_p \approx -\frac{1}{\rho}\left(\frac{\rho_\infty - \rho}{T_\infty - T}\right) \tag{3.162}$$

This approximation represents an equation of state relating density to the temperature of the gas.

As a result, we obtain the following mass, momentum, and energy equations for laminar, free convection boundary layer flow along a vertically oriented surface:

$$\frac{\partial u}{\partial x} + \frac{\partial v}{\partial y} = 0 \tag{3.163}$$

$$\rho \frac{\partial (uu)}{\partial x} + \rho \frac{\partial (uv)}{\partial y} = \mu \frac{\partial^2 u}{\partial y^2} - \rho g \beta (T - T_\infty) \tag{3.164}$$

$$\rho c_p u \frac{\partial T}{\partial x} + \rho c_p v \frac{\partial T}{\partial y} = k \frac{\partial^2 T}{\partial y^2} \tag{3.165}$$

For an ideal gas, it can be readily shown that the thermal expansion coefficient, β, can be written as the reciprocal of temperature. Also, since buoyancy represents a force, these effects are confined to the momentum equation, and the mass and energy equations remain the same as problems involving forced convection. Also, it is usually assumed that viscous dissipation is negligibly small in comparison to other terms in the energy equation for free convection problems.

In dimensionless form, the free convection boundary layer equations become

$$\frac{\partial u^*}{\partial x^*} + \frac{\partial v^*}{\partial y^*} = 0 \tag{3.166}$$

$$\rho^* u^* \frac{\partial u^*}{\partial x^*} + \rho^* v^* \frac{\partial u^*}{\partial y^*} = \left(\frac{1}{Re_L}\right) \frac{\partial^2 u^*}{\partial y^{*2}} + Gr_L T^* \tag{3.167}$$

$$u^* \frac{\partial T^*}{\partial x^*} + v^* \frac{\partial T^*}{\partial y^*} = \left(\frac{1}{Re_L Pr}\right) \frac{\partial^2 T^*}{\partial y^{*2}} \tag{3.168}$$

where the reference velocity for nondimensionalization is ν / L and $T^* = (T - T_\infty)/(T_s - T_\infty)$. From these expressions, it can be observed that the average Nusselt number depends on the Reynolds number, Re_L, Prandtl number, Pr, and Grashof number, Gr_L (or Rayleigh number, $Ra_L = Gr_L Pr$).

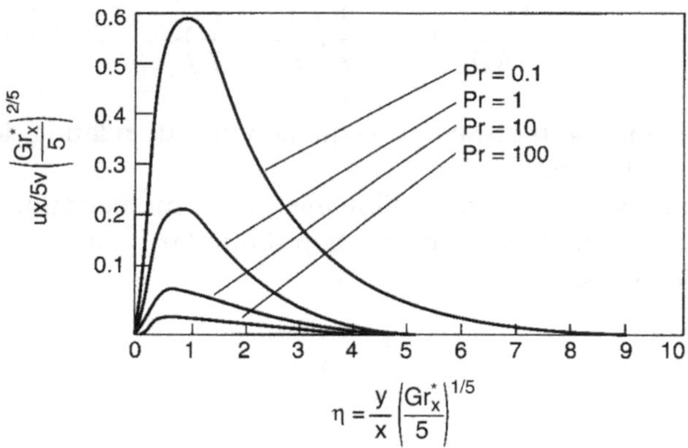

FIGURE 3.11
Dimensionless velocities for free convection boundary layers (flat plate, constant wall heat flux).

The solution of these free convection equations may be obtained by various methods, including approximate methods (described in the next section) or similarity methods. The results, from a similarity solution of the free convection boundary layer equations, are based on the similarity variable of $\eta = y(Gr_x/5)^{1/5}/x$. The results are shown in Figure 3.11. From the figure, it can be observed that the maximum velocity occurs inside the boundary layer. Also, the boundary layer thickness increases when the Prandtl number, Pr, decreases.

From the analytical solution of the free convection equations for both specified (constant) wall heat flux and constant wall temperature cases, we have

$$Nu_x = \left(\frac{Gr_x}{4}\right)^{1/4} g(Pr) \tag{3.169}$$

$$\overline{Nu}_L = \frac{4}{3}\left(\frac{Gr_L}{4}\right)^{1/4} g(Pr) \tag{3.170}$$

where

$$g(Pr) = \frac{0.75Pr^{1/2}}{(0.609 + 1.221Pr^{1/2} + 1.238Pr)^{1/4}} \tag{3.171}$$

Furthermore, based on experimental data over the laminar and turbulent flow regimes,

$$\overline{Nu}_L = \left[0.825 + \frac{0.387 Ra_L^{1/6}}{(1 + (0.492/Pr)^{9/16})^{8/27}}\right]^2 \quad \text{(Churchill, Chu)} \tag{3.172}$$

Boundary layer transition to turbulence occurs at approximately $Ra_x \approx 10^9$ for flow along a vertical flat plate. Furthermore, the previously described flat plate correlations may be applied to wide vertical cylinders when surface curvature effects become negligible (i.e., $\delta << D$; $D > 35L/Gr_L^{1/4}$).

Example 3.8.1.2

Optimal Fin Spacing for Cooling of an Electronic Assembly.

A set of rectangular fins with a length of 22 mm (each spaced 30 mm apart) is used for cooling of an electronic assembly. The total base width is 280 mm, and the ambient surrounding air temperature is 25°C. The fins must dissipate a total of 15 W to the surrounding air by free convection alone. What recommendations would you suggest to enhance the heat removal for designers who have selected a fin height of 8.6 cm? Also, determine the necessary fin thickness to allow the fins to dissipate 15 W of heat to the surrounding air, based on a fin surface temperature of 70°C.

It will be assumed that radiation effects are negligible. Also, the surrounding ambient air is motionless and the fins are isothermal. If the spacing of the fins is too close, then the boundary layers on the adjoining surfaces coalesce and the convective heat transfer decreases. On the other hand, if the spacing between the fins is too large, then the resulting exposed surface area decreases, and thus the total heat transfer also decreases. As a result, it is desirable to find the optimal height and spacing of the fins to remove heat as effectively as possible from the electronic enclosure.

From the similarity solution results for free convection along a vertical plate (see Figure 3.11), the edge of the boundary layer is defined by

$$\eta = \left(\frac{\delta}{H}\right)\left(\frac{Gr_H}{5}\right)^{1/5} \approx 5 \tag{3.173}$$

Substituting numerical values outlined in this problem,

$$5 \approx \frac{0.015}{H}\left[\frac{9.8 \times (1/320)45H^3}{5(18 \times 10^{-6})^2}\right]^{1/5} \tag{3.174}$$

where twice the boundary layer thickness at the base of the fin has been used as the value for δ. This thickness represents the optimal height of the plate since this value corresponds to merging of the boundary layers at that location. Solving Equation (3.174) yields an optimal height of approximately

$H_{opt} = 1.4$ cm. Since the designers have selected $H = 8.6$ cm, it is recommended to reduce this height to provide better thermal effectiveness of the fins (that is, if this reduction is feasible, in terms of cost effectiveness and manufacturing).

Subsequent calculations will be based on the initially selected fin height of 8.6 cm. The heat transfer from all fins excluding the base regions is given by

$$q = 2N\bar{h}(HL)(T_s - T_\infty) \tag{3.175}$$

where $N = w/(s + t)$ is the number of fins, based on a spacing, s, between each fin and a thickness, t, of each fin. The average convection coefficient, $\bar{h}$, is determined by the appropriate correlation for natural convection, with the result that

$$t = 2\frac{w}{q}\left[\frac{k}{H_{opt}}\frac{4}{3}\left(\frac{Gr_{H_{opt}}}{4}\right)^{1/4}\frac{0.75Pr^{1/2}}{(0.609 + 1.221Pr^{1/2} + 1.238Pr)^{1/4}}\right]H_{opt}L(T_s - T_\infty) - s \tag{3.176}$$

Based on the previously selected fin height, the Grashof number is computed to be 2.7×10^7 (i.e., laminar flow correlations may be adopted). Substituting numerical values and appropriate thermophysical properties into Equation (3.176), we obtain a required fin spacing of $s = 5.3$ mm. It is anticipated that a further increase of total fin surface area from a reduction in fin spacing would increase the heat removal from the electronic enclosure. Further optimization studies that consider the trade-offs between more surface area and the resulting boundary layer structure could reveal other beneficial modifications of the fin design.

3.8.2 Approximate Analysis of Free Convection

An approximate analysis of free convection for various geometrical configurations, based on an integral method, has been presented by Raithby and Hollands (1975). For integral methods with zero degrees of freedom, the profile of the integrated quantity (such as the integrated velocity profile yielding a mass flow rate) is solely determined by a solution of an appropriate conservation equation. In this way, zero degrees of freedom arises because the integrand is governed by the appropriate conservation equation, rather than empiricism alone.

Example 3.8.2.1

Free Convection along a Vertical Plate.

Consider steady-state heat transfer due to natural convection between a fluid at a uniform temperature of T_f and a vertical plate at T_w. Upon close examination of this problem, we notice the existence of an inner region of the boundary layer, where conduction is dominant, and an outer region (see Figure 3.12). We will define the inner region as extending from the wall ($y = 0$) to the position of the maximum velocity ($y = y_m$). Also, the outer region extends from this position out to a value of y where $u \approx 0$ (quiescent ambient conditions). Inertial effects with fluid acceleration arise from buoyancy forces in the outer region. However, these inertial effects may be de-coupled from the energy equation in the inner region as viscous effects dampen the fluid motion near the wall.

From an appropriate scaling or dimensional analysis, we find that the governing equations in the inner region can be represented as follows:

$$\mu \frac{\partial^2 u}{\partial y^2} = -(\rho(y) - \rho_f) g_x; \quad 0 \le y \le y_m \tag{3.177}$$

$$k \frac{\partial^2 T}{\partial y^2} = 0; \quad 0 \le y \le y_m \tag{3.178}$$

where g_x refers to the component of the gravitational acceleration in the x direction. Based on this component tangent to the surface, the analysis will

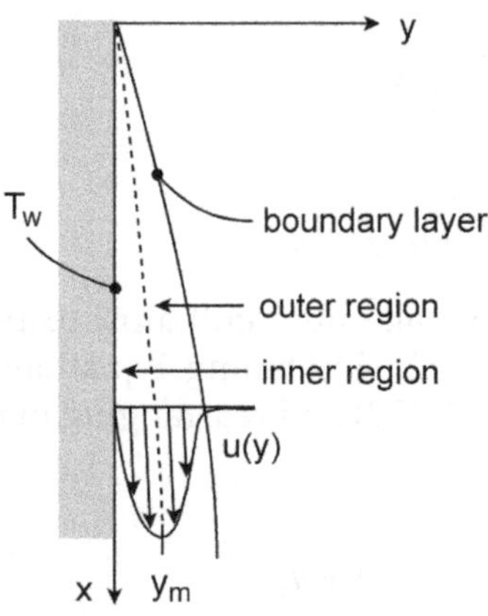

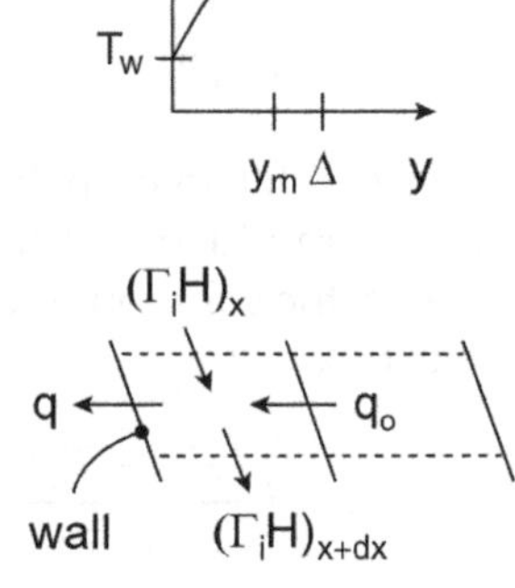

FIGURE 3.12
Schematic of free convection boundary layer.

allow us to consider convection along an inclined plate, or axisymmetric surface, oriented at any arbitrary angle with respect to the x axis.

Solving Equation (3.178), subject to $T(y_m) = T_m$, introducing the thermal expansion coefficient relating temperature and density differences, and substituting that result into the solution from Equation (3.177), we obtain

$$u(y) = \frac{\rho \beta g_x}{\mu}(T_f - T_w)\left[y y_m - \frac{y^2}{2} - \frac{y_m^2 y}{2\Delta} + \frac{y^3}{6\Delta} \right] \tag{3.179}$$

where Δ refers to the location where the extrapolated temperature from the inner region equals the ambient fluid temperature. Its value will be obtained by an interface condition eliminating the unknown values at that position.

We can now find the mass flow rate (per unit width of surface) in the inner region for subsequent use in other energy and mass balances,

$$\Gamma_i = \int_0^{y_m} \rho u dy = \frac{\rho^2 \beta g_x}{\mu}(T_f - T_w)\frac{y_m^3}{3}\left(1 - \frac{5M}{8}\right) \tag{3.180}$$

where the unknowns in the analysis now become y_m and $M = y_m/\Delta$. Since the velocity profile is already available from Equation (3.179), we can refer to the current integral solution as a method involving zero degrees of freedom since the integrand in Equation (3.180) was obtained from the solution of the momentum equation. In this way, we have not imposed any empirical curve fitting for the velocity profile in the integrand.

Furthermore, writing the conservation of mass and total energy for a differential control volume in the inner region (see Figure 3.12), we have

$$m_i'' + \Gamma_i = \left(\Gamma_i + \frac{d\Gamma_i}{dx} dx\right) \tag{3.181}$$

$$q_o + m_i'' H_m + (\Gamma_i \overline{H})_x = (\Gamma_i \overline{H})_{x+dx} + q \tag{3.182}$$

where H and $\overline{H}$ refer to enthalpy and average fluid enthalpy, respectively. Also, q_o and q are illustrated in Figure 3.12. Combining Equations (3.181) and (3.182) with the solution of Equation (3.178) and rearranging in terms of the wall heat flux,

$$q = k\frac{(T_m - T_w)}{y_m} = \frac{d\Gamma_i}{dx}(H_m - \overline{H}_i) + q_o \tag{3.183}$$

Furthermore, similar mass and energy balances can be applied in the outer region, i.e.,

$$m_o'' - m_i'' = \frac{d\Gamma_o}{dx} \tag{3.184}$$

$$m_o''(H_f - \overline{H}_o) = q_o + m_i''(H_m - \overline{H}_o) \tag{3.185}$$

Combining Equations (3.183) to (3.185) and rearranging terms,

$$\frac{k}{c_p B(x)} = y_m \frac{d\Gamma_i}{dx} \tag{3.186}$$

where

$$B(x) = \frac{T_f - \overline{T}_i}{T_m - T_w} + \left(\frac{d\Gamma_o/dx}{d\Gamma_i/dx}\right)\frac{T_f - \overline{T}_o}{T_m - T_w} \tag{3.187}$$

Based on earlier observations that the convective boundary layer exhibits self-similarity, we find that $B = B(M)$ only. Also, combining Equation (3.180) with Equation (3.187) and eliminating y_m,

$$\Gamma_i^{1/3}\frac{d\Gamma_i}{dx} = \frac{1}{B(M)}\left(1 - \frac{5M}{8}\right)^{1/3}\frac{k}{c_p}\left[\frac{\rho^2 g\beta(T_f - T_w)}{3\mu}\right]^{1/3}\left(\frac{g_x}{g}\right)^{1/3} \tag{3.188}$$

Assuming that M does not vary with x, applying the boundary condition of $\Gamma_i(0) = 0$, and integrating,

$$\Gamma_i = \frac{(1 - 5M/8)^{1/4}}{B(M)^{3/4}}\left(\frac{4k}{3c_p}\right)^{3/4}\left[\frac{\rho^2 g\beta(T_f - T_w)}{3\mu}\right]^{1/4}\left[\int_0^x \left(\frac{g_x}{g}\right)^{1/3} dx\right]^{3/4} \tag{3.189}$$

We can now determine y_m explicitly from Equations (3.180) and (3.189).

The wall heat flux is proportional to the temperature difference in the inner region, $T_m - T_w$, divided by y_m, or alternatively, the total temperature difference, $T_f - T_w$, divided by y_m/M. As a result, we can find the local Nusselt number as follows:

$$Nu_x = \frac{hx}{k} = \frac{q''\cdot x}{(T_f - T_w)k} \tag{3.190}$$

$$Nu_x = \frac{(kM(T_f - T_w)/y_m)x}{(T_f - T_w)k} = \frac{Mx}{y_m} \tag{3.191}$$

Substituting the expression for y_m obtained after combining Equations (3.180) and (3.189) into Equation (3.191) and simplifying,

$$Nu_x = f(M) \cdot Ra_x^{1/4} \cdot \frac{(g_x/g)^{1/3}}{\left[\frac{1}{x}\int_0^x (g_x/g)dx\right]^{1/4}} \tag{3.192}$$

where

$$f(M) = M\left[\left(1 - \frac{5M}{8}\right)\frac{B}{4}\right]^{1/4} \tag{3.193}$$

and the Rayleigh number is $Ra = Pr \cdot Gr$.

Alternatively, integrating the heat flux over the surface, we obtain the following result for the average Nusselt number:

$$\overline{Nu_s} = \frac{4}{3}\, f(M) \cdot Ra_s^{1/4} \cdot \left[\frac{1}{S}\int_0^S \left(\frac{g_x}{g}\right)^{1/3} dx\right]^{3/4} \tag{3.194}$$

Since M depends on the Prandtl number, it is estimated that

$$f(M) = 0.48\left(\frac{Pr}{0.861 + Pr}\right)^{1/4} \tag{3.195}$$

based on a comparison between the forms of Equation (3.194) and the known exact (similarity) solution for natural convection along a vertical plate. Solution errors of predicted $\overline{Nu_s}$ values from Equation (3.194) are normally within $\pm 1\%$ of the exact results. Good agreement with experimental data is achieved for flat surfaces under laminar flow conditions. However, at high Ra numbers and other geometrical configurations, larger discrepancies are observed, partly due to turbulence and surface curvature effects.

The method of separating inner and outer regions of a boundary layer (as discussed in the previous example) is a useful technique for de-coupling the momentum and energy equations. It can be applied to free convection, as well as other types of boundary layer problems.

3.8.3 Effects of Geometrical Configuration

In this section, correlations for free convection in other important configurations (i.e., spheres, concentric spheres, enclosures) will be outlined. Additional material describing other free convection flows is documented in books by Kays and Crawford (1980) and Kreith and Bohn (2001).

3.8.3.1 Spheres

For free convective heat transfer from the external surface of an isothermal sphere,

$$\overline{Nu}_D = 1 + \frac{0.589 Ra_D^{1/4}}{[1 + (0.469/Pr)^{9/16}]^{4/9}} \tag{3.196}$$

which is recommended in the range of $Pr \geq 0.7$ and $Ra_D \leq 10^{11}$ by Churchill (1983). This result is developed for an immersed (external flow) geometry, whereas subsequent correlations (i.e., concentric spheres and enclosures) refer to internal flow configurations. The following internal flows are constrained by the physical presence of solid boundaries on all sides of the flow.

3.8.3.2 Concentric Spheres

Free convective heat transfer in the region between concentric spheres exhibits a combination of features observed in earlier geometries, including immersed flow over the inner sphere. Raithby and Hollands (1975) define an effective thermal conductivity of free convection,

$$\frac{k_{eff}}{k} = 0.74 (Ra_s^*)^{1/4} \left(\frac{Pr}{0.861 + Pr} \right)^{1/4} \tag{3.197}$$

where

$$Ra_s^* = \frac{L \cdot Ra_L}{D_1^4 D_2^4 (D_1^{-7/5} + D_2^{-7/5})^5} \tag{3.198}$$

The variables D_1 and D_2 refer to the diameters of the inner and outer spheres, respectively, and L is the radial gap width. Then, in the range of $100 \leq Ra_s^* \leq 10,000$, the following expression is recommended to obtain the total heat transfer rate:

$$q = k_{eff} \left(\frac{\pi D_1 D_2}{L} \right) (T_1 - T_2) \tag{3.199}$$

The effective thermal conductivity, k_{eff}, represents a thermal conductivity that a stationary fluid must have to produce the same amount of heat transfer as the actual moving fluid.

3.8.3.3 Tilted Rectangular Enclosures

Free convection within rectangular enclosures is often encountered in engineering systems. Some examples include solar collectors, channels within electronic enclosures, and heat transfer through windowpanes in buildings. Consider heat transfer across a rectangular enclosure with an aspect ratio of H/L. Two opposing walls are maintained at different temperatures, and the remaining surfaces are well insulated. The following correlation is recommended by Hollands et al. (1976) for large aspect ratios ($H/L \geq 12$) and tilt angles less than the critical value of τ^* :

$$\overline{Nu}_L = 1 + 1.44\left[1 - \frac{1708}{Ra_L cos(\tau)}\right]^* \left[1 - \frac{1708(sin 1.8\tau)^{1.6}}{Ra_L cos(\tau)}\right] + \left[\left(\frac{Ra_L cos(\tau)}{5830}\right)^{1/3} - 1\right]^* \quad (3.200)$$

In this correlation, τ is the tilt angle (with respect to the horizontal plane) and [ ]* means that if the quantity in brackets is negative, set it to zero. These correlations provide close agreement with experimental data when applied below the following critical tilt angles: 25° for $H/L = 1$, 53° for $H/L = 3$, 60° for $H/L = 6$, 67° for $H/L = 12$, and 70° for $H/L > 12$.

3.9 Second Law of Thermodynamics

The Second Law of Thermodynamics describes a physical constraint on the upper limits and performance of engineering systems. It requires that entropy is produced, but never destroyed, in an isolated system. Thus, $\dot{\mathscr{P}}_s \geq 0$ for an isolated system, where $\dot{\mathscr{P}}_s$ refers to the entropy production rate. In this section, two useful aspects of the second law in convection analysis (namely, the entropy transport equation and entropy production minimization) will be considered.

3.9.1 Entropy Transport Equation

Specific entropy is a scalar quantity (denoted by s) that is transported with the heat flow rate, $\mathbf{q}$, to yield the following flux of entropy (associated with $\mathbf{q}$):

$$F^s = -\nabla \cdot \left(\frac{\mathbf{q}}{T}\right) \quad (3.201)$$

Based on this entropy flux due to heat flow, as well as entropy transported by the mass flow, we have the following overall balance of entropy for a differential control volume within a fluid stream:

$$\rho \frac{Ds}{Dt} = -\nabla \cdot \left(\frac{\mathbf{q}}{T}\right) + \dot{\mathscr{P}}_s \tag{3.202}$$

where Ds/Dt is the substantial derivative of entropy.

Also, differentiating the Gibbs equation for a simple compressible substance,

$$T \frac{Ds}{Dt} = \frac{D\hat{e}}{Dt} + p \frac{Dv'''}{Dt} \tag{3.203}$$

Combining this Gibbs equation with continuity, Equation (3.6), comparing with Equation (3.202), and using the quotient rule of calculus and Fourier's law of heat conduction (i.e., to express **q** in terms of temperature), we obtain the following result:

$$\dot{\mathscr{P}}_s = \frac{k(\nabla T \cdot \nabla T)}{T^2} + \frac{\mu \Phi}{T} \geq 0 \tag{3.204}$$

This result gives an explicit, positive-definite expression for the entropy production rate, thereby confirming the Second Law of Thermodynamics. Applying the second law to heat transfer problems can be a very useful and powerful tool in engineering design because it gives a quantitative measure of irreversibilities throughout the system. Regions exhibiting high entropy production can be clearly identified for design modifications to improve the overall system efficiency and performance.

3.9.2 Entropy Production Minimization

A certain quality of fluid energy is lost when entropy increases. For example, the ability of a fluid stream to perform useful work is reduced when friction irreversibilities convert mechanical energy into thermal energy through viscous dissipation. Optimizing engineering systems by minimizing thermofluid irreversibilities can provide substantial benefits leading to direct cost savings and more efficient use of energy resources. In particular, minimizing the entropy production can reduce system irreversibilities and provide a systematic approach for achieving optimal performance and efficiency (Bejan, 1996).

Exergy refers to the energy availability or maximum possible work that a system can deliver as it undergoes a reversible process from a specified initial state to the state of the environment (i.e., "dead state" at standard temperature/pressure (STP), 101 kPa, 25°C). Exergy can be interpreted in terms of work potential, entropy production, or loss of energy availability

(Camberos, 2000). For example, consider air in a tank, closed by a valve, at the same temperature and pressure as the surrounding air outside the tank (note: neglecting chemical potentials). In this case, there would be a zero thermomechanical availability of air in the tank since there would be no potential for work to be done on the external system (surroundings). On the other hand, if an underground thermal energy source contains pressurized fluid at a temperature and pressure higher than STP, then the availability of energy is greater than zero since this fluid has a potential to perform work. For example, the pressurized fluid could flow through a turbine to generate power. Minimizing entropy production becomes equivalent to minimizing the destruction of exergy.

Example 3.9.2.1

External Flow Past a Submersed Body.

Optimizing the geometry of bodies in external flow is an important consideration in the effective design of surfaces in heat exchangers, electronic assemblies, aircraft, and other applications. Consider an external flow past an arbitrarily shaped body. Find an expression for the entropy production through fluid flow and heat transfer past the object. Apply this expression to a flat plate to find the optimal plate length based on entropy production minimization.

In this example, we will define the body as enclosed within a stream tube, meaning that the external streamline (boundaries of the stream tube) is selected so that it is not affected by the presence of the body (see Figure 3.13). In other words, an incoming fluid stream at U_∞ and T_∞ across the body gives the boundary conditions along the edges of this stream tube. The body is assumed to be sufficiently far from the edge of this stream tube. As a result, we can define a control volume enclosing the immersed body while retaining freestream conditions along the outer boundaries (except at the outlet where the pressure changes).

The purpose of this example is to determine the optimal shape in the external flow, based on a minimization of the combined irreversibilities due to heat transfer and friction. The body is maintained at a temperature of T_s. A fixed mass flow rate, $\dot{m} = A_{tube}\rho U_\infty$, is specified for the flow through the stream tube.

In this case, the first and second laws of thermodynamics, together with the Gibbs equation, are written in the following form:

$$q''A_s = \dot{m}(h_{out} - h_{in}) = \bar{h}A_s(T_s - T_\infty) \tag{3.205}$$

$$\dot{\mathscr{P}}_s = \dot{m}(s_{out} - s_{in}) - \frac{q''A_s}{T_s} \geq 0 \tag{3.206}$$

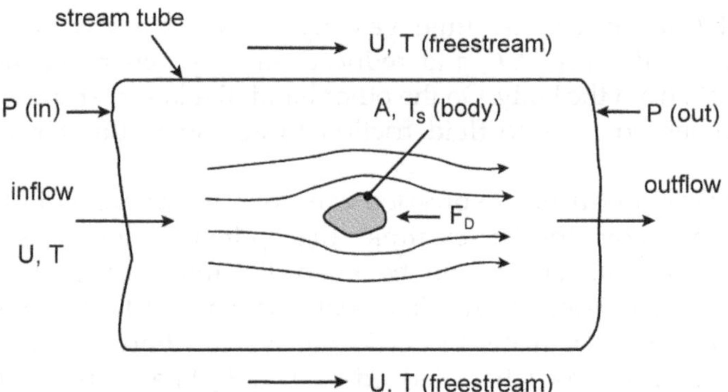

FIGURE 3.13
External flow with definition of a stream tube.

$$\Delta h = T\Delta s + \frac{\Delta P}{\rho} \tag{3.207}$$

where $\overline{h}$ refers to the heat transfer coefficient and h_{in} and h_{out} refer to inlet and outlet enthalpies, respectively. The subscript s refers to entropy in $\dot{\mathscr{P}}_s$, but it refers to surface in the other variables. Combining the previous expressions,

$$\dot{\mathscr{P}}_s = q''A_s\left(\frac{1}{T_\infty} - \frac{1}{T_s}\right) - \frac{\dot{m}(P_{out} - P_{in})}{T_\infty \rho} \tag{3.208}$$

From a balance of linear momentum on the stream tube, we find that the latter term in the previous equation (involving the pressure difference) can be rewritten in terms of the drag force, F_D. In particular, the pressure difference is given by the drag force on the body divided by the cross-sectional area of the stream tube, since the inlet and outlet momentum fluxes are identical. As a result,

$$\dot{\mathscr{P}}_s = q''A_s\left(\frac{T_s - T_\infty}{T_s T_\infty}\right) + \frac{F_D U_\infty}{T_\infty} \tag{3.209}$$

Furthermore, the first term (left side of the previous equation) can be simplified after assuming that $\Delta T << T$, so that

$$\dot{\mathscr{P}}_s = \left(\frac{q''}{T_\infty}\right)^2 \frac{A_s}{\overline{h}} + \frac{F_D U_\infty}{T_\infty} \tag{3.210}$$

The first term indicates the entropy production rate due to heat transfer. This thermal irreversibility decreases with U_∞. An increased convection coefficient (due to a larger fluid velocity) entails a lower surface–fluid temperature difference, ΔT, and reduced entropy generation for a fixed heat flux, q'', from the body. On the other hand, the latter term describes the entropy generation due to fluid friction (drag force); this term increases with U_∞.

The entropy production expression can now be optimized in terms of a specific body geometry. For example, the optimal length of a plate in a finned heat exchanger can be obtained by minimizing the previous expression, while using appropriate heat transfer and friction correlations for flows along a flat plate. In particular, we can find the optimal plate length for external flow (characterized by U_∞, T_∞) past a plate of length L and width W and heated uniformly at q''. For this case, we can use the following correlations:

$$Nu_x = \frac{hx}{k} = 0.458 Pr^{1/3} Re_x^{1/2} \tag{3.211}$$

$$c_{f,x} = \frac{\tau_w}{\rho U_\infty^2/2} = 0.664 Re_x^{-1/2} \tag{3.212}$$

Integrating Equation (3.210) in reference to a flat plate,

$$\dot{\mathscr{P}}_s = \left(\frac{q''}{T_\infty}\right)^2 \int_0^L \frac{W dx}{h} + \frac{U_\infty}{T_\infty} \int_0^L \tau_w W dx \tag{3.213}$$

Substituting Equations (3.211) and (3.212) into Equation (3.213) and integrating,

$$\dot{\mathscr{P}}_s = \left(\frac{q''^2 L^2 W}{k T_\infty^2}\right) \left(1.456 Pr^{-1/3} Re_L^{-1/2} + 0.664 \frac{U_\infty^2 \mu k T_\infty}{q''^2 L^2} Re_L^{1/2}\right) \tag{3.214}$$

This result outlines the irreversibilities of heat transfer and fluid friction as they contribute to the total entropy production rate. Optimizing L by setting the derivative of $\dot{\mathscr{P}}_s$ with respect to Re_L equal to zero, we obtain

$$Re_{L,opt} = 2.193 Pr^{-1/3} \left(\frac{q'^2}{U_\infty^2 \mu k T_\infty}\right) \tag{3.215}$$

where q' refers to the fixed heat transfer per unit length of plate. The optimal Reynolds number can be rewritten in terms of the optimal plate length.

Although the previous example considers laminar flow, most applications of industrial relevance involve fluid turbulence. Fundamental aspects of turbulence modeling will be presented in the following section.

3.10 Turbulence Modeling

In comparison to laminar flows, turbulence usually enhances momentum transport and heat transfer. Turbulence is composed of fine-scale velocity fluctuations superimposed on a mean velocity. Physical features of turbulence can be effectively observed experimentally by nonintrusive, laser-based techniques, such as particle image velocimetry (Naterer and Glockner, 2001). Turbulent flows are inherently three-dimensional. Although they can provide certain benefits, such as enhanced heat transfer rates in a heat exchanger, other undesirable consequences may arise, i.e., higher wall friction and pumping power requirements. In this section, the general nature of turbulence, definitions, spectrum, and modeling of turbulence will be described.

3.10.1 Nature of Turbulence

In turbulent flows, there is some organization within the flow, particularly at low frequencies. Also, there are specific scales that are systematic or have some order within the turbulence, such as the structure of vortices behind a cylinder in a cross-flow. As the following example indicates, turbulence generally arises from instabilities in a laminar flow.

Example 3.10.1.1

Laminar Instability in a Boundary Layer Flow.

The purpose of this example is to describe some physical aspects of instabilities in laminar flows leading to turbulence. Consider an incoming airflow past a cylinder. Along the upstream side of the cylinder the boundary layer remains attached to the wall. However, with an increasing pressure on the back side in the streamwise direction, the boundary layer separates from the wall. This adverse pressure gradient leads to a flow instability and boundary layer separation, with mixing and turbulence in the downstream wake.

Also, consider the characteristics of turbulence development in a boundary layer when a fluid (such as water) flows across a plate. As viewed from above the plate, the section near the leading edge exhibits a stable, laminar character. Shortly thereafter (assuming that the Reynolds number exceeds the approximate transition value of 10^5), a fluid instability arises near the upper portion of the boundary layer as the inertial motion of fluid eddies overrides any viscous dampening near the wall. The smooth motion then breaks down into axial vortices, which then break down into a turbulent "spot" (region of concentrated mixing) and corresponding downstream wake. Finally, the flow becomes fully turbulent further downstream. This example indicates that many stages of development occur during the transition from laminar to turbulent flow.

The analysis of turbulence is a multiscaled problem that often requires a statistical approach. In many cases, a single scale (involving velocity or length) inadequately describes the actual turbulence. For fully developed turbulent flow in a pipe, a single scale (such as the pipe diameter) works reasonably well. In flows where the turbulence diffusion is governed primarily by sufficiently large eddies, single-scale models may be adequate.

Furthermore, we can observe that turbulence is not a self-sustaining phenomenon. Energy must be added to the flow in order to maintain a given level of turbulence. For example, consider a situation in which a projectile passes through an otherwise motionless fluid. The Reynolds number associated with this flow is within the turbulence regime ($Re \approx 10^6$). In this example, the turbulent wake behind the object returns to a laminar state after the projectile passes away. In general, the turbulence is not self-induced and self-sustained, since it needs to be supported by kinetic energy extracted from the mean flow (i.e., moving projectile).

In giving a mathematical description of turbulence, we need to consider the analysis of instabilities in the laminar flow. The mathematics and physics associated with these instabilities are highly nonlinear. As a result, simplified theories are often based on linearized equations, in view of the difficulties in precisely dealing with the vast scales of the actual turbulence.

3.10.2 Turbulence Definitions

Turbulent flows are characterized by flow properties consisting of a mean component and fluctuations about this mean component. The mean component is obtained by integration over a characteristic time period associated with the flow. For example, the mean quantity corresponding to the scalar B is given by

$$\overline{B} = \frac{1}{\Delta t}\int_t^{t+T} B dt; \qquad B = \overline{B} + B' \tag{3.216}$$

where T represents a time scale that adequately covers the time range over which the turbulence occurs. Also, B' is the fluctuating component and the components of turbulence intensity are defined as follows:

$$I_x = \frac{\sqrt{\overline{u'^2}}}{\overline{u}}; \quad I_y = \frac{\sqrt{\overline{v'^2}}}{\overline{v}}; \quad I_z = \frac{\sqrt{\overline{w'^2}}}{\overline{w}} \tag{3.217}$$

We refer to *stationary turbulence* as occurring when the probability distribution of the turbulence fluctuations becomes independent of time. Also, *homogeneous turbulence* occurs when the probability distribution becomes independent of position. For example, this may occur near the outer edge of a flat plate boundary layer, but not near the surface, because the velocity fluctuations are dampened by the presence of the wall. *Isotropic turbulence* is a special case of homogeneous turbulence, and it is restricted to small-scale motions. It occurs when the intensity of turbulence in each coordinate direction is identical.

3.10.3 Turbulence Spectrum

The Kolmogorov spectrum of turbulent kinetic energy gives a graphical summary (see Figure 3.14) that describes the exchange of turbulence at various scales. Turbulent kinetic energy, k, is defined in the following manner:

$$k = \frac{1}{2}(\overline{u'^2} + \overline{v'^2} + \overline{w'^2}) \tag{3.218}$$

There are three fundamental concepts in the Kolmogorov spectrum of

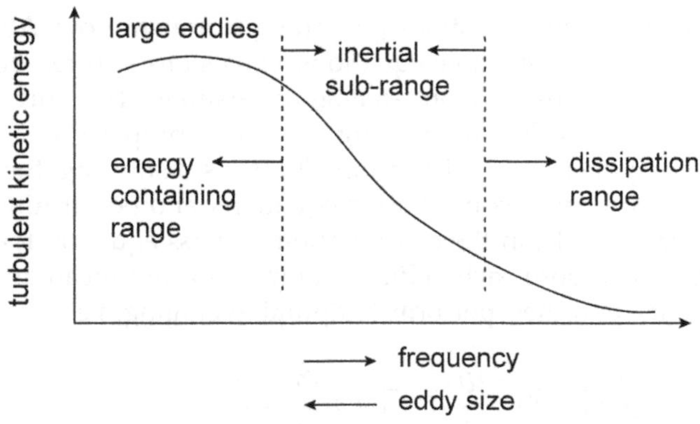

FIGURE 3.14
Kolmogorov turbulence spectrum.

turbulence. First, turbulent kinetic energy is extracted from the mean flow by the large eddies (i.e., large swirling motions) in the flow, which occurs at low frequencies. It is transferred to higher frequencies and smaller eddies until it is finally dissipated by viscous action into thermal energy and heat. Second, all turbulent motion possesses "local isotropy" in the higher frequency (smaller eddy) range of the spectrum. Thirdly, in this local isotropic region the turbulence properties of the flow are uniquely determined by viscosity, ν, and the dissipation rate, ϵ (energy dissipation rate per unit mass).

Based on this spectrum, we can calculate the following Kolmogorov scales of turbulence for length (L_{ko}), velocity (V_{ko}), and time (T_{ko}):

$$L_{ko} = \left(\frac{\nu^3}{\epsilon}\right)^{1/4} \tag{3.219}$$

$$V_{ko} = (\nu\epsilon)^{1/4} \tag{3.220}$$

$$T_{ko} = \left(\frac{\nu}{\epsilon}\right)^{1/2} \tag{3.221}$$

We find that the Reynolds number (based on these Kolmogorov scales) equals 1, which indicates that the scales are applicable to the small-eddy range in the viscous dissipation region of the spectrum.

3.10.4 Modeling of Turbulence

Let us now consider the equations governing turbulent flow. Although we will derive the two-dimensional equations, extensions to three-dimensional flows can be developed in an analogous manner. The turbulent flow equations are obtained by subdividing each flow property into mean and fluctuating components, i.e., Equation (3.216), substituting these expressions into the appropriate conservation equations, and performing temporal averaging. For example, in the conservation of mass equation for turbulent flow we first decompose each velocity component into mean and fluctuating components, and then perform temporal averaging, i.e.,

$$\frac{\partial}{\partial x}(\overline{\overline{u}+u'}) + \frac{\partial}{\partial y}(\overline{\overline{v}+v'}) = \frac{\partial \overline{u}}{\partial x} + \frac{\partial \overline{v}}{\partial y} = 0 \tag{3.222}$$

The averaged momentum equation in the x direction becomes

$$\frac{\partial(\rho\overline{u})}{\partial t}+\frac{\partial(\rho\overline{u}\overline{u})}{\partial x}+\frac{\partial(\rho\overline{u}\overline{v})}{\partial y}=-\frac{\partial\overline{p}}{\partial x}+\mu\left(\frac{\partial^2\overline{u}}{\partial x^2}+\frac{\partial^2\overline{u}}{\partial y^2}\right)-\frac{\partial\overline{\rho u'u'}}{\partial x}-\frac{\partial\overline{\rho u'v'}}{\partial y} \tag{3.223}$$

A similar result is obtained for the y momentum equation.

For incompressible flow in Equation (3.12), recall that the fifth and six terms in Equation (3.223) refer to the derivatives of molecular (laminar) viscous stresses. In an analogous manner, it can be observed that the last two terms of Equation (3.223) have a similar form, i.e., derivatives of turbulent, rather than laminar (molecular), viscous stresses,

$$\tau_{xx}^t=-\overline{\rho u'u'} \tag{3.224}$$

$$\tau_{yx}^t=-\overline{\rho u'v'} \tag{3.225}$$

In other words, we can interpret the total stresses on the fluid particle as composed of a laminar part and a turbulent part. The action of turbulent stresses can be viewed as a diffusion type process arising from turbulent mixing, rather than momentum exchange at a molecular level.

The time-averaged temperature equation for incompressible turbulent flow is obtained in a manner similar to that in the previous momentum equation, with the following result:

$$\rho c_p\frac{\partial\overline{T}}{\partial t}+\rho c_p\frac{\partial(\overline{u}\overline{T})}{\partial x}+\rho c_p\frac{\partial(\overline{v}\overline{T})}{\partial y}$$
$$=-\frac{\partial}{\partial x}\left[-k\frac{\partial\overline{T}}{\partial x}+\rho c_p\overline{u'T'}\right]-\frac{\partial}{\partial y}\left[-k\frac{\partial\overline{T}}{\partial y}+\rho c_p\overline{v'T'}\right]+\nu\overline{\Phi} \tag{3.226}$$

In an analogous manner, when compared with the turbulent stresses, it can be observed from the right side of this equation that the total heat flux can be decomposed into a laminar component, $\mathbf{q}^l$, and a turbulent component, $\mathbf{q}^t$, i.e.,

$$q_x^l=-k\frac{\partial\overline{T}}{\partial x} \tag{3.227}$$

$$q_y^l=-k\frac{\partial\overline{T}}{\partial y} \tag{3.228}$$

$$q_x^t=\rho c_p\overline{u'T'} \tag{3.229}$$

$$q_y^t=\rho c_p\overline{v'T'} \tag{3.230}$$

As a result, the unknowns arising in the time-averaged flow equations become the turbulent stresses (τ^t_{xx}, τ^t_{yx}, τ^t_{xy}, τ^t_{yy}) and heat fluxes (q^t_x, q^t_y).

Although it may be possible to develop an exact turbulence model by discretizing into very fine scale equations, it would be difficult in view of the wide range of scales occurring in turbulence. For example, about 10^{10} grid points for predicting turbulent flow in a duct (10 cm × 10 cm × 2 m) would be required in resolving certain turbulence scales (about 0.1 mm). This represents an excessive storage requirement for computers. Also, designers are often more interested in average flow characteristics than in all of the detailed turbulence information. Similar concerns arose earlier in the conservation equations when we formulated the continuum equations, rather than microscopic tracking of molecular motion. In that case, μ, k, and other thermophysical properties were used for closure of the shear stress and heat flux in terms of velocity and temperature, respectively. In the current modeling, we will seek models for turbulent stress and heat flux, due to the turbulence transport, in terms of the mean flow quantities.

A common approach for turbulence modeling of boundary layers and many other flows is the turbulent viscosity approach. Recall the following definition of laminar stress for incompressible flow:

$$\tau^l_{ij} = \mu\left(\frac{\partial u_i}{\partial x_j} + \frac{\partial u_j}{\partial x_i}\right) \tag{3.231}$$

By analogy, we have the following relationship for turbulent flow:

$$\tau^t_{ij} = \mu_t\left(\frac{\partial \overline{u_i}}{\partial x_j} + \frac{\partial \overline{u_j}}{\partial x_i}\right) = -\rho\overline{u_i'u_j'} \tag{3.232}$$

Alternatively, $\nu_t = \mu_t/\rho$ refers to turbulent kinematic viscosity. In laminar flows, ν represents a property of the fluid, whereas ν_t represents a property of the flow in turbulent flow conditions.

Similarly, the turbulent heat flux is written as

$$q^t_x = -k_t\frac{\partial \overline{T}}{\partial x} = \rho c_p\overline{u'T'} \tag{3.233}$$

$$q^t_y = -k_t\frac{\partial \overline{T}}{\partial y} = \rho c_p\overline{v'T'} \tag{3.234}$$

where $k_t = \mu_t c_p/\sigma_t$ refers to the turbulent conductivity and σ_t is the turbulent Prandtl number (usually taken as constant or 1). Thus, it only remains to find suitable models for the turbulent viscosity. Zero-, one-, and two-equation models of the turbulent viscosity will be considered in this section.

3.10.4.1 Eddy Viscosity (Zero-Equation Model)

In the eddy viscosity approach, the enhanced diffusion transport due to turbulence is expressed in terms of an eddy viscosity, μ_t, and eddy diffusivity, α_t. For example, considering momentum transport and heat transfer, respectively,

$$\tau_{tot} = (\mu + \mu_t)\frac{\partial \bar{u}}{\partial y} \tag{3.235}$$

$$q_{tot} = -\rho c_p(\alpha + \alpha_t)\frac{\partial \overline{T}}{\partial y} \tag{3.236}$$

where y refers to the cross-stream direction (i.e., perpendicular to the wall) and the total stress and diffusivity are comprised of laminar and turbulent components.

In the zero equation approach, a single-length scale is used to estimate the turbulent viscosity. For example,

$$\nu_t = V_{ref} l = \mu_t/\rho \tag{3.237}$$

where the mixing length, l, refers to a representative scale of the turbulent eddies. It is usually an order of magnitude smaller than the characteristic flow dimension. Also, V_{ref} is a characteristic or reference velocity of the turbulence.

Example 3.10.4.1.1

Turbulent Free Jet.

Select a suitable characteristic length scale of turbulence for a free jet formed by a flow exiting from a confined channel.

In this case, V_o and V_{max} will denote the ambient and maximum velocities, respectively, after the jet leaves the channel. An important characteristic of this free jet is the self-similar profiles of velocity that form along the downstream direction. The gap width is denoted by L, and the characteristic length for this problem is selected to be the half-gap width in the channel opening. A reference velocity is selected as $V_{max} - V_o$. The eddy viscosity approach is a successful model here since the flow mainly depends on a single length or velocity scale.

An alternative approach in turbulent viscosity-based models is selecting the reference velocity to be the square root of the turbulent kinetic energy (called the *Prandtl–Kolmogorov model*), i.e.,

$$V_{ref} \approx \sqrt{k} \tag{3.238}$$

Also l can be related to mean flow quantities through algebraic relations.

3.10.4.2 Mixing Length (Zero-Equation Model)

The mixing length model was initially proposed by Prandtl in 1925. In this approach, the turbulent viscosity is written in terms of a mixing length. Consider the turbulent fluctuations of a fluid particle about an imaginary plane (i.e., $y = 0$ plane). A fluid particle arrives at the y-plane from a distance l away with a velocity that is approximately $l \cdot d\bar{u}/dy$ different. The velocity fluctuation, u', is proportional to this difference. However, from continuity, v' is proportional to u' so that

$$\tau_{yx}^{t} = -\rho\overline{u'v'} \approx \rho l_m^2 \left|\frac{d\bar{u}}{dy}\right| \frac{d\bar{u}}{dy} \tag{3.239}$$

where the proportionality constants have been absorbed into a single mixing length, l_m, and the modulus is needed to ensure a positive turbulent stress.

Thus, we have the following mixing length hypothesis:

$$\tau_{yx}^{t} = \mu_t \frac{d\bar{u}}{dy} \tag{3.240}$$

where

$$\mu_t = \rho l_m^2 \left|\frac{d\bar{u}}{dy}\right| \tag{3.241}$$

The choice of a particular mixing length depends on location. For example, the mixing length is approximately the distance to the wall (when close to the wall), whereas the mixing length depends on the large-scale turbulence away from the wall. Difficulties in this approach include a lack of knowledge of the mixing length beforehand in "new" problems. Also, we may obtain a zero turbulent viscosity (or zero turbulent diffusivity) due to $d\bar{u}/dy$, even though the velocity profile around the velocity maximum may be nonzero and nonsymmetrical.

3.10.4.3 Turbulent Near-Wall Flow (Zero-Equation Model)

The importance of accurate turbulence modeling in the boundary layer region is widely evident. For example, in airplane flight the thin viscous region near the aircraft surface is primarily responsible for the lift and drag forces. Thus, an accurate understanding of the turbulent boundary layer is essential in aircraft design. As a result, a special focus on analytical and empirical expressions is given for turbulent near-wall flow. From the boundary layer equations, we have the total wall stress, τ_w, consisting of laminar and turbulent parts, i.e.,

$$\tau_w = \mu \frac{\partial \bar{u}}{\partial y} - \rho \overline{u'v'} \tag{3.242}$$

In addition, we can subdivide the boundary layer into two distinct regions: the inner viscous layer (where laminar stresses dominate) and the outer layer (where turbulent stresses dominate). Although there is an overlap layer, where both stresses are important, we will identify only the inner and outer regions for the purposes of this turbulence modeling. Molecular diffusion dominates in the viscous wall layer, and laminar stresses are dominant, since this layer is very thin and turbulent eddy motion is suppressed by the wall. However, in the outer region eddy motion becomes more significant, so that turbulent stresses are usually much larger than molecular (laminar) stresses.

Outside the viscous sublayer, observations confirm that the mixing length is approximately proportional to the distance from the wall. As a result, we obtain the following *law of the wall* as a widely used correlation for the velocity profile:

$$\frac{u}{u_\tau} = \frac{1}{\kappa} ln\left(\frac{yu_\tau}{\nu}\right) + C \tag{3.243}$$

where $u_\tau = \sqrt{\tau_w/\rho}$ is the friction velocity, and extensive measurements show $\kappa \approx 0.41$ (von Karman constant) and $C \approx 5.2$. If the surface is rough, then the roughness elements will break up and affect the wall layer. In this case, the wall shear stress depends on pressure-related drag forces on the roughness elements, rather than just molecular action alone. From experimental data in this case, the previous log law is modified so that $\kappa = 0.4$, $C \approx 3.5$, and the fraction u_τ/ν (in the log term) is replaced by $1/y_r$, where y_r refers to the average roughness element height.

In terms of the mixing length, the following van Driest and exponential models are often adopted:

$$l = \kappa y \left[1 - exp\left(\frac{-yu_\tau}{\nu A^+} \right) \right] \quad (inner \;\; viscous \;\; layer) \tag{3.244}$$

$$l = \frac{\delta\kappa}{n} \left[1 - \left(1 - \frac{y}{\delta} \right)^n \right] \quad (outer \;\; layer) \tag{3.245}$$

where $A^+ = 26$, δ refers to the boundary layer thickness, and $n = 5$ for boundary layer flow.

3.10.4.4 One-Equation Model (k)

In one-equation models, the turbulent kinetic energy (k) in the Prandtl–Kolmogorov equation is calculated from a differential transport equation for k (one-equation closure), rather than from an algebraic relationship to mean flow variables (zero-equation closure). In the present case, let us take the dot product between the velocity field and the momentum equations, perform averaging, and divide by 2 to yield the averaged mechanical energy equation in terms of k. Furthermore, assume that the turbulent diffusion of this kinetic energy occurs down its gradient in a manner analogous to molecular diffusion (called the *gradient diffusion hypothesis*). We then obtain the following *equation for the transport of turbulent kinetic energy*:

$$\rho \frac{Dk}{Dt} = \frac{\partial}{\partial y} \left[\left(\mu + \frac{\mu_t}{\sigma_k} \right) \frac{\partial k}{\partial y} \right] + \mu_t \left(\frac{\partial \overline{u}}{\partial y} \right)^2 - \frac{c_D k^{3/2}}{l_t} \tag{3.246}$$

where commonly adopted values of the constants are $c_D \approx 0.09$ and $\sigma_k \approx 1$ (empirical constants).

Equation (3.246) indicates that the rate of change of k with time equals molecular and turbulent diffusion of k, plus production of k by velocity gradients in the mean flow, minus dissipation of k to internal energy in small-scale motion ($\epsilon \geq 1$). This model applies well to shear layer flows; i.e., flows are incorporated within the turbulent diffusion term. The selection of the turbulent length scale, l_t, depends on the problem under consideration. For example, l_t/δ remains approximately constant in free shear layer flows and wakes, where δ refers to the wake thickness. In wall-bounded flows, the turbulent length scale is approximately equal to the mixing length, as described in the previous secion.

3.10.4.5 Two-Equation Model ($k-\epsilon$)

In calculating the turbulent viscosity as the product of the reference velocity, V_{ref}, and length, l, a differential transport equation can be established for more widely applicable predictions of the length scale and turbulent kinetic energy. The Prandtl–Kolmogorov model allowed us to estimate the reference velocity as approximately the square root of the turbulent kinetic energy. Alternatively, we can use the differential equation for k and any product of k and l, such as $k \cdot l$ or $k^{3/2}/l$. The dissipation of k is defined as $\epsilon = c_D k^{3/2}/l$, which has the proper units of a suitable product term. Using this definition, the turbulent viscosity becomes

$$\nu_t = V_{ref} l \tag{3.247}$$

$$\nu_t = c_\mu \frac{k^2}{\epsilon} \tag{3.248}$$

The *equation for the transport of dissipation rate* (ETDR) consists of the following four essential processes: convection, diffusion (viscous, turbulent, and pressure), production, and dissipation. Furthermore, we will consider the gradient diffusion hypothesis for the dissipation rate, two-dimensional incompressible flow, steady state, and negligible effects of curvature. Then the convection of ϵ balances the diffusion plus production of ϵ, minus the destruction of ϵ, plus any additional terms, such as buoyancy (denoted by E), i.e.,

$$\bar{u}\frac{\partial \epsilon}{\partial x} + \bar{v}\frac{\partial \epsilon}{\partial y} = \frac{\partial}{\partial y}\left[\left(\mu + \frac{\mu_t}{\sigma_\epsilon}\right)\frac{\partial \epsilon}{\partial y}\right] + c_{\epsilon 1} P \frac{\epsilon}{\kappa} - c_{\epsilon 2}\frac{\epsilon^2}{k} + E \tag{3.249}$$

Upon combining Equations (3.246) and (3.249), we have the $k-\epsilon$ model of turbulence. Typical values of the empirical constants are $\sigma_\epsilon = 1.3$, $\sigma_k = 1$, $c_\mu = 0.09$, $c_{\epsilon 1} = 1.44$, and $c_{\epsilon 2} = 1.92$. Also

$$P = \nu_t \left(\frac{\partial \bar{u}}{\partial y}\right)^2 \tag{3.250}$$

This widely used $k-\epsilon$ model is a two-equation closure solving Equations (3.246) and (3.249) for k and ϵ, thereby giving ν_t and the turbulent stresses in the momentum equations. Finally, the turbulent Prandtl number is used to relate ν_t to the turbulent diffusivity for subsequent solutions of the turbulent heat transfer equations. Various complex flows have been successfully predicted by $k-\epsilon$ models. Furthermore, the addition of a buoyancy term, $G = \beta g(\nu_t/\sigma_k)(\partial T/\partial x_i)$, to Equation (3.246) can permit the analysis of turbulent free convection flows, such as buoyant jets and plumes.

A shortfall in the $k - \epsilon$ model is that empirical correlations (such as the log law) are usually required in the near-wall region for numerical procedures. These correlations involve certain limitations (i.e., invalid beyond boundary layer separation points), thereby limiting the applicability of the overall approach. Furthermore, the magnitudes of the turbulent stresses are assumed be equal (or isotropic) in all directions. Although they are beyond the scope of this book, other alternative turbulence models are available in the technical literature: Reynolds stress models (Rodi, 1984), algebraic stress models (Hanjalic and Launder, 1972), and renormalization group theory-based models (Kirtley, 1992).

References

E. Achenbach. 1978. "Heat transfer from spheres up to Re = 6 × 106," in *Proceedings of the Sixth International Heat Transfer Conference*, Vol. 5, Washington, D.C.: Hemisphere.

A. Bejan. 1996. *Entropy Generation Minimization*, Boca Raton, FL: CRC Press.

J.A. Camberos. 2000. "Revised interpretation of work potential in thermophysical processes," *AIAA J. Thermophys. Heat Transfer* **14**: 177–185.

S.W. Churchill. Free convection around immersed bodies, in *Heat Exchanger Design Handbook*, Sec. 2.5.7, 1983, Hemisphere, New York.

K. Hanjalic and B.E. Launder. 1972. "A Reynolds stress model of turbulence and its application to thin shear flows," *J. Fluid Mech.* **52**: 609–638.

K.G.T. Hollands et al. 1976. "Free convective heat transfer across inclined air layers," *J. Heat Transfer* **98**: 189.

F.P. Incropera and D.P. DeWitt. *Fundamentals of Heat and Mass Transfer*, 3rd ed., John Wiley & Sons, New York, 1990.

W.M. Kays and M.E. Crawford. *Convective Heat and Mass Transfer*, McGraw-Hill, New York, 1980.

K. Kirtley. 1992. "Renormalization group based algebraic turbulence model for three-dimensional turbomachinery flows," *AIAA J.* **30**: 1500–1506.

F. Kreith and M.S. Bohn. *Principles of Heat Transfer*, 6th ed., Brooks/Cole Thomson Learning, 2001.

G.F. Naterer and P.S. Glockner. Pulsed Laser PIV Measurements and Multiphase Turbulence Model of Aircraft Engine Inlet Flows, paper presented at AIAA 31st Fluid Dynamics Conference and Exhibit, AIAA Paper 2001–2032, Anaheim, CA, June 11–14, 2001.

G.D. Raithby and K.G.T. Hollands. A general method of obtaining approximate solutions to laminar and turbulent free convection problems, in *Advances in Heat*

Transfer, T.F. Irvine Jr. and J.P. Hartnett, eds., Vol. 11, New York: Academic Press, 1975, pp. 265–315.

H. Schlichting. *Boundary Layer Theory*, McGraw-Hill, New York, 1979.

W. Rodi. 1984. *Turbulence Models and Their Application in Hydraulics*, Brookfield, VT: Brookfield.

S. Whitaker. 1972. "Forced convection heat transfer correlations for flow in pipes, past flat plates, single cylinders, single spheres, and flow in packed beds and tube bundles," *AIChE J.* **18**: 361–371.

A. Zhukauskas. 1972. "Heat transfer from tubes in cross flow," in *Advances in Heat Transfer*, J.P. Hartnett and T.F. Irvine, Jr., eds., Vol. 8, New York: Academic Press.

Problems

1. Moist air at 22°C flows across the surface of a body of water at the same temperature. If the surface of the body of water recedes at 0.2 mm/h, what is the rate of evaporation of liquid mass from this body? Explain how this evaporation rate can be used to determine the rate of convective heat transfer from the water to the moist air. It may be assumed that the body of water is a closed system, with the exception of mass outflow due to evaporation.
2. In a manufacturing process a surface of area 0.7 m^2 loses heat (without evaporative cooling) to a surrounding airstream at 20°C with an average convective heat transfer coefficient of 20 W/m^{2}K. The surface temperature of the component is 30°C. The same manufacturing process is performed on another, more humid day (50% relative humidity) at an ambient air temperature of 32°C. In this case, the surface is saturated, and a surface liquid film is formed at 36°C. Using the same convection coefficient for both cases, what proportion of the total heat loss is due to evaporative cooling during the humid day? Use a diffusion coefficient (air–H_2O) of 2.6×10^{-5} m^2/sec.
3. In this problem, Bernoulli's equation is derived by integrating the mechanical energy equation (or total energy minus internal energy) along a streamline within a duct. Two-dimensional steady flow through a duct of varying cross-sectional area (unit depth) is considered. Start with the total energy equation and convert the external work term (work done by gravity) to a potential energy form. Integrate the resulting equation over a cross-sectional area, $A(ds)$, where A refers to the local cross-sectional area. Then integrate from the inlet (1) to the outlet (2), and subtract the analogous integrated form of

the internal energy equation to obtain Bernoulli's equation for compressible flow. You may obtain the same result by performing the integration of the mechanical energy equation (across $A(ds)$ and from the inlet to the outlet) instead. Assume uniform velocity profiles at the inlet and the outlet. What expression is obtained for the head loss term?

4. Determine the nondimensional functional relationship for the total heat transfer from a convectively cooled solid material. Express your result in terms of the Biot number and nondimensional time and temperature variables.
5. In this problem, the similarity solution method (developed in this chapter) will be applied to a problem involving transient heat conduction. Consider a problem whereby a semi-infinite body (or finite body at early stages of time) at a temperature of T_o suddenly has its surface temperature changed to T_s and maintained at that level. In this situation, a "thermal wave" propagates into the material and self-similarity of temperature profiles is maintained over time. Find a similarity solution of this problem using the similarity variable of $\theta = (T - T_s)/(T_o - T_s)$ as a function of $\eta = xg(t)$.
6. Consider frictional heating within a laminar boundary layer flow where the freestream flow (at T_∞, U_∞) is parallel to an adiabatic wall. We will assume that the Blasius velocity solution may be adopted in the fluid flow problem.
   (a) Perform a scaling analysis to find the relevant boundary layer energy equation and explain the meaning of each resulting scaling term. Show that the increase of wall temperature scales as U_∞^2/c_p and $U_\infty^2\mu/k$ for cases where $Pr \gg 1$ and $Pr \ll 1$, respectively.
   (b) Find a similarity solution of this problem based on the following similarity transformations:

$$\eta = y\sqrt{\frac{U_\infty}{\nu x}} \tag{3.251}$$

$$\psi = \sqrt{U_\infty \nu x} f(\eta) \tag{3.252}$$

$$\theta(\eta) = \frac{T - T_\infty}{U_\infty^2/(2c_p)} \tag{3.253}$$

Give appropriate boundary conditions for this problem and find the temperature rise in the boundary layer as a function of η and the Prandtl number.

(c) The dimensionless wall temperature rise in part (b) can now be evaluated numerically by using an analytical approximation for $f(\eta)$ from the Blasius similarity results. Compare these results with the earlier scaling results in part (a). In particular, use the scaling analysis to show that θ is of order $\mathcal{O}(Pr)$ and $\mathcal{O}(1)$ in the limit as $Pr \to 0$ and $Pr \to \infty$, respectively.

7. Liquid propane at 27°C flows along a flat surface with a velocity of 1 m/sec. At what distance from the leading edge of the surface does the velocity boundary layer reach a thickness of 6 mm?
8. Water at 27°C enters the inlet section between two parallel flat plates at a velocity of 2 m/sec. If each plate length is 10 cm, then what gap spacing between plates should be used so that the hydrodynamic boundary layers merge in the exit plane?
9. Air at 17°C flows past a flat plate at 77°C with a velocity of 15 m/sec.
   (a) Find the convective heat transfer coefficient at the midpoint along the plate of length 1 m.
   (b) What plate width is required to provide a heat transfer rate of 1 kW from the plate to the surrounding airstream?
10. Heated air passes vertically through a vertical channel consisting of two parallel plates, where one plate is well insulated and the other plate is maintained at a temperature of 20°C. The air enters the passage at 200°C with a uniform velocity of 8 m/sec. The width and height of each plate are 80 and 60 cm, respectively.
   (a) What spacing between the plates should be used to reach a fully developed flow condition at the channel outlet?
   (b) At what position within the channel does the mean temperature fall 5°C below its inlet value?
11. A hot-wire anemometer is a device that can measure gas velocities over a wide range of flow conditions. The most common materials for anemometers are platinum and nickel alloy wires since resistivity variations with temperature are very small for these materials. Consider a fine platinum wire with a diameter, surface temperature, and resistivity of 0.9 mm, 400 K, and 268.8 Ω/m, respectively (note: 1 W = 1 $A^2\Omega$). The wire is exposed to an airstream at a temperature of 300 K.
   (a) If the gas velocity is 2 m/sec, calculate the current flow through the wire.
   (b) Find a relationship between small changes in the gas velocity, ΔV, and the current, ΔI. In other words, consider two experiments, namely, parts (a) and (b). Describe a technique that determines the gas velocity difference in terms of the current difference between both experiments.

12. Air at $-5°C$ flows past an overhead power transmission line at 10 m/sec. A 5-mm-thick ice layer covers the 1.9-cm-diameter cable. The cable carries a current of 240 A with an electrical resistance of 4×10^{-4} Ω/m of cable length. Estimate the surface temperature of the cable beneath the ice layer.
13. A pipeline carries oil above the ground over a distance of 60 m where the surrounding air velocity and temperature are 6 m/sec and 2°C, respectively. The surface temperature of the 1-m-diameter pipeline is 17°C. What thickness of insulation ($k = 0.08$ W/mK) is required to reduce the total heat loss by 90% (as compared with an uninsulated pipe with the same surface temperature)?
14. Metallic spheres (diameter of 2 cm) at 330°C are suddenly removed from a heat treatment furnace and cooled by an airstream at 30°C flowing past each sphere at 8 m/sec. Find the initial rate of temperature change of each iron sphere. Does this rate of change vary appreciably throughout the sphere? Neglect radiative heat exchange with the surroundings.
15. In a reaction chamber, spherical fuel elements are cooled by a nitrogen gas flow at 30 m/sec and 210°C. The 2-cm-diameter elements have an internal volumetric heating rate of 9×10^7 W/m^3. Determine the surface temperature of the fuel element.
16. Engine oil flows through the inner tube of a double-pipe heat exchanger with a mass flow rate of 1 kg/sec. An evaporating refrigerant in the annular region around the inner tube absorbs heat from the oil and maintains a constant wall temperature, T_w. The thermal conductivity of the inner tube is 54 W/mK, and the inner and outer radii are 3 and 3.2 cm, respectively. The oil enters the exchanger at a temperature of 320 K and flows over a distance of 80 m through the tube. Find the required saturation temperature of the refrigerant, T_w, to cool the oil to 316 K at the outlet of the tube.
17. A long steel pipe with inner and outer diameters of 2 and 4 cm, respectively, is heated by electrical resistance heaters within the pipe walls. The electrical resistance elements provide a uniform heating rate of 10^6 W/m^3 within the pipe wall. Water flows through the pipe with a flow rate of 0.1 kg/sec, and the external sides of the electrically heated walls are well insulated. Is the water flow fully developed at the pipe outlet? If the mean temperatures of the water at the inlet and outlet are 20 and 40°C, respectively, then find the inner wall temperature at the outlet.
18. Oil transport through underground pipelines poses certain problems in the petroleum industry. For example, pipe heat losses may significantly increase pumping power requirements since the oil viscosity increases when the oil temperature decreases. Consider a single phase (liquid oil)

flow within a long underground insulated pipe. Assume fully developed conditions and constant thermophysical properties for oil ($\rho = 90$ kg/m^3, $c_p = 2$ kJ/kgK, $\nu = 8.5 \times 10^{-4}$ m^2/sec, and $k = 0.14$ W/mK), insulation ($k = 0.03$ W/mK), and soil ($k = 0.5$ W/mK).

(a) Develop an expression for the mean oil temperature as a function of distance, x, along the pipe. (Hint: you may use a shape factor of $2\pi L/ln(4z/D)$ for a cylinder buried in a semi-infinite medium.)

(b) Find the mean oil temperature at the pipe outlet, where $x = L =$ 250 km. The mean oil inlet temperature, mass flow rate, ground surface temperature (above the pipe), pipe depth, and insulated pipe radii (inner and outer) are 110°C, 400 kg/sec, 5°C, 6 m, 1.2 m, and 1.5 m, respectively.

19. Pressurized water at 200°C is pumped from a power plant to a nearby factory through a thin-walled circular pipe at a mass flow rate of 1.8 kg/sec. The pipe diameter is 0.8 m, and a layer of 0.05-m-thick insulation with a conductivity of 0.05 W/mK covers the pipe. The pipe length is 600 m, and the pipe is exposed to an external air cross-flow with a velocity of 5 m/sec and $T_\infty = -10$°C. Develop an expression for the mean water temperature as a function of distance, x, along the pipe. Also, find the mean water temperature at the outlet of the pipe.

20. The purpose of this problem is to derive the Nusselt number for fully developed slug flow through an annulus with inner and outer diameters of d_i and d_o, respectively. A uniform heat flux, q_{wi}, is applied at the inner surface, and the outer surface is maintained at a constant temperature, T_{wo}. In thermal analysis of slug flows, it may be assumed that the velocity is approximately constant (u_m; mean velocity) across the pipe (or duct).

(a) Explain the assumptions adopted in obtaining the following governing energy equation:

$$\rho c_p u_m \frac{\partial T}{\partial x} = \frac{k}{r}\frac{\partial}{\partial r}\left(r\frac{\partial T}{\partial r}\right)$$

Solve this equation (subject to appropriate boundary conditions) to obtain the fluid temperature distribution within the annulus.

(b) Explain how the Nusselt number (as a function of r_o/r_i) can be obtained from the results in part (a) (without finding the explicit closed-form solution for Nu).

21. Air flows at 0.01 kg/sec through a pipe with a diameter and length of 2.5 cm and 2 m, respectively. Electrical heating elements around the pipe provide a constant heat flux of 2 kW/m^2 to the airflow in the pipe. The air inlet temperature is 27°C. Assume that thermally and hydrodynamically developed conditions exist throughout the pipe.

(a) Estimate the wall temperature at the pipe outlet.
(b) Will additional pipe wall roughness increase or decrease the values of the convection coefficient, h, and the friction factor, f? Explain your responses.
(c) Various options are available for enhancing the heat transfer in this problem, including longitudinal internal fins or helical ribs. What adverse effects would these schemes exhibit in terms of the related fluid mechanics of this problem?

22. A vertical composite wall in a building consists of a brick material (8 cm thick with $k = 0.4$ W/mK) adjoined by an insulation layer (8 cm thick with $k = 0.02$ W/mK) and plasterboard (1 cm thick with $k = 0.7$ W/mK) facing indoors. Ambient air temperatures outside and inside the wall are -20 and 25°C, respectively. If the wall is 3 m high, find the rate of heat loss through the wall per unit width of the wall.
23. An assessment of heat losses from two rooms on the side of a building is required. The walls of rooms A and B have lengths of 12 and 8 m, respectively, and a height of 4 m with a wall thickness of 0.25 m. The effective thermal conductivity of the wall is approximately 1 W/mK. Outside air flows parallel to the walls at 8 m/sec with an ambient temperature of -25°C. The air temperatures in rooms A and B are 25 and 18°C, respectively, and the air velocity in each room is assumed to be zero. Calculate the total heat transfer rate through both external walls of rooms A and B of the building.
24. A section of an electronic assembly consists of a board rack with a vertically aligned circuit board and transistors mounted on the surface of each circuit board. Four lead wires ($k = 25$ W/mK) conduct heat between the circuit board and each transistor case. The circuit board and ambient air temperatures are 40 and 25°C, respectively. The transistor case and wire lead dimensions are 5 mm (width) $\times$ 9 mm (height) $\times$ 8 mm (depth) and $3 \times 11 \times 0.2$ mm, respectively. The transistor case is mounted with a 1-mm gap above the board. The radiation exchange and heat losses from the board, and case edges may be neglected.
    (a) Explain how the thermal resistances are assembled together in a thermal circuit for this problem.
    (b) The transistor temperature should not exceed 70°C for reliable and effective performance. Consider the options of a (i) stagnant air layer ($k_a = 0.028$ W/mK) or a (ii) filler paste material ($k_p = 0.1$ W/mK) in the gap between the transistor case and the board. Which option would permit the highest transistor heat generation without exceeding the temperature limit?
25. Water at a mean temperature of 70°C with a mean velocity of 1 m/sec flows from a heat exchanger through a long copper pipe (diameter of 1

cm) inside a building. The poorly insulated pipe loses heat by free convection into the room air at 20°C. Find the rate of heat loss per meter length of pipe. It may be assumed that the insulation and pipe wall resistances are small in comparison to the thermal resistances associated with convection.

26. Perform a discrete scaling analysis for natural convection from a vertical plate. Consider a flat plate maintained at $T = T_s$ adjacent to a quiescent freestream at a temperature of $T = T_\infty$. Find expressions for the boundary layer thickness, in terms of the Prandtl number, as well as the average heat transfer coefficient and Nusselt number, in terms of the Prandtl and Rayleigh numbers. Compare your result with other available correlations for natural convection.
27. In this problem we will consider an integral analysis (zero degrees of freedom) for free convective heat transfer from a sphere.
    (a) Derive expressions for the local and average Nusselt numbers, based on an integral analysis, for natural convective heat transfer from a sphere.
    (b) How does surface curvature affect the results in part (a)?
    (c) Does your result approach the correct limits as $Ra \to \infty$ and $Ra \to 0$?
28. An inlet section of an airflow tunnel of total length L consists of a central duct carrying the main (inner) inflow with a surrounding (outer) flow that independently carries a separate inflow stream. The inner velocity, u, or corresponding mass flow rate must be found in order to transport a fixed outer flow across a specified pressure difference (i.e., $\Delta p = p_2 - p_1$ across the tunnel section). Use an integral analysis to find an expression for the required inner velocity in terms of the inlet pressure, outlet pressure, and duct areas (inner and outer). In the analysis, velocity variations in the cross-stream (y) direction may be neglected.
29. In a manufacturing process, a six-sided metal rod is heated through laminar free convection. The rod is oriented with the corner facing upward, and the vertical sides have twice the length ($2L$) of the other sides (L). The ambient air temperature is T_∞, and the wall temperature is T_w (where $T_w < T_\infty$).
    (a) Calculate the average Nusselt number, $\overline{Nu}_L$, for this configuration in terms of the side length, L. What average heat transfer rate (per unit depth of rod) is obtained when $T_\infty = 70°C$, $T_w = 20°C$, and $L = 6$ cm? Explain how and why you would expect $\overline{Nu}_L$ to change if the vertical sides are lengthened (while maintaining the same total surface area).

(b) How would the heat transfer be altered if surface curvature effects were included (i.e., higher or lower $\overline{Nu}_L$ in comparison to part (a))? Explain your response.

30. Small indentations along a vertically oriented surface are proposed for more effective convective cooling of an electronic assembly. The lengths of the vertical and indented surfaces are L_1 and L_2, respectively. The indented triangular cavity consists of both surfaces inclined at 45° with respect to the vertical direction. The ambient air temperature is T_∞, and the wall temperature is T_w (where $T_w > T_\infty$).
    (a) Calculate the average Nusselt number, $\overline{Nu}_S$, for natural convection in this configuration, where the total side length, S, corresponds to N sets of vertical surface and cavity sections. Find the total heat flow from the surface (per unit depth) when $T_\infty = 17°C$ and $T_w = 57°C$. Express your answer in terms of L_1, L_2, and N.
    (b) Find the percentage increase of heat flow due to surface indentations with $L_1 = 1$ cm $= L_2$, in comparison to no surface indentations, while maintaining the same gap spacing between the vertical sections. How does your result depend on the relative magnitudes of L_1 and L_2? Explain your response.
31. Liquid benzene is heated by free convection by a vertical plate of 10 cm in height with a surface temperature of 30°C. If the lowest allowed benzene temperature is 20°C, then what minimum plate width is required to provide at least 300 W of heat transfer to the liquid benzene?
32. An electrically heated rod of 4 cm in height with an electrical resistance of 0.3 Ω/m is immersed in a liquid acetic acid bath at 15°C. The rod's diameter is 3 cm. Estimate the electrical current required to provide an average surface temperature of 45°C along the cylinder. Assume that the heat transfer occurs predominantly by free convection.
33. Thin metal plates are suspended in air at 25°C before processing in a manufacturing operation. What initial plate surface temperature is required to produce an initial rate of temperature change of 0.7 K/sec for a single plate? The same length of each 1-mm-thick square iron plate is 40 cm. Neglect heat exchange by radiation.
34. Optimizing the parameters within a pipe flow has important implications in the design of heat exchangers, underground pipelines carrying oil, and other applications. Consider steady, fully developed pipe flow with a fixed mass flow rate, $\dot{m}$, and rate of heat transfer (per unit length), q', to the pipe. Find the optimal pipe diameter based on the fixed values of $\dot{m}$ and q' and the method of entropy generation minimization.

35. Determine the optimal sphere diameter (based on entropy generation minimization) for external flow of a fluid at U_∞ and T_∞ past a sphere of diameter D heated by a fixed amount q. Express your answer in terms of q, D, U_∞, T_∞, and thermophysical properties.
36. A heat exchanger design involves turbulent flow of water at 310 K (average fluid temperature) through a tube while requiring a mass flow rate of 12 kg/sec and a temperature rise of 6 K/m. Find the optimal diameter based on entropy generation minimization.
37. Air flows at 20 m/sec and 400 K across a plate heated uniformly at 100 W/m. Find the optimal plate length based on the method of entropy generation minimization.
38. Turbulent flow of air past a plate of length L and width W is encountered in an electronic assembly. Under these conditions, the following correlations for heat transfer (Nusselt number) and friction coefficient are adopted:

$$Nu_x = 0.029 Pr^{1/3} Re_x^{4/5} \tag{3.254}$$

$$c_{f,x} = \frac{\tau_x}{\rho_\infty U_\infty^2/2} = 0.0576 Re_x^{-1/5} \tag{3.255}$$

where U_∞ and T_∞ refer to freestream velocity and temperature, respectively. The plate is subjected to a uniform wall heat flux, q''. Explain why an optimal plate length, L_{opt}, exists for this external turbulent flow and find this optimal length (based on entropy generation minimization). Express your answer in terms of q'' (or $q' = q''L$), U_∞, T_∞, Pr, and relevant thermophysical properties.
39. A brass cylindrical fin is joined to the top surface of an electronic component to enhance convective cooling of the component. Correlations for heat transfer and drag coefficient, c_d, in regard to external flow past the fin, are known as follows:

$$Nu_d = 0.683 Pr^{1/3} Re_d^{0.466}; \quad c_d = 1.2$$

where U_∞, Nu_d, Pr, and Re_d refer to incoming velocity and Nusselt, Prandtl, and Reynolds numbers, respectively. The freestream temperature is T_∞, and the fin is required to transfer a fixed heat flow, q, from the base to the freestream air.

(a) Find the optimal fin length, L, based on the method of entropy generation minimization. Express your answer in terms of T_∞, q, fin diameter (d), U_∞, and thermophysical properties.

(b) Give a physical interpretation to explain how this optimum could be used to deliver better convective cooling of the microelectronic system.

4

Radiative Heat Transfer

4.1 Introduction

Thermal radiation is a form of energy emitted as electromagnetic waves (or photons) by all matter above a temperature of absolute zero. The emissions are due to changes in the electron configurations of the constituent atoms or molecules. In this chapter, the mechanisms and governing equations of thermal radiation will be described.

Consider the cooling of a hot solid object in a vacuum chamber. The vacuum chamber is an evacuated space containing a very low pressure and resulting negligible mass. Despite the absence of conduction or convection modes of heat transfer, energy is still transferred from the object to the vacuum chamber walls. There are two main theories that explain how this heat transfer occurs, based on *quantum theory* (Planck) or *electromagnetic theory* (Maxwell).

In the former case (Planck, 1959), it is known that the energy is transported by radiation in the form of *photons* (or energy packets), which travel at the speed of light. In terms of the photon energy, e, and the frequency of radiation, ν,

$$e = \hbar \nu \tag{4.1}$$

where $\hbar = 6.63 \times 10^{-34}$ J·s refers to Planck's constant. Alternatively (Maxwell), radiation is interpreted to be transported in the form of electromagnetic waves traveling at the speed of light. In this interpretation, the speed of light, c, is related to the wavelength, λ, in the following way:

$$c = \lambda \nu \tag{4.2}$$

The speed of light is 3×10^8 m/sec in a vacuum.

4.2 Fundamental Processes and Equations

Radiation occurs across an electromagnetic spectrum (see Figure 4.1), i.e., over a wide range of wavelengths and corresponding frequencies. From very small wavelengths, below 10^{-5} μm (or very high frequencies), to the longest wavelengths in the microwave region (above 10^2 μm), the spectrum identifies the characteristics of the transmitted radiation. Many common everyday experiences can be explained by the electromagnetic spectrum. For example, the color of a rainbow is related to the radiative properties of the atmosphere in the visible range of the spectrum. The development of the human eye is remarkable in view of how it interprets incoming waves only within a certain range (between 0.4 μm and 0.7 μm). All other electromagnetic waves are not visually detected. The basis of many science fiction novels is often linked to interesting possibilities inherent within the electromagnetic spectrum. For example, in a manner similar to that in effectively scattering radar signatures in advanced military aircraft, the future may perhaps hold the opportunity to mask an object through special material properties that manipulate incoming electromagnetic waves in the visible range alone.

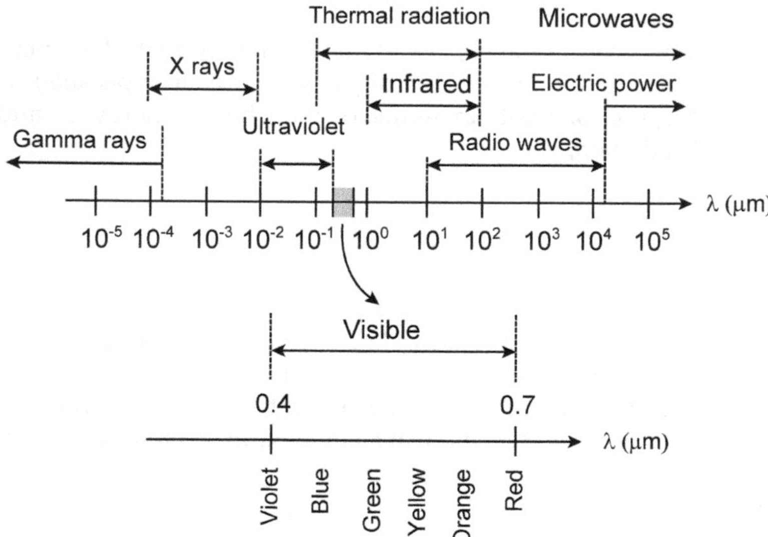

FIGURE 4.1
Electromagnetic spectrum of radiation.

Gamma rays are mainly of interest to astronomers and astrophysicists since these waves are generally encountered only in the transmission of signals in deep space. Between about 10^{-4} μm and 10^{-2} μm x-rays are encountered; these rays often arise in nuclear and medical applications. Further up the scale (between about 0.01 and 0.4 μm), ultraviolet rays are experienced. It has been speculated that incoming ultraviolet rays from the sun are harmful to the human body, particularly skin. As a result, depletion of the ozone in the atmosphere, due to the release of certain chemicals into the atmosphere (such as chlorofluorocarbons (CFCs) from refrigerants in air conditioning systems), is viewed to have a harmful effect, since the ozone absorbs incoming solar radiation in the ultraviolet part of the spectrum. Developing more environmentally friendly refrigerants is thus an important challenge facing engineers in the future.

The next range is visible radiation, between 0.4 and 0.7 μm, which can be further subdivided into regions interpreted by human vision as white, violet, blue, green, yellow, and red. When an object is heated, its energy increases and thus the frequency of emitted radiation increases (i.e., wavelength decreases). This process explains why objects initially become red and eventually turn white while they are heated. The remaining ranges are the infrared (between 0.7 and 100 μm) and microwave (above 100 μm) regions. Applications involving radio wave propagation and electrical engineering often arise in the microwave region. Thermal radiation refers to the range between 0.1 and 100 μm (encompassing ultraviolet, visible, and infrared regions) since the majority of thermal engineering applications occur in this range. Thermal radiation that is emitted, absorbed, or reflected by a surface varies with temperature, wavelength, and direction. For example, emitted radiation normal to a surface is anticipated to be larger than the emission nearly tangent to the surface since the roughness elements along a surface scatter and obstruct the outgoing radiation (i.e., establishing a directional distribution).

The properties of radiation are typically compared with an ideal radiator, called a *blackbody*. A blackbody absorbs all incident radiation, including radiation at all wavelengths and directions. The ratio of an actual surface's radiative absorption to a blackbody's absorption is called the absorptivity, α (note: the absorptivity of a blackbody is $\alpha = 1$). No surface can emit more energy than a blackbody at a given temperature and wavelength (characterized by an emissivity of $\epsilon = 1$). Furthermore, the radiative emission from a blackbody is independent of direction; in other words, a blackbody is a *diffuse emitter*. This diffuse property includes both surface-emitting radiation equally in all directions and incoming radiation (called *irradiation*), absorbed equally from all directions at a particular point on the surface of the blackbody.

A blackbody can refer to the property of a particular configuration of surfaces, rather than only a single surface. For example, an isothermal cavity containing a small hole, through which radiation passes or enters, can be considered to be a blackbody since it possesses all of the previously mentioned blackbody properties. There is complete absorption of the incident radiation by the cavity, regardless of the condition of the surfaces comprising the inner walls of the cavity. Since all irradiation is absorbed, the entire cavity is seen by another object to behave as a blackbody, even though the surfaces within the cavity may not be black. Furthermore, irradiation within the cavity is diffuse and emission from the cavity through the hole is diffuse. As a result, this cavity satisfies the properties required to show blackbody behavior.

The energy emitted by radiation varies with wavelength and the temperature of the emitting body. The precise nature of this transport is a subject of statistical thermodynamics. An object above a temperature of absolute zero contains molecules with electrons that are situated at discrete energy levels (called quantized energy states), rather than a continuous range of levels. For example, a hot object would have more electrons located at higher energy levels than a cold object. As these electrons fluctuate back and forth between different quantum states, due to temperature fluctuations of the surface, each fluctuation generates electromagnetic waves that are emitted from that location in characterizing the quantum disturbance.

Based on statistical methods, it can be shown that there is an exponential probabilistic distribution where electrons are more likely to occupy a configuration having more sites at certain quantum energy levels. The exponential probabilistic decay of blackbody spectral emissive power, $E_{\lambda,b}(\lambda, T)$, at a temperature T and wavelength λ is given by *Planck's law*:

$$E_{\lambda,T}(\lambda, T) = \frac{C_1}{\lambda^5(exp(C_2/\lambda T) - 1)} = \pi I_{\lambda,b}(\lambda, T) \tag{4.3}$$

where $I_{\lambda,b}$ is the spectral radiation intensity. The factor π is obtained due to angular integration of the spectral distribution over the hemispherical range of radiative emission by a surface. Also, $C_1 = 3.742 \times 10^{-16}$ Wm^2 and

$$C_2 = \frac{\hat{h}c}{B} = 0.01439 \text{ W/mK} \tag{4.4}$$

where $\hat{h}$ is the Planck constant, $B = 1.3805 \times 10^{-23}$ J/K is the Boltzmann constant, and c is the speed of light. The emissive power of a blackbody is shown in Figure 4.2.

The quantization of energy, as graphically depicted through Planck's law, shows that the emissive power varies with wavelength in a continuous

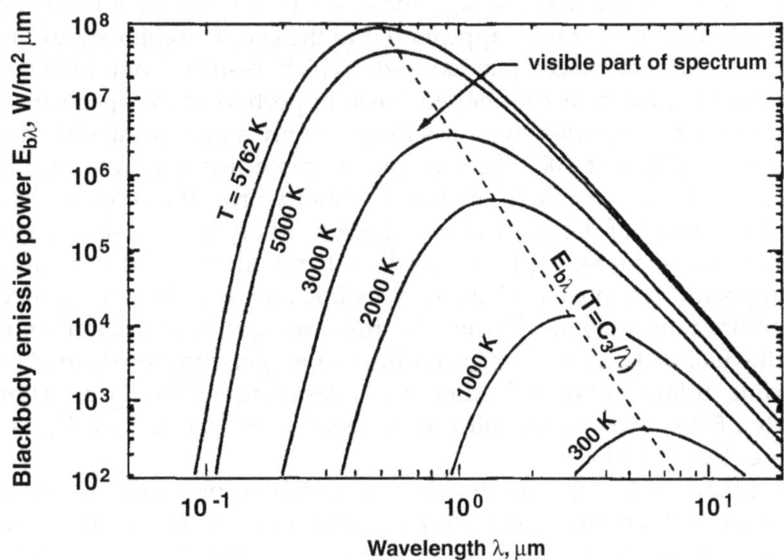

FIGURE 4.2
Variation of blackbody emissive power with wavelength. (From Hewitt, G.F., Shires, G.L., and Polezhaev, Y.V., Eds., *International Encyclopedia of Heat and Mass Transfer*, Copyright 1997, CRC Press, Boca Raton, FL. Reprinted with permission.)

manner. For example, the spectral distribution of solar radiation can be approximated by blackbody radiation at 5800 K with a maximum emissive power in the visible region of the spectrum. Also, at a fixed wavelength, E_λ decreases with surface temperature. For surface temperatures above approximately 700 K, a portion of the spectral emission lies within the visible range, whereas the spectra are shifted rightward outside the visible range below this temperature. Another interesting observation is that the spectral distribution of emissive power decreases on both sides of the maxima. Most electron activity is centered about a specific quantum level corresponding to a given wavelength and frequency of electromagnetic wave emission. Fewer fluctuations between energy states at other quantum levels lead to the decrease of spectral power away from the peak value.

The locus of maxima of E_λ at different wavelengths is described by *Wein's displacement law*. In order to find the wavelength where the emissive power is maximized at a given surface temperature, Planck's law is differentiated with respect to λ. Setting the resulting expression to zero and rearranging,

$$\lambda_{max} T = 0.002898 \text{ mK} \tag{4.5}$$

For example, $\lambda_{max} \approx 0.5$ μm for solar radiation. As the surface temperature decreases, λ_{max} increases, i.e., λ_{max} increases to 2.9 μm for a blackbody at 1000 K. At 1000 K the object appears red to the eye. The color should not be confused with an object painted red, which results from reflection of radiation from the object in the red (visible) portion of the spectrum.

If the spectral distribution passes through only a part of the visible range (such as an object at 1000 K), the color is discerned through the highest frequency photons visible to the eye. In other words, the color is observed from the tail end of the spectral distribution. Another example is tungsten, which is the material typically used in lamp filaments due to its high melting point. A tungsten filament at 2900 K appears white to the eye and λ_{max} occurs at approximately 1 μm. In this case, significant radiant emission from the filament occurs over the entire visible spectrum (mixture of colors, including bright yellow), leading to a resulting white light. After the tungsten filament is turned off and it cools down below 1300 K, its light becomes barely visible.

The fundamental equation for thermal radiation problems, given by the Stefan–Boltzmann law, is obtained by integration of the spectral distribution, $E_{\lambda,T}$, over all wavelengths. Changing variables by $x = C_2/(\lambda T)$ and integrating Equation (4.3),

$$E_b = \int_0^{\infty} \left(\frac{C_1 T^4}{C_2^4} \right) \frac{x^3 dx}{(e^x - 1)} = \sigma T^4 \tag{4.6}$$

where $\sigma = 5.67 \times 10^{-8}$ W/m^{2}K^4 is the Stefan–Boltzmann constant. This expression for E_b represents the total blackbody emissive power, or the radiative heat transfer, q_s, from the blackbody.

Based on the previous results for a blackbody, the following form of the Stefan–Boltzmann law is obtained for the radiative heat flux, q''_{rad}, emitted by a real surface of emissivity ϵ:

$$q''_{rad} = \epsilon \sigma T_s^4 \tag{4.7}$$

Furthermore, radiation exchange between two surfaces or objects can be approximated (linearized) and written in terms of an effective radiation heat transfer coefficient, h_r, as follows:

$$q''_{rad} = h_r (T_1 - T_2) \tag{4.8}$$

where

$$h_r = \epsilon \sigma (T_1 + T_2)(T_1^2 + T_2^2) \tag{4.9}$$

In this way, radiation and convection coefficients can be combined in an analysis involving both modes of heat transfer.

The emissive power represents the energy emitted per unit of actual (unprojected) surface area. A related concept, called the total blackbody radiation intensity, I_b, is defined on the basis of projected area. A beam of radiation originates from a point on the surface and travels in some direction, denoted by **s**, and upon arriving at another point, such as P, forms a circular cross-sectional plane, dA_n, with a projected unit normal vector, **n**. The growing conical (three-dimensional) beam is enclosed by a solid angle of $d\omega$, where

$$dw = \frac{dA}{s^2} \tag{4.10}$$

The units of this solid angle are steradians (sr).

If we then consider that d^2Q is an amount of radiant energy that passes (one way) past dA_n in the direction of $d\omega$, the resulting intensity of radiation at **s** in the direction of **d** is defined as

$$I(\mathbf{s},\ \mathbf{d}) = \frac{d^2Q}{dA_n d\omega} \tag{4.11}$$

The solid angle, $d\omega$, is defined in a way to conveniently identify the heat flux or radiation intensity across a particular projected area before it arrives at a surface.

For example, for radiation emanating from a flat surface in the direction of θ_1 (measured relative to the perpendicular direction to the plate) and arriving at a second horizontal surface located at a distance of L from the first surface,

$$d\omega = \frac{dA_2 cos(\theta_2)}{L^2} \tag{4.12}$$

where dA_2 and θ_2 refer to the surface area and the directional angle of the incoming radiation measured with respect to the vertical direction. In general, if all outgoing beams of radiation (leaving surface 1) are considered, then integration over the hemispherical range of angular directions gives

$$q_{rad} = \pi I_b \tag{4.13}$$

This result indicates that the radiative heat flux and the total blackbody radiation intensity are equivalent within a factor of π.

The emitted radiation in a specified band of the electromagnetic spectrum is a frequently required quantity. In particular, $F_{(0 \to \lambda)}$ is defined to be the fraction of total blackbody radiation in the wavelength range from $0 \to \lambda$.

This fraction can be computed from the integral of $E_{b,\lambda}$ over the appropriate wavelength range, i.e.,

$$F_{(0\to\lambda)} = \frac{\int_0^{\lambda} E_{b,\lambda} d\lambda}{\int_0^{\infty} E_{b,\lambda} d\lambda} = \int_0^{\lambda T} \left(\frac{E_{b,\lambda}}{\sigma T^5}\right) d(\lambda T) = f(\lambda T) \tag{4.14}$$

Since $E_{b,\lambda}$ is a function of λT only (using Planck's law), the band emission is written as a function of this product alone.

Two important properties of the blackbody functions are

$$F_{(\lambda_1\to\lambda_2)} = F_{(0\to\lambda_2)} - F_{(0\to\lambda_1)} \tag{4.15}$$

and

$$F_{(\lambda\to\infty)} = 1 - F_{(0\to\lambda)} \tag{4.16}$$

Sample values of $F_{(0\to\lambda)}$ at various products of λT are shown in Table 4.1.

Based on these blackbody functions, the fraction of emitted energy in any particular wavelength region can be determined. For example, consider the problem of finding the amount of visible energy emitted in the visible range (between 0.4 μm and 0.7 μm) by various sources, such as a tungsten filament (temperature of 2900 K) and the sun (temperature of about 5800 K). Given the required wavelength range, $\lambda_1 \leq \lambda \leq \lambda_2$, and temperature, the products $\lambda_1 T$ and $\lambda_2 T$ can be computed. Then the band emission table (Table 4.1) can be used to find the blackbody functions $F_{(0\to\lambda_1)}$ and $F_{(0\to\lambda_2)}$.

TABLE 4.1

Radiation Functions for a Blackbody

λT(μm·K)	$F_{(0\to\lambda)}$	λT(μm·K)	$F_{(0\to\lambda)}$
100	0.0000	6,000	0.738
1,000	0.0003	6,500	0.776
1,500	0.014	7,000	0.808
2,000	0.067	7,500	0.834
2,500	0.162	8,000	0.856
3,000	0.273	9,000	0.890
3,500	0.383	10,000	0.914
4,000	0.481	12,000	0.945
4,500	0.564	15,000	0.970
5,000	0.634	20,000	0.986
5,500	0.691	40,000	0.998

Finally, subtracting these two values gives the fraction of emitted energy in the specified range of wavelengths. For example, 37% of emitted solar energy and 7% of energy emitted by the tungsten filament are emitted in the visible range of the spectrum. Further clarification of these concepts is outlined in the following example.

Example 4.2.1

Energy Emitted in the Visible Range.

The purpose of this example is to find what wavelength range contains 80% of emitted solar energy. The appropriate range can be determined by leaving off the top and bottom 10% of the spectral distribution for a blackbody at 5800 K. The fraction of 0.9 is obtained for the blackbody function of $F_{(0\to1.19\ \mu m)}$, and a fraction of 0.1 is obtained for $F_{(0\to0.38\ \mu m)}$. As a result, 80% of the emitted solar energy lies between 0.38 μm and 1.19 μm. Other ranges could be obtained by leaving off different percentages from the bottom and top parts of the spectral distribution.

Radiative properties of the surface describe how an actual surface emits (ϵ), reflects (ρ), and transmits (τ) radiative energy. The radiative properties, ϵ (emissivity), ρ (reflectivity), α (absorptivity), and τ (transmissivity), vary with temperature, wavelength, and direction (Wood et al., 1964). Some properties represent the surface's characteristics relative to a blackbody. For example, ϵ represents the ratio of actual energy emitted by the surface to the energy emitted by a blackbody at the same temperature. Other definitions apply to ρ (fraction of irradiation reflected), α (fraction of irradiation absorbed), and τ (fraction of irradiation transmitted).

As discussed earlier, the spectral distribution of emissive power for a blackbody is given by Planck's law. This distribution is smooth with respect to increasing wavelength, but $E_\lambda(\lambda, T)$ for an actual surface (not a blackbody) has a lower magnitude and typically fluctuates with λ. Similar fluctuations are observed for radiative properties of materials. Various factors, such as surface roughness, surface coating, and material density, can lead to variations of surface scattering at different wavelengths, thereby yielding an emissivity of less than unity for an actual surface.

For a diffuse radiator, energy is emitted equally in all directions, but angular variations of emissivity are typically observed with real surfaces. For example, surface roughness elements block emissions along the direction parallel to the surface, and so the emissivity typically varies from 0 (along the surface direction) to a maximum value in the direction normal to the surface. The surface characteristics can be effectively designed or controlled to take advantage of angular variations of the radiative properties.

An example of how the directional distribution can be controlled is a lamp shade. The cover focuses the emitted radiation in a particular angular range. Similarly, a solar absorber can be designed with a corrugated surface to allow preferential directional properties with respect to incoming solar radiation. The proposed surface would absorb well in the direction of incoming radiation and poorly in other directions. For example, $\alpha_{\lambda,\theta} = \epsilon_{\lambda,\theta}$ is high for $\theta < 20°$, yet $\alpha_{\lambda,\theta}$ and $\epsilon_{\lambda,\theta}$ are low for $\theta \geq 20^o$. This surface emits or loses less energy than a smooth surface emitting well in all angular directions. In this way, more heat can be retained by the solar collector, thereby increasing its effectiveness.

The structure of the material also largely influences the radiation properties of the surface. Nonconductors are generally observed to have much higher surface emissivities than conductors (such as metals), except in a direction normal to the surface. This can be partly attributed to the faceted surfaces of nonconductors (such as ceramic surfaces) being smoother than nonfaceted surfaces (i.e., metals). A faceted interface typically arises with a low entropy of fusion during solidification of the material, thereby tending to reduce or close interatomic gaps. On the other hand, a high entropy of fusion during solidification of a nonfaceted material (such as metals) requires that the solidifying material absorbs crystals more easily from the solidifying liquid. A lower tendency to reduce interatomic gaps increases the surface roughness, thereby giving some evidence as to why conductors emit energy less effectively than nonconductors, except in the direction normal to the surface, where surface roughness is a less significant factor. In most engineering calculations, surface properties that represent directional averages will be used.

More specifically, the radiative properties will be defined as follows. The spectral emissivity, ϵ_λ, is defined as the actual surface emission at the surface temperature, T, divided by the surface emission at T for an ideal radiator (blackbody), i.e.,

$$\epsilon_\lambda(\lambda, T) = \frac{E_\lambda(\lambda, T)}{E_{b,\lambda}(\lambda, T)} = \frac{\int_0^{2\pi}\int_0^{\pi/2} \epsilon_{\lambda,\theta} cos(\theta) sin(\theta) d\theta d\phi}{\int_0^{2\pi}\int_0^{\pi/2} cos(\theta) sin(\theta) d\theta d\phi} \tag{4.17}$$

For example, the emissivity of stainless steel at 800 K decreases with wavelength beyond 0.4 μm. On the other hand, ϵ_λ remains approximately constant in the visible range for alumina at 1400 K, whereas it increases, reaches a maximum value, and subsequently decreases for wavelengths beyond 0.7 μm.

Also, the total emissivity, ϵ, is obtained by integrating the radiative emission across all wavelengths in the spectrum, i.e.,

$$\epsilon(T) = \frac{\int_0^\infty \epsilon_\lambda E_{b,\lambda} d\lambda}{\int_0^\infty E_{b,\lambda} d\lambda} = \frac{E(T)}{E_b(T)} = \frac{E(T)}{\sigma T^4} \tag{4.18}$$

In addition to surface emission, radiation can be reflected, absorbed, or transmitted through an object (see Figure 4.3). Energy absorbed by the object due to the incident electromagnetic waves may be subsequently transmitted through the object in the form of heat conduction within the solid.

We have used ϵ to denote emissivity, while ρ, α, and τ will refer to the reflectivity (fraction of radiation reflected by the object), absorptivity (fraction of radiation absorbed by the object), and transmissivity (fraction transmitted through the object), respectively. These quantities are defined analogously to Equation (4.18), but in terms of the incident radiation, G_λ, i.e.,

$$\rho(T) = \frac{\int_0^\infty \rho_\lambda G_\lambda d\lambda}{\int_0^\infty G_\lambda d\lambda} \tag{4.19}$$

$$\alpha(T) = \frac{\int_0^\infty \alpha_\lambda G_\lambda d\lambda}{\int_0^\infty G_\lambda d\lambda} \tag{4.20}$$

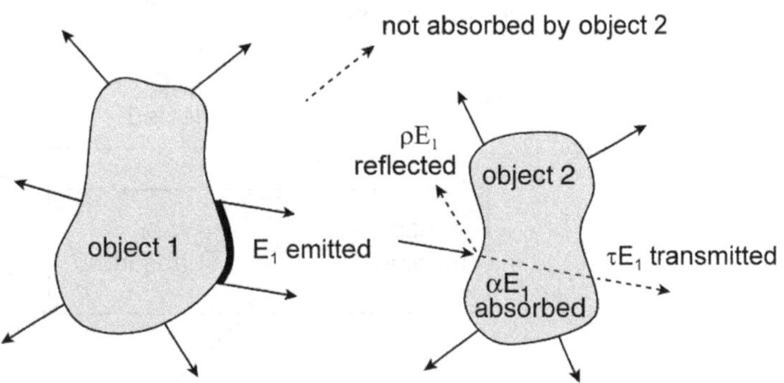

FIGURE 4.3
Radiation exchange between two objects.

$$\tau(T) = \frac{\int_0^\infty \tau_\lambda G_\lambda d\lambda}{\int_0^\infty G_\lambda d\lambda} \tag{4.21}$$

Based on conservation of energy for object 2 in Figure 4.3, the sum of reflectivity, absorptivity, and transmissivity relating to that object must equal unity (note: $\rho + \alpha + \tau = 1$ to be derived shortly).

The incident radiation on a surface is called the *irradiation*, G_λ. The sum of emitted radiation from a surface and the reflected irradiation is called *radiosity*, J_λ, where

$$J_\lambda = E_\lambda + \rho_\lambda G_\lambda \tag{4.22}$$

For a transparent medium, a portion of radiation ($\alpha_\lambda G_\lambda$) may be absorbed as heat in the object (see Figure 4.4). Furthermore, the remaining parts of the incoming radiation that are not reflected or absorbed may be transmitted through the object (in the amount of $\tau_\lambda G_\lambda$, where τ_λ refers to the spectral transmissivity). In other words, the total irradiation may be decomposed by

$$G_\lambda = G_{\lambda,ref} + G_{\lambda,abs} + G_{\lambda,tran} \tag{4.23}$$

Integrating these spectral quantities over all wavelengths,

$$G = G_{ref} + G_{abs} + G_{tran} \tag{4.24}$$

In addition, the respective parts of irradiation may be obtained by the total irradiation multiplied by the respective radiative property, i.e.,

$$G_\lambda = \rho_\lambda G_\lambda + \alpha_\lambda G_\lambda + \tau_\lambda G_\lambda \tag{4.25}$$

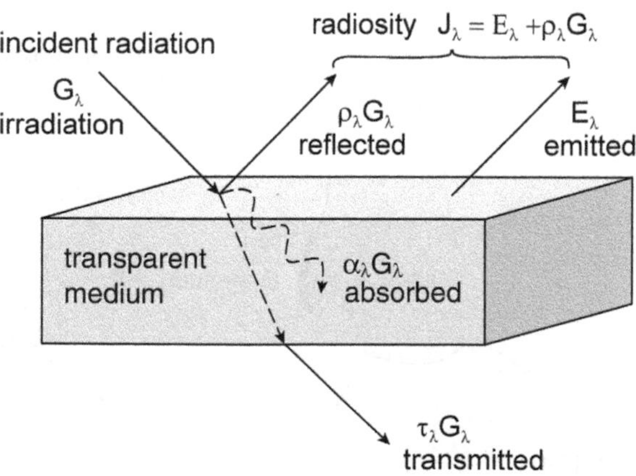

FIGURE 4.4
Schematic of irradiation and radiosity.

where ρ_λ, α_λ, and τ_λ refer to the spectral reflectivity, spectral transmissivity, and spectral absorptivity, respectively. Dividing Equation (4.25) by G_λ yields the familiar result that

$$\rho_\lambda + \alpha_\lambda + \tau_\lambda = 1 \tag{4.26}$$

Alternatively, integrating across all wavelengths,

$$\rho + \alpha + \tau = 1 \tag{4.27}$$

where

$$\alpha = \frac{\int_0^\infty \alpha_\lambda G_\lambda d\lambda}{\int_0^\infty G_\lambda d\lambda} \tag{4.28}$$

and the other integrated radiative properties are defined in an analogous manner.

In general, ρ_λ, α_λ, and τ_λ are weak functions of temperature. They are more dependent on the wavelength, or spectral distribution of the irradiation. For an *opaque medium*, $\tau_\lambda = \tau = 0$ (i.e., no transmission through the object). The spectral absorptivity is dependent on the surface's ability to absorb radiation in the wavelength range corresponding to the irradiation; in this way, α is dependent on the irradiation rather than the surface emission. On the other hand, ϵ_λ depends strongly on the surface temperature (i.e., surface emission), but not on the spectral distribution of the irradiation. These concepts are important in regard to evaluation of the radiative properties at the appropriate temperature (i.e., source or surface).

The spectral properties of materials and gases have many important practical implications. For example, hydrocarbon emissions from internal combustion engines and power plants have a high absorptivity (low τ_λ) in the long wavelength region of the spectrum. This feature is largely responsibly for the so-called *greenhouse effect*, where emissions from objects inside the atmosphere (relatively low temperatures, large λ) are effectively trapped because of the high α_λ of the atmospheric gases in the long wavelength region.

Another example of selective radiative properties is tinted glass, which is transparent in the visible range, but opaque in the infrared and ultraviolet regions. In terms of reflectivity, an aluminum film is almost completely reflective across all wavelengths of the spectrum, whereas white paint reflects well only between about 0.4 and 1 μm, and poorly otherwise. A surface covered with black paint is a poor reflector in all wavelength regions. Red paint alternates between good, moderate, and poor reflectivities across various ranges of the spectrum. Its properties are similar to those of brick materials consisting of clay, shale, silicon carbide, and other materials. The decrease of ρ_λ between 1 and 5 μm occurs partly because clay and shale in the material exhibit a high absorptivity in that range. A surface covered with black paint is a poor reflector, but it is a good absorber and emitter. Conversely, polished aluminum absorbs and emits radiation weakly above 0.4 μm as a result of its high reflectivity in that range. The following example applies these concepts to heat transfer from an object in a vacuum chamber.

Example 4.2.2

Small Object in a Vacuum Chamber.

A small hot object of mass M is cooled by radiation exchange with walls of an evacuated chamber at temperatures of T_w. The purpose of this example is to find the governing equation describing heat transfer from the object. It may be assumed that the Biot number is small (i.e., $Bi < 0.1$; lumped capacitance approximation valid) and the cavity can be represented as a blackbody.

In this problem, the net incident radiation onto the object is the amount of irradiation (from the walls at T_w) absorbed by the object less the net radiation emitted by the surface of the object. Based on an energy balance, which requires that the rate of energy decrease of the object balances the net outflow of heat by radiation from the object,

$$Mc_p \frac{dT}{dt} = A_s[\alpha G(T_w) - \epsilon E_b(T)] = A_s[\alpha \sigma T_w^4 - \epsilon \sigma T^4] \tag{4.29}$$

where A_s refers to the surface area of the object. This governing equation is a first-order nonlinear differential equation that can be solved to find $T(t)$ once the initial temperature of the object is specified (assuming constant radiative properties).

An important question arises from close consideration of the previous example: *Under what conditions is* $\alpha = \epsilon$ *true?* In other words, the question may be expressed in the following manner:

$$\epsilon_\lambda \equiv \frac{\int_0^{2\pi}\int_0^{\pi/2} \epsilon_{\lambda,\theta} cos(\theta) sin(\theta) d\theta d\phi}{\int_0^{2\pi}\int_0^{\pi/2} cos(\theta) sin(\theta) d\theta d\phi} ?=? \frac{\int_0^{2\pi}\int_0^{\pi/2} \alpha_{\lambda,\theta} I_\lambda cos(\theta) sin(\theta) d\theta d\phi}{\int_0^{2\pi}\int_0^{\pi/2} I_\lambda cos(\theta) sin(\theta) d\theta d\phi} \equiv \alpha_\lambda \tag{4.30}$$

$$\epsilon = \frac{\int_0^\infty \epsilon_\lambda E_{b,\lambda}(\lambda, T) d\lambda}{\int_0^\infty E_{b,\lambda}(\lambda, T) d\lambda} = \frac{\int_0^\infty \alpha_\lambda G_\lambda(\lambda) d\lambda}{\int_0^\infty G_\lambda(\lambda) d\lambda} = \alpha \tag{4.31}$$

A starting point of assessing the validity of the above equalities is *Kirchhoff's law*, which states that the emissivity at a given wavelength, λ, and direction, θ, must equal the absorptivity at that given wavelength and direction. In mathematical terms,

$$\epsilon_{\lambda,\theta} = \alpha_{\lambda,\theta} \tag{4.32}$$

For *diffuse irradiation* (i.e., I_λ is independent of direction) or a *diffuse surface* (i.e., surface emission is independent of direction), Equation (4.32) becomes $\epsilon_\lambda = \alpha_\lambda$. In other words, the spectral properties are identical if either the surface or irradiation is diffuse (independent of direction).

In addition to these diffuse properties that yield directional independence, two further conditions provide spectral independence:

1. Irradiation arrives from a blackbody at the same temperature as the incident surface. In this way, $G_\lambda = E_{b,\lambda}$ and the previously defined integrals in Equation (4.31) become identical.
2. Alternatively, for a *gray surface*, ϵ_λ and α_λ are both constant. If ϵ_λ is constant, then $\epsilon = \epsilon_\lambda$. Similarly, if α_λ is constant, then $\alpha = \alpha_\lambda$, which requires that $\epsilon = \alpha$ since $\epsilon_\lambda = \alpha_\lambda$.

Under either condition (1) or (2), together with the diffuse property, we have the result that $\epsilon = \alpha$ for a *diffuse gray surface*. A diffuse gray surface is a surface exhibiting property characteristics with both directional and wavelength (spectral) independence. In other words, ϵ and α are identical at all angles and wavelengths. A *gray surface* is a surface where ϵ_λ and α_λ are independent of λ over the dominant spectral regions of G_λ and E_λ.

These results may be summarized in the following way:

- Kirchhoff's law requires that $\epsilon_{\lambda,\theta} = \alpha_{\lambda,\theta}$ for any surface.
- For a diffuse surface, $\epsilon_\lambda = \alpha_\lambda$.
- For a diffuse gray surface, $\epsilon = \epsilon_\lambda = \alpha_\lambda = \alpha$.

These definitions are further clarified through the following example.

Example 4.2.3

Gray Surfaces Exposed to Solar Radiation.

Consider the radiative properties of the following four surfaces:

1. $\alpha_\lambda = 0.4$ within $3 < \lambda < 6$ μm and 0.9 otherwise.
2. $\alpha_\lambda = 0.7$ for $\lambda < 3$ μm and 0.5 otherwise.
3. $\alpha_\lambda = 0.2$ for $\lambda < 3$ μm, $\alpha_\lambda = 0.6$ for $\lambda > 6$ μm, and $\alpha_\lambda = 0.8$ otherwise.
4. $\alpha_\lambda = 0.7$ within $3 < \lambda < 6$ μm and 0.1 otherwise.

The purpose of this problem is to determine which diffuse surfaces at a temperature of 300 K are gray when they are exposed to solar radiation.

From the electromagnetic spectrum, approximately 98% of solar irradiation occurs at $\lambda < 3$ μm, whereas 96% of the emission from a surface at 300 K occurs at $\lambda > 6$ μm (with the peak occurring at approximately 10 μm). Thus, in comparing the various surfaces, a gray surface is identified when the absorption of the incoming energy over the spectral range of the irradiation matches the emission from the source of irradiation (i.e., sun) over the spectral range of the source.

The surfaces possess the following characteristics:

- $\alpha_\lambda = 0.9$ for G_λ and $\epsilon_\lambda = 0.9$ for E_λ (gray).
- $\alpha_\lambda = 0.7$ for G_λ and $\epsilon_\lambda = 0.5$ for E_λ (not gray).
- $\alpha_\lambda = 0.2$ for G_λ and $\epsilon_\lambda = 0.6$ for E_λ (not gray).
- $\alpha_\lambda = 0.1$ for G_λ and $\epsilon_\lambda = 0.1$ for E_λ (gray).

From the previous results, it can be observed that the property values of gray surfaces must match each other in the appropriate regions of the spectrum where the radiative transfer occurs.

4.3 Radiation Exchange between Surfaces

In many practical problems involving radiation heat transfer, only a portion of radiation emitted from a surface arrives at another surface. The *view factor*, F_{ij}, is defined as the fraction of radiation leaving surface i that is intercepted by surface j, i.e.,

$$F_{ij} = \frac{q_{ij}}{A_i J_i} \tag{4.33}$$

where J_i is the radiosity of surface i. The view factor is often called the radiation *shape factor* or the *configuration factor*. It will be used to calculate the appropriate radiation exchange between surfaces simultaneously emitting, absorbing, reflecting, or transmitting radiation.

Consider a surface of area dA_i at a temperature of T_i emitting a beam of radiation toward an element of surface area dA_j at T_j and a distance R away. The normal vectors to each surface are at angles of θ_i and θ_j, respectively, with respect to the line joining each of the differential surface elements (see Figure 4.5). The solid angle formed by a beam of radiation leaving surface i and spreading and arriving at surface j is given by

$$d\omega_{ij} = \frac{dA_j cos(\theta_j)}{R^2} \tag{4.34}$$

Then the radiative energy arriving at surface j is written as follows:

$$dq_{ij} = I_i cos(\theta_i) dA_i d\omega_{ji} \tag{4.35}$$

Using Equation (4.13), the intensity of radiation, I_i, in Equation (4.35) may be written as the radiosity, J_i (consisting of radiative emissions and reflections), divided by π.

Then, combining Equations (4.34) and (4.35), the total radiation emitted from surface i that arrives at surface j becomes

$$q_{ij} = \int dq_{ij} = J_i \int_{A_i}\int_{A_j} \frac{cos(\theta_i)cos(\theta_j)}{\pi R^2} \, dA_i dA_j = F_{ij} A_i J_i \tag{4.36}$$

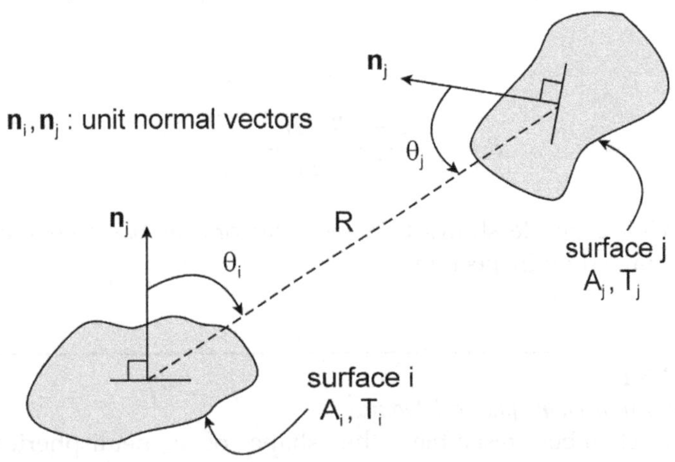

FIGURE 4.5
Schematic of view factor between two surfaces.

Thus, the view factor can be expressed in the following form:

$$F_{ij} = \frac{1}{A_i} \int_{A_i} \int_{A_j} \frac{cos(\theta_i)cos(\theta_j)}{\pi R^2} dA_i dA_j \tag{4.37}$$

which represents an integral involving geometrical parameters only (see Figure 4.5).

In a similar way, F_{ji} refers to the radiation leaving surface j and arriving at surface i, where

$$F_{ji} = \frac{1}{A_j} \int_{A_i} \int_{A_j} \frac{cos(\theta_i)cos(\theta_j)}{\pi R^2} dA_i dA_j \tag{4.38}$$

Comparing Equations (4.37) and (4.38), we obtain the following *reciprocity relation*:

$$A_i F_{ij} = A_j F_{ji} \tag{4.39}$$

which conveniently allows us to write one shape factor in terms of the other factor based on the respective area ratio between both surfaces.

View factors for basic geometries can be derived analytically or by other means (Howell, 1982). For example, the view factor between coaxial parallel disks (i.e., disk 1 of radius R_1 at a spacing of H below another disk of radius R_2) can be expressed as

$$F_{12} = \frac{1}{2}\left[B - \sqrt{B^2 - 4(R_2/R_1)^2}\right] \tag{4.40}$$

where

$$B = 1 + \frac{1 + (R_2/H)^2}{(R_1/H)^2} \tag{4.41}$$

The following example shows that view factors for basic geometries can often be obtained by inspection.

Example 4.3.1

View Factors in a Hemispherical Dome.

A heating chamber resembles the shape of a hemispherical dome subdivided into three separate surfaces. The first surface is the base section of the dome (surface 1), and surfaces 2 and 3 are the upper halves of the hemispherical chamber. Both upper halves have the same shape and size.

The purpose of this example is to find the view factors for radiation exchange between each of the surfaces.

By inspection, it can be observed that all of the radiation leaving surface 1 arrives at the other surfaces, i.e., $F_{1-23} = 1$. Also, the radiation leaving the base (surface 1) is equally distributed among the upper surfaces due to symmetry, so $F_{12} = 1/2$ and $F_{13} = 1/2$.

None of the radiation leaving surface 1 arrives back at that surface (excluding reflections from other surfaces) since it is a flat surface. The view factors for flat and convex surfaces, with respect to themselves, are zero, but concave surfaces may involve radiation arriving back upon itself. If the base surface was concave, then the previously described view factors of 1/2 would have to be smaller since some of the radiation would be intercepted by surface 1. The remaining view factors in this problem can be inferred from the previously obtained view factors, based on the reciprocity relation between respective surfaces.

Other useful relations can be obtained for radiation exchange between surfaces inside enclosures. Considering an enclosure with n surfaces (see Figure 4.6), an energy balance for surface 1 is given by

$$J_1 A_1 = q_{11} + q_{12} + \ldots + q_{1n} \tag{4.42}$$

The first term may be nonzero if it is a concave surface (as discussed in previous example). The conservation of energy in Equation (4.42) states that the energy leaving surface 1 arrives at surface 1, surface 2, etc. up to surface

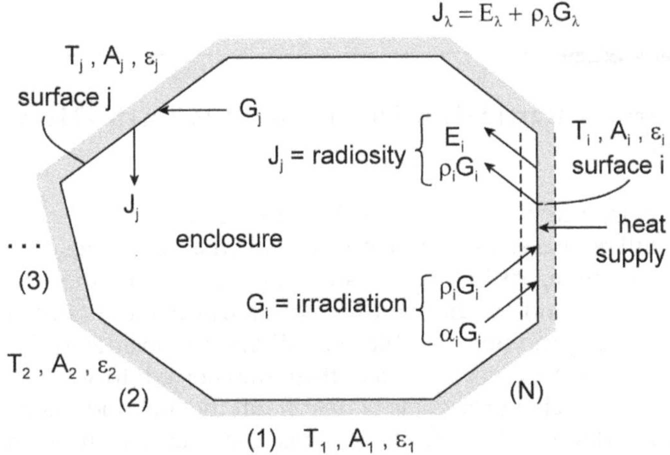

FIGURE 4.6
Radiation exchange in an enclosure.

n. Dividing Equation (4.42) by $J_i A_i$ and generalizing from surface 1 to surface i,

$$1 = \sum_{j=1}^{n} \left(\frac{q_{ij}}{A_i J_i} \right) \tag{4.43}$$

In Equation (4.43), $i = 1, 2, 3, \ldots n$. The lower limit in the summation represents radiation leaving surface i and arriving at surfaces $j = 1, 2, 3, \ldots n$.

Observing that the parenthetical term in Equation (4.43) is the view factor, we obtain the following *summation relation*:

$$\sum_{j=1}^{n} F_{ij} = 1 \tag{4.44}$$

Alternatively, Equation (4.44) is called the *enclosure relation*, which is valid only for radiation exchange occurring within enclosures. It should be noted that $F_{ii} \neq 0$ for concave surfaces, but $F_{ii} = 0$ for plane or convex surfaces. As described earlier, a convex surface does not intercept any of its outgoing radiation. The total heat flow is obtained once the respective view factors are known. In the following section, heat transfer within enclosures is further discussed, particularly for enclosures with internal diffuse gray surfaces.

4.4 Thermal Radiation in Enclosures with Diffuse Gray Surfaces

In this section, radiation exchange between diffuse gray surfaces in an enclosure will be presented. It will be assumed that the enclosure consists of isothermal, opaque, diffuse gray surfaces (i.e., $\tau = 0$ and $\alpha = \epsilon$). Also, uniform radiosity and irradiation over each surface and a nonparticipating (i.e., nonscattering and nonabsorbing) medium are considered. The purpose of this section is to determine the heat transferred to each surface (or temperature of each surface) as a result of the net radiation exchange between all surfaces in the enclosure. Since each surface emits and absorbs radiation simultaneously in conjunction with other surfaces, it is anticipated that a system of simultaneous, discrete equations describing this exchange will need to be solved.

Consider an enclosure consisting of n surfaces (see Figure 4.6). The temperature, area, and emmisivity of surface i are denoted by T_i, A_i and ϵ_i, respectively. Based on an energy balance for this surface,

$$q_i = A_i(J_i - G_i) \tag{4.45}$$

where J_i, G_i, and q_i refer to the radiosity, irradiation, and heat transfer required to maintain surface i at T_i, respectively. The energy balance in Equation (4.45) requires that the energy needed to maintain surface i at T_i is the net energy leaving the surface (i.e., radiosity minus irradiation).

The radiosity consists of the sum of emitted radiation, E_i, and reflected radiation, $\rho_i G_i$ (see Figure 4.6). Also, the irradiation may be decomposed into a reflected component, $\rho_i G_i$, and an absorbed part, $\alpha_i G_i$, since none of the incident radiation is transmitted through the surface. Since each surface is diffuse gray ($\alpha_i = \epsilon_i$) and opaque ($\tau_i = 0$),

$$\rho_i + \alpha_i = \rho_i + \epsilon_i = 1 \tag{4.46}$$

Using this result, the radiosity can be expressed in the following form:

$$J_i = E_i + \rho_i G_i = \epsilon_i E_{bi} + (1 - \epsilon_i) G_i \tag{4.47}$$

The blackbody radiation emitted, i.e., E_{bi} in Equation (4.47), can be evaluated by the Stefan–Boltzmann law based on the surface temperature of T_i.

Substituting the expression for G_i from Equation (4.47) into Equation (4.45) and rearranging terms,

$$q_i = \frac{E_{bi} - J_i}{(1 - \epsilon_i)/(\epsilon_i A_i)} \tag{4.48}$$

This result is given in a convenient form since it appears analogous to earlier analyses of thermal networks where thermal resistances were written in the form of a temperature difference divided by a thermal resistance. In Equation (4.48), the heat transfer to or from surface i is represented as a potential difference divided by the surface resistance. The surface resistance represents the real surface behavior (J_i), opposed to a blackbody surface (E_{bi}), at the same temperature. It approaches the correct limiting behavior of zero resistance as $\epsilon \to 1$, and conversely, a low emissivity correctly yields a high surface resistance to radiative emission and heat transfer. Using this resistance, thermal circuits involving radiation can be constructed in series or parallel similarly to previously described thermal circuit analyses.

More specifically, let us consider the radiation exchange between individual surfaces inside the enclosure (see Figure 4.7). The radiosity

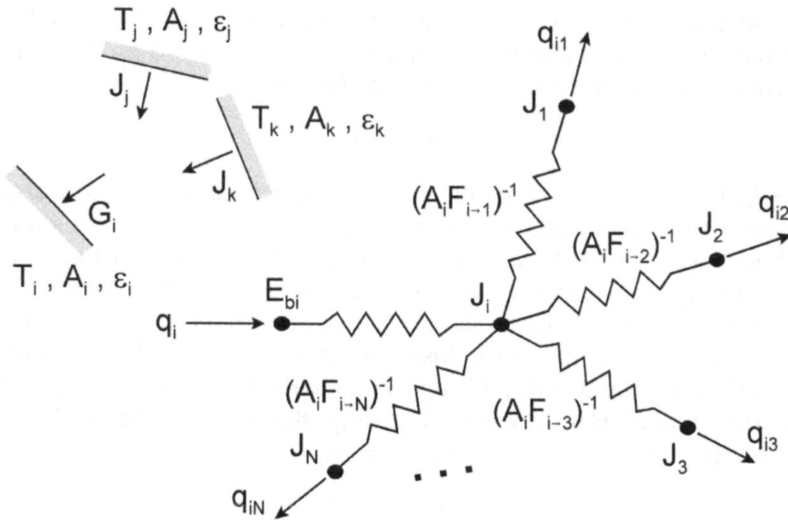

FIGURE 4.7
Radiation exchange between surfaces inside an enclosure.

leaving a particular surface eventually becomes part of the irradiation on another surface within the enclosure. In particular, considering all irradiation arriving on a specific surface,

$$A_i G_i = \sum_{j=1}^{n} F_{ji}(A_j J_j) \tag{4.49}$$

This equality suggests that the irradiation arriving on surface i consists of radiation leaving all other surfaces (i.e., surfaces $j = 1, 2, \ldots n$). The fraction of radiation arriving from surface j is multiplied by the area and radiosity of that surface.

Using the reciprocity relation in Equation (4.49) to write the energy balance in terms of the area A_i and then dividing by this area,

$$G_i = \sum_{j=1}^{n} F_{ij} J_j \tag{4.50}$$

This result indicates that the irradiation arriving on surface i consists of the sum of view factor-weighted radiosity contributions from all surfaces within the enclosure. Furthermore, combining Equations (4.47) to (4.50),

$$J_i = \epsilon T_i^4 + (1 - \epsilon_i) \sum_{j=1}^{n} F_{ij} J_j \tag{4.51}$$

which means that the radiosity leaving surface i includes the emitted radiation and the reflected irradiation.

The heat transfer to or from surface i can be expressed in the following form, based on Equations (4.45) and (4.50):

$$q_i = A_i J_i - \sum_{j=1}^{n} A_i F_{ij} J_j \tag{4.52}$$

Thus, the heat required to maintain surface i at T_i balances the radiation leaving surface i minus the radiation arriving at surface i from all other surfaces. This energy balance still describes the same principle of radiation exchange as Equation (4.45), but in terms of radiosity and view factors instead, since the specific exchange between individual surfaces is outlined in detail.

Using the summation relation with Equation (4.52),

$$q_i = A_i \left[\sum_{j=1}^{n} F_{ij} J_i - \sum_{j=1}^{n} F_{ij} J_j \right] = A_i \sum_{j=1}^{n} F_{ij}(J_i - J_j) \equiv \sum_{j=1}^{n} q_{ij} \tag{4.53}$$

This result shows that the net heat loss or gain by surface i balances the sum of all net radiative heat exchanges between surface i and all other surfaces. It can be written in a more convenient form as follows:

$$q_i = \sum_{j=1}^{n} \frac{J_i - J_j}{1/(A_i F_{ij})} \tag{4.54}$$

which indicates that the heat transfer is given by the sum of potential differences divided by spatial differences arising from radiation exchange with each surface (see Figure 4.7).

The standard form of Equation (4.54) allows a thermal circuit involving radiation exchange to be constructed. For example, the heat flow into surface i (node i of network) experiences a surface resistance of $(1 - \epsilon_i)/(\epsilon_i A_i)$ with a radiosity of J_i, and then a group of spatial resistances, $(A_i F_{ij})^{-1}$, in parallel, corresponding to a heat flow of q_{ij} between surfaces i and j. If any of the surfaces are blackbodies, then the same previously described equations may be applied, but J_i is replaced with E_{bi} and $\epsilon_i = \alpha_i = 1$. The following example shows how the previously described results can be used to find the heat transfer in a simplified two-surface enclosure.

Example 4.4.1

Two Diffuse Gray Surfaces in an Enclosure.

An enclosure is approximated by two diffuse gray surfaces. The first and second surfaces are maintained at temperatures T_1 and T_2, respectively. If the surface emissivities of surfaces 1 and 2 are ϵ_1 and ϵ_2, respectively, find the heat transfer from surface 1 to 2.

The heat exchange between surfaces 1 and 2 may be written as the potential difference between these surfaces divided by the sum of resistances, i.e.,

$$q_1 = q_{12} = q_2 = \frac{E_{b1} - E_{b2}}{\left(\frac{1-\epsilon_1}{A_1\epsilon_1}\right) + \left(\frac{1}{A_1F_{12}}\right) + \left(\frac{1-\epsilon_2}{A_2\epsilon_2}\right)} \tag{4.55}$$

where the denominator includes the sum of surface and space resistances. Rearranging this result,

$$q_{12} = \frac{\epsilon_1 A_1 \sigma(T_1^4 - T_2^4)}{(1-\epsilon_1) + \left(\frac{\epsilon_1}{F_{12}}\right) + \left(\frac{A_1\epsilon_1}{A_2\epsilon_2}\right)(1-\epsilon_2)} \tag{4.56}$$

This general result may be used for any two diffuse gray surfaces forming an enclosure.

Special cases can be obtained from the previously described example:

1. For large parallel plates (surfaces 1 and 2), the view factor is $F_{12} = 1$ and $A_1 = A_2$. Substituting these results into Equation (4.56),

$$q_{12} = \frac{\epsilon_1 A_1 \sigma(T_1^4 - T_2^4)}{1 + \epsilon_1/\epsilon_2 - \epsilon_1} \tag{4.57}$$

2. For long concentric cylinders (of radii r_1 and r_2), the view factor is $F_{12} = 1$ and $A_1/A_2 = r_1/r_2$. The heat exchange is obtained as

$$q_{12} = \frac{\epsilon_1 \sigma A_1 (T_1^4 - T_2^4)}{1 + \left(\frac{r_1\epsilon_1}{r_2\epsilon_2}\right)(1-\epsilon_2)} \tag{4.58}$$

3. For a small object (surface 1) in an enclosure (surface 2), we have $A_1 \ll A_2$, $F_{12} = 1$, and

$$q_{12} = A_1 \epsilon_1 \sigma (T_1^4 - T_2^4) \tag{4.59}$$

The previous equations have allowed us to determine either the temperature or heat transfer to each surface. In summary, the following equations and procedures may be adopted for the analysis of radiative exchange between diffuse gray surfaces in an enclosure:

$$J_i = \epsilon_i \sigma T_i^4 + (1 - \epsilon_i) \sum_{j=1}^{n} F_{ij} J_j \tag{4.60}$$

$$q_i = \sum_{j=1}^{n} A_i F_{ij} (J_i - J_j) \tag{4.61}$$

where $i = 1, 2, 3 \ldots n$. Equation (4.60) requires that the radiosity must balance the emission and reflection of irradiation from the surface. In Equation (4.61), the net heat loss or gain balances the sum of radiation exchanges between surface i and the other surfaces.

Various solution methods can now be summarized for three types of problems as follows:

1. All T_i are given, and all q_i must be found. Use Equation (4.60) to find all J_i by a simultaneous solution of the linear algebraic equations. Then use Equation (4.61) to find the resulting q_i quantities.
2. All q_i are given, and all T_i must be found. Use Equation (4.61) to compute all J_i simultaneously. Then the temperatures, T_i, can be obtained from Equation (4.60).
3. Some q_i and T_i are given, and the remaining q_i and T_i must be found. Use the procedures in problems 1 and 2 as they apply to a given surface.

Although a thermal network becomes more complicated with many surfaces, it is a useful approach for analyzing enclosures with fewer than five surfaces. In any case, the resulting system of n linear equations may be solved by a computer algorithm or a type of iterative method (such as a Gauss–Seidel scheme). Further details are available in books by Siegel, Howell (1993), and Incropera and DeWitt (1996).

In some applications, it is desirable to reduce the net radiation heat transfer between surfaces. A *radiation shield* is a surface constructed from a low ϵ (high ρ) material to reduce the radiation exchange between the surfaces. Some typical examples include a screen or cover on a fireplace or a microwave oven. A radiation shield can be effectively designed with different emissivities on both sides of the shield. If a radiation shield is placed between two objects, the net radiation heat transfer between the objects consists of the potential difference, $E_{b1} - E_{b2}$, divided by the sum of appropriate thermal resistances, including surface and shape resistances.

For example, the net heat transfer rate, q_{12}, between two parallel plates of equal area, separated by a radiation shield, is given by

$$q_{12} = \frac{A_1 \sigma (T_1^4 - T_2^4)}{1/\epsilon_1 + 1/\epsilon_2 + (1 - \epsilon_{s1})/\epsilon_{s1} + (1 - \epsilon_{s2})/\epsilon_{s2}} \tag{4.62}$$

where the subscripts 1, 2, $s1$, and $s2$ refer to plate 1, plate 2, shield (side 1), and shield (side 2), respectively. It can be observed that smaller ϵ_{s1} and ϵ_{s2} values lead to an increased total thermal resistance in the denominator, thereby reducing the net heat transfer between the objects. As a result, a radiation shield can effectively reduce the radiation exchange by independently controlling the shield emissivities.

4.5 Solar Energy

4.5.1 Components of Solar Radiation

Solar radiation is an essential source of energy for life on Earth. Also, it constitutes an important source of energy in many engineering technologies. The amount of solar radiation arriving at the top of the Earth's atmosphere is called the *solar constant*, $G_s = 1353 \text{ W/m}^2$. This amount can be readily derived from an energy balance between concentric spheres (i.e., outer edge of the sun and a sphere encompassing the Earth's orbit). The energy emitted across the surface of the sun's spherical edge balances the energy passing across a much larger spherical surface area formed around the orbit of the Earth. The spectral distribution of this extraterrestrial irradiation (outside of the Earth's atmosphere) can be closely approximated by the blackbody emissive power of an emitting body at a temperature of 5800 K.

The solar constant represents the energy flux (per unit surface area) in the direction of the incoming radiation from the sun. However, the radiation component actually passing through the top of the atmosphere, G_o, is the component normal to the relevant surface. In other words, G_s must be multiplied by $cos(\theta)$, where θ refers to the angle between the incident radiation and the tangent to the top surface of the atmosphere. This can be envisioned by a plate of an area of 1 m^2 held at the top of the atmosphere. In this case, the plate is nearly parallel to the sun's incoming radiation near the north and south poles of the Earth, and thus $cos(\theta) \approx 0$ correctly yields little or no heat passing downwards through the atmosphere there. On the other hand, near the equator, $cos(\theta) \approx 1$, and the full amount of solar radiation passes through the top of the atmosphere.

In free space (i.e., evacuated space outside of the atmosphere), the intensity of emitted radiation, I_λ, from the sun remains constant in a particular path of travel since little or no scattering or absorption of radiation occurs there. However, once the radiation passes through the top of the atmosphere, absorption and scattering of radiation by dust particles, moisture, etc. reduces the radiant intensity with distance traveled. If 100% of the incident radiation arrives at the top of the atmosphere, the remaining parts of the radiation are distributed as follows:

- 1–6% is scattered back to space.
- 11–23% is absorbed by the atmosphere (i.e., gases such as O_3, O_2, H_2O, and CO_2, dust, and aerosols).
- 5–15% is diffuse (scattered throughout the sky and arriving on the surface of the Earth).
- 56–83% is radiation arriving directly on the Earth's surface.

The combined direct and diffuse radiation is called the *global irradiation* since only this amount arrives at the Earth's surface. For the diffuse component, sky radiation may be approximated as radiation emitted by a blackbody (i.e., $\epsilon = 1$) at a sky temperature between 230 and 285 K.

Radiation emitted from the Earth's surface (at a temperature of $250 \text{ K} < T < 320 \text{ K}$) occurs predominantly in the long wavelength part of the spectrum, $4 \leq \lambda \leq 40$ μm. Using the Stefan–Boltzmann law to find the Earth's emitted radiation, E,

$$E = \epsilon \sigma T^4 \tag{4.63}$$

where typical values of emissivity, ϵ, are 0.97 for water and 0.93–0.96 for soil. For the atmospheric emission, represented by G_{sky},

$$G_{sky} = \sigma T_{sky}^4 \tag{4.64}$$

In this case, the irradiation occurs predominantly in the ranges of $5 \leq \lambda \leq 8$ μm and $\lambda \geq 13$ μm.

The previously cited 11–23% absorption by the atmosphere can be further understood by considering the spectral distribution of the sun's radiation as seen through the Earth's atmosphere. Outside the atmosphere, the solar radiation closely resembles a blackbody at 5800 K, but within the atmosphere, radiative absorption renders it more closely to a gray body with a reduced effective emissivity. Absorption by ozone is strong in the ultraviolet region. This partly explains why ozone depletion by certain chemicals (such as escaped CFCs from refrigeration systems) is speculated to be harmful, since this depletion allows more direct solar ultraviolet irradiation to pass through the atmosphere.

Unlike the continuous, idealized spectral distribution of a blackbody, the distribution of solar radiation is banded (i.e., occurring in discrete wavelength bands). This is partly explained by individual constituents in the atmosphere, such as H_2O and CO_2, whose radiative properties vary widely with wavelength. A detailed understanding of the banding phenomenon may be gained through consideration of electron activity at the quantum level. For a particular constituent in the atmosphere (such as H_2O), vibrational molecular motion emits electromagnetic waves in a certain frequency range, whereas rotational motions may emit electromagnetic waves over another discrete range of frequencies. In between these discrete frequency ranges, there may be little or no electromagnetic activity, thereby leaving a band in the electromagnetic spectrum. The quantization of energy provides explanations for the observed bands in the spectral distribution of radiation.

In view of solar radiation having a notably different spectral distribution than the Earth's emissions, the gray surface assumption is rarely justifiable. The solar irradiation, $G_{\lambda,sol}$, is largely located below wavelengths of 4 μm and centered at about 0.5 μm, whereas the Earth's emission, $E_{\lambda,earth}$, is mainly located above 4 μm and centered at about 10 μm. As a result, $\alpha_s \neq \epsilon$ since the spectral distributions of solar irradiation and the Earth's emissions are located in different spectral regions. Instead of the gray surface assumption, α_s/ϵ ratios are typically tabulated and listed for different surfaces. As an example for a solar collector or absorber plate, it would be desirable to use a material with a high α_s/ϵ ratio since this type of material would absorb well, while losing relatively less heat by radiation due to a low surface emissivity.

On the other hand, the gray surface assumption is reasonable for many surfaces involving sky irradiation, $G_{\lambda,sky}$, and the Earth's emission, $E_{\lambda,earth}$. Most sky irradiation is located within the same wavelength range as the Earth's emission since the temperatures involved in both cases are close in comparison to the previously described solar irradiation and Earth's emissions. As a result, $\alpha_{sky} = \epsilon$ is a reasonable assumption when considering the radiative exchange between the sky and the Earth's surface.

Example 4.5.1.1

Heated Surface Exposed to Solar and Atmospheric Irradiation.

A 10-cm-square metal plate with an electrical heater on its back side is placed firmly against the ground in a region where the Earth temperature and effective sky temperature are both 285 K. The plate is exposed to direct solar irradiation of 800 W/m^2 and an ambient airstream at 295 K flowing at 5 m/sec along the plate. The plate emissivity is $\epsilon_\lambda = 0.8$ for $0 < \lambda < 2$ μm and $\epsilon_\lambda = 0.1$ for $\lambda > 2$ μm. What is the electrical power required to maintain the plate surface at a temperature of 345 K?

In this problem, the following assumptions will be adopted: (i) steady-state conditions and (ii) an isothermal, diffuse plate surface. The plate surface emission and atmospheric irradiation both occur in the long wavelength part of the spectrum, whereas the solar irradiation is centered at approximately 0.5 μm (low wavelength part of the spectrum). Performing an energy balance on the plate,

$$\alpha_s G_s + \alpha_a G_a + q''_{elec} = \epsilon \sigma T_s^4 + q''_{conv} \tag{4.65}$$

where the terms represent (from left to right) solar irradiation, atmospheric irradiation, electrical heat supplied, surface emission, and convective heat losses.

The absorptivity of solar radiation, α_s, is computed by

$$\alpha_s = \frac{\int_0^\infty \alpha_\lambda G_{\lambda,sun} d\lambda}{\int_0^\infty G_{\lambda,sun} d\lambda} = \frac{\int_0^\infty \alpha_\lambda E_{b,\lambda}(5800K) d\lambda}{\int_0^\infty E_{b,\lambda}(5800K) d\lambda} \tag{4.66}$$

$$\alpha_s = \frac{\int_0^2 \alpha_\lambda E_{b,\lambda} d\lambda}{\int_0^\infty E_{b,\lambda} d\lambda} + \frac{\int_2^\infty \alpha_\lambda E_{b,\lambda} d\lambda}{\int_0^\infty E_{b,\lambda} d\lambda} = 0.8 F_{(0\to 2\ \mu m)} + 0.1[1 - F_{(0\to 2\ \mu m)}] \tag{4.67}$$

Using the blackbody function table (Table 4.1) with $\lambda T = 2 \times 5800 = 11{,}600$ μmK, we find that $F_{(0\to\lambda T)} = 0.94$. Then Equation (4.67) yields an absorptivity of $\alpha_s = 0.758$. Thus, the surface absorbs about 76% of the incident solar energy.

The absorptivity for atmospheric irradiation and the plate emissivity are computed in a similar manner. In particular, for atmospheric irradiation,

$$\alpha_{atm} = \frac{\int_0^{\infty} \alpha_\lambda G_{\lambda,atm} d\lambda}{\int_0^{\infty} G_{\lambda,atm} d\lambda} = 0.8F_{(0\to 2\ \mu m)} + 0.1[1 - F_{(0\to 2\ \mu m)}] \tag{4.68}$$

In this case, the product λT in the blackbody function is obtained as $\lambda T = 2 \times 285 = 570$ μmK, yielding $F_{(0\to\lambda T)} = 0.000$ and $\alpha_{atm} = 0.1$ in Equation (4.68).

Similarly, assuming a diffuse surface for the computation of the plate emissivity,

$$\epsilon = \frac{\int_0^{\infty} \epsilon_\lambda E_{b,\lambda}(345K) d\lambda}{\int_0^{\infty} E_{b,\lambda} d\lambda} = \frac{\int_0^{\infty} \alpha_\lambda E_{b,\lambda}(345K) d\lambda}{\int_0^{\infty} E_{b,\lambda} d\lambda}$$
$$= 0.8F_{(0\to 2\ \mu m)} + 0.1[1 - F_{(0\to 1\ \mu m)}] \tag{4.69}$$

From the blackbody table (Table 4.1) and Equation (4.69), using $\lambda T = 2 \times 345 = 690$ μmK, we obtain $F_{(0\to\lambda T)} = 0.000$ and $\epsilon = 0.1 = \alpha_{atm}$. Thus, the plate is gray with respect to G_{atm}, but since $\epsilon \neq \alpha_s$, it is nongray with respect to G_{sun}.

For air at a film temperature of 320 K, $\nu = 17.8 \times 10^{-6}$ m²/sec, $k = 0.028$ W/mK, and $Pr = 0.703$. Then the Reynolds number based on the length of the plate, L, is

$$Re_L = \frac{VL}{\nu} = \frac{5 \times 0.1}{19.9 \times 10^{-6}} = 2.8 \times 10^4 \tag{4.70}$$

The flow is laminar, and the following correlation may be adopted:

$$\overline{Nu}_L = \frac{\overline{h}L}{k} = 0.664(Re_L)^{1/2}(Pr)^{1/3} \tag{4.71}$$

Substituting the relevant numerical values, we find $\overline{Nu}_L = 99$ and $\overline{h} \approx 27.7$ W/m²K.

Thus, the convective heat flux from the plate is estimated as follows:

$$q''_{conv} = \overline{h}(T_p - T_\infty) = 27.7(345 - 295) = 1,385\ \text{W/m}^2 \tag{4.72}$$

Using this result and the previously obtained radiative properties in Equation (4.65),

$$0.758(800) + 0.1\sigma(285)^4 + q''_{elec} = 0.1\sigma(345)^4 + 1,385 \tag{4.73}$$

Solving Equation (4.73), we obtain a required electrical heating of approximately 821.9 W/m^2.

From the previous example, it can be observed that a diffuse surface or a diffuse irradiation assumption leads to the results that $\epsilon_\lambda = \alpha_\lambda$ and $\epsilon = \alpha_{atm}$.

4.5.2 Solar Angles

The Earth's position in its orbit influences the solar radiation at a particular location and any resulting energy balances involving solar radiation (i.e., solar collectors). The following angles are defined in reference to a point *P* on the Earth's surface and, in particular, an inclined surface at that location (see Figure 4.8):

- δ = declination angle (angle between the north pole and the axis normal to the sun's incoming rays)
- λ = latitude (degrees north or south of the equator)
- ω = hour angle (relative to the meridian of the plane of the sun's incoming rays)
- θ_s = zenith angle of sun (angle between the normal to the Earth's surface and the sun's incoming rays at point *P*)

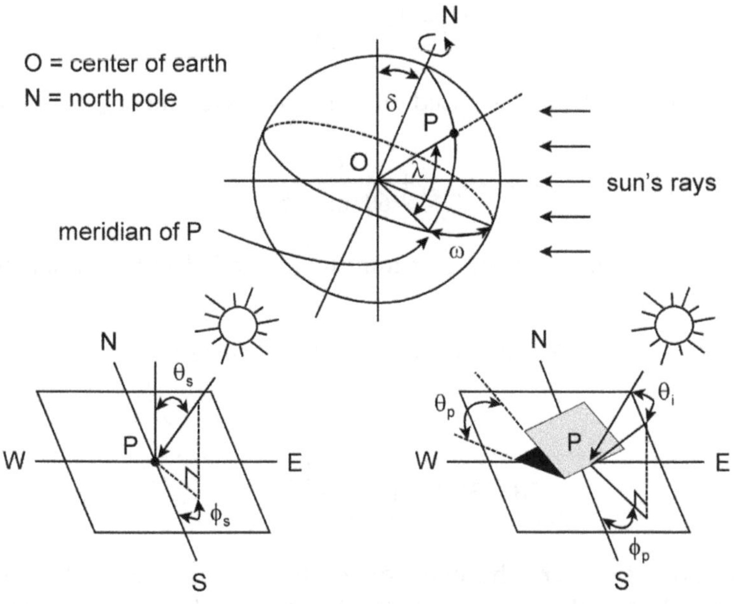

FIGURE 4.8
Solar angles.

- α = solar altitude angle = $\pi/2 - \theta_s$
- θ_p = inclination (tilt) angle of the surface
- θ_i = incident angle of surface (between the normal to the inclined surface and the sun's incoming rays at point P on the surface)
- ϕ = azimuth angle

The azimuth angle is defined with respect to some direction. For example, ϕ_s is the azimuth angle formed between the direction south of point P and the sun's incoming rays at point P.

A trigonometric analysis can be performed to find various relationships between the previously defined angles. In particular, the declination angle can be written in terms of the day of the year, n, as follows:

$$sin(\delta) = -sin(23.45)cos\left[\frac{360(n+10)}{365.25}\right] \tag{4.74}$$

Also, the zenith angle can be determined from

$$\cos(\theta_s) = cos(\lambda)cos(\delta)cos(\omega) + sin(\lambda)sin(\delta) \tag{4.75}$$

The hour angle is given by

$$\omega = (t_{sol} - 12\ \text{h})\frac{360°}{24\ \text{h}} \tag{4.76}$$

where the solar time, t_{sol}, is based on the local civic time, $t_{loc,civ}$, as follows:

$$t_{sol} = t_{loc,civ} + \frac{E_t}{60\ \text{min/h}} \tag{4.77}$$

In Equation (4.77), E_t (in units of minutes) is called the equation of time, where

$$E_t = 9.87sin(2B) - 7.53cos(B) - 1.5sin(B) \tag{4.78}$$

and

$$B = 360° \times \left(\frac{n-81}{364}\right) \tag{4.79}$$

for day n of the year. The equation of time in Equation (4.78) gives a correction of the solar time (in units of minutes) due to variations of the Earth's orbit around the sun throughout the year. It provides a correction of the local civic time in Equation (4.77).

The local civic time in Equation (4.77) differs from standard time by 4 min (1/15 h) for each degree of difference in longitude from the reference meridian, such as the meridian dividing Eastern Standard Time (EST) and Central Standard Time (CST). Some examples of standard time zones are shown as follows:

- Pacific Standard Time (PST) at 120° W
- Mountain Standard Time (MST) at 105° W
- Central Standard Time (CST) at 90° W
- Eastern Standard Time (EST) at 75° W

In addition, the azimuth and incident angles can be calculated from

$$sin(\phi_s) = \frac{cos(\delta)sin(\omega)}{sin(\theta_s)} \tag{4.80}$$

$$cos(\theta_i) = sin(\theta_s)sin(\theta_p)cos(\phi_s - \phi_p) + cos(\theta_s)cos(\theta_p) \tag{4.81}$$

The declination angle varies throughout the year due to the seasonal variations of the Earth's position in its orbit. For example, on June 21 in the summer solstice, the Earth's declination is 23.5° and this corresponds to summer in the northern hemisphere. The autumnal equinox begins on September 21. The declination angle becomes –23.5° (i.e., north pole facing away from the sun's incoming rays) on December 21 in the winter solstice (winter in the northern hemisphere). On the vernal equinox (March 23) and autumnal equinox (September 21), the sun appears directly overhead from an observer's perspective at the equator. During these various stages, transition of the declination angle occurs between a negative angle (winter in the northern hemisphere) and a positive angle (summer in the northern hemisphere). The following example illustrates how the declination angle is used in the calculation of various solar angles.

Example 4.5.2.1

Solar Angles for an Inclined Solar Collector.

The purpose of this problem is to find the incidence angle for a solar collector inclined at 50° and facing south in Winnipeg, Canada, at 10:30 A.M. on January 12. 10:30 The location of Winnipeg is 49° 50′ N and 97° 15′ W.

Winnipeg is located about 7.25° W of the CST meridian. Thus, the local civic time is

$$t_{loc,civ} = 10{:}30 - 7.25(4) = 10{:}01 \text{ A.M.} \tag{4.82}$$

Since the Earth turns 360° in 24 h, 1° of rotation takes about 4 min (24(60)/360). Then, on January 12 ($n = 12$),

$$B = 360\left(\frac{12-81}{364}\right) = -68.2° \tag{4.83}$$

$$E_t = 9.87sin(2B) - 7.53cos(B) - 1.5sin(B) = -8.2 \text{ min} \tag{4.84}$$

As a result, the local solar time is given as follows:

$$t_{sol} = 10{:}01 - 0{:}08 = 9{:}53 \text{ A.M.} \tag{4.85}$$

which yields the following hour angle:

$$\omega = (12{:}00 - 9{:}53)15° = 2.1 \times 15 = 31.5° \tag{4.86}$$

(note: 15°/h based on 360° in 24 h). The positive sign on the hour angle signifies morning. Thus,

$$sin(\delta) = -sin(23.45)cos\left[\frac{360(12+10)}{365.25}\right] = -0.37 \tag{4.87}$$

which gives a declination angle of $\delta = -21.7°$.

Also,

$$sin(\alpha) = sin\left(49\frac{50}{60}\right)sin(-21.7) + cos(49.83)cos(31.5)cos(-21.7) \tag{4.88}$$

which yields a solar altitude angle of 13.45° at this time. Furthermore, the azimuth angle is obtained by

$$sin(\phi_s) = \frac{cos(-21.7)sin(31.5)}{cos(13.45)} = 0.5 \tag{4.89}$$

and so $\phi_s = 30°$ (east of south). Finally, the incidence angle is determined by

$$cos(\theta_i) = cos(30-0)cos(13.45)sin(50) + sin(13.45)cos(50) = 0.7945 \tag{4.90}$$

which gives $\theta_i = 37.4°$.

It can be observed that the times of sunset and sunrise in the previous example could be determined by setting $\alpha = 0$ in Equation (4.88) and solving the resulting hour angles. Substituting these hour angles into Equation (4.76) would then give the corresponding times of sunset and sunrise. The *total daily extraterrestrial radiation* is obtained by integrating $cos(\theta_s)$ from sunrise to sunset and multiplying by the solar constant (I_0).

4.5.3 Direct, Diffuse, and Reflected Solar Radiation

Incoming solar radiation at the top of the Earth's atmosphere is apportioned into direct, diffuse, absorbed, and scattered components (see Figure 4.9). Consider a surface located at point P on the Earth's surface and inclined at an angle of θ_p with respect to the horizontal plane (ground). The total incoming solar radiation on the surface, I_p, is the sum of direct solar radiation, $I_{dir,p}$, diffuse sky irradiation due to scattering, $I_{dif,p}$, and radiation reflected from the ground and other surrounding surfaces, $I_{ref,p}$, i.e.,

$$I_p = I_{dir,p} + I_{dif,p} + I_{ref,p} \tag{4.91}$$

As discussed earlier, some incoming radiation from the sun is scattered within the atmosphere, both scattered back to space and scattered eventually to reach the Earth's surface ($I_{dif,p}$). This component is distinguished from the reflected portion ($I_{ref,p}$), which refers to radiation reflected off the ground and other surfaces on the Earth, rather than scattered within the atmosphere. Various factors affect the diffuse (scattered) component of radiation, $I_{dif,p}$, including local weather conditions (i.e., cloud cover), surface orientation, and other factors. Different components of solar radiation and their effects on solar collectors have been extensively analyzed and documented by Duffie, Beckman (1974) and Hsieh (1986).

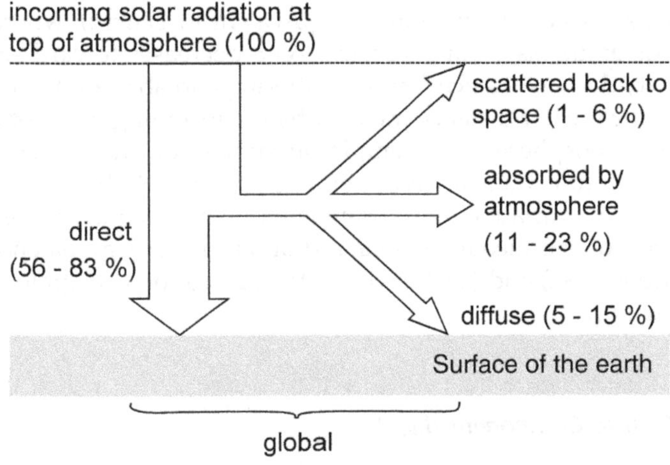

FIGURE 4.9
Components of solar radiation.

4.5.3.1 Direct Solar Radiation ($I_{dir,p}$)

The direct component of incident solar radiation can be expressed in terms of the incidence angle, θ_i, and solar constant, I_0, as follows:

$$I_{dir,p} = I_{dir} cos(\theta_i) = I_0 \tau_{atm} \tag{4.92}$$

where τ_{atm} refers to the transmissivity of the atmosphere. This transmissivity is dependent on the path length of the incoming beam of radiation through the atmosphere.

For clear skies without pollution, the transmissivity can be approximated by

$$\tau_{atm} \approx 0.5[e^{-0.095m(z,\alpha)} + e^{-0.65m(z,\alpha)}] \tag{4.93}$$

where z designates the elevation above sea level. Also,

$$m(z, \alpha) = m(0, \alpha) \frac{P_{atm}(z)}{P_{atm}(0)} \tag{4.94}$$

where $P_{atm}(z)$ is the atmospheric pressure at an elevation of z and

$$m(0, \alpha) = [(614 sin(\alpha))^2 + 1229]^{1/2} - 614 sin(\alpha) \tag{4.95}$$

This coefficient, m, can be physically interpreted as the ratio of the actual distance traveled by the incoming beam of radiation through the atmosphere to the distance traveled when the sun is directly overhead of the point P (i.e., solar noon). This ratio is equivalent to $1/sin(\alpha)$, where α is the solar altitude angle (as defined earlier). As α decreases toward sunset, the transmissivity decreases since further scattering and absorption of radiation occurs over the greater distance traveled by the incoming beam of radiation through the atmosphere. For example, at sunrise or sunset, $\alpha = 0$, which yields a minimum transmissivity of $\tau_{atm} = 0.018$. Although a spectral distribution (i.e., variation with wavelength and frequency) affects the characteristics of the incoming solar radiation, the energy balances involving Equations (4.91) and (4.92) give the total radiation (integrated over the entire spectrum).

4.5.3.2 Diffuse Component ($I_{dif,p}$)

The diffuse (scattered) component of solar radiation can be expressed in terms of the total horizontal radiation, I_h, and extraterrestrial radiative flux on a horizontal surface, $I_{0,h}$, as follows:

$$I_{dif,p} = I_{dif,h}\left[\frac{1+cos(\theta_p)}{2}\right] = (I_h - I_{dir,h})\left[\frac{1+cos(\theta_p)}{2}\right] \tag{4.96}$$

where

$$\frac{I_h}{I_{0,h}} = 0.8302 - 0.03847m(z,\ \alpha) - 0.04407(CC) + 0.011013(CC)^2$$
$$- 0.01109(CC)^3 \tag{4.97}$$

and CC refers to the cloud cover (i.e., 0 = clear sky, 1 = fully overcast).

For the diffuse component, the trigonometric factor in brackets in Equation (4.96) represents a view factor between the sky and the inclined surface (tilt angle of θ_p), which collects radiation at point P on the Earth's surface. For example, this view factor is $F = 1$ at $\theta_p = 0°$ (upward facing) and $F = 0$ at $\theta_p = 180°$ (downward facing). Also, it should be noted that the total horizontal radiation (i.e., incident radiation on the surface when it is horizontally oriented), I_h, refers to the incident radiation after it has passed through the atmosphere and arrives at the Earth's surface (in contrast to the total solar flux of $I_{0,h}$ prior to passage through the atmosphere). Also, the extraterrestrial flux on a horizontal surface, $I_{0,h}$, is equivalent to $I_0 cos(\theta_{i,h})$, where $\theta_{i,h}$ refers to the incident angle for a horizontal surface.

4.5.3.3 Reflected Component ($I_{ref,p}$)

The remaining component in Equation (4.91) is the following reflected component of solar radiation:

$$I_{ref,p} = \rho_g I_h\left[\frac{1-cos(\theta_p)}{2}\right] \tag{4.98}$$

where ρ_g refers to the ground (or surrounding surface) reflectivity. The last expression, in brackets, in Equation (4.98) represents a view factor between the ground and the inclined surface (tilt angle of θ_p), which collects the incoming solar radiation.

Example 4.5.3.3.1

Incident Solar Flux on an Inclined Solar Collector.

Estimate the direct, diffuse, and reflected components of incident solar radiation arriving on a solar collector that faces south in Winnipeg, Canada, on February 8. The ground reflectivity is $\rho_g = 0.68$. Also, the hour angle is $\omega = -40°$ (afternoon), and the solar collector is inclined at 60° (with respect to the horizontal plane) during the partially cloudy day (20% cloud cover).

On February 8, $n = 38$, and so

$$sin(\delta) = -sin(23.45)cos\left[\frac{360(38 + 10)}{365.25}\right] = -0.27 \tag{4.99}$$

which yields a declination angle of $\delta = -15.7°$. Thus,

$$sin(\alpha) = sin(49.8)sin(-15.7) + cos(49.8)cos(-15.7)cos(-40) \tag{4.100}$$

The solar altitude angle becomes approximately $\alpha = 15.7°$.

As a result, the azimuth angle is computed by

$$sin(\phi_s) = \frac{cos(-15.7)sin(-48.8)}{cos(15.7)} = -0.75 \tag{4.101}$$

which yields $\phi_s = -48.6°$. Furthermore,

$$cos(\theta_i) = cos(-48.6 - 0)cos(15.7)sin(60) + sin(15.7)cos(60) \tag{4.102}$$

and so $\theta_i = 46.6°$. In terms of the horizontal plane,

$$cos(\theta_{i,h}) = cos(-48.6 - 0)cos(15.7)sin(0) + sin(15.7)cos(0) = 0.27 \tag{4.103}$$

which gives $\theta_{i,h} = 74.3°$.

The incoming solar flux consists of three components: direct, diffuse, and reflected. For the first component (direct),

$$m(0,\ \alpha) = [(614sin(15.7))^2 + 1229]^{1/2} - 614sin(15.7) = 3.66 \tag{4.104}$$

$$\tau_{atm} \approx 0.5[exp(-0.095 \times 3.65) + exp(-0.65 \times 3.65)] = 0.4 \tag{4.105}$$

$$I_{dir,p} = I_{dir}cos(\theta_i) = I_0\tau_{atm}cos(\theta_i) = 1353(0.4)cos(46.7) = 371.2\ \text{W/m}^2 \tag{4.106}$$

For the second component (diffuse), under a partial cloud cover (CC = 0.2),

$$\frac{I_h}{I_{0,h}} = 0.8302 - 0.03847(3.65) - 0.04407(0.2) + 0.011013(0.2)^2 - 0.01109(0.2)^3$$
$$= 0.68 \tag{4.107}$$

where

$$I_{0,h} = I_0cos(\theta_{i,h}) = 1353(0.27) = 365.3\ \text{W/m}^2 \tag{4.108}$$

As a result,

$$I_h = 0.68(365.3) = 248.4 \text{ W/m}^2 \tag{4.109}$$

Also,

$$I_{dir,h} = I_0 \tau_{atm} cos(\theta_{i,h}) = 1353(0.4)0.27 = 146.1 \text{ W/m}^2 \tag{4.110}$$

which yields the following diffuse radiation arriving on a horizontal surface:

$$I_{dif,h} = I_h - I_{dir,h} = 248.4 - 146.1 = 102.3 \text{ W/m}^2 \tag{4.111}$$

Thus, for the inclined surface,

$$I_{dif,p} = 102.3\left[\frac{1 + cos(60)}{2}\right] = 76.7 \text{ W/m}^2 \tag{4.112}$$

Finally, the reflected portion of the incoming solar flux is

$$I_{ref,p} = 0.68\left[\frac{1 - cos(60)}{2}\right]248.4 = 42.2 \text{ W/m}^2 \tag{4.113}$$

Combining the direct, diffuse, and reflected portions of the incoming radiation, a total hourly averaged incident solar flux of 490.2 W/m^2 on the solar collector is obtained.

The calculation of the incident solar flux (as shown in the previous example problem) is an important part of effectively designing solar collectors. Further details are described in books by Howell et al. (1982) and Meinel and Meinel (1976). The following section considers various detailed aspects of the design of solar collectors.

4.5.4 Design of Solar Collectors

The design of solar collectors involves absorption, storage, and distribution of energy collected by solar radiation. A solar collector is essentially a heat exchanger that absorbs the incident solar radiation and transfers it to a working fluid, such as an ethylene glycol–water solution, thereby raising the fluid temperature for purposes such as space or water heating in buildings. The main parts of a solar collector include the top glass cover(s) above an absorptive collector plate with a surface coating (see Figure 4.10). Tubes that carry the working fluid are bonded and joined to the absorber plate. The back and bottom sides of the solar collector are usually well insulated to reduce heat losses from the collector, so that most of the incoming solar energy is transferred to the working fluid. By adding the top glass cover(s), heat losses such as convective heat losses to the environment

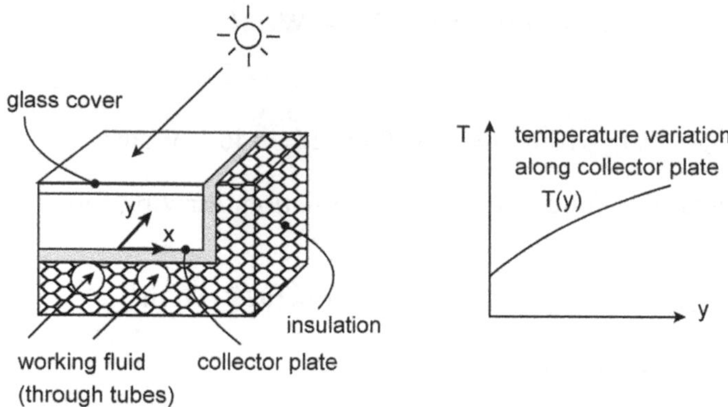

FIGURE 4.10
Flat plate solar collector.

can be reduced, thereby increasing the net amount of heat absorbed by the working fluid. A *single-glazed collector* refers to a single glass cover, whereas a *double-glazed collector* refers to two glass covers above the absorber plate.

Based on an overall energy balance for a solar collector,

$$I_p A_c \tau_s \alpha_s = q_{loss} + q_u \tag{4.114}$$

i.e., the energy absorbed by the absorber plate (left side) must balance the (undesired) heat losses to the surroundings due to reflected radiation and convective losses, q_{loss}, and the (desired) energy gained by the working fluid, q_u. The incident radiation at point P in Equation (4.114), denoted by I_p, is multiplied by the collector surface area, A_p, transmissivity of the glass cover, τ_s, and the absorptivity of the collector plate, α_s. The incident radiation is first transmitted through the glass covers, and then absorbed by the collector plate.

The instantaneous collector efficiency, η_c, is defined as the ratio of energy gained by the working fluid to the total incoming solar energy, i.e.,

$$\eta_c = \frac{q_u}{A_c I_p} \tag{4.115}$$

Integrating the parts of Equation (4.115) over a characteristic time period, T, such as an entire day, yields the following expression for the average collector efficiency:

$$\overline{\eta}_c = \frac{\int_0^T q_u dt}{\int_0^T A_c I_p dt} \tag{4.116}$$

The heat losses, q_{loss}, and energy gained by the working fluid, q_u, can be effectively examined by a thermal network involving the various thermal resistances. Consider a solar collector with two glass covers (top and lower covers), absorber plate, collector tubes beneath the absorber plate, and insulation at the bottom side of the solar collector. The incident radiation arriving at the top glass cover, I_p, is either reflected (combined convection and radiation resistance, R_5, between the top cover at T_{g1} and the ambient air at T_a), absorbed by the top cover, or transmitted to the air gap beneath the top cover. Similarly, the radiation that passes through the top cover, $\tau_s I_p$ (i.e., based on the transmissivity of the top cover), is either reflected back (combined convection and radiation resistance, R_4, between T_{g1} and the lower cover at T_{g2}), absorbed by the lower cover at T_{g2}, or transmitted through to the absorber plate at T_c.

At the cover (temperature of T_c), the incoming radiation of $\tau_s I_p$ is either gained by the working fluid, q_u, or lost back to the surroundings (i.e., combined convection and radiation resistance, R_3, between T_c and T_{g2}). Below the collector plate, a series sum of resistances, including R_1 (between T_c and the base of the collector at T_b) and R_2 (between T_b and T_a), is included as the final part of the overall thermal network. In summary, the overall thermal network at node T_c (collector plate) includes an incoming radiation flux, $\tau_s I_p$, outgoing energy that is gained by the working fluid, and a total resistance, R_{tot}, between the collector plate and the surrounding ambient air. This total resistance includes heat transfer from both the top (convection to air, reflected and emitted radiation) and bottom (convection to air and working fluid, conduction) sides of the collector.

Thus, the total conductance for the overall thermal network is

$$U_c = \frac{1}{R_{tot}} = \frac{1}{R_1 + R_2} + \frac{1}{R_3 + R_4 + R_5} \tag{4.117}$$

which allows the heat losses to the surroundings to be written as follows:

$$q_{loss} = U_c A_c (T_c - T_a) \tag{4.118}$$

Heat losses from the insulated bottom of the collector, expressed in the first fraction on the right side of Equation (4.117), will be neglected in this analysis.

A separate calculation of total heat losses is required in view of its influence on the energy balance in Equation (4.114) and the resulting collector efficiency in Equation (4.115). Based on the portion of the thermal network between the absorber plate and bottom glass cover (see Figure 4.10),

$$q_{top,loss} = A_c\bar{h}_{c2}(T_c - T_{g2}) + \frac{\sigma A_c(T_c^4 - T_{g2}^4)}{1/\epsilon_{p,i} + 1/\epsilon_{g2,i} - 1} = \frac{T_c - T_{g2}}{R_3} \tag{4.119}$$

On the right side of Equation (4.119), the terms represent heat losses by natural convection and radiation (based on view factors between two parallel plates facing each other), respectively.

Similarly, between the two glass covers,

$$q_{top,loss} = A_c\bar{h}_{c1}(T_{g2} - T_{g1}) + \frac{\sigma A_c(T_{g2}^4 - T_{g1}^4)}{1/\epsilon_{g2,i} + 1/\epsilon_{g1,i} - 1} = \frac{T_{g2} - T_{g1}}{R_4} \tag{4.120}$$

where the first and second terms on the right side of Equation (4.120) refer to heat losses by natural convection and radiation, respectively. Alternatively, writing the heat loss from the top of the collector based on the thermal network above the top glass cover,

$$q_{top,loss} = A_c\bar{h}_{c,a}(T_{g1} - T_{air}) + \epsilon_{g1,i}\sigma A_c(T_{g1}^4 - T_{sky}^4) = \frac{T_{g1} - T_{air}}{R_5} \tag{4.121}$$

It can be observed that Equations (4.119) to (4.121) must be solved iteratively since values of R_3, R_4, and R_5 depend on the unknown temperatures T_c, T_{g1}, and T_{g2}. In particular, the resistances include linearized radiation coefficients, based on definitions analogous to Equation (4.9), which are dependent on temperature.

In contrast to an iterative solution of Equations (4.119) to (4.121), Klein's method provides an empirical alternative for a direct solution. The following variables are defined in this method:

- N = number of glass covers
- θ_p = collector inclination with respect to the horizontal plane
- ϵ_g and ϵ_p = emissivity of glass cover and absorber plate, respectively
- $C = 365(1 - 0.00883\theta_p + 0.00013\theta_p^2)$
- $f = (1 - 0.4\, h_{c,a} + 0.0005\, h_{c,a}^2)\,(1 + 0.091N)$
- $h_c[\text{W/m}^2\,\text{K}] \approx 5.7 + 3.8\ V\ [\text{m/sec}]$

Based on these variables, Klein's method predicts that

$$q_{top,loss} = \frac{A_c(T_c - T_a)}{NT_c\left(\frac{T_c - T_a}{N + f}\right)^{0.33} / C + 1/h_c} + \frac{\sigma A_c(T_c^4 - T_a^4)}{1/[\epsilon_p + 0.05N(1 - \epsilon_p)] + (2N + f - 1)/\epsilon_g - N} \quad (4.122)$$

This approach gives a direct solution as an alternative to solving Equations (4.119) to (4.121) iteratively.

Once the heat losses are determined in Equation (4.114), the remaining heat flow term is the energy gained by the fluid, q_u. Since tubes are spaced at some distance apart, heat is absorbed from incident solar radiation arriving directly above a tube, as well as heat transfer along the absorber plate (effectively acting as a fin) between the tubes. Heat gained by the absorber plate between the tubes is transferred by conduction through the plate to the section of plate directly above the tubes carrying the working fluid. Thus, q_u will be computed as the sum of heat transfer through the fin and solar energy absorbed directly above each tube.

In order to find the fin heat transfer, the temperature variation in the absorber plate (x direction) is required. Considering an energy balance across a differential section of absorber plate (thickness t) between the tubes (see Figure 4.11),

$$\alpha_s I_s dx + [-ktq''_x] = U_c(T_c - T_a)dx + [-ktq''_{x+dx}] \quad (4.123)$$

where

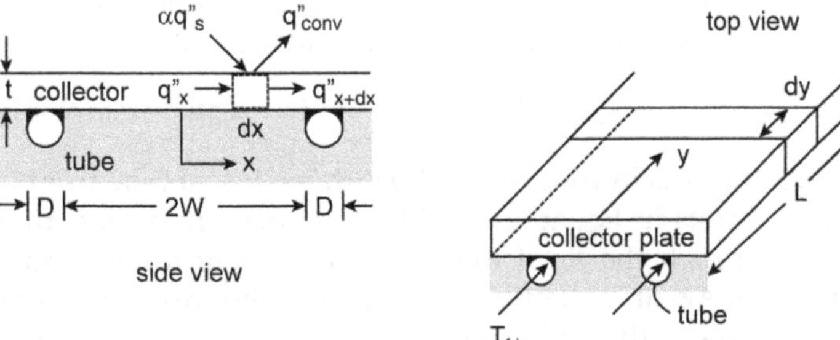

FIGURE 4.11
Heat balance for a solar collector.

$$q''_x = -k\frac{dT_c}{dx} \tag{4.124}$$

(Fourier's law). Also, expanding the heat flux at position $x + dx$ about the heat flux at x using a Taylor series expansion,

$$q''_{x+dx} = q''_x + \frac{dq''_x}{dx}dx + \ldots \tag{4.125}$$

Higher order terms will be neglected in view of the differential thickness of the control volume.

Substituting Equations (4.124) and (4.125) into Equation (4.123) and rearranging,

$$\frac{d^2T_c}{dx^2} = \frac{U_c}{kt}\left[T_c - \frac{\alpha_s I_s}{U_c} - T_a\right] \tag{4.126}$$

which is subject to $T_c = T_b$ at $x = W$ (above tube) and $dT_c/dx = 0$ at $x = 0$ (midway between the tubes). Equation (4.126) can be solved to yield

$$T_c(x) = T_a + \frac{\alpha_s I_s}{U_c}\left(T_b - T_a - \frac{\alpha_s I_s}{U_c}\right)\frac{cosh(mx)}{cosh(mW)} \tag{4.127}$$

where $cosh(mW) = (e^{mW} + e^{-mW})/2$ is the hyperbolic cosine function and $m^2 = U_c/(kt)$.

Differentiating Equation (4.127) and evaluating at $x = W$ (above tube, per unit depth),

$$q_{fin} = -kt\left.\frac{dT_c}{dx}\right|_W = -ktm\left(T_b - T_a - \frac{\alpha_s I_s}{U_c}\right)tanh(mW) \tag{4.128}$$

where $tanh(mW)$ refers to the hyperbolic tangent function. Alternatively, based on the definition of fin efficiency in Equation (2.41),

$$q_{fin} = \eta_{fin} q_{max} = 2W\eta_{fin}[\alpha_s I_s - U_c(T_b - T_a)] \tag{4.129}$$

where the factor of 2 is introduced for heat transfer to a tube from two sides. The term in brackets in Equation (4.129) is the maximum heat transfer that would occur through the fin if the entire fin was at a temperature of T_b. With a fin entirely at a temperature of T_b, the resulting temperature difference (between the fin and air) would be maximized, thereby maximizing the convective heat transfer between the absorber plate and surrounding air. For solar collectors, large convective losses to the surrounding air are undesirable.

Dividing Equations (4.128) and (4.129),

$$\eta_{fin} = \frac{tanh(mW)}{mW} \tag{4.130}$$

which indicates that the fin efficiency increases with decreasing total conductance, increasing plate conductivity, or increasing plate thickness. Since heat absorbed by the absorber plate between the tubes is largely transferred by conduction through the fin to the tube and working fluid, appropriate steps should be taken in the design of the solar collector to increase this heat transfer. Examples include a reduced plate reflectivity (i.e., lower total conductance), while increasing the plate's thermal conductivity (lower m). The fin performance can be graphically depicted in a conventional form (i.e., Figure 2.7) to show the trends of η_{fin} with increasing values of mW.

Then, combining the heat flows from the fin, Equation (4.129), and solar energy gained directly above the tube (with area equal to diameter, D, multiplied by a unit depth),

$$q_u = (2W\eta_{fin} + D)[\alpha_s I_s - U_c(T_b - T_a)] \tag{4.131}$$

Alternatively, based on an energy balance on the working fluid (inside each tube) in terms of the convective heat transfer coefficient,

$$q_u = \dot{m}c_p \frac{dT_f}{dy} = \pi D\bar{h}_c(T_b - T_f(y)) \tag{4.132}$$

where $\dot{m}$ refers to the mass flow rate of working fluid through the tube. The performance of solar collectors in Equation (4.132) is often expressed in terms of the following two factors: (i) collector efficiency factor and (ii) collector heat removal factor.

4.5.4.1 Collector Efficiency Factor

Since the base temperature, T_b, is often unknown in the design, Equations (4.131) and (4.132) can be combined to eliminate the appearance of T_b. Once T_b is eliminated, an alternative expression for the heat gain of the working fluid is obtained, i.e.,

$$q_u = (D + 2W)F'[\alpha_s I_s - U_c(T_f - T_a)] \tag{4.133}$$

where the collector efficiency factor, F', is given by

$$F' = \frac{1/U_c}{(D+2W)}\left[\frac{1}{U_c(D+2W\eta_{fin})}+\frac{1}{\bar{h}_c \pi D_i}\right] \tag{4.134}$$

The collector efficiency represents a ratio of the thermal resistance between the collector and environment to the thermal resistance between the working fluid and environment. In Equation (4.134), the terms within square brackets refer to the fin resistance and fluid resistance to heat transfer, respectively. The average convection coefficient, $\bar{h}_c$, in Equation (4.134), applies to the flow of liquid within the tubes. Typical values of the total conductance, U_c, are 4 W/m^{2}K (two glass covers) and 8 W/m^{2}K (one glass cover). A working fluid of water with typical absorber plate materials of steel or copper is commonly used in solar collectors.

4.5.4.2 Collector Heat Removal Factor

In Equation (4.133), the local rate of heat transfer was given since the fluid temperature, T_f, was used at a specific location, y. However, this fluid temperature varies from $T_{f,in}$ at the tube inlet to $T_{f,out}$ at the tube outlet. Thus, in order to find the total heat removed, the entire variation of T_f with position y is required. This permits spatial integration of the heat removal rate from the inlet to the outlet. Based on an overall energy balance for the working fluid from the inlet to the outlet of a single tube (see Figure 4.11),

$$q_u = \dot{m}c_p(T_{f,out} - T_{f,in}) \tag{4.135}$$

It can be observed that the total temperature difference of the fluid between the inlet and outlet is required. This temperature difference will be obtained by first deriving the variation of fluid temperature with position within the tube, and then integrating this expression appropriately from the inlet to the outlet.

Considering a differential segment of thickness dy within the tube, an energy balance for this control volume (see Figure 4.11) yields

$$\dot{m}c_p\frac{dT_f}{dy} = q_u = (D+2W)F'[\alpha_s I_s - U_c(T_f - T_a)] \tag{4.136}$$

which represents Equations (4.132) and (4.133). Separating variables and integrating from $y = 0$ ($T_f = T_{f,in}$) to $y = L$ ($T_f = T_{f,out}$),

$$\frac{T_{f,out} - (T_a + \alpha_s I_s/U_c)}{T_{f,in} - (T_a + \alpha_s I_s/U_c)} = exp\left[-\frac{U_c F' G}{c_p}\right] \tag{4.137}$$

where the mass flux, G, is given by

$$G = \frac{\dot{m}}{L(D + 2W)} \tag{4.138}$$

The fluid temperature difference can be isolated by adding and subtracting $T_{f,in}$ in the numerator of Equation (4.137). Then, substituting this temperature difference into Equation (4.135),

$$q_u = A_c F_r [\alpha_s I_s - U_c (T_{f,in} - T_a)] \tag{4.139}$$

where the heat removal factor, F_r, is

$$F_r = \frac{Gc_p}{U_c}\left[1 - exp\left(-\frac{U_c F'}{Gc_p}\right)\right] \tag{4.140}$$

The heat removal factor represents a coefficient that describes the ability of the working fluid to remove heat from the fin and collector.

An alternative way of writing Equations (4.115) and (4.139) is

$$\eta_c = \frac{q_u}{A_c I_p} = F_r \alpha_s \tau_s - F_r U_c \left(\frac{T_{f,in} - T_a}{I_p}\right) \tag{4.141}$$

This result is often used for the testing and evaluation of solar collector performance. If η_c is measured based on the first equality in Equation (4.141) at various values of $(T_{f,in} - T_a)/I_p$, then corresponding values of F_r and U_c can be determined experimentally. The variation of η_c typically becomes approximately linear with respect to $(T_{f,in} - T_a)/I_p$ (called X_T, independent variable along the horizontal axis) since $\alpha_s \tau_s$, F_r, and U_c are often constant for a fixed collector design.

The objectives of effective solar collector design include selecting a system with a high $\alpha_s \tau_s$, high F_r, and low U_c. These choices would increase the collector efficiency, thereby delivering a higher energy gain by the working fluid for a given amount of incident solar radiation. The collector performance can then be depicted by an approximately linear decrease of η_c with X_T. The intercept of this efficiency curve is $F_r \alpha_s \tau_s$ at $X_T = 0$, and the slope (negative) is $F_r U_c$. Thus, a high intercept with a shallow slope would provide a high collector efficiency over a wide range of fluid and incident radiation conditions. Although a curve with a steep slope may provide a higher η_c at lower values of I_p, it would drop off to lower η_c values at higher I_p. Since X_T varies with the time of day, surface orientation, and other factors (in view of I_p in the denominator), the former trends are often more desirable (i.e., shallow slope of $\eta_c - X_T$ curve).

Example 4.5.4.2.1

Solar Collector with Aluminum Fins and Tubes.

A solar collector panel is 2.2 m wide and 3.1 m long. It is constructed with aluminum fins and tubes with a tube-to-tube centered distance of 10 cm, fin thickness of 0.4 mm, tube diameter of 1 cm, and fluid–tube heat transfer coefficient of 1100 W/m^{2}K. If the collector heat loss coefficient is 8 W/m^{2}K, find the required water flow rate yielding a heat removal factor of 0.74. Also, find the required absorptivity of the collector plate for a collector efficiency of 0.13. The cover transmissivity to solar radiation is 0.92, and the solar flux is $I_p = 400$ W/m^2. Also, $T_a = 15°C$ and $T_{f,\,in} = 47°C$.

The fin efficiency of the collector plate is computed as follows:

$$m = \sqrt{\frac{U_c}{kt}} = \sqrt{\frac{8}{237(0.0004)}} = 9.19/\text{meter} \tag{4.142}$$

$$\eta_{fin} = \frac{tanh(mW)}{mW} = \frac{tanh(9.19 \times 0.045)}{9.19 \times 0.045} = 0.95 \tag{4.143}$$

Then the collector efficiency and heat removal factors are

$$F' = \frac{1/8}{0.10\left[\dfrac{1}{8(0.01 + 0.09 \times 0.95)} + \dfrac{1}{1100\pi(0.01)}\right]} = 0.934 \tag{4.144}$$

$$F_r = 0.74 = \frac{\dot{m} \times 4200}{2.2(3.1)8}\left[1 - exp\left(-\frac{8 \times 0.934}{4200\dot{m}/(2.2 \times 31)}\right)\right] \tag{4.145}$$

which can be solved iteratively to give a mass flow rate of $\dot{m} \approx 0.025$ kg/sec.

Also, from the collector efficiency relation,

$$\eta_c = \frac{q_u}{A_c I_p} = F_r\left[\alpha_s \tau_s - U_c\left(\frac{T_{f,in} - T_a}{I_p}\right)\right] \tag{4.146}$$

$$0.13 = 0.74\left[\alpha_s(0.92) - 8\left(\frac{320 - 288}{400}\right)\right] \tag{4.147}$$

which gives a required absorptivity of $\alpha_s \approx 0.89$.

The previous sections have described certain thermal aspects of solar energy processes. A more complete treatment of solar energy engineering, including further applications in practice, is provided in each of the following references: Duffie and Beckman (1974), Hsieh (1986), Garg (1982), and Howell and Bereny (1979).

4.5.5 Solar Based Power Generation

In addition to solar energy collection for space and water heating, solar-based systems can also be used for the production of electricity. An example is a 10-mW solar power plant in Barstow, California. A working fluid (typically water) is pumped up into a solar receiver, which can be a type of enclosure surrounded by heat-conducting panels with a boiler inside. A solar concentrator (collector subsystem) consists of a series of mirrors and heliostats that carefully focus the incident solar radiation onto the receiver to convert the water into superheated steam. The steam is then directed down the *solar power tower* to drive a steam turbine and electric generator in the regular fashion of a Rankine cycle (see Figure 4.12). A thermal storage subsystem can be used to retain heat over nights and cloudy days.

In this method of power generation, the solar heat is stored in the form of a heated working fluid. However, in some solar energy applications, such as thermal control of spacecraft, it may be less costly and less bulky to store the energy in phase change materials (PCMs), rather than the working fluid alone. Energy is released during solidification of the PCM and absorbed during melting. For example, an outer layer of a solar receiver tube (such as $LiF-CaF_2$ material) is a PCM that stores incident solar energy or releases this heat to a working fluid that passes through the tube. Heat storage occurs in the PCM during the melting cycle (i.e., high incident solar radiation period), and then heat is released by solidification during the low

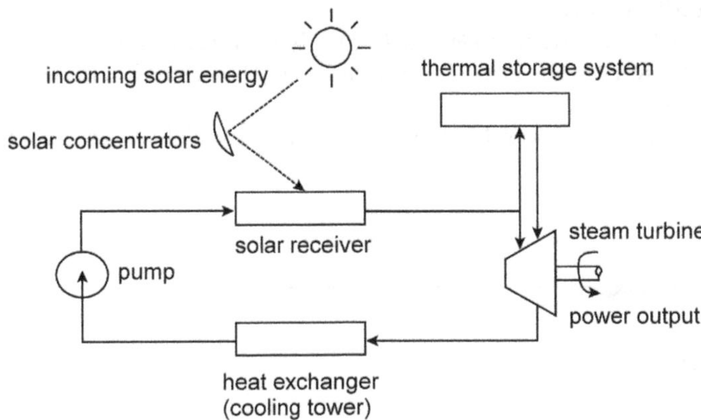

FIGURE 4.12
Solar power generation.

incoming solar radiation period (i.e., overnight). This heat transfer can occur essentially isothermally since the latent heat is released (or absorbed) over a small temperature range for pure PCMs. For multiconstituent materials, the latent heat may be released over a discrete temperature range (discussed further in Chapters 5 and 8).

References

J.A. Duffie and W.A. Beckman. 1974. *Solar Energy Thermal Processes*, New York: John Wiley & Sons.

H.P. Garg. 1982. *Treatise on Solar Energy*, New York: John Wiley & Sons.

J.R. Howell. 1982. *A Catalog of Radiation Configuration Factors*, New York: McGraw-Hill.

J.R. Howell, R.B. Bannerot, and G.C. Vliet. 1982. *Solar-Thermal Energy Systems, Analysis and Design*, New York: McGraw-Hill.

Y. Howell and J.A. Bereny. 1979. *Engineer's Guide to Solar Energy*, San Mateo, CA: Solar Energy Information Services.

J.S. Hsieh. 1986. *Solar Energy Engineering*, New York: Prentice Hall, Inc.

F.P. Incropera and D.P. DeWitt. 1996. *Fundamentals of Heat and Mass Transfer*, 4th ed., New York: John Wiley & Sons.

A.B. Meinel and M.P. Meinel. 1976. *Applied Solar Energy: An Introduction*, Reading, MA: Addison-Wesley.

M. Planck. 1959. *The Theory of Heat Radiation*, New York: Dover Publications.

R. Siegel and J.R. Howell. 1993. *Thermal Radiation Heat Transfer*, 3rd ed., New York: Hemisphere.

W.D. Wood, H.W. Deem, and C.F. Lucks. 1964. *Thermal Radiative Properties*, New York: Plenum Press.

Problems

1. The tungsten filament in a light bulb reaches a temperature of 2800 K.
   (a) Find the wavelength corresponding to the maximum amount of radiative emission at the given surface temperature.

(b) What range of wavelengths contains 90% of the emitted radiation? Find the range containing the largest portion of the visible range.

2. Find the amount of radiation emitted by a surface at 1600 K with the following spectral emissivities: (i) $\epsilon = 0.4$ for $0 < \lambda < 1$ μm, (ii) $\epsilon = 0.8$ for $1 = \lambda < 4$ μm, and (iii) $\epsilon = 0.3$ for $\lambda = 4$ μm. Express your answer in units of W/m^2.
3. A flat exterior spacecraft surface consists of a 12-cm-thick layer of material with a thermal conductivity of 0.04 W/mK. The Biot number involving convection on the inner surface is 90 (based on the wall thickness). The exterior wall emissivity is 0.1. Find the required air temperature inside the spacecraft to provide a net heat transfer rate of 20 W/m^2 from the exterior wall into space at 0 K.
4. The top side of a horizontal plate is exposed to solar and atmospheric irradiation. The back side of the opaque plate is electrically heated. A layer of insulation covers this back side below the resistance heating elements. The ambient air and effective sky temperatures are 10 and 0°C, respectively, and the convection coefficient is 20 W/m^{2}K. Assume that the plate is diffuse with a hemispherical reflectivity of 0.1 below wavelengths of 1 μm, and $\rho_\lambda = 0.8$ for $\lambda > 1$ μm. Find the solar irradiation if 60 W/m^2 of electrical power is required to keep the plate surface temperature at 40°C.
5. An oil-fired furnace heats a diffuse, cylindrical iron ingot with a diameter and length of 0.9 and 3 m, respectively. The furnace wall temperature is 1700 K, and hot combustion gases at 1400 K flow across the ingot at 4 m/sec from a burner inlet to a flue outlet through the furnace. The ingot is elevated so that the gas flow approximates a symmetric cross-flow past a cylinder. The iron surface emissivity is 0.3 for wavelengths below 2 μm and $\epsilon_\lambda = 0.15$ for $\lambda > 2$ μm. Find the steady-state ingot temperature. Assume that the combustion gas properties may be approximated by air properties and the furnace walls are large, compared to the ingot.
6. A steel sheet emerges from a hot-roll section of a steel mill at a temperature of 1100 K with a thickness of 4 mm and uniform properties ($\rho = 7900$ kg/m^3, $c_p = 640$ J/kgK, and $k = 28$ W/mK). In the spectral range below a wavelength of 1 μm, the sheet has an emissivity of $\epsilon =$ 0.55. The spectral emissivity values are $\epsilon = 0.35$ and $\epsilon = 0.25$ in the ranges 1 μm $\leq \lambda \leq 6$ μm and $\lambda \geq 6$ μm, respectively. The convection coefficient between the air and the strip is $h = 10$ W/m^{2}K with $T_\infty = 300$ K. Neglect conduction effects in the longitudinal (x) direction.

(a) Determine whether significant temperature gradients exist in the strip in the transverse (y) direction.

(b) Calculate the initial cooling rate (i.e., rate of temperature change with time) of the strip as it emerges from the hot-roll section.

7. A horizontal black plate (area of 7 cm^2) is located beneath and between two vertical plates (plate 2 of area of 8 cm^2 and plate 3). Both vertical plates emit 33 kW/m^2sr in the normal direction. The angles between each surface normal (first subscript) and the beam of radiation connecting it with an adjacent surface (second subscript) are given as follows: $\theta_{12} = 60°$, $\theta_{13} = 50°$, $\theta_{21} = 40°$, and $\theta_{31} = 70°$.
   (a) Find the solid angle subtended by plate 1 and the required distance between plates 1 and 2 if surface 1 intercepts 0.03 W of irradiation from surface 2. Assume that the surfaces are sufficiently small and that they can be approximated as differential surface areas.
   (b) Find the required surface area of plate 3 to yield the same irradiation (0.03 W) from that surface onto plate 1. Plate 3 is located the same distance from plate 1 as determined for plate 2.
8. A small object is cooled by radiation exchange with the walls of an enclosure around the object. Find the view factors between the two surfaces represented by the enclosure walls (surface 1) and the small object (surface 2).
9. A cubical cavity with side lengths of 6 cm has a 3-cm-diameter hole in a side wall. Find the view factor representing the total radiation emitted from the walls to (through) the single hole.
10. A radiation shield is used in a spectrometer instrument for controlling temperature and measuring the composition of certain gas molecules in the Earth's troposphere. The radiation shield encloses a cube-shaped inner metal component. The configuration consists of an upper half cube (shield, 4 × 4 × 2 cm) with a base that intersects the midplane of the enclosed inner metal cube (2 × 2 × 2 cm). Find the following radiation view factors: F_{12}, F_{21}, F_{11}, and F_{13}, where surfaces 1, 2, and 3 refer to the outer shield, inner metal piece, and surroundings formed by the gap between the shield and the metal piece, respectively.
11. A black spherical object (diameter of 4 cm) is heated in a furnace oven containing air at a temperature of 160°C. Find the required oven wall temperature to produce a sphere surface temperature of 60°C with a net rate of heat flow of 60 W to the sphere. The convection coefficient between the sphere and surrounding airflow is 50 W/m^2K. Assume that the wall reflectivity and view factor (wall–wall) are 0.1 and 0.9, respectively.
12. A special coating on a disk surface is heated and cured by a heater located 30 cm above the lower disk. The two parallel disks are placed in a large room at 300 K, and their radii are 30 cm (absorber surface) and 40 cm (heater), respectively. The emissivity and heat loss from the

absorber surface are 0.42 and 3 kW, respectively. The heater emissivity is 0.94, and its surface temperature is 1100 K. Find the absorber surface temperature and the radiative heat transfer rate from the heater.

13. A radiation shield is placed between two large plates of equal area. Find the steady-state temperature of the shield in terms of the temperatures and emissivities of both plates.
14. The cross-sectional area of a solar collector consists of a right-angled triangle with two equal-sided absorber plates and a glass cover along the longest side. A heat flux of 300 W/m^2 passes through the glass cover to yield a cover temperature of 30°C. The surface emissivity is 0.8. The absorber plates are approximated as blackbody surfaces that are insulated on both back sides. If the temperature of the upper absorber plate is 70°C, calculate the surface temperature of the other plate resulting from radiation exchange between all surfaces in the enclosure. It may be assumed that the cover plate is opaque with respect to radiation exchange with the absorber plates.
15. Solar collectors are used to deliver heated service water in a building. Consider a single solar collector that is facing S 20° E during a clear day in Toronto on April 20 (79° 30′ W, 43° 40′ N). What tilt angle (with respect to the horizontal plane) is required to provide a direct solar radiation flux of 800 W/m^2 on the inclined collector at 11:00 A.M., EST (note: CST at 90°, EST at 75°)?
16. A solar collector is located in Winnipeg, Canada (latitude 49° 50′ N, longitude 97° 15′ W). It is tilted 45° up from the horizontal plane, and it is facing S 15° E at 9:30 A.M., CST, on August 17.
    (a) Find the solar altitude and azimuth angles.
    (b) Estimate the total incident solar radiation on a clear day.
17. A solar collector with a single glass cover (surface area of 3 m^2) and insulated back and bottom sides is tilted at 50° with respect to the horizontal plane. The spacing between the plate ($\epsilon_p = 0.25$) and glass cover ($\epsilon_g = 0.85$) is 4 cm. The average plate temperature is 70°C, and the air temperature and wind speed are 15°C and 5 m/sec, respectively. Using the method of Klein and the exact (iterative) method, find the heat loss from the solar collector.
18. The copper tubes of a single-glazed solar collector are connected by a 2-mm-thick copper plate with a conductivity of 400 W/m^{2}K. Water flows through the tubes (0.5-mm-thick wall) with a heat transfer coefficient of 440 W/m^{2}K. The total conductance is $U_c = 10$ W/m^{2}K. If the tube-to-tube distance between inner edges of adjacent tubes is 15 cm, what tube diameter is required to give a collector efficiency factor of 0.87?
19. A solar collector is designed with the following characteristics: one glass cover, tilt angle of 60°C, overall conductance of $U_c = 6$ W/m^2, $T_a = 18$°C, and $I_p = 760$ W/m. The cover transmissivity is $\tau = 0.8$, and

the plate absorptivitiy is $\alpha = 0.9$. The copper plate thickness is 0.06 cm, and each tube has an inner diameter and wall thickness of 1.5 and 0.06 cm, respectively. The tube-to-tube centered distance is 10 cm. What is the plate temperature above each tube, T_b, when the collector efficiency reaches 0.35?

20. Performance testing of a single-glazed flat plate solar collector yields the following result for collector efficiency, η_c, in terms of incident radiative flux, I_p, ambient temperature, T_a, and incoming fluid temperature (within collector), $T_{f,in}$:

$$\eta_c \approx 0.82 - 7.0\left(\frac{T_{f,in} - T_a}{I_p}\right)$$

The transmissivity of the glass cover is 0.92, and the collector heat removal factor is 0.94.

(a) Find the collector surface absorptivity, α_s, and heat loss conductance, U_c, for this solar collector.

(b) The net energy absorbed by a solar collector over a surface area of 3 m^2 is $q_u = 1600$ W. If the incident radiative flux and ambient air temperature are 800 W/m^2 and 4°C, respectively, calculate the incoming fluid temperature, $T_{f,in}$, into the solar collector.

(c) Find the collector efficiency factor, F', under conditions with a water mass flow rate of 8 kg/min through the collector tubes.

5

Phase Change Heat Transfer

5.1 Introduction

In addition to heat exchange within a single phase, many engineering technologies involve multiphase systems. In these cases, additional consideration of heat transfer during phase transition may be required. In this chapter, the main physical processes arising during phase change will be described.

An essential characteristic of problems dealing with phase change is the movement of a phase interface with the release or absorption of latent heat at this interface. These types of problems are usually highly nonlinear, and thermophysical properties are typically different on each side of the phase interface. As a result, analytical solutions are generally available only for a limited class of one-dimensional problems, typically pure materials or fluids in infinite or semi-infinite domains.

In this chapter, phase change heat transfer in pure liquids and materials will be considered, as well as multicomponent systems. In general multiphase systems (i.e., solid, liquid, or gas phases) involving one or more components, the *Gibbs phase rule* identifies the number of phases present, P, as follows:

$$P + F = C + N \tag{5.1}$$

where C, N, and F refer to the number of components (i.e., two components in a binary alloy), number of noncompositional variables, and the number of degrees of freedom, respectively. Examples of noncompositional variables are pressure and temperature in a liquid–vapor system. The degrees of freedom refer to the number of externally controlled variables.

Example 5.1.1
Number of Phases in a Binary Component System.
The purpose of this example is to find the number of phases that coexist in equilibrium for a binary component mixture containing solid and liquid

phases. Use the Gibbs phase rule to find the number of independent thermodynamic variables in the pure phase and two-phase regions.

In the phase equilibrium diagram of a binary component system, $P+F=2+1$, where 1 refers only to temperature on a temperature–composition diagram for a solid–liquid system (see Figure 1.6). A value of $N=2$ would be realized for a liquid–vapor system since temperature and pressure would be the noncompositional variables. In the case of a solid–liquid system, $P=1$ (a pure phase region), and so $F=3-1=2$, which means that two degrees of freedom (or two externally controlled variables), such as temperature and material composition, are specified. In the two-phase region, both phases coexist in equilibrium. In this region, $P=2$, and so $F=3-2=1$. This implies that only temperature needs to be specified, since the individual phase compositions are identified by the phase equilibrium diagram for that particular system. In a solid–liquid system, these individual compositions are the solidus and liquidus lines, respectively (see Figure 1.6).

Furthermore, if $P=3$ (i.e., eutectic point), then $F=3-3=0$ since a single point is identified on the phase equilibrium diagram. If $P=3$ and $F>0$, this corresponds to a nonequilibrium condition (i.e., not in accordance with the phase equilibrium diagram).

Although the governing equations vary for each type of multiphase system (i.e., solid–liquid, liquid–gas, etc.), the overall features of the physical and mathematical models are analogous. These continuum models are generally written based on a *mixture formulation* or an *interface tracking formulation*. In the mixture approach, the control volume consists of a homogeneous mixture encompassing both (or all) phases. In this way, the control volume equation can be written in terms of a single mixture quantity, such as mixture velocity, rather than independently varying phase quantities. This approach is advantageous since it reduces the resulting number of equations, as well as their complexity. However, the spatial averaging procedure may lead to a certain loss of information regarding the interphase transport processes. Further supplementary relations, such as cross-phase interactions, are generally required in the absence of this detailed interface resolution.

In some classes of problems, such as two-phase (liquid–gas) flows, the alternative interface tracking approach is required instead. In this approach, the interface is explicitly resolved and the conservation equations are applied on both sides of the interface with appropriate matching conditions to equate the results at the phase interface. The final form of the equations is dependent on the phases on each side of the interface. Mixture and interface tracking formulations are presented in this chapter. Subsequent chapters will present detailed formulations for each specific type of multiphase system.

5.2 Processes of Phase Change

5.2.1 Fundamental Definitions

Various definitions are given here as a basis for the study of phase change processes. A *phase* is considered to be any homogeneous aggregation of matter. Liquids take the shape of their container and are essentially incompressible, whereas solids retain their original shape unless deformed by external forces. On the other hand, gases are characterized by their low density and viscosity, compressibility, lack of rigidity, and optical transparency.

In characterizing the structure of solids, a *unit cell* is used to identify a geometric configuration for a grouping of atoms in the solid. *Allotropy* refers to the existence of two or more molecular or crystal structures with the same chemical composition. This group is repeated many times in space within a crystal. *Interstitial diffusion* refers to migration of atoms through microscopic (interstitial) openings in the lattice structure of a solid. A *peritectic reaction* is a reaction whereby a solid goes to a new solid plus a liquid on heating and the reverse on cooling. A *peritectoid* refers to transforming from two solid phases to a third solid phase upon cooling. In a *polyphase* material, two or more phases are present. The *primary phase* is the phase that appears first upon cooling. The *solubility* of a multi-component mixture refers to the maximum concentration of atoms of a particular constituent that can dissolve into the mixture. Gases can dissolve into liquids. This solubility generally increases with pressure, but decreases with temperature.

In the following sections, the transport phenomena of specific phase change processes will be outlined.

5.2.2 Boiling and Condensation

During phase change with boiling, vapor bubbles initially form within cavities along the heated surface. The bubbles grow when liquid is evaporated around the bubble. In particular, bubble growth along the heated surface occurs by evaporation of a liquid *microlayer* beneath the bubble. Bubbles typically expand due to heat transfer through the base of each bubble. Heat is also transferred from the wall around the cavity to the liquid beneath the bubble (liquid microlayer). This heat transfer leads to evaporation of the bubble above the microlayer and movement of the vapor–liquid interface toward the heated wall until it reaches the wall and the bubble detaches from the surface cavity. Additional liquid comes into contact with the heating surface to sustain the boiling process. Due to its higher heat capacity and thermal conductivity, the local heat transfer rate is enhanced when liquid comes into contact with the surface. Velocity and temperature diffusion layers at a micro-scale are formed around the surface of each departing bubble.

Once a bubble departs from the heated surface, a microconvection process occurs whereby liquid is entrained into the cavity where the vapor emerged. During the departure of the previous bubble, a vapor remnant was retained in the cavity, and liquid flow over the cavity traps this remnant to initiate formation of another bubble at this location. If further heating occurs, the number of nucleation sites increases and isolated bubbles can merge into continuous columns or channels of vapor. During this time, liquid flows downward toward the surface between these vapor columns.

A variety of transport processes occurs simultaneously during the bubble growth. Predictive models of bubble growth rates have been developed by various researchers, such as Plesset and Zwick (1954). As the microlayer evaporates beneath the bubbles, latent heat is extracted, while colder liquid flows into the near-wall region. In subcooled liquids, evaporation on the hot side of the bubble, together with conduction across the bubble due to small temperature differences, may lead to varying surface tension along the edge of the bubble. Microlayer evaporation is enhanced with the convective flow associated with this process. Due to the difficulties of temperature and other measurements at such small scales associated with these processes, the relative interactions between these physical processes are not well understood.

Surface orientation is an important factor in the boiling process. In comparison to upfacing horizontal surfaces, inclined surfaces can promote more active microlayer agitation and mixing along the surface. It has been shown that more frequent bubble sweeps along the surface lead to more effective heat transfer in downward-facing orientations (Naterer et al., 1998). For inclined surfaces, bubbles forming in a cavity begin to slide along and away from the cavity's upper edge as heat is transferred from the surface. After a bubble expands and departs from the cavity, the lower edge of the trailing liquid–vapor interface moves into the cavity. Other ascending bubbles approach this cavity and merge with the remaining gas phase in the cavity. A new bubble is formed through this merging process. The bubble motion along the inclined surface is continued, and the wake of the passing bubble mixes with the microlayer forming over the cavity. This sequence of stages during the bubble growth is repeated throughout the boiling process. Thus, boiling heat transfer consists of several distinct stages.

As mentioned earlier, the boiling process begins when individual bubbles form and grow along the heated surface. The liquid microlayer beneath a growing bubble is evaporated, leading to departure of the newly formed bubble. Following the bubble's departure (before the next bubble emerges), heat is transferred by transient conduction from the heated surface to the superheated liquid layer in contact with the surface. This process may be approximated through a combination of liquid and vapor periods where heat conduction solutions can be applied at both successive stages (Nishikawa et al., 1984; Naterer et al., 1998). The average heat flux over both periods is anticipated to have a close dependence on the frequency of bubble departure (Mikic and Rohsenow, 1969). The changing liquid layer thickness beneath a bubble is implicitly coupled with the heat conduction process in the liquid and latent heat released in the adjacent vapor phase.

In contrast to boiling heat transfer, condensation arises when vapor is brought into contact with a cooled surface below the saturation temperature. If a quiescent vapor comes in contact with a cooled surface, two fundamental modes of condensation can occur, namely, dropwise and film condensation. The liquid does not fully cover or wet the surface in dropwise condensation, whereas a liquid film covers the entire surface in film condensation. A lower rate of heat transfer between the vapor and wall usually arises in film condensation, due to the added thermal resistance of the liquid film.

It is useful to compare the approximate ranges of condensation heat transfer coefficients under various thermal conditions. For condensation of steam on surfaces about 3 to 20°C below the saturation point, the average heat transfer coefficient, $\bar{h}$, ranges between 11,000 and 23,000 W/m^{2}K for horizontal tubes (outer diameter of 25 to 75 mm) and between 5,700 and 11,000 W/m^{2}K for vertical surfaces (3 m high). However, for condensation of ethanol along a horizontal tube (outer diameter of 50 mm, $5 < \Delta T <$ 20°C), $\bar{h}$ varies between 1700 and 2600 W/m^{2}K. Various physical and other parameters affect the variations in these values. For example, the condensation heat transfer coefficient depends considerably on thermophysical properties, as well as the geometrical configuration, thereby partly explaining the previously reported differences in values between steam and ethanol. Detailed analysis of condensation heat transfer is documented by Collier (1981).

The heat released or absorbed during liquid–vapor phase change (boiling or condensation) can be expressed as the amount of mass multiplied by the latent heat of vaporization. For example, the rate of heat transfer during boiling, q, can be determined based on

$$\frac{q}{\dot{m}} = h_{fg} \tag{5.2}$$

where $\dot{m}$ and h_{fg} refer to the mass of liquid boiled per unit time and the latent heat of vaporization, respectively. Similarly, under this approach for condensation of pure substances, $\dot{m}$ would refer to the rate of vapor condensation on a cooled surface.

However, a subtle correction of the right side of Equation (5.2) is observed in view of the phase change that occurs at a fixed temperature for pure (single constituent) fluids (Bejan, 1993). Consider that the enthalpy of vaporization in Equation (5.2) is evaluated at a fixed temperature. A change of phase (liquid or vapor) at a fixed temperature implies a constant pressure, since pressure and temperature are mutually dependent for liquid–vapor mixtures at equilibrium. This correction of Equation (5.2) is clarified in the following example.

Example 5.2.2.1

Release of Latent Heat by Phase Change.

A rigid tank initially consists of a specified mass of liquid and vapor in equilibrium, together with a relief valve at the top of the tank to maintain a constant pressure during the phase change (see Figure 5.1). The problem may be considered in view of condensation (i.e., heat lost from the saturated vapor) or boiling (i.e., heat gained by the saturated liquid). Determine an expression for the release of latent heat during the change of phase, particularly a correction to Equation (5.2), that appears as a result of the phase change at a constant pressure.

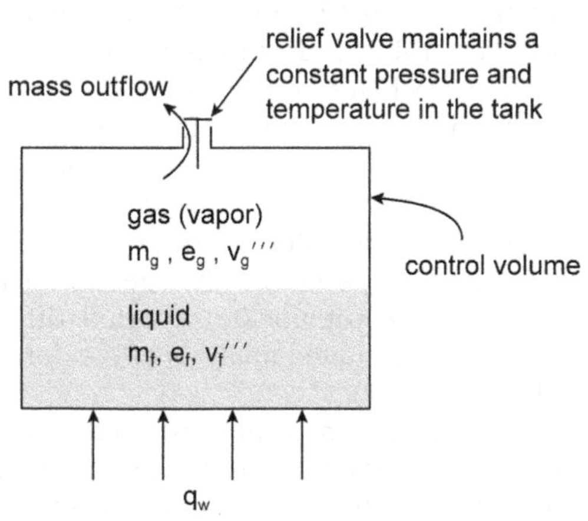

FIGURE 5.1
Liquid–vapor phase change in a rigid tank.

Based on an energy balance on the control volume,

$$\frac{d\hat{E}}{dt} = q - \dot{m}h_g \tag{5.3}$$

where the positive q refers to heat addition (i.e., boiling). An analogous result for condensation can be obtained by switching to a negative q. The energy balance in Equation (5.3) states that the rate of change of internal energy in the tank is equal to the heat addition minus the energy loss by vapor carried out of the tank to maintain an isobaric process. Expanding the internal energy in Equation (5.3) in terms of individual specific internal energies of the liquid and vapor phases,

$$\frac{d\hat{E}}{dt} = \frac{d}{dt}(m_f\hat{e}_f + m_g\hat{e}_g) = \dot{m}_f\hat{e}_f + \dot{m}_g\hat{e}_g \tag{5.4}$$

where the subscripts f and g refer to fluid (liquid) and gas (vapor), respectively.

Also, conservation of mass requires that the rate of mass change within the tank balances the flow rate of escaping steam, i.e.,

$$\frac{d}{dt}(m_f + m_g) = -\dot{m} \tag{5.5}$$

During the phase change process, the total volume remains constant, so that

$$\mathcal{V} = m_f v_f''' + m_g v_g''' \tag{5.6}$$

which can be differentiated to yield

$$\dot{m}_f v_f''' + \dot{m}_g v_g''' = 0 \tag{5.7}$$

Combining Equations (5.5) and (5.7),

$$\dot{m}_f = -\dot{m}\left(\frac{v_g'''}{v_g''' - v_f'''}\right) \tag{5.8}$$

$$\dot{m}_g = \dot{m}\left(\frac{v_f'''}{v_g''' - v_f'''}\right) \tag{5.9}$$

Furthermore, substituting Equations (5.8) and (5.9) into Equation (5.4) and then Equation (5.3) and rearranging,

$$\frac{q}{\dot{m}} = h_g - \frac{\hat{e}_f v_g''' - \hat{e}_g v_f'''}{v_g''' - v_f'''} = h_{fg} + \left(h_f - \frac{\hat{e}_f v_g''' - \hat{e}_g v_f'''}{v_g''' - v_f'''} \right) \tag{5.10}$$

where the bracketed term represents the appropriate correction of Equation (5.2). The enthalpy of the saturated liquid, h_f, was subtracted from h_g in Equation (5.2) (note: $h_{fg} = h_g - h_f$), whereas a correction of h_f involving internal energies and specific volumes was provided by the last fraction in Equation (5.10). It can be shown that the right side of Equation (5.10) deviates only slightly from the right side of Equation (5.2), except near the critical point.

The correction discussed in the previous example has arisen due to the change of internal energy required to maintain a constant pressure and temperature during the liquid–vapor phase change. The appropriate expression for $q/\dot{m}$, namely, Equation (5.2) or Equation (5.10), is selected based on the specific problem conditions. The right side of Equation (5.2) is usually adopted in this textbook, but it requires the indicated modification in proximity of the critical point. In the next section, another type of phase change process (between solid and liquid phases) will be described.

5.2.3 Solidification and Melting

Solidification and melting occur in many engineering applications (Hewitt et al., 1997), including materials processing (i.e., casting solidification, extrusion, injection molding), ice formation on aircraft and other structures, and thermal energy storage with phase change materials (PCMs). During solid–liquid phase change, various physical processes (including heat conduction and convection) affect the structure and movement of the phase interface. A solid or liquid phase will refer to a portion of the system where the properties and composition of the material are homogeneous. In a manner similar to that described in the previous section, the heat transfer, q, due to phase change between solid and liquid phases can be written as

$$q = \dot{m} h_{sf} \tag{5.11}$$

where $\dot{m}$ and h_{sf} refer to the rate of mass change due to the phase transition and the latent heat of fusion (sometimes denoted by L), respectively. In this section, the fundamental processes leading to solid–liquid phase change will be examined.

Heat transfer has important effects on interfacial processes during phase change. Consider a process involving solidification of a liquid initially held at a uniform temperature. Then the liquid is suddenly exposed to a cooled boundary at a temperature below the phase change temperature. Solidification is initiated and a planar solid–liquid interface moves into the liquid over time. The boundary is maintained below the phase change temperature and remains parallel to the planar phase interface. In order to understand whether the interface remains planar over time, consider that the interface becomes slightly perturbed so that a small section of it moves further from the heat sink (cooled boundary) than the undisturbed parts of the interface. Since this perturbed part is further from the heat sink, its temperature gradient becomes smaller than that of the surrounding undisturbed parts of the phase interface. Thus, less heat flows from the disturbed section, thereby requiring it to grow slower than the undisturbed regions. In this case, the perturbation will disappear and the phase interface is restored to its initially stable and planar form.

During this heat extraction, the energy of the solid and liquid phases is changed in two ways: (i) sensible cooling with a temperature change and (ii) a change of energy (equivalent to the latent heat of fusion) arising from the change of phase. Despite a temperature decrease due to the heat removal from the cooling boundary, latent heat is released simultaneously from the solidifying material. Once this latent heat is released, it is transferred by conduction (and other modes of heat transfer) to the surrounding solid or liquid. However, if the rate of release of latent heat exceeds the rate at which heat is removed from the solidifying crystal and the cooling boundary, a local rise of temperature (called *recalescence*) may be observed. In some circumstances, these phenomena can explain physical processes such as repetitive remelting of solidified crystals.

The Gibbs free energy, G, has special importance in solid–liquid phase change. It was defined in Chapter 1 as

$$G = H - TS \tag{5.12}$$

where H, T, and S refer to enthalpy (extensive), temperature, and entropy (extensive), respectively. Furthermore, the chemical potential, μ, is defined as the rate of change of Gibbs free energy with respect to a change in the number of moles of the mixture. For example, considering component A in a multicomponent mixture,

$$\mu_A = \left.\frac{\partial G_A}{\partial n_A}\right|_{T,p} \tag{5.13}$$

where the derivative is evaluated at constant pressure and temperature. It can be observed that uniformity of the Gibbs free energy, i.e., $dG = 0$, is required at thermodynamic equilibrium since no temperature or pressure gradients would exist therein.

A useful result that follows from the definition of the Gibbs free energy is the calculation of the entropy of fusion due to the solid–liquid phase change. Consider a solid and liquid separately, and define the Gibbs free energy of each phase as it varies with temperature (below and above the phase change temperature), i.e.,

$$G^l = H^l - TS^l \tag{5.14}$$

$$G^s = H^s - TS^s \tag{5.15}$$

where the superscripts l and s refer to liquid and solid, respectively. It should be noted that these two curves intersect at the phase change temperature, T_m.

The change in Gibbs free energy, ΔG_{mix}, due to mixing of the two phases at any temperature is

$$\Delta G_{mix} = \Delta H_{mix} - T\Delta S_{mix} \tag{5.16}$$

Thus, at the phase change temperature ($T = T_m$),

$$\Delta G = 0 = \Delta H - T_m \Delta S \tag{5.17}$$

which suggests that the entropy of fusion can be approximated by

$$\Delta S_f \approx \frac{L}{T_m} \approx R \tag{5.18}$$

where L represents the latent heat of fusion (ΔH_{sf}) and $R \sim 8.4$ J/mol K is approximately constant for most metals (called *Richard's rule*). The entropy of fusion is a useful way to characterize different materials, i.e., $\Delta S_f/R$ is about 1 for metals, 2 or 3 for semiconductors, and 20 to 100 for complex molecular structures and polymers.

In a manner similar to that discussed for nucleate boiling, phase change from liquid to solid phases is initiated by a nucleation process. Nucleation occurs when the probability of atoms arranging themselves on a crystal lattice is high enough to form a solid crystal from the liquid. *Homogeneous nucleation* refers to solidification of liquid initiated by undercooling alone (i.e., without particle impurities in the liquid to assist in crystal formation). On the other hand, *heterogeneous nucleation* (more common case) occurs when the walls of a container or particle impurities exist to assist in providing nucleation sites for the solidifying crystals. Since heterogeneous nucleation is often characterized in terms of the undercooling of its corresponding homogeneous nucleation, the latter type (homogeneous nucleation) will be further described here.

The critical condition for the onset of homogeneous nucleation occurs when the decrease in the Gibbs free energy of the crystal becomes (Kurz, Fisher, 1984)

$$\Delta G = \sigma A + \Delta g \Delta V \tag{5.19}$$

where Δg and ΔV refer to Gibbs free energy difference between the liquid and solid (per unit volume) and spherical volume enclosed by the crystal (surface area of A, characterized by a radius of r_o), respectively. Also, σ refers to the specific interface energy (J/m^2). Equation (5.19) states that the decrease in the Gibbs free energy due to phase change balances the work required to keep the initial crystal bonds in the lattice structure from melting back to the liquid (first term) plus the change in the Gibbs free energy in changing from liquid to solid phases (second term).

The first term on the right side of Equation (5.19), the work term required to maintain the crystal bonds together without melting, is characterized by the interface energy, σ. This interface energy is a property of the material. For example, it can be written in terms of the Gibbs–Thomson coefficient, Γ_{gt}, for metals, i.e.,

$$\Gamma_{gt} = \frac{\sigma}{\Delta s_f} \approx 1 \times 10^{-7} \text{ Km} \tag{5.20}$$

where Δs_f is the entropy of fusion per unit volume (J/m^{3}K).

Using Equation (5.18) while considering phase change at T_m, the Gibbs free energy per unit volume becomes

$$\Delta g = \Delta h - T\Delta s = L - T\frac{L}{T_m} \tag{5.21}$$

where the enthalpy and entropy changes with temperature (i.e., due to undercooling) in the middle expression of Equation (5.18) are assumed to be small in comparison to the respective changes associated with the phase transformation.

Combining both terms in Equation (5.21),

$$\Delta g = \left(\frac{L}{T_m}\right)\Delta T = \Delta s_f \Delta T \tag{5.22}$$

where $\Delta T = T_m - T$ is the undercooling level, i.e., temperature deviation below the phase change temperature. Substituting Equation (5.22) into Equation (5.19) with a radius of r to represent a characteristic spherical volume, $r = r_o$, encompassing the nucleating crystal,

$$\Delta G = \sigma 4\pi r^2 + \frac{4}{3}\pi r^3 \Delta s_f \Delta T \tag{5.23}$$

The critical condition for the onset of nucleation occurs when the driving force for solidification is equal to the respective force for melting. In particular, this occurs when G is minimized or, equivalently, ΔG from the initial liquid state is maximized. This extremum corresponds to the activation energy that must be overcome in order to form a crystal nucleus from the liquid. This critical condition is outlined in the following example.

Example 5.2.3.1

Critical Radius of a Crystal for the Onset of Nucleation.

A liquid is cooled to a temperature below its equilibrium phase change temperature. Use the Gibbs free energy to determine the critical radius of a crystal that initiates homogeneous nucleation in the solidifying liquid.

The maximized ΔG occurs when

$$\frac{d(\Delta G)}{dr} = 0 \tag{5.24}$$

For a spherical volume, the Gibbs free energy is given by Equation (5.23).

Differentiating the expression in Equation (5.23) with respect to r, substituting into Equation (5.24), and solving for the resulting critical undercooling,

$$\Delta T = \frac{2\sigma}{r_o \Delta s_f} \tag{5.25}$$

which gives the critical condition to initiate homogeneous nucleation. Alternatively, in terms of the critical radius,

$$r_o = \frac{2\sigma}{\Delta T \Delta s_f} \tag{5.26}$$

The radius r_o refers to the critical nucleation radius for homogeneous nucleation (i.e., a larger radius would not solidify as a stable lattice structure).

Without particle impurities in the liquid or a rough-walled container to prematurely initiate the phase transition, the undercooling level in Equation (5.25) is required to sustain the nucleation site. This result can represent a large level of undercooling for homogeneous nucleation. For example, liquid nickel can be theoretically held at $\Delta T \approx 250$ K before solidifying. In addition to the required undercooling for the initial nucleation, another important parameter is the rate of ongoing nucleation once it begins. The variable $I(T)$ (units of m^3/sec) is used to refer to the nucleation rate as a function of temperature, T. It denotes the number of nuclei created per unit time. The rate of nucleation, or rate of forming new crystals on the lattice structure, increases with the level of undercooling below the phase change temperature.

For example, the rate of nucleation for a slightly undercooled liquid is less than the rate of nucleation for a material that is substantially undercooled below the phase change temperature because the former (warmer) material possesses more internal energy to possibly melt crystals back into the liquid, thereby reducing the rate of nucleation. This increase of nucleation rate with undercooling level (up to some maximum point, I_{max}) can be expressed as follows:

$$I(T) = exp\left(\frac{-\Delta G_n^o}{\kappa T}\right) \tag{5.27}$$

where ΔG_n^o and κ refer to Gibbs activation energy (material dependent) and Boltzmann's constant ($\kappa \approx 1.38 \times 10^{-23}$ J/K), respectively.

At a low undercooling level, there is a relatively high energy barrier to sustain the nucleus formation. The exponential term in Equation (5.27) indicates the sufficient amount of activation energy required for nucleation of a critical number of clustered atoms in the crystal lattice. However, at very high levels of undercooling, the nucleation rate begins to decrease as a result of the decrease in the rate of atomic diffusion as the temperature approaches absolute zero (0 K). As a result, the nucleation rate reaches a maximum value, I_{max}, and then decreases with further reduction in the undercooling level.

For materials exhibiting a higher density in the solid phase (such as most metals and alloys), the liquid occupies a larger volume than the solid phase at the phase change temperature. As a result, the atomic packing is greater in the solid phase, which implies that there is more freedom of movement of atoms in the liquid. At the stage of phase transition between liquid and solid phases, many close-packed clusters of atoms in the liquid phase are temporarily in the same crystalline array as the solid phase. Equation (5.27) may be interpreted alternatively as follows:

$$n_r = n_o exp\left(\frac{-\Delta G_n^o}{\kappa T}\right) \tag{5.28}$$

where n_r and n_o refer to the average number of newly formed spherical clusters of radius r and the total number of atoms, respectively. As described earlier, ΔG_n^o represents a thermal activation barrier over which a solid cluster must pass to become a stable nucleus.

In Equation (5.28), ΔG_n^o increases with r, and so n_r decreases when the spherical radius cluster increases. In other words, the probability of forming a cluster from the liquid phase decreases very rapidly as the cluster size increases. A typical limit is $n_{r,max} \approx 100$ atoms, which represents a required number for a reasonable probability of sustaining a stable solid crystal from the liquid. At small undercooling levels, the critical radius of nucleation is large, which means that there is a small probability of forming a stable nucleus from the liquid under small undercooling conditions.

Based on Equations (5.25) to (5.28), the speed at which solid nuclei appear in the liquid (denoted by $\dot{n}_{hom}$) at a given undercooling level can be better understood. Equation (5.25) was rearranged to predict the critical nucleation radius, r_o, required at the critical nucleation undercooling. Also, the left side of Equation (5.28) (with units of clusters/m^3) can be interpreted as the number of clusters reaching the critical radius r_o. It is related to the speed of solid nuclei formation as follows:

$$\dot{n}_{hom} = f_o n_r \tag{5.29}$$

where f_o is the frequency of adding atoms to the stable, newly formed solid cluster. This frequency is dependent on the interface energy, surface tension at the phase interface, and areas of the nuclei.

A certain time is required for nucleation of an undercooled liquid. This duration is analogous to the previously described nucleation rate. The time required for nucleation is large at small undercooling levels, but then decreases to a minimum point with further undercooling and increases thereafter as the temperature approaches absolute zero, due to the reduced atomic diffusion at very low temperatures. If temperatures in the subcooled liquid are dropped rapidly (i.e., rapid solidification), an *amorphous solid* may be formed, which is characterized by a lack of any crystalline structure. In rapid solidification, little time is allowed for regular atomic ordering on the crystal lattice during phase transition.

Most solidification processes usually occur at undercooling levels lower than those outlined by the previous homogeneous nucleation analysis. In heterogeneous nucleation, the formation of the initial solid is assisted by particle impurities in the liquid, surface roughness, and other assisting factors to nucleation. The following example demonstrates how relevant forces affect the onset of heterogeneous nucleation.

Example 5.2.3.2

Heterogeneous Nucleation of Crystals.

A hemispherical-shaped solid nucleus (subscript s) with a spherical radius of curvature r is formed in contact with a smooth surface (mold, subscript m). The angle of contact between the tangent to the surface and the solid–liquid interface is θ (note: subscript l will refer to liquid). Using appropriate force balances at the interfaces between the solid and liquid, explain how to find the critical undercooling required for the onset of heterogeneous nucleation in terms of the angle of contact, θ.

Based on a force balance involving surface tensions, γ, at the junction between the solid, wall (mold), and liquid,

$$\gamma_{ml} = \gamma_{sm} + \gamma_{sl} cos\theta \tag{5.30}$$

The relevant areas of the solid–liquid and solid–mold interfaces, as well as the solid volume, respectively, can be shown to be

$$A_{sl} = 2\pi r^2 (1 - cos\theta) \tag{5.31}$$

$$A_{sm} = \pi r^2 sin^2\theta \tag{5.32}$$

$$V_s = \frac{1}{3}\ \pi r^3 (2 + cos\theta)(1 - cos\theta)^2 \tag{5.33}$$

The Gibbs free energy required to sustain this solid nucleus along the mold wall without remelting is

$$\Delta G_{het} = \gamma_{sl} A_{sl} + \gamma_{sm} A_{sm} + V_s G_s - A_{sm}\gamma_{ml} - V_s G_l \tag{5.34}$$

The first three terms on the right side of Equation (5.34) refer to the energies required to create the new solid–liquid interface during the nucleation process. These energies involve surface tension to sustain the new phase interface and the Gibbs free energy to form a spherical cluster of nuclei. The latter two terms in Equation (5.34) are subtracted as a result of the destruction of the previous mold–liquid interface and liquid that previously occupied the volume of the newly formed solid.

The surface tension expressions in Equation (5.34) may be combined in terms of the required interface energy, σ, to sustain the new phase interface. Also, the difference of Gibbs energies can be written in terms of the entropy of fusion based on Equation (5.22). In this way, Equation (5.34) becomes

$$\Delta G_{het} = \left(\sigma 4\pi r^2 + \frac{4}{3}\ \pi r^3 \Delta s_f \Delta T\right) S(\theta) \tag{5.35}$$

where

$$S(\theta) = \frac{1}{4}(2 + cos\theta)(1 - cos\theta)^2 \quad (5.36)$$

is the shape factor in this particular example of heterogeneous nucleation. In a fashion similar to that of the differentiation performed in Equation (5.24), the Gibbs free energy difference in Equation (5.35) is maximized when the derivative of ΔG_{het} with respect to r is set to zero. Performing this differentiation, the same critical radius as that in Equation (5.26) is obtained.

Comparing Equations (5.23) and (5.35), it can be observed that the Gibbs free energy for heterogeneous nucleation is the corresponding ΔG for homogeneous nucleation multiplied by a shape factor, $S(\theta)$, in Equation (5.36). Since the shape factor is less than unity, ΔG_{het} is less than ΔG_{hom} (generally much less than the Gibbs free energy difference for homogeneous nucleation). The critical undercooling can now be obtained through a procedure similar to that carried out in Equations (5.24) and (5.25).

Although the previous analysis was performed for heterogeneous nucleation along a flat mold wall, the critical nucleation radius analysis holds for any nucleation geometry, such as rough surfaces or solid impurities in the liquid. The initiation of heterogeneous nucleation is important in various practical instances, such as the addition of inoculants to industrial chemicals to act as agents reacting with a component in the liquid to form a solid nucleus. The effectiveness of an inoculant is dependent on the nucleation geometry, particularly θ, interface energy, σ, and entropy of fusion of the solidified material.

The structure of the phase interface can be characterized by the nature of the surface area bonding at the interface. This can be explained by the number of sides attached to the lattice structure in a cube-shaped crystal. The interface growth is determined by the probability that a molecule will reach the interface and remain there until it bonds to the interface (without melting back to the liquid). This probability increases with an increasing number of neighbors in the crystal. For example, a type 3 atom is halfway in the solid (three sides bonded to the solid) and halfway in the liquid (three sides exposed to liquid). A certain sequence of phase transition can then be envisioned (see Figure 5.2). Type 3 atoms are added until a row along the interface is completed. Then type 2 atoms (i.e., two sides bonded to the solid) are needed to start the next row, which is adjacent to the previous row at the solid–liquid interface. This requires more undercooling than a type 3 formation. Finally, an entirely new row is initiated by a type 1 atom (i.e., only one side attached to the solid). This formation requires the most

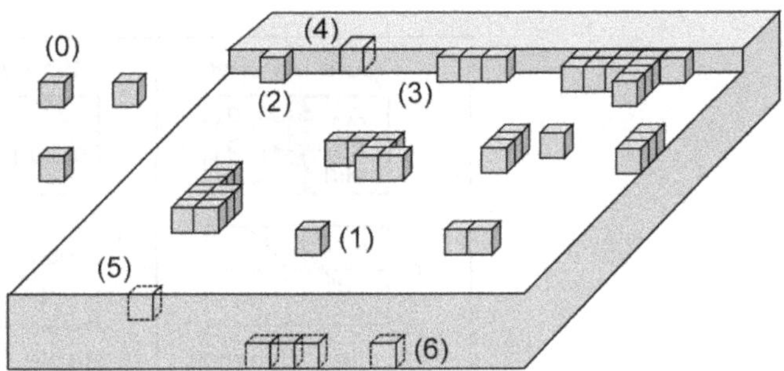

FIGURE 5.2
Stages of crystal and solid formation.

undercooling since the remaining five sides are exposed to liquid and possible remelting.

Another way that the phase interface can be characterized is by its faceted structure. A *faceted interface* between the liquid and solid can be described as atomically smooth (i.e., smooth at a 0.5-nm scale), yet microscopically jagged (i.e., jagged at a 10-μm scale). Comparatively few bonds are exposed to atoms in the liquid, and there is a tendency to close atomic gaps at the phase interface. It is usually formed under high undercooling levels in materials with a high entropy of fusion as a result of a large difference in microscopic structures between the liquid and solid phases. Minerals are considered to have faceted interfaces during phase transition. The large difference in atomic structures in both phases leads to a diminished tendency to incorporate new atoms into the solidified crystal structure, thereby preserving a microscopically smooth phase interface.

On the other hand, a *nonfaceted interface* is characterized as atomically rough, yet microscopically smooth. In comparison to faceted interfaces, there are more sites available for atom attachment from the liquid phase during solidification. Nonfaceted interfaces are usually formed under low undercooling levels and materials with low entropy of fusion values, such as metals and metal alloys. For example, $\Delta S_f/R$ is about 1 for metals and 2 or 3 for semiconductors. Anisotropy in the material properties, such as interface energy, is commonly experienced in these materials. The dendritic growth usually occurs along preferred crystallographic directions, and there is a small difference in structure and bonding between the phases.

The morphological stability affects the structure of the solid–liquid interface. For a pure material, a stable interface is called *columnar*, whereas an unstable interface is *equiaxed*. In the stable, columnar interface, consider the onset and advancement of a wavy phase interface in the positive x direction due to solidification and heat extraction through the solid (see

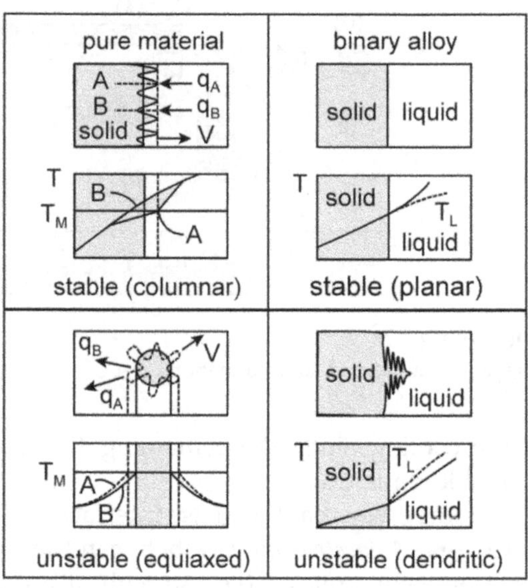

FIGURE 5.3
Stability of phase interface.

Figure 5.3). The peak and valley of the wavy interface (where the peak refers to the farthest extent into the liquid) will be denoted along planes A and B, respectively. The corresponding heat flows are q_A (through the solid) and q_B (into the solid from the liquid), respectively. The temperature gradient in the liquid is higher along plane A since the phase change temperature is reached at the edge of the solid perturbation. As a result, q_A is larger than q_B at that location, so that the perturbation is melted and the interface remains stable (planar).

On the other hand, consider the wavy interface, specifically an interface formed along the outer edge of a spherically shaped crystal that grows radially outward. In this case, latent heat is released radially outward from the solidified crystal into the liquid. The crystal grows radially outward until it impinges on other nuclei. The heat flows along the peak (plane A) and valley (plane B) of the wavy interface are denoted by q_A and q_B, respectively. In this case, the temperature gradient is again higher along plane A, which suggests that the tip of the wavy interface rejects more heat. As a result, the perturbation at this tip grows faster and the waviness of the interface increases (i.e., unstable interface). The resulting structure is called equiaxed due to the sustained growth of interface perturbations along certain crystal axis directions.

The morphological stability of the solid–liquid interface is affected by the presence of other constituents in the material. The solubility limit refers to the maximum concentration of solute that can dissolve in the solvent to form a solid solution. This solubility limit affects the coupled heat and mass transfer in phase change processes. For example, consider a binary alloy that consists of lead and tin (both in solid and liquid phases). During phase transition, the concentration of solute (tin) in the solid remains less than that of the original mean solute composition due to the difference in solubilities of both constituents within each phase. The balance of the solute from that original concentration remains in the liquid.

This transfer of solute (called solute "rejection") at the phase interface creates a solute diffusion layer that reduces the equilibrium liquidus temperature (see Figure 1.6). The diffusive flux of solute is based on the local solute concentration gradient through *Fick's law*. For instance, considering a one-dimensional diffusive flux, denoted by J_x, in the x direction,

$$J_x = -D\frac{dC}{dx} \tag{5.37}$$

where D refers to the mass diffusivity (analogous to the thermal conductivity used in Fourier's law). The solute diffusion layer at the phase interface affects the liquidus temperature. Values of mass diffusivity in Equation (5.37) are typically much lower than thermal diffusivities in heat conduction. For example, $D \approx 9 \times 10^{-8}$ m^2/sec for carbon–iron steels at a temperature of 1500 K, whereas the thermal diffusivity is about 3×10^{-5} m^2/sec at the same temperature. In aluminum alloys (i.e., Mg–Al) at 500 K, the thermal diffusivity is about 2×10^{-4} m^2/sec, whereas $D \approx 10^{-14}$ m^2/sec for mass transfer.

Example 5.2.3.3

Solute Diffusion with Solidification in a Binary Component Mixture.

Find the governing equation for diffusion and transfer of solute at an advancing solid–liquid interface under one-dimensional conditions. In the analysis, consider diffusion in the liquid and solute redistribution at the phase interface; solid diffusion and convection may be neglected. Express the result in terms of a uniform interface velocity, U, and the liquid mass diffusivity, D_l.

A control volume of thickness Δx is defined ahead of the advancing interface. Within this control volume, an incremental change of solid fraction, df_s, is observed with the interface advance. The constant interface velocity and mass flux of solute at position x are denoted by U and J_x, respectively. Then, based on conservation of solute in the control volume and a Taylor series approximation of the mass outflux,

$$\frac{d}{dt}(\rho C \Delta x) = J_x - \rho U C_l + \rho U C_s - \left(J_x + \frac{dJ_x}{dx} \Delta x \right) \tag{5.38}$$

Solid diffusion is assumed to be appreciably less than the corresponding liquid diffusion over the same time scale. The change of interface concentration with time on the left side of Equation (5.38) is assumed to be small in view of the constant solute concentration in the bulk liquid.

The terms on the right side of Equation (5.38) refer to the liquid diffusion at position x, rate of solute loss in the newly formed solid volume ($df_s \Delta V$), rate of solute gain in the newly formed solidified volume, and liquid diffusive flow across $x + \Delta x$, respectively. Convection effects have not been included in this analysis. Then, dividing Equation (5.38) by Δx, taking the limit as $\Delta x \to 0$, and applying Fick's law,

$$D_l \frac{d^2 C_l}{dx^2} + U \frac{dC_l}{dx} = 0 \tag{5.39}$$

which can be solved, subject to appropriate boundary conditions, to find the spatial profile of solute concentration ahead of the advancing phase interface (note: see end-of-chapter problems).

If the actual temperature in the liquid ahead of the solid–liquid interface is higher than T_l (liquidus temperature; see Figure 1.6), then there would be no constitutional supercooling due to the solute diffusion layer. In this case, an initially planar phase interface is stable and would remain planar. On the other hand, if the actual temperature ahead of the interface is less than T_l, then the constitutional supercooling from the interfacial solute rejection would lead to an unstable, solidified dendritic formation ahead of the phase interface (see Figure 5.3). A dendritic structure refers to solidification ahead of the interface into main, secondary, and other branched arms to eliminate the constitutional supercooling caused by the interfacial solute diffusion layer.

A certain sequence usually occurs in interface transition from a stable, planar interface to an unstable interface. This transition usually proceeds in the following fashion for an increasing degree of constitutional supercooling: planar, cellular, cellular–dendritic, dendritic, and finally globulitic (i.e., formation of isolated globule-type structures resembling dendrites). For instabilities at the phase interface (pure material or multicomponent material), the rate of growth of the perturbation is dependent on the wavelength of the instability, λ. This wavelength refers to the distance between successive peaks of the wavy phase interface. The ratio of this rate of growth, $d\epsilon/dt$, to the perturbation height, ϵ, will be called the perturbation growth ratio (PGR). It is useful to examine the variation of

PGR with λ since this distinguishes stable interface growth from unstable growth.

For example, a solidifying aluminum–copper (2%) alloy with an interface velocity of 0.1 mm/sec exhibits a negative but increasing PGR up to $\lambda_i \sim 2$ μm (limit of stability), where PGR = 0. At these small wavelengths, the perturbations disappear due to the high curvature of the phase interface. As a result, dendrite arms grow into each other to eliminate the interfacial instability. Above a wavelength of λ_i, the value of PRG increases to reach a maximum value of nearly 100/sec at λ_o, where diffusion of solute across the waves of the wavy interface leads to accelerated instability growth. Beyond λ_o, the value of PRG decreases since less solute diffusion across the wavy peaks of the perturbed interface at larger λ means that each instability is not accelerated. The ratio of PRG asymptotically reaches 0 at high wavelengths since $\lambda \to 0$ represents an effectively nonwavy (nonperturbed) interface.

As discussed, when the actual liquid temperature is less than T_l (liquidus temperature) ahead of the interface, the resulting constitutional supercooling leads to an interface instability. It also leads to the formation of cells along the phase interface since any perturbation forming on the interface would be supercooled and not melted back into the liquid. These resulting cell formations grow with their axes parallel and opposite to the direction of heat flow without regard to the actual crystal orientation in the solid. An asymmetry of the dendritic cells can be caused by asymmetry in the thermal field. Due to anisotropy of properties such as σ (interface energy), the dendrites will typically grow in the preferred crystallographic direction, which is closest to the heat flow direction. For example, the dendrite orientation for aluminum (face-centered cubic (FCC)) is [1 0 0], whereas the orientation for tin (body-centered tetragonal) is [1 1 0]. The tip of a dendrite may be closely approximated as a paraboloid of revolution. A dendrite that has a small tip radius, R_s, tends to increase its radius due to interaction with side branches of the full dendrite. Dendritic growth often occurs at a critical $R_s \approx \lambda_i$ (limit of stability), since λ_i represents the shortest wavelength perturbation that would cause the dendrite tip to become unstable.

The range of interface structures during solidification may be graphically summarized to depict the variation of interface velocity, V (mm/sec), with liquid temperature gradient at the phase interface, $G = \partial T / \partial x$ (K/mm). Typical trends are summarized in the list below and illustrated in Figure 5.4.

- For low temperature gradients, the following structures are typically observed: (i) cellular (V of order of 10^{-4} mm/sec), (ii) dendritic (V of order of 10^{-2} mm/sec), and (iii) equiaxed dendritic (V of order of 10^{1} mm/sec).

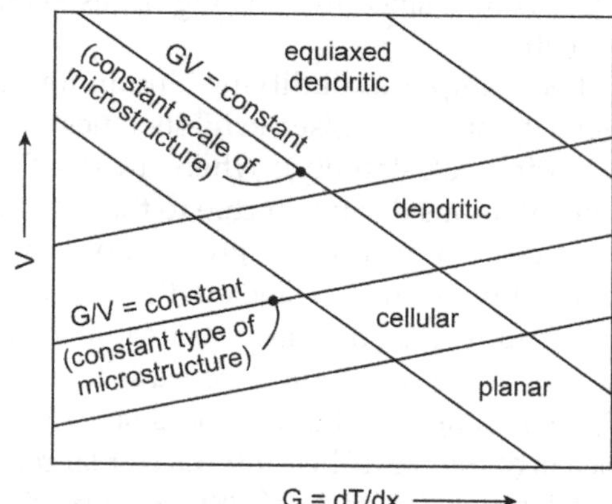

FIGURE 5.4
Interface structures at varying temperature gradients.

- For high liquid temperature gradients, the following structures are typically observed: (i) planar (V of order of 10^{-4} mm/sec), (ii) cellular (V of order of 10^{-2} mm/sec), (iii) dendritic (V of order of 10^{1} mm/sec), and (iv) equiaxed dendritic (V of order of 10^{2} mm/sec).

Transitions between each type of interface structure usually occur across lines of constant G/V ratios. Also, transitions between different scales of interface structures in the previous classifications are typically identified across lines of constant GV products.

Unlike solidification, which typically requires some undercooling below the phase change temperature for the onset of nucleation, melting processes do not require superheating. Phase transition from a solid to a liquid phase occurs at the phase change temperature. Superheating is not required in melting since no additional interface energy is required to hold a specific molecular structure together in the liquid phase. For phase transition from solid to liquid phases in metals, an increase in the vacancy concentration is experienced. A newly melted solid may be considered an array of voids surrounded by loose regions of disordered crystal. The crystallographic order of the solid disappears when a series of dislocations breaks up the closely packed structure during the melting process.

This section has described certain transport phenomena during solidification and melting processes. Additional sources of information, including various practical applications, are provided in Kurz and Fisher (1984) and Chalmers (1964).

5.2.4 Evaporation and Sublimation

Evaporation is distinguished from boiling since it can be solely driven by mass transfer processes. For example, a concentration gradient of water or vapor alone can lead to evaporation from the surface of liquid water. Evaporation refers to the transfer of molecules from the surface of a liquid to the vapor phase. Only liquid molecules with sufficient energy are capable of overcoming the cohesive forces that hold the molecules in the liquid phase along the surface. Since the higher energy molecules leave the liquid surface, the resulting liquid temperature is lower (thus called evaporative cooling). In an enclosed space, vapor can be added until the pressure reaches a maximum when the vapor becomes saturated.

The evaporative heat loss may be approximated as follows:

$$q''_{evap} = h_{fg} m''_A \tag{5.40}$$

where m''_A and h_{fg} refer to the evaporative mass flux (component A of a mixture, such as water vapor) and latent heat of vaporization, respectively. Under steady-state conditions, the latent energy lost by the liquid due to evaporation is supplied by heat transfer to the liquid from the surroundings. For example, if this heat is supplied by convection alone, the heat gain balances the evaporative loss from Equation (5.40) at steady state, i.e.,

$$h(T_\infty - T_s) = h_{fg} h_m [\rho_A(T_s) - \rho_{A,\infty}] \tag{5.41}$$

where the vapor density at the liquid surface, $\rho_A(T_s)$, refers to the density associated with saturated conditions at T_s. Also, h_m is the convective mass transfer coefficient.

Using Equation (3.88) to express the ratio of convective heat to mass transfer coefficients, together with the ideal gas law, Equation (5.41) becomes

$$T_\infty - T_s = \frac{h_{fg}}{\rho c_p Le^{2/3}} \left[\frac{P_{A,sat}(T_s)}{RT_s} - \frac{P_{A,\infty}}{RT_\infty} \right] \tag{5.42}$$

where the thermophysical properties (ρ, c_p, and Le) are evaluated at the mean temperature of the thermal boundary layer.

The gas constant, R, is evaluated for the particular constituent (i.e., water vapor). Alternatively, the gas constant can be written in terms of the universal gas constant ($R_u = 8.315$ kJ/kmolK) divided by the molecular weight of the constituent of interest. Representing T_s and T_∞ by an average temperature of T_{av}, Equation (5.42) becomes

$$T_\infty - T_s = \frac{\mathcal{M}_A h_{fg}}{\rho c_p R_u Le^{2/3} T_{av}} [P_{A,sat}(T_s) - P_{A,\infty}] \tag{5.43}$$

where $\mathcal{M}_A$ is the molecular weight of constituent A in the mixture (i.e., water vapor).

Also, from the ideal gas law,

$$P = \rho\left(\frac{R_u}{\mathcal{M}_B}\right)T_{av} \tag{5.44}$$

which can be substituted into Equation (5.43) for T_{av} to give the following result for an air (component A)–vapor (component B) mixture:

$$T_\infty - T_s = \frac{0.622h_{fg}}{c_p Le^{2/3} P}[P_{A,sat}(T_s) - P_{A,\infty}] \tag{5.45}$$

The factor of 0.622 is obtained from the ratio of the molecular weights of water vapor and dry air.

From Equations (5.40) and (5.41), it can be observed that the right side of Equation (5.41) gives q''_{evap} (evaporative heat flux). This balances the left side of Equation (5.41), which can be obtained by the convection coefficient, h, multiplied by the ΔT that was obtained in Equation (5.45). Thus, the evaporative heat flux can be written as

$$q''_{evap} = \chi_{evap}[P_{A,sat}(T_s) - P_{A,\infty}] \tag{5.46}$$

where

$$\chi_{evap} = \frac{0.622h(h_{fg})}{c_p Le^{2/3} P} \tag{5.47}$$

A less accurate (but more convenient) approximation is obtained by assuming that the partial pressure ($P_{A,\infty}$) is negligible relative to the saturation pressure ($P_{A,sat}$). The previously derived result was obtained for water vapor in air. However, an analogous result could be obtained for other gas mixtures, except that 0.622 would be replaced by the appropriate ratio of molecular weights of the two constituent gases.

In contrast to evaporation, sublimation is a phase change process where a material is transformed from the solid to vapor phase without passing through a liquid state. All solid materials will sublime below their triple point. The triple point is the thermodynamic state when solid, liquid, and gas phases can all coexist in equilibrium. Sublimation is encountered in practical applications such as dry ice (i.e., solid carbon dioxide that is used as a refrigerant for transporting perishables). Also, iodine and its compounds are used as antiseptics, fungicides, and light-sensitive materials (silver iodide) in photography, as well as in the production of dyes. Iodine forms black crystals that readily sublime to violet vapor. Another practical application involving sublimation is dehydration, or drying to remove water from a substance. For example, modern freeze drying of foods retains their texture and flavor. Freeze drying is performed through sublimation of ice from frozen foods under a vacuum.

Heat transfer by sublimation can be calculated similarly as Equations (5.40) to (5.47), except that the latent heat of vaporization, h_{fg}, is replaced by the latent heat of sublimation, h_{sg}. In other words, the sublimation heat flux is calculated by replacing χ_{evap} with χ_{sub} in Equation (5.46) and replacing h_{fg} with h_{sg} in Equation (5.47). These results apply to sublimation over ice. Other solids and gases would follow a procedure analogous to that in Equations (5.40) to (5.47), except that the appropriate ratio of molecular weights of the two constituents would replace 0.622 (obtained for vapor/air and ice) in Equation (5.45) and thereafter.

5.3 Mixture and Two-Fluid Formulations

In the previous section, the overall physical processes of phase change heat transfer were discussed. In this section, the relevant governing equations for these processes will be outlined. Two broad categories of techniques can be identified for the solutions of problems involving moving phase boundaries. These two methods are based on *mixture modeling* (or other spatial averaging techniques such as *two-fluid modeling*) and *interface tracking*. In general, these different approaches are based on either spatial averaging of the governing equations or tracking of individual phases.

Interface tracking is based on solutions of the single phase equations in each phase separately, after the position of the phase interface has been identified. In particular, the governing equations are solved separately in each phase, and iterations are performed at the anticipated location of the phase interface until convergence of certain interfacial balances is obtained. These interfacial balances include balances of mass, momentum, and energy. This approach allows the position of the phase interface to be clearly identified and used to separate the solutions of the governing equations on both sides of the interface.

On the other hand, mixture models or other similar methods based on spatial averaging techniques (such as two-fluid formulations) do not explicitly identify the microscopically detailed interface between phases. Instead, spatial averaging of the governing equations is performed over both phases simultaneously (mixture model) or individual phases (two-fluid model) within a multiphase control volume. In this sense of spatial averaging, mixture and two-fluid formulations are analogous. For example, a spatially averaged heat balance is applied over a collection of droplets and air separately, where each assembly is considered to be a separate fluid in the two-fluid model.

Alternatively, the heat balance could be applied individually within each phase, and then the results from all phases could be combined (mixture approach). This approach requires some input regarding the cross-phase interactions, in view of the spatial averaging performed across the phase interface. The mixture and two-fluid approaches are similar since they are both based on spatial averaging of conservation equations over an assembly of smaller elements. As a result, the following section will consider this spatial averaging in a broad sense (based on general mixture modeling) in anticipation that the main concepts are outlined for other types of spatial averaging techniques (such as two-fluid modeling).

5.3.1 General Scalar Transport Equation

Consider the transport of a general scalar quantity, B_k, associated with phase k in a multiphase mixture within a differential control volume (i.e., two phases occupying control volume in Figure 3.2). The control volume may contain a single phase or a mixture of phases at a differential level. The general conservation equation for B_k may be written as

$$\frac{\partial(f_k \rho_k B_k)}{\partial t} + \nabla \cdot (f_k \rho_k \mathbf{v}_k B_k) = -\nabla \cdot (f_k \mathbf{J}_k) + f_k S_k \tag{5.48}$$

where f_k and $\mathbf{J}_k$ refer to phase fraction and diffusive flux in phase k, respectively. Also, the subscript k refers to the phase. For example, $k = 1$ and $k = 2$ can refer to solid and liquid phases, respectively.

In this formulation, it will be assumed that the differential surface area of phase k, denoted by dA_k, associated with the control volume is equal to $f_k dA$. The component terms in Equation (5.48) represent the transient accumulation of B_k in the control volume occupied by phase k, the net advective flow of B_k across the phase k portion of the control surface, the net diffusive flow into phase k across the interface, and the source of B_k in the phase k portion of the control volume, respectively.

It can be observed that Equation (5.48) is essentially identical to equations developed in earlier chapters (i.e., convection equations in Chapter 3) since a value of k substituted into Equation (5.48) yields a single-phase equation. The mixture multiphase equation is obtained by summing the separate phase equations over all phases within a control volume and rewriting the variables in terms of mixture variables. For example, if the liquid equation is added to the solid phase equation, the respective portions of each phase are included through a phase fraction that appears in the summed two-phase equation. This approach permits a solution strategy similar to that of single-phase systems. However, final closure of the system of equations requires suitable correlations for the interphase processes arising from spatial averaging at the phase interface within the control volume.

A general mixture quantity, such as B, is defined as the mass fraction-weighted sum of individual phase components of B. For example, in a two-phase system,

$$B = \sum_{k=1}^{2} f_k B_k \tag{5.49}$$

where f_k refers to the mass fraction of phase k in the multiphase mixture within the given control volume. For example, $\rho = f_l \rho_l + f_s \rho_s$ is the mixture density and $\mathbf{v} = f_l \mathbf{v}_l + f_s \mathbf{v}_s$ is the mixture velocity. In these cases, the subscripts 1 and 2 refer to solid, s, and liquid, l, respectively. Specific forms of the homogeneous mixture equations will now be considered by using specific values of B in Equation (5.48).

Whether an equation is written for a scalar quantity, B, in a single phase or as a mixture, its physical properties can be characterized by the mathematical form of the equation. For example, consider the following reference equation:

$$c_1 \frac{\partial^2 B}{\partial x^2} + c_2 \frac{\partial^2 B}{\partial x \partial y} + c_3 \frac{\partial^2 B}{\partial y^2} = 0 \tag{5.50}$$

If $c_2^2 - 4c_1c_3 < 0$, the equation is called *elliptic* and the boundary conditions must be specified on the entire boundary enclosing the domain. This type of problem is called a *boundary value problem*. An example is Laplace's equation for steady-state heat conduction or incompressible, potential flow. On the other hand, if $c_2^2 - 4c_1c_3 = 0$, the equation is called *parabolic* and the boundary conditions are specified on the boundary, including some inlet boundary. An example of this type is the transient heat conduction equation, which is a combined boundary value problem and *initial value problem* (i.e., initial conditions in time are required). The third classification is a *hyperbolic* equation, which occurs when $c_2^2 - 4c_1c_3 > 0$. In this case, precise boundary conditions are required only at certain portions of the domain boundary. Examples of hyperbolic equations are the wave equation and compressible gas flow in a nozzle where outflow boundary conditions are not required under choked flow conditions. The classification of the governing equations into these categories is useful, particularly for the solution techniques described in the upcoming sections and chapters.

5.3.2 Mass and Momentum Equations

For conservation of mass, substitute $B_k = 1$, $J_k = 0$, and $S_k = \dot{M}_k$ into Equation (5.48). Then, adding the individual phase conservation equations over all phases, the following familiar form of the continuity equation is obtained for the previous example of liquid–solid systems:

$$\frac{\partial \rho}{\partial t} + \nabla \cdot (\rho \mathbf{v}) = 0 \tag{5.51}$$

where ρ and $\mathbf{v}$ refer to mixture density and velocity, respectively.

The x momentum equation for phase k can be obtained from Equation (5.48) with $B_k = u_k$, $\mathbf{J}_k = p_k \mathbf{i} - \tau_{kx}$, and $S_k = \rho_k F_{b,kx} + G_{kx}$. The source term, $\rho_k F_{b,kx}$, represents the x component body force on phase k, and G_{kx} accounts for momentum production due to cross-phase interactions. An analogous approach is taken for the y momentum equation.

For certain types of multiphase systems, such as solid–liquid phase change, the mixture velocity can be effectively expressed by a mass fraction-weighted sum of phase velocities. In these cases, the following mixture momentum equation can be obtained by summing over individual phase equations:

$$\frac{\partial (\rho \mathbf{v})}{\partial t} + \nabla \cdot \left(\sum_{k=1}^{2} f_k \rho_k \mathbf{v}_k \mathbf{v}_k \right) = -\nabla p + \nabla \cdot \left(\sum_{k=1}^{2} \mu_k \nabla (f_k \mathbf{v}_k) \right) + \mathbf{F}_b + \mathbf{G} \tag{5.52}$$

This result can be written in terms of a single mixture velocity by decomposing the advective flux of momentum into a mean mixture component and a resulting difference between mixture and phase quantities (moved to the right side of the equation). Also, the interphase terms for a given multiphase system, $\mathbf{F}_b$ and $\mathbf{G}$, must be specified (typically empirically). In Equation (5.52), Newtonian constitutive relationships are used within the stress tensor.

The specific form of interphase correlation is obtained based on the physical transport phenomena associated with each type of multiphase system. For example, in solid–liquid phase change (subscript sl), examples of body forces are thermal and solutal buoyancy, i.e.,

$$\mathbf{F}_{b,sl} = \mathbf{g} \beta_T (T - T_0) + \mathbf{g} \beta_C (C - C_0) \tag{5.53}$$

where β_T and β_C refer to thermal and solutal expansion coefficients, respectively. The values of T_0 and C_0 refer to reference temperature and concentration, respectively. Similar expressions for body forces can be obtained for other types of multiphase systems.

Also, the last term on the right side of Equation (5.52), **G**, represents the production of momentum due to interactions between phases along the interfacial boundary. In the example of solid–liquid systems, it includes fluid acceleration through pockets or channels of the dendritic solid matrix. Other supplementary relations, such as Darcy's law for flow through a porous medium, may be used for closure of the phase interaction terms. Darcy's law for liquid flow through a permeable dendritic solid matrix may be written as

$$K_x G_{x,sl} = \nu_l (f_l u_r) \tag{5.54}$$

where the coefficient K_x denotes the permeability in the x direction and $u_r = u_l - u_s$ represents the x component relative phase velocity (note: similar expressions are obtained in the y direction). A typical trend in permeability would be $K_x \approx C/f_s$, where C and f_s refer to a constant of proportionality (i.e., $C \approx 10^{-8}$) and solid fraction, respectively. Then K_x becomes small when $f_s \to 1$, whereas $K_x \to +\infty$ when $f_s \to 0$ (i.e., single-phase flow equations are retained). In certain porous medium applications, such as groundwater flow problems, the phase interactions, $G_{x,sl}$ in Equation (5.54), are often grouped together with the pressure gradient.

In gas–solid systems (subscript *gs*), such as multiphase flows with particles (or droplets), the body forces ($\mathbf{F}_b$) and cross-phase interactions (**G**) are dependent on whether the flow involves free, mixed, or mainly forced convection. For free convection, the body forces can be based on the Boussinesq approximation (as described in Chapter 3), whereby the density includes a mass fraction weighting of gas and particles. Particles would remain dispersed within the gas flow for sufficiently fine particles. However, they may settle out under low thermal buoyancy, or if large particles were contained in the mixture (i.e., weight largely exceeding the upward buoyancy of the gas flow). In the latter case (large particles), a particle tracking method would be recommended (rather than the current mixture formulation) since the particles would not be homogeneously dispersed while following the gas–solid flow.

For forced and mixed convection of gas–solid flows, higher inertia in the gas phase will more likely carry particles with the flow. However, particles can settle out of the flow (i.e., sedimentation along the bottom of a domain) if their individual (or clustered) weights exceed the inertia provided by the main flow. Thus, the homogeneous mixture approach is best suited to flows involving small particle diameters or high gas velocities. In many practical applications, the particles (or droplets) closely follow the main (gas) flow in the freestream away from physical boundaries, but depart appreciably from the main flow characteristics near the solid boundaries of the domain. Thus, continuum mixture modeling is more effective in the freestream than in the boundary layer equations.

Similar observations arise in other types of multiphase systems, such as liquid–gas flows. In the previous discussions regarding solid–liquid phase change, including Equations (5.53) and (5.54), macroscopic phenomena have been modeled. However, since the homogeneous mixture does not track the explicit location of the phase interface, certain microscopic phenomena (such as particle settling and crystal sedimentation) are not directly modeled. In this case, the mixture approach assumed either a zero solid phase velocity (i.e., stationary dendrite crystals) or crystals following the liquid flow. Otherwise, tracking of crystal (or particle) trajectories is required separately from the liquid flow. This tracking of trajectories would be essentially equivalent to the method of interface tracking (discussed in the next section).

The cross-phase interactions in gas–solid systems (with particles or droplets), $\mathbf{G}_{gs}$ in Equation (5.52), require a detailed understanding of the loss or gain of momentum due to interphase forces exerted across the phases. In particular, the drag force, F_d, exerted between the particles (or droplets) and the gas phase depends on the local particle Reynolds number, Re_p. The following correlation provides a commonly used expression for drag force that can be adopted in the cross-phase interactions of the mixture momentum equation (Hewitt et al., 1997):

$$F_d = \frac{\pi}{8} \rho_g u_p^2 d_p^2 c_d \tag{5.55}$$

The drag coefficient, c_d, can be approximated by

$$c_d = \frac{24}{Re_p}(1 + 0.15 Re_p^{0.687}) \tag{5.56}$$

where the subscripts g and p refer to gas and particle (or droplet), respectively. It can be observed that the drag force is proportional to the square of velocity and diameter. The drag force increases with $u_p d_p$. In general, the correlations involving drag force depend on a variety of factors, such as the Reynolds number, turbulence, and other factors. As a result, the specific expression adopted for $\mathbf{G}_{gs}$ in Equation (5.52) requires a correlation that is well suited to the flow conditions under consideration.

For liquid–gas (vapor) systems, including two-phase flows with boiling, other fluid processes lead to appropriate expressions for the body forces, $\mathbf{F}_{b,lg}$, and cross-phase interactions, $\mathbf{G}_{lg}$. For example, in pool boiling (i.e., boiling of initially quiescent liquid), vapor bubbles rise by buoyancy since the density of vapor is much less than that of the liquid. These buoyancy forces in a two-phase control volume are dependent on the frequency of bubble departure and bubble departure diameter. If it were adopted in a mixture model, the mixture velocity might require a phase fraction weighting of the velocity of ascending bubbles and the liquid motion induced by bubble mixing. Vapor buoyancy is also included within body forces for problems involving forced convection with boiling. However, the decomposition of mixture velocities into separate phase velocities becomes difficult in boiling problems due to the various possible flow regimes and wide (unknown) discrepancy between phase velocities. As a result, some type of explicit interface tracking (to be discussed in the following section) is typically required in these types of problems.

Similar considerations are given to the cross-phase interactions, $\mathbf{G}_{lg}$, in liquid–gas (vapor) flows. In addition to buoyancy in boiling problems, the momentum equations include viscous forces between bubbles and the surrounding liquid. The net force on the bubbles leads to the acceleration of vapor and removal of heat from the boiling surface. Similarly, droplets that condense out of a vapor flow encounter viscous drag in a manner similar to that discussed earlier with gas–solid (particle) flows. In surface condensation (i.e., condensation of saturated vapor along a cooled surface), the liquid and vapor phases are essentially separated from each other. As a result, interfacial balances (to be discussed in the following section) are required to find the position of the phase interface, in addition to mixture modeling of the main (gas) flow. Macroscopic transport processes can often be handled effectively through mixture modeling. However, microscopic analysis of a nondispersed phase and its phase interface, such as a nondispersed condensate film, rather than a dispersed droplet, requires some form of detailed interfacial balances (to be discussed in upcoming sections).

5.3.3 Energy Equation

In continuation of the earlier cited example of solid–liquid systems, the conservation of energy equation can be obtained from Equation (5.48) with $B_k = e_k$, $\mathbf{J}_k = -k_k \nabla T$, and $S_k = \dot{E}_k$ (internal heat generation). Summing these phase conservation equations, while neglecting viscous dissipation and heat sources, the following mixture energy equation is obtained for the two-phase system:

$$\frac{\partial}{\partial t}\left(\sum_{k=1}^{2} f_k \rho_k e_k\right) + \nabla \cdot \left(\sum_{k=1}^{2} f_k \rho_k \mathbf{v}_k e_k\right) = \nabla \cdot (k \nabla T) \tag{5.57}$$

where

$$e_k(T) = \int_{T_0}^{T} c_{r,k}(\zeta)d\zeta + e_{r,k}(T) \tag{5.58}$$

In Equation (5.58), $c_{r,k}(T)$ represents the effective specific heat of phase k. Also, (T) denotes the functional dependence on temperature and the subscript r refers to reference value.

The advection term in Equation (5.57) can be decomposed into mean and relative phase motion components. Then, the final equation can be written in terms of a mixture energy, rather than individual phase quantities, and thus it can be solved in the regular fashion for a single scalar quantity, e (rather than e_k). For solid–liquid systems, the following result is obtained for the mixture energy equation:

$$\frac{\partial(\rho e)}{\partial t} + \nabla \cdot (\rho \mathbf{v} e) = \nabla \cdot (k \nabla T) + \nabla \cdot \left(\rho \mathbf{v} e - \sum_{k=1}^{2} \rho_k \mathbf{v}_k e_k \right) \tag{5.59}$$

A supplementary equation of state is required so that Equation (5.59) can be written and solved in terms of temperature alone. For example, using a piecewise linear approximation,

$$e_k = e_{r,k} + c_{r,k}(T - T_{r,k}) \tag{5.60}$$

where the subscripts r and k refer to reference value and phase, respectively. Also, $c_{r,k}$ represents the effective specific heat, which includes both sensible and latent heat portions, since it incorporates the change of energy between phases undergoing phase transition.

For example, $c_{r,k}$ represents the regular specific heats in the solid ($k = 1$) and liquid ($k = 3$), but it includes the latent heat of fusion throughout the phase transition region ($k = 2$). The reference energies and temperatures reflect a linear rise of $e_{r,1}$ in the solid, an abrupt increase of $e_{r,2}$ at the phase change temperature, and a subsequent linear increase with temperature in the liquid. For pure materials, the change of energy during phase transition occurs over a very small temperature range. However, for multicomponent systems, the phase transition occurs over a discrete temperature range, as specified through the reference values used in Equation (5.60).

In the mixture energy equation (5.57), the cross-phase interactions are largely modeled through the change in phase fraction, such as f_s (solid fraction). The governing equation is based on an energy balance within a control volume, and individual phase equations are summed together to yield the mixture equation. In mixture modeling, both phases are usually considered to be at the same temperature within the control volume. During solidification, the release of latent heat due to phase transition within the control volume is transferred accordingly to surrounding material based on the relevant heat balance. For example, a local heating effect (called recalescence) occurs when the release of latent heat in a control volume exceeds the rate of external cooling that initiated the solidification process. These energy interactions are accommodated through the equation of state in Equation (5.60).

In the homogeneous mixture approach for modeling of multiphase systems, the phase interface can be described in detail only to the level of resolution of the control volume. Within a sufficiently fine control volume (i.e., differential scale) at the phase interface, it is often assumed that equilibrium conditions exist therein. In other words, interfacial variables such as temperature (and concentration for multicomponent systems) are based on values expected under equilibrium conditions. However, under some circumstances, such as rapid solidification of metal alloys, phase change occurs too quickly for the solute to be completely diffused throughout a control volume. In the mixture approach, these types of nonequilibrium conditions can be modeled through the phase fraction dependence on other problem variables in Equation (5.59). Rather than tracking the explicit location of the phase interface, the mixture approach predicts the interfacial phenomena by treating the material as a homogeneous mixture within the control volume.

Results similar to those of Equations (5.59) and (5.60) can be derived for other multiphase systems. For example, if the mixture velocity, $\mathbf{v}$, is computed for liquid–gas (vapor) systems, the individual phase velocities must be decomposed from this mixture velocity for adoption in Equation (5.59). This decomposition can be effectively handled under certain conditions, such as boiling of liquid in a porous material, since the liquid flow can be approximated separately from the dynamics of the full two-phase flow. Then, similarly with the equation of state in Equation (5.60), the mixture energy equation can be solved to obtain temperature. However, in more complicated two-phase flows, such as forced convection of liquid with boiling in nuclear reactors, the behavior of each fluid stream is independently based on the appropriate momentum balances in each phase. The governing equations are typically solved separately in each phase (such as two-fluid models) with matching conditions at the moving phase interface. The individual phase equations refer to Equation (5.57) without the summation (i.e., within a single phase, k).

Similar processes are observed for gas–solid (particle) systems, where cross-phase interactions would represent heat transfer between the gas stream and particles. For example, the particles would be heated by the gas stream until the temperatures of both phases are equal. Without phase change, an equation of state is not required in the full form of Equation (5.60). In this case with incompressible flows, the differential of energy in Equation (5.59) can be written directly as specific heat multiplied by differential of temperature. However, it must properly accommodate any variations of specific heat with temperature when these variations become significant (i.e., at high temperatures).

Under certain conditions when the particles are sufficiently fine and dispersed in the two-phase flow, it can be assumed that the particles closely follow the main (gas) flow stream. In this way, the mixture and individual phase velocities become equal in Equation (5.59). The homogeneous mixture approach for two-phase modeling can be well suited to problems when the particle and gas stream temperatures are equal. Heating (or cooling) of the mixture occurs similarly to single-phase flows, with the exception of thermophysical properties that are based on the phase fraction-weighted sum of individual phase properties. However, if differences between the gas and particle temperatures are important aspects of the overall problem, individual particle tracking may be required, since a mixture approach uses a single temperature for both phases within a control volume. Alternatively, Eulerian modeling in a two-fluid model could utilize the spatial averaging technique in the droplet and gas phases separately.

Particle tracking refers to tracking of individual particle trajectories throughout the flow field, whereas mixture modeling approximates flow quantities based on mixture quantities (rather than individual phase quantities). Particle tracking represents a type of *interface tracking* that will be described in upcoming sections. It may be considered a type of Lagrangian approach (rather than Eulerian). For larger particles (not following the gas stream), tracking of individual trajectories or Eulerian two-fluid modeling is recommended over mixture modeling, unless particle velocities can be clearly extracted from mixture velocities. These particle–mixture velocity relationships may be available under certain conditions, such as large particles falling at their terminal velocity through a slowly moving gas.

5.3.4 Second Law of Thermodynamics

For an individual phase (denoted by subscript k) in a multiphase mixture, the second law of thermodynamics can be written as

$$\dot{P}_{s,\,k} \equiv \frac{\partial S_k}{\partial t} + \nabla \cdot \mathbf{F}_k \geq 0 \tag{5.61}$$

where $\dot{P}_{s,\,k}$ refers to the entropy production rate. Also, S_k and F_k represent the thermodynamic entropy and entropy flux in phase k, respectively.

In Equation (5.61), the inequality refers to irreversible processes, whereas the equality is a limiting case of reversible processes. The extensive variable,

S_k, can be written in terms of the entropy per unit mass, s_k, as follows:

$$S_k = \rho_k s_k \tag{5.62}$$

The entropy flux consists of both advective and diffusive components. For a multicomponent system, the entropy flux can be expanded so that Equation (5.61) becomes

$$\frac{D(f_k \rho_k s_k)}{Dt} = \nabla \cdot \left(\frac{f_k k_k \nabla T}{T} \right) + \nabla \cdot \left(\sum_{c=1}^{N} \frac{f_k g_{c,k} \mathbf{j}_{c,k}}{T} \right) + \dot{P}_{s,k} \tag{5.63}$$

where $g_{c,k} = \partial e_k / \partial C_{c,k}$ refers to the chemical potential of constituent c in phase k.

The chemical potential, $g_{c,k}$ in Equation (5.63), may be interpreted as the additional increase of work potential in the fluid if $dC_{c,k}$ of constituent c is added to the mixture. The summation over N constituents in Equation (5.63) includes each chemical potential from $c = 1$ to $c = N$ constituents; i.e., $N = 2$ for a binary alloy. The diffusive mass flux, $\mathbf{j}_{c,k}$, for constituent c in phase k, is written in terms of the species concentration, $C_{c,k}$, using Fick's law (analogous to Fourier's law for heat conduction). The substantial derivative on the left side of Equation (5.63) includes both transient and advection terms. It can be observed that the entropy flux, $\mathbf{F}_k$ in Equation (5.61), includes an advective component from the substantial derivative on the left side of Equation (5.63), as well as diffusive (heat and species) components in the first two terms on the right side of Equation (5.63).

Also, the Gibbs equation provides a relationship between entropy and the conservation variables. For a multicomponent mixture, the Gibbs equation in phase k is given by

$$ds_k = \frac{d\hat{e}_k}{T} + \frac{p dv_k'''}{T} - \sum_{c=1}^{N} \frac{g_{c,k} dC_{c,k}}{T} \tag{5.64}$$

where v''' represents specific volume. Since Equation (5.64) is written within phase k alone, a latent heat term is not shown.

For an incompressible substance in each phase, Equation (5.64) can be written in terms of substantial derivatives as follows:

$$T \frac{D(f_k \rho_k s_k)}{Dt} = \frac{D(f_k \rho_k \hat{e}_k)}{Dt} - \sum_{c=1}^{N} g_{c,k} \frac{D(f_k \rho_k C_{c,k})}{Dt} \tag{5.65}$$

Substituting the thermal energy equation from Chapter 3 into Equation (5.65),

$$\frac{D(f_k\rho_k s_k)}{Dt} = \frac{1}{T}\{k_k\nabla^2 T + \tau : \nabla \mathbf{v}_k + \dot{P}_{e,k}\} - \frac{1}{T}\sum_{c=1}^{N} g_{c,k}\{-\nabla\cdot \mathbf{j}_{c,k} + \dot{P}_{c,k}\} \quad (5.66)$$

Expanding the divergence terms in Equation (5.66) using the product rule of calculus and comparing the result with Equation (5.63), the following result for the entropy production in phase k is obtained:

$$\dot{P}_{s,k} = \frac{f_k k_k(\nabla T)^2}{T^2} + \frac{\mu\Phi}{T} + \frac{\dot{P}_{e,k}}{T} - \frac{1}{T}\sum_{c=1}^{N} \mathbf{j}_{c,k}\cdot\nabla g_{c,k} + \frac{1}{T^2}\sum_{c=1}^{N} g_{c,k}\mathbf{j}_{c,k}\cdot\nabla T - \sum_{c=1}^{N}\frac{g_{c,k}\dot{P}_{c,k}}{T} \quad (5.67)$$

On the right side of Equation (5.67), the production terms become zero after summation over all phases because production or destruction of energy or species concentration in one phase is accompanied by destruction or production in the other phases. However, viscous dissipation and fluid mixing are entropy-producing processes within individual phases, and thus a zero summation of entropy production terms does not apply to the second law.

The entropy production rate is a positive-definite quantity. After summing over both solid and liquid phases, using the Gibbs equation, Equation (5.67) becomes

$$\dot{P}_s = \sum_{k=1}^{2}\frac{f_k k_k(\nabla T\cdot\nabla T)}{T^2} + \frac{\mu\Phi}{T} - \frac{1}{T}\sum_{c=1}^{N} \mathbf{j}_{c,k}\cdot\nabla g_{c,k} + \frac{1}{T^2}\sum_{c=1}^{N} g_{c,k}\mathbf{j}_{c,k}\cdot\nabla T \quad (5.68)$$

where Φ refers to the viscous dissipation function. In the second-law calculations, an entropy equation of state is required to relate entropy to other thermodynamic variables (such as temperature). In the following example, a sample equation of state is constructed using the Gibbs equation.

Example 5.3.4.1

Entropy Equation of State for a Two-Phase Mixture.

Find a piecewise linear approximation for the entropy of a two-phase (solid and liquid) mixture in equilibrium. Use the Gibbs equation and express the answer in terms of temperature and solute concentration for the binary component mixture.

A piecewise linear approximation based on Equation (5.64) is

$$s_k = s_{r,k} + c_{r,k}\log\left[\frac{T}{T_{r,k}}\right] \quad (5.69)$$

where the subscript r refers to values at a reference state. For a multicomponent solid–liquid mixture, entropy varies with temperature and solute concentration. Also, the entropy of fusion is released during phase transition.

In the solid phase ($k = 1$), the following reference values are specified:

- $s_{r,1} = 0$
- $c_{r,1} = c_s$
- $T_{r,1} = T_e$ (eutectic)

In the two-phase region ($k = 2$, solid and liquid phases coexisting in equilibrium), the reference specific heat must include both sensible and latent entropy contributions. The reference entropy is specified at the solidus line (see Figure 1.6), and the reference values become:

- $s_{r,2} = c_s log(T_{sol}/T_e)$
- $c_{r,2} = \left(\frac{c_s T_{liq} - c_l T_{sol}}{T_{liq} - T_{sol}}\right) + \frac{c_l - c_s + \Delta s_f}{log(T_{liq}/T_{sol})}$
- $T_{r,2} = T_{sol}$

In the liquid phase ($k = 3$), the Gibbs equation, Equation (5.64), is integrated across the phase transition region. A linear variation of liquid fraction with temperature is assumed in this integration. Performing the integration and substituting $s_s = c_s log\ (T_{sol}/T_e)$ from the previous results in the solid phase:

- $s_{r,3} = \left(\frac{c_s T_{liq} - c_l T_{sol}}{T_{liq} - T_{sol}}\right) log(T_{liq}/T_{sol}) + c_s log(T_{sol}/T_e) + c_l - c_s + \Delta s_f$
- $c_{r,3} = c_l$
- $T_{r,3} = T_{liq}$

These variables give the reference values for the entropy equation of state in Equation (5.69). Entropy and the Second Law have important implications in the study of liquid–solid systems (Naterer, 2000, 2001).

The previous derivation has referred to solid–liquid systems, but similar mixture formulations can be derived for other multiphase systems. A procedure similar to that in Equations (5.61) to (5.68) can be used to obtain the rate of entropy production for other multiphase mixtures. However, some steps in the previous derivation may require modifications. For example, in a liquid–gas (vapor) mixture, Equation (5.64) is applicable, but compressible flow in the gas phase would require the specific volume to be retained. An analogous expression to Equation (5.67) is obtained. An entropy equation of state in the form of Equation (5.69) is required for phase change problems. However, the reference entropy is not essential in two-phase flows without phase change since this reference value is identical between two thermodynamic states in the flow.

The previously obtained results for a single phase (i.e., without summation over k) can often be adopted for gas–solid (particle) flows without phase change. However, it would be required that the previously adopted thermophysical properties would become the phase fraction-weighted sum of individual phase properties. For example, the effective thermal conductivity would be approximated as the appropriate weighting (based on mass fraction) of solid (particle) and gas conductivities. Since the solutions of the conservation equations are obtained before the application of the second law, the mixture velocities can be determined and used in the entropy transport equations, such as Equation (5.61).

Cross-phase interactions, such as interfacial resistance forces between the particles and gas stream, affect the momentum (and resulting velocity) of the flow, as well as the entropy produced by viscous shear action. In a mixture formulation, this interfacial entropy production is predicted through resulting velocity gradients in the main (gas) flow, i.e., viscous dissipation in Equation (5.67). Thus, with sufficient accuracy in predicting the cross-phase momentum interactions in a mixture model, this approach is expected to yield equivalent results to entropy produced by shear action in particle tracking of individual particle trajectories through the main (gas) flow.

5.4 Interface Tracking

In the previous section, the governing equations were based on modeling of a homogeneous mixture in a multiphase, differential control volume. In certain cases, the two-phase fluid cannot be well modeled as a homogeneous mixture, since differences in certain flow quantities (such as temperature or velocity) within the control volume can be significant, yet unknown and essential to the dynamics of the system. In these cases, a method of *interface tracking*, or tracking of individual phases (such as particles) throughout the flow field, is often required. In addition to the conservation equations (described earlier), further interfacial balances must be satisfied at the phase interface. These balances are important in establishing the location, temperature, and motion of the interface between different phases. In this section, these interfacial balances will be considered. Although references are initially made to solid–liquid systems, analogous principles and results are obtained afterward for other multiphase systems.

5.4.1 Interfacial Mass and Momentum Balances

During a phase change process, mass and momentum balances across the phase interface are essential in predicting the location and movement of the interface. For example, in the case of solid–liquid systems, the difference between the liquid and solid flows out of and into the interface, respectively, must balance the rate of mass change due to differences in phase densities (i.e., material shrinkage during solidification). Let us denote n, v_1, and v_2 as the normal to the interface and the individual phase velocities normal to the interface, respectively. For phase change under one-dimensional conditions with different phases designated by subscripts 1 and 2, the interfacial mass balance requires that

$$\rho_1 v_1 - \rho_2 v_2 = (\rho_1 - \rho_2)\frac{dn}{dt} \tag{5.70}$$

A similar result can be obtained in multidimensional flow, which is typically more representative of actual phase change processes involving two independently moving phases.

The interface velocity, dn/dt in Equation (5.70), is based on the movement of the interface over a distance of dn in a time increment of dt. Although it is the unique velocity at the phase interface, differences in individual phase velocities (corresponding to the density changes) near the interface are required to sustain this interface advancement. If there is little or no density difference between both phases, then Equation (5.70) indicates that the resulting velocities of both phases are equal on both sides of the interface. In many practical applications, such as porosity formation in solidified metal alloys, the mass change (shrinkage) effect due to differences in phase density plays a significant role in engineering design.

It can be observed that the interfacial mass balance is based on a change of mass within a differential control volume encompassing both sides of the phase interface. This change of mass and phase within an interfacial control volume, in conjunction with the difference of phase densities, requires a corresponding change of mass flow into or out of the interface. Without phase change, this change of mass flow rate is not observed since the mass of each phase remains constant at a particular temperature.

Similar interfacial balances arise in other types of multiphase systems. In another example of forced convection with condensation in a cooled pipe, vapor flow in the core region (center of the pipe) leads to formation of a condensate film along the wall. Although the rate of condensation is largely controlled by heat transfer from the vapor to the wall, the growth of the condensate film in the axial direction is based on entrainment of condensed vapor and accumulated mass from the vapor phase. The contribution of accumulated vapor mass to the thicker condensate film is largely based on the density difference between these phases. This mass exchange can be expressed by an analogous multidimensional form of Equation (5.70).

The momentum balance at the interface between two different phases generally requires a matching between shear stresses on both sides of the interface. Furthermore, the velocity at the interface between fluids (or between a gas and solid particle) is unique, which requires that both phase velocities are equal to each other at the phase interface. At a stationary wall, this corresponds to a no-slip condition whereby the fluid velocities are both zero. In other circumstances, particularly multiphase flows with particles or droplets, the interfacial momentum balances are effectively handled by tracking of particle trajectories throughout the flow field based on Newton's law of motion. For example, Newton's law represents the momentum equation of a particle that balances net forces on the particle (such as viscous drag and weight) and particle acceleration (with respect to the gas stream). An example will be given in the following section with reference to mass, momentum, and energy balances for an evaporating droplet.

5.4.2 Interfacial Energy Balance

For phase change problems, the heat balance at the phase interface largely affects the movement and position of the phase interface. For example, in solid–liquid systems, the heat transfer from the liquid phase into the phase interface, HT_l, consists of conduction and advection components:

$$HT_l = -k_l dA \left.\frac{dT}{dn}\right|_l dt + \rho_l v_l \hat{e}_l dA dt \tag{5.71}$$

where $\hat{e}$ refers to specific internal energy. A similar expression, HT_s, is obtained in the solid phase.

A control volume at the phase interface is selected with a thickness of dn. Then the change of energy that accompanies the advance of the interface arises due to the energy difference between the initially liquid volume, occupying $dAdn$, and a final solid volume, i.e.,

$$dE \equiv HT_l - HT_s = \rho_l \hat{e}_l dAdn - \rho_s \hat{e}_s dAdn \tag{5.72}$$

Using Equation (5.71), as well as the analogous result for the heat flux in the solid phase, and substituting those results in Equation (5.72), the following result is obtained:

$$(\rho_l \hat{e}_l - \rho_s \hat{e}_s) \cdot \frac{dn}{dt} = -k_l \frac{dT}{dn}\bigg|_l + k_s \frac{dT}{dn}\bigg|_s + \rho_l v_l \hat{e}_l - \rho_s v_s \hat{e}_s \tag{5.73}$$

Also, combining Equation (5.70) with Equation (5.73) and rearranging terms,

$$-k_l \frac{dT}{dn}\bigg|_l + k_s \frac{dT}{dn}\bigg|_s = -v_s \rho_s \Delta\hat{e}_f + \rho_s \Delta\hat{e}_f v_i \tag{5.74}$$

where $\Delta\hat{e}_f = \hat{e}_l - \hat{e}_s$ is the latent heat of fusion and $v_i = dn/dt$ refers to the interface velocity. If a stationary solid phase is assumed, then the first term on the right side of Equation (5.74) is neglected.

Equation (5.74) represents the heat balance at a solid–liquid interface. In practical solidification and melting problems, the interfacial balances are solved in conjunction with the governing conservation equations (i.e., the mixture equations described in previous sections), appropriate phase diagram(s), and initial and boundary conditions for the relevant problem variables, such as velocity and temperature. In subsequent chapters, these governing equations and supplementary relations will be examined in greater detail for various types of multiphase systems.

The previous heat balances were applied to a solid–liquid interface. Similar results are obtained for liquid–gas (vapor) systems, such as problems involving boiling or condensation heat transfer. For heat transfer with condensation, the latent heat of vaporization is released at the phase interface and typically conducted through the liquid film. The vapor typically condenses at or near the saturation temperature. Latent heat is released and primarily transferred from the phase interface through the liquid due to a lower vapor conductivity and decreasing temperature through the liquid. In the presence of convection with superheated vapor (above the saturation temperature), the rate of heat transfer to the interface is typically enhanced, thereby increasing the rate of growth of the condensate film. In single-phase problems with convection (Chapter 3), the wall heat flux was expressed in terms of the convective heat transfer coefficient multiplied by the wall–fluid temperature difference. Similarly, convection at the phase interface in two-phase flows can be expressed through the resulting steepened temperature gradient, analogous to the left side of Equation (5.74). This contribution to the heat exchange would not arise if the freestream vapor was at the saturation temperature (i.e., same as the temperature at the phase interface).

In boiling heat transfer, latent heat is absorbed in the liquid to complete the change of phase. Heat is transferred by conduction, convection, or radiation from the heating surface and extracted by liquid at the phase interface during boiling of the liquid. This absorption of latent heat from the liquid occurs in contrast to the release of latent heat from the vapor during condensation. The additional mode of radiation heat transfer is not shown in Equation (5.71), but could be readily incorporated based on material discussed in Chapter 4. In film boiling, a continuous interface is often identified between the liquid and vapor. However, interfacial modeling of nucleate and transition boiling may be restricted in some sense, such as analysis near the heating boundary, since bubble coalescence, elongation, stretching, and other deformations largely complicate the analysis of bubble flow within the liquid. At the heating boundary, heat transfer can be modeled by alternating periods of liquid and vapor bubbles in contact with the wall (Naterer et al., 1998).

For gas–solid (particle) flows without phase change, the previous interfacial heat balances can be modified appropriately, but without the latent heat terms. A convective type of boundary condition is encountered at the interface between a gas and solid (particle). For example, the relative motion between soot particles and a hot gas stream in a combustion chamber leads to transfer of heat between the gas and particles. As discussed in Chapter 2, the temperature remains approximately uniform within each particle if the particle Biot number is sufficiently small (less than 0.1).

In other cases involving gas–solid (particle) flows with phase change, such as chemically reacting flows, the interfacial heat balances are also analogous to Equations (5.71) to (5.74). In these previous equations, the energy balance was constructed at both sides of the phase interface. However, in multiphase flows with particles, various flow regimes are observed (discussed in Chapters 7 and 9). For example, in cases where particles are densely packed in the flow, some particles may be deposited in a sedimentation layer along the bottom boundary. Then a clearly identified boundary exists between both phases, and interfacial balances can be well applied. However, if particles are dispersed and fully mixed in the main flow (rather than segregated), it may be more practical to apply mixture modeling (or two-fluid modeling) by treating the flow as a mixture with properties based on the fraction of each phase. Applying interfacial heat balances for each particle could become intractable due to the substantial computational effort in tracking a large number of particles individually.

Other examples are multiphase flows with droplets, such as convective droplet evaporation in combustion systems and convective droplet solidification in atmospheric icing of structures. The following example considers the interfacial mass, momentum, and energy balances for an evaporating droplet.

Example 5.4.2.1

Interfacial Balance for Evaporating Droplets.

Find expressions for the interfacial mass, momentum, and heat balances at the liquid–vapor interface of an evaporating droplet in a microgravity environment. It may be assumed that the liquid and gas have constant thermophysical properties and the droplet evaporates with a constant mass flux $\dot{m}$. Also, it is assumed that the surrounding gas is at a uniform temperature T_∞.

In this example, the subscripts l, g, and i represent liquid, gas, and interface (on the droplet surface), respectively. The multidimensional form of Equation (5.70) corresponds to the following interfacial mass balance:

$$\dot{m} = \rho_l(\mathbf{v}_l - \mathbf{v}_i)\cdot\mathbf{n} = \rho_g(\mathbf{v}_g - \mathbf{v}_i)\cdot\mathbf{n} \tag{5.75}$$

In the case of fluid momentum, the change of momentum across the phase interface during evaporation must balance the difference of forces across the interface due to shear stresses and pressure, i.e.,

$$\dot{m}(\mathbf{v}_g - \mathbf{v}_l)\cdot\mathbf{n} = [\mathbf{n}\cdot(\tau_g - \tau_l)]\cdot\mathbf{n} \tag{5.76}$$

where τ is the stress tensor that includes both viscous and pressure components. An additional term would be added to the right side of the above equation to accommodate the effects of surface tension on the interface shape. In practice, τ_g may be neglected, since the gas viscosity is much smaller than the liquid viscosity. Also, the effects of fluctuations in the ambient gas pressure may be neglected in comparison to the effects of pressure fluctuations in the liquid. Further detailed dynamics of these momentum interactions for evaporating and oscillating droplets in a zero gravity environment are examined by Mashayek (2001).

For interfacial heat transfer, the latent heat evolved by the evaporating droplet must balance the rate of heat transfer from the surrounding gas, i.e.,

$$\dot{m}h_{fg} = k_g\nabla T_g\cdot\mathbf{n} \tag{5.77}$$

where h_{fg} refers to the enthalpy of vaporization, and the spatial gradients of temperature within the droplet have been neglected (i.e., isothermal droplet, $Bi < 0.1$).

In some cases, nonequilibrium or meta-stable conditions appreciably influence the interfacial heat balance. For example, supercooled droplets in clouds often exist in a liquid phase below the equilibrium phase change temperature in the absence of a suitable nucleation site to initiate freezing. A nonequilibrium criterion is required to initiate the phase change. Then the interfacial heat balance is expected to approach the previously described equilibrium conditions once the phase change has commenced. For reasons similar to those cited earlier for gas–particle flows, some type of spatial averaging (mixture or two-fluid approach) can be adopted for the prediction of multiphase flows with droplets, rather than tracking of individual droplets.

5.4.3 Entropy and the Second Law

At the moving phase interface, entropy is produced in accordance with the second law of thermodynamics. In the previously discussed example of solid–liquid systems, the entropy transfer from the liquid phase into the interface (designated by ET_l) consists of diffusive and advective parts, i.e.,

$$ET_l = -k_l \frac{dA}{T_l} \frac{dT}{dn}\bigg|_l dt + \rho_l V_l s_l dAdt \tag{5.78}$$

where the subscript l refers to the liquid phase. Similarly, in the solid phase (subscript s),

$$ET_s = -k_s \frac{dA}{T_s} \frac{dT}{dn}\bigg|_s dt + \rho_s V_s s_s dAdt \tag{5.79}$$

A change of entropy, dS, accompanies the phase transition between liquid and solid phases in the interfacial control volume, $dAdn$ (area multiplied by distance normal to the interface), i.e.,

$$dS = \rho_l s_l dAdn - \rho_s s_s dAdn \tag{5.80}$$

The difference between Equations (5.78) and (5.79), together with the entropy produced, $P_{s,i}$, at the moving phase interface, balances the entropy change, i.e.,

$$dS = ET_l - ET_s + \rho_l P_{s,i} dAdn \tag{5.81}$$

The entropy production, $P_{s,i}$, designates the entropy produced due to shear action along the dendrite arms as the dendrite (or other microscopic solidified structure) moves a distance of dn during the time interval dt.

Combining the previous equations and rearranging,

$$(\rho_l s_l - \rho_s s_s)\frac{dn}{dt} = -\frac{k_l}{T_l}\frac{dT}{dn}\bigg|_l + \frac{k_s}{T_s}\frac{dT}{dn}\bigg|_s + \rho_l V_l s_l - \rho_s V_s s_s + \rho_l P_{s,i}\frac{dn}{dt} \tag{5.82}$$

The interfacial entropy constraint is then obtained after combining Equation (5.82) with continuity, Equation (5.70), yielding

$$P_{s,i} = \frac{\rho_s}{\rho_l}\left(\Delta s_f - \frac{h_{sf}}{T}\right) + \frac{k_l}{\rho_l V_i}\left.\frac{dT}{dn}\right|_l\left(\frac{1}{T_l} - \frac{1}{T_s}\right) \tag{5.83}$$

where $\Delta s_f = s_l - s_s$ refers to the entropy of fusion. As mentioned earlier, the entropy of fusion for most metals and alloys is approximately constant, i.e., $\Delta S_f \approx 8.4$ J/mol K. It was outlined earlier that the entropy of fusion is approximately equal to the heat of fusion divided by the phase change temperature (called *Richard's rule*).

In addition to the entropy transport equations and interfacial constraints, two additional properties characterize entropy within a particular phase: *downward concavity* and *compatibility*. The downward concavity property of entropy may be written as

$$S_{,\phi\phi} < 0 \tag{5.84}$$

where the comma subscript notation refers to differentiation and ϕ refers to any general scalar (conserved) quantity, such as energy or mass. Equation (5.84) indicates that $S_{,\phi\phi}$ is a negative definite matrix since the second law requires that entropy is produced through irreversible processes. In other words, entropy is bounded from above as it attains a maximum value at a state of equilibrium. Also, for an irreversible process, conservation of entropy yields the following compatibility requirement of entropy:

$$\mathbf{F}_{,\phi} = S_{,\phi}\mathbf{f}_{,\phi} \tag{5.85}$$

where $\mathbf{F}_{,\phi}$ refers to the entropy flux derivative matrix (second-order tensor). Also, $\mathbf{f}_{,\phi}$ is a third-order tensor since it describes a derivative of flux terms in three directions with respect to the conservation variables.

The properties of downward concavity and compatibility can be applied within individual phases of a multiphase mixture. The previous development of interfacial entropy constraints was applied to solid–liquid systems, but analogous results are obtained for other types of multiphase systems. For example, instead of entropy produced by viscous action of an advancing dendrite arm (solid–liquid), an analogous irreversibility in liquid–gas systems is viscous shear stresses along the phase interface due to vapor flow during forced convection with boiling. In pool boiling, bubbles typically ascend into the quiescent liquid, and the entropy produced in Equation (5.81) would correspond to viscous drag at the vapor–liquid interface. The remaining terms in the entropy balances are similar to those in the previously derived equations, except that the subscripts would be replaced by g (gas instead of liquid) and l (liquid instead of solid). Also, the latent entropy of fusion in Equation (5.83) becomes the latent entropy of vaporization in the analysis of liquid–gas (vapor) systems. The latent entropy of vaporization is available from the difference between entropy values on the saturation lines for the state diagram of the given liquid. For example, the difference of $s_f - s_g$ from the saturated water tables indicates the entropy difference arising from vaporization (or condensation) at a specified saturation pressure.

Condensing flows are another example of liquid–gas (vapor) systems. Condensation may occur in the presence of a quiescent vapor exposed to a cool surface (direct contact condensation). Alternatively, it may occur by flow of vapor in contact with a cool surface or suddenly exposed to a pressure change such that vapor condenses out as droplets that become dispersed in the gas phase as a type of fog (called homogeneous condensation). In the case of quiescent vapor, entropy flows in Equations (5.78) and (5.79) occur predominantly due to conduction, but the flow components are included when forced (or free) convection accompanies the flow in the vapor phase. If the vapor temperature is at the saturated point, the entropy flow is based on heat conduction from the liquid side of the phase interface. In the general case including convection with vapor above the saturation temperature, $P_{s,i}$ in Equation (5.81) includes viscous drag of vapor along the moving phase interface (direct contact condensation) or along the suspended droplets in the gas phase (homogeneous condensation). The prediction of entropy production with condensing droplet formation has significance in many engineering systems. For example, reduced entropy production of flow through a steam turbine with condensing droplets at the turbine outlet can largely increase power output and generation of electricity in power plants.

Other examples involving interfacial entropy balances are multiphase flows with particles (gas–solid) or droplets (gas–liquid–solid). Viscous drag between the droplet (or particle) and gas stream is responsible for entropy produced at the interface in Equation (5.81). This entropy production arises with or without phase change, and in the latter case, a latent entropy of phase change between the two appropriate phases must be included in Equation (5.83). Relating the interfacial entropy production due to viscous drag in Equation (5.81) with flow quantities in multiphase flows with particles (or droplets) can be accommodated through Equation (3.23). The viscous dissipation function in Equation (3.23) describes the conversion of kinetic energy to internal energy through frictional effects in the fluid. Mechanical energy is dissipated in this manner to produce entropy as described in Equation (3.204). In the present case of multiphase flows, the relative motion between particles or droplets and the gas stream, or temperature gradients therein, is responsible for interfacial entropy production as described in Equations (5.81) to (5.83). Computing and analyzing entropy in this manner allows designers to reduce flow irreversibilities, thereby transporting multiphase flows with reduced pressure losses, while providing the desired heat transfer.

References

A. Bejan. 1993. *Heat Transfer*, New York: John Wiley & Sons.

B. Chalmers. 1964. *Principles of Solidification*, New York: John Wiley & Sons.

J.G. Collier. 1981. *Convective Boiling and Condensation*, 2nd ed., New York: McGraw-Hill.

International Encyclopedia of Heat and Mass Transfer, G.F. Hewitt, G.L. Shires, and Y.V. Polezhaev, eds., Boca Raton, FL: CRC Press, 1997

W. Kurz and D.J. Fisher. 1984. *Fundamentals of Solidification*, Switzerland: Trans Tech Publications.

N. Lion. 2000. "Melting and freezing," in *CRC Handbook of Thermal Engineering*, F. Kreith, ed., Boca Raton, FL: CRC Press.

F. Mashayek. 2001. "Dynamics of evaporating drops. Part I: formulation and evaporation model," *Int. J. Heat Mass Transfer* **44**: 1517–1526.

B.B. Mikic and W.M. Rohsenow. 1969. "A new correlation of pool-boiling data including the effect of heating surface characteristics," *J. Heat Transfer*, 245–250.

G.F. Naterer. 2000. "Predictive entropy based correction of phase change computations with fluid flow — part 1: second law formulation," *Numerical Heat Transfer B* **37**: 393–414.

G.F. Naterer. 2001. "Establishing heat-entropy analogies for interface tracking in phase change heat transfer with fluid flow," *Int. J. Heat Mass Transfer* **44**: 2903–2916.

G.F. Naterer, et al. 1998. "Near-wall microlayer evaporation analysis and experimental study of nucleate pool boiling on inclined surfaces," *ASME J. Heat Transfer* **120**: 641–653.

K. Nishikawa, et al. 1984. "Effects of surface configuration on nucleate boiling heat transfer," *Int. J. Heat Mass Transfer* **27**: 1559–1571.

M.S. Plesset and S.A. Zwick. 1954. "The growth of vapor bubbles in superheated liquids," *J. Appl. Phys.* **25**: 693–700.

L.S. Tong. 1965. *Boiling Heat Transfer and Two-Phase Flow*, New York: Wiley.

P.B. Whalley. 1987. *Boiling, Condensation and Gas-Liquid Flow*, Oxford: Clarendon Press.

Problems

1. The number of phases that coexist in equilibrium for a ternary component mixture is considered. Use the Gibbs phase rule to find the number of independent thermodynamic variables in the pure phase and two-phase regions of the ternary mixture as it undergoes phase transition.
2. Consider a phase change process with boiling involving bubble detachment from a heated horizontal surface. Alternating periods of a liquid- and vapor-covered surface are observed after the bubble detachment and during the bubble growth, respectively. An important parameter is the fraction of time with a particular phase covering the surface, such as f_v (fraction of time that a bubble covers the surface at a specific position). If f_v is well known or measured accurately, explain how the boiling heat flux to the liquid could be determined based on f_v and methods described in Chapter 2. Neglect the effects of convective motion and the region of influence encompassing the vortices and wakes of the departing bubble.
3. During a boiling process, bubbles grow outward from small indentations or cavities along a heated horizontal wall and detach from the surface once their size becomes sufficiently large. Beneath the bubble (surrounding the cavity), a thin microlayer of liquid is evaporated gradually as the bubble expands. This process may be approximated by one-dimensional downward movement of the gas–liquid interface until the vapor front reaches the wall.
   (a) Using a scaling analysis, estimate the transient variation of heat flux by conduction from the wall to the liquid during the microlayer evaporation. Express your answer in terms of q_w (wall heat flux), T_w (wall temperature), thermophysical properties, time, and the initial microlayer thickness, δ_0.
   (b) If the initial microlayer thickness is $\delta_0 \approx 3.2 \times 10^4 q_w^{-1.5}$, how much time will elapse before the microlayer evaporates entirely? What value is obtained for water with $q_w = 800\ \text{kW/m}^2$?
4. A bubble grows spherically outward along a heated wall during a boiling process. The rate of phase change with time is assumed to be constant (i.e., bubble radius grows linearly in time). The pressure within the bubble is denoted by p_v, whereas the pressure in the liquid surrounding the bubble is denoted by p_l.

(a) Use the one-dimensional form of the continuity equation to determine the liquid velocity surrounding the bubble (note: due to bubble expansion alone, not other convective motion).

(b) Perform an energy balance on an expanding bubble to find the pressure difference between the phase interface and the surrounding liquid over time (i.e., $p_i - p_l$).

(c) Perform a force balance to estimate the pressure difference between the vapor (within the bubble) and the phase interface over time (i.e., $p_v - p_i$).

5 Steam flows through a pipe (3.3-cm inner diameter, 3.8-cm outer diameter) with a mean velocity, temperature, and convection coefficient of 2 m/sec, 105°C, and 400 W/m^{2}K, respectively. Heat losses to the surrounding air at 20°C lead to condensation of steam along the inner pipe walls. The convection coefficient at the outer surface is 15 W/m^{2}K, and the thermal conductivity of the 1.3-cm-thick layer of insulation around the pipe is 0.03 W/mK.

(a) Over what length of pipe will the rate of condensation reach 0.1 g/sec?

(b) What is the change of phase fraction of steam over this distance of pipe length?

6 The wall of a tank containing liquid is heated electrically to provide a constant heat flux during boiling of the liquid along the wall. Due to alternating periods of liquid and vapor covering the heated wall, the temperature of the wall, T_w, varies with time. Explain how the various stages of phase change along the wall contribute to this variation of wall temperature with time. How would thermocapillarity effects influence this variation with time?

7 During phase change with boiling, a bubble grows spherically outward from a small indentation along a heated wall. Two distinct regions of fluid motion are observed: (i) viscous boundary layer in the liquid microlayer beneath the bubble and (ii) motion outside the boundary layer ahead of the outer bubble edge.

(a) Give the governing equations and boundary conditions for laminar fluid flow in each of these regions.

(b) Under suitable simplifying assumptions (to be stated in your analysis), solve the one-dimensional form of these equations to find the radial component of velocity outside the boundary layer. Assume that the phase interface moves with inverse proportionality to the square root of time (i.e., $V \approx A/\sqrt{t}$ where A depends on thermophysical properties and other factors).

8. Liquid methanol at its saturation temperature (65°C) is suddenly exposed to a heated vertical wall at 75°C. Measurements are taken

during boiling and bubble formation during phase change with the following results: (i) average microlayer thickness of 0.025 mm beneath the bubble and (ii) characteristic bubble diameter of 1 mm after 15 msec from its onset of formation. Can this data provide an indication of the relative significance of viscous forces acting on the liquid (associated with bubble expansion)? The following properties may be adopted for methanol: $\rho_l = 751$ kg/m^3, $\rho_v = 1.2$ kg/m^3, and $\mu_l = 3.3 \times 10^{-4}$ kg/msec.

9. Surface roughness can have a substantial effect on various forces, such as surface tension, contributing to phase change (liquid–vapor or liquid–solid) at a fixed wall. How could the surface roughness be characterized? Give an example and physical explanation of how its effects could be included within the phase change predictions.
10. Heat conduction from the wall through the liquid beneath a growing bubble affects the phase change process during boiling. Consider the two-dimensional cross-sectional area of the liquid region between a flat, smooth wall and an adjoining stationary bubble. This region may be transformed to a planar channel region by the following Mobius transformation:

$$\left(\frac{z - z_1}{z - z_3}\right) \cdot \left(\frac{z_2 - z_3}{z_2 - z_1}\right) = \frac{w - w_1}{w_2 - w_1}$$

The complex variables are $z = x + iy$ and $w = u + iv$, where z_1, z_2, and z_3 refer to points on two concentric circles (joined at a point) that are mapped to corresponding points w_1, w_2, and w_3 in the channel region.

(a) Use the Mobius mapping to find the temperature distribution beneath the bubble at a given time (note: wall can be characterized by a large radius of the outer circle).

(b) Based on the result in part (a), determine the wall heat flux.

11. In the previous problem, conformal mapping was used to transform the temperature solution from a simplified geometry (planar channel) to the actual complicated domain (between the wall and the bubble). Assume that the liquid temperature ahead of the phase interface is T_{sat} (saturation temperature) and the vapor temperature is linear in the simplified domain. Express your answer in terms of T_w, T_{sat}, δ_0 (initial microlayer thickness), and thermophysical properties.

(a) Perform a heat balance at the phase interface to estimate the time required to evaporate the liquid beneath the bubble entirely.

(b) What value of time is obtained in part (a) for water heated by a wall at a temperature of 20°C above T_{sat}? The initial microlayer thickness is 0.01 mm. The vapor density and thermal conductivity are 0.6 kg/m^3 and 0.025 W/mK, respectively.

12. A liquid–vapor phase change is encountered in a porous medium under one-dimensional, diffusion-dominated conditions (i.e., no convective motion). The liquid temperature is initially T_0 (below the phase change temperature, T_{sat}). Then the wall temperature at $x = 0$ is suddenly raised to T_s (above T_{sat}), vapor is generated, and the phase interface moves into the medium over time. Use a similarity solution to find the change of interface position with time. The similarity variable is $\eta = x/\sqrt{4\alpha t}$, and the phase interface is assumed to move according to $\delta = 2\gamma\sqrt{\alpha t}$, where γ must be determined.
13. A balance of pressure and surface tension forces in droplet formation can be used to find the pressure difference across the phase interface. An elongated droplet of saturated liquid with radii of R_1 and R_2 in the xz and yz planes, respectively, is in thermal and mechanical equilibrium with the surrounding subcooled vapor. Show that the pressure difference across the liquid–vapor interface can be written as follows:

$$P_v - P_{sat} = \gamma\left(\frac{1}{R_1} + \frac{1}{R_2}\right)$$

where γ, P_v, and P_{sat} refer to surface tension, the pressure of subcooled vapor outside the droplet, and the pressure of saturated liquid, respectively. The parenthetical term on the right side of the above equation is called the mean curvature of nonspherical surfaces.
14. A new technique has been considered to overcome water shortage problems in a particular country. Icebergs could be towed southward from the north Atlantic and Arctic oceans to that country. An iceberg with a relatively flat base (700 m long, 600 m wide) and a volume of 9×10^7 m^3 is specifically considered. What power is required to tow this iceberg to its destination (7200 km from its origin) if at least 90% of the ice mass must be retained (i.e., not melted) once it arrives at its destination? It may be assumed that the average temperature of ocean water during the voyage is 13°C. Consider only friction and heat transfer along the flat base in this approximate analysis. State the assumptions adopted in your solution.
15. In Equations (5.27) and (5.28), it was shown that the nucleation rate was based on the exponential of the Gibbs activation energy. This nucleation rate can be viewed as the frequency of adding atoms to a stable cluster multiplied by the number of clusters reaching a critical size to sustain the phase change. Derive Equation (5.27) or (5.28) by using the statistical definition of entropy to find the number of equally probable quantum states of an atom in the stable cluster.

16. What differences are encountered between solidification of metals, compared with polymers?
17. A thin liquid film covers an inclined surface in an industrial process. Evaporative cooling of the surface film occurs under steady-state conditions, together with convective heating by air at 35°C flowing past the surface. What is the saturated vapor pressure of the liquid when the measured surface temperature is 10°C? The film properties are given as follows: $h_{fg} = 180$ kJ/kg, $D_{AB} = 0.6 \times 10^{-5}$ m^2/sec, and $R = 0.1$ kJ/kg.
18. Under what conditions of pressure and temperature will sublimation occur?
19. In the liquid region of a binary component mixture ahead of an advancing solid–liquid interface during phase change, the mean concentration of solute is given by C_0. The ratio of solute concentrations in the solid and liquid phases is κ. Show that the solute concentration decreases exponentially with distance ahead of the phase interface. Apply suitable boundary conditions and express your answer in terms of C_0, κ, interface velocity, R, and the liquid mass diffusivity, D. How would this analysis be altered if a ternary mixture was considered?
20. The mixture form of the conservation of mass was obtained in Equation (5.51) as a special case of the general scalar conservation equation. Derive this mixture equation of mass conservation for a control volume occupied by two phases (such as solid and liquid) by performing a mass balance in each phase individually and then summing over both phases. In this analysis, use an averaged interfacial velocity, $\overline{V}_i$, and a volume occupied by phase k, V_k, which is approximated by the phase fraction, f_l, multiplied by the total volume, V.
21. A piecewise linear equation of state was presented in Equation (5.60) for energy in terms of temperature alone. By integrating the Gibbs equation through the phase change region (where the liquid fraction is assumed to vary linearly with temperature), derive this result for the equation of state of a binary component mixture.

6

Gas (Vapor)–Liquid Systems

6.1 Introduction

Heat transfer in two-phase flows involving gas and liquid phases is characterized by the presence of a moving and deforming phase interface. In this chapter, we will consider the fundamental physical processes in gas–liquid flows, as well as a variety of modeling and solution techniques, including analysis of boiling and condensation heat transfer. Many engineering technologies involve boiling and condensation. Examples include boiler operation in a power plant, nuclear reactors, evaporator performance in refrigeration systems, quenching in materials processing, and condensers in power and refrigeration systems.

6.2 Boiling Heat Transfer

6.2.1 Physical Mechanisms of Pool Boiling

Pool boiling refers to boiling along a heated surface submerged in a large volume of quiescent liquid. Liquid motion arises from free convection and mixing due to bubble growth and detachment from the heated surface. Pool boiling arises under two types of conditions. First, electrical heating (called 'power-controlled heating') allows the heat flux to be calculated based on measurements of the applied current and voltage. The power setting and heat flux are independent variables, whereas temperature is a dependent variable. Second, in 'thermal heating' (see Figure 6.1), the surface temperature can be set independently of the heat flux.

Saturated pool boiling arises when the temperature of the liquid pool is maintained at the saturation temperature, T_{sat}, or closely thereabout. Bubbles formed at the heated surface are propelled upward through the liquid by buoyancy. In *subcooled pool boiling*, the temperature of the liquid

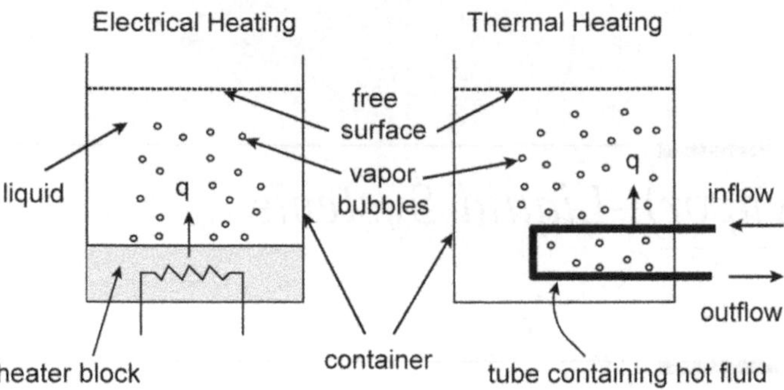

FIGURE 6.1
Electrical and thermal heating.

pool is less than the saturation temperature, and bubbles formed near the heater surface may later condense in the liquid.

In pool boiling problems, the fluid is initially quiescent near the heating surface, and subsequent fluid motion arises from free convection and the circulation induced by bubble growth and detachment. A main parameter in boiling problems is the degree of wall superheating, ΔT. It is defined as the difference between the wall and the bulk liquid saturation temperature at the local pressure. As ΔT increases, free convection, nucleate, transition, and film boiling modes along the *boiling curve* may be observed (see Figure 6.2).

Up to point A, free convection occurs since there is insufficient vapor to cause active boiling. Small temperature differences exist in the liquid, and heat is removed by free convection to the free surface. At point A, isolated bubbles initially appear along the heating surface (onset of nucleate

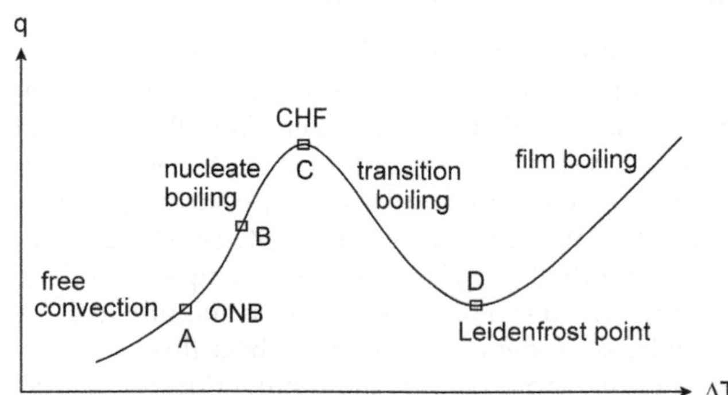

FIGURE 6.2
Boiling curve.

boiling). Nucleate boiling occurs between points A and C. Isolated bubbles appear, and heat is transferred mainly from the surface to the liquid. As ΔT increases (B–C), more nucleation sites become active and bubbles coalesce, mix, and ascend as merged jets or columns of vapor. At point C, the maximum heat flux, or critical heat flux (CHF), occurs.

Between points C and D, transition boiling (temperature controlled) occurs. An unstable (partial) vapor film forms on the heating surface, and conditions oscillate between nucleate and film boiling. Intermittent vapor formation blocks the liquid (higher conductivity) from contacting the surface (lowering the surface heat flux). Film boiling occurs beyond point D. In addition to conduction and convection, heat transfer by radiation is important at these high wall superheat levels. A stable vapor film covers the surface in this regime.

Differences in thermophysical properties lead to various boiling curves for different fluids. Phase change of water is widely encountered in many engineering applications. Selected thermophysical properties of saturated water at various temperatures are shown in Appendix A6. Based on those data, it can be observed that the thermophysical properties vary widely with temperature and pressure, thereby leading to changes in the boiling curve at different saturation pressures. Other properties of fluids and gases are outlined in the appendices.

6.2.2 Nucleate Pool Boiling Correlations

During nucleate boiling, vapor bubbles initially emerge from cavities along the heating surface where a gas or vapor phase already exists. After their formation, the bubbles grow when liquid is evaporated and heat is extracted from the surface by the near-wall fluid. A thin layer of liquid beneath the expanding bubble (called the *microlayer*) is evaporated. The vapor bubble expands upon further heating and eventually emerges and departs from the cavity under the influence of buoyancy. Each bubble transfers heat by convection as it moves away from the heated surface. In pool boiling, bubbles ascend and carry away the latent heat of evaporation, while liquid between the bubbles continues to absorb heat by natural convection from the surface.

Bubble coalescence and interactions between vapor columns affect the convective flow of liquid returning to the heating surface following the bubble departure. Furthermore, the effects of surface tension and pressure on the bubble motion influence the convective heat removal rate. The surface forces acting on the bubble, including buoyancy, weight, and surface tension at the nucleation surface site, are responsible for the bubble's departure frequency. The surface tension acts along the surface of contact where the bubble forms within the surface cavity. From a simplified balance of these forces, it may be readily shown that the departure diameter of a bubble is inversely proportional to the square roots of gravitational acceleration and phase density difference, as well as directly proportional to the square root of fluid surface tension.

From a dimensional analysis of all relevant variables arising in nucleate boiling, the Buckingham–Pi theorem yields 5 dimensionless pi groups. Based on these pi groups, the overall form of the heat transfer correlation becomes

$$\frac{(q_w''/\Delta T)}{k} = \overline{C}\left(\frac{\rho g(\rho_l - \rho_v)L^3}{\mu^2}\right)^a \left(\frac{c_p \Delta T}{h_{fg}}\right)^b \left(\frac{\mu c_p}{k}\right)^c \left(\frac{g(\rho_l - \rho_v)L^2}{\sigma}\right)^d \tag{6.1}$$

where L, q_w'', σ, $\overline{C}$, a, b, c, and d refer to surface length, wall heat flux, surface tension, and coefficients to be determined from the Buckingham–Pi analysis, respectively.

After these coefficients are determined, from both the dimensional analysis and experimental data (over a wide variety of conditions), the following widely used correlation is obtained (Rohsenow, 1952):

$$q_w'' = \mu_l h_{fg}\left[\frac{g(\rho_l - \rho_v)}{\sigma}\right]^{1/2}\left(\frac{c_{p,l}\Delta T}{c_{s,f} h_{fg} Pr^n}\right)^3 \tag{6.2}$$

In Equation (6.2), the empirical coefficients are dependent on the fluid–surface combination. Some typical values of the coefficient $c_{s,f}$ for various liquid–surface combinations are shown as follows:

- $c_{s,f} = 0.013$, $n = 1.0$ (water–copper)
- $c_{s,f} = 0.006$, $n = 1.0$ (water–brass)
- $c_{s,f} = 0.0132$, $n = 1.0$ (water–mechanically polished stainless steel)
- $c_{s,f} = 0.101$, $n = 1.7$ (benzene–chromium)
- $c_{s,f} = 0.027$, $n = 1.7$ (ethyl alcohol–chromium)
- $c_{s,f} = 0.00225$, $n = 1.7$ (isopropyl alcohol–copper)
- $c_{s,f} = 0.015$, $n = 1.7$ (n-pentane–chromium)

The addition of other constituents into the working fluid can largely alter the heat transfer characteristics of the liquid–surface combination. For example, at atmospheric pressure, the ratio of the average heat transfer coefficient for diluted water to pure water, h/h_w, can vary appreciably with constituent concentration, i.e., $h/h_w = 0.61$ (24% NaCl, 76% water), $h/h_w = 0.87$ (20% sugar, 80% water), and $h/h_w = 0.53$ (100% methanol, no water).

In these examples, the modified thermophysical properties, as well as the resulting changes in bubble dynamics, produce variations in the heat transfer coefficient. In addition, the conditions of the surface, including its roughness, surface coating, oxidation, and fouling, affect the bubble formation and dynamics. The effects of these surface conditions may be incorporated through modified empirical exponents in Rohsenow's correlation (Vachon et al., 1968). On the other hand, the effects of subcooling on the bubble formation process are generally small. The formation of intermittent bubble and vapor film regions will be considered in the following section involving transition boiling.

6.2.3 Critical and Minimum Heat Flux

The maximum heat flux during nucleate boiling depends on various thermophysical properties, i.e., latent heat of evaporation, vapor density, and surface tension. This maximum heat flux (CHF) occurs at the transition between nucleate boiling and film boiling. Kutateladze (1948) provided a dimensional analysis to estimate the dependence of the CHF on various operating parameters. The prediction of the CHF is important because of potential dangers associated with operating equipment near or beyond the CHF (see Figure 6.2), particularly when the heat source is operated independently of temperature. For example, if the heat flux is raised above the CHF, then a large sudden increase in system temperature can lead to severe system damage, such as melting of components or overheating and explosive reactions. In other types of equipment, operating near the CHF leads to risks of boiling in the transition region (i.e., unstable changes in the output heat flux).

Applying dimensional analysis with the Buckingham–Pi theorem at the maximum heat flux, we have five variables and four primary dimensions yielding $5-4=1$ pi group. As a result, we obtain

$$q''_{max} = 0.149 h_{fg} \rho_v \left[\frac{\sigma g(\rho_l - \rho_v)}{\rho_v^2} \right]^{1/4} \tag{6.3}$$

where the coefficient 0.149 was obtained by comparisons with experimental data (Lienhard et al., 1973). This result holds for pure liquids (saturated) on horizontal, upfacing surfaces. In Equation (6.3), it can be observed that the maximum heat flux depends on pressure, mainly through the dependence of thermophysical properties on the saturation pressure.

The Critical Heat Flux during subcooled pool boiling arises when the liquid pool temperature is less than the saturation temperature. The maximum heat flux for subcooled conditions, $q''_{max,sub}$ can be expressed in terms of the maximum heat flux for a saturated liquid, $q''_{max,sat}$, as follows:

$$q''_{max,sub} = q''_{max,sat}(1 + B\Delta T_{sub}); \quad B = 0.1 c_{p,l} \frac{(\rho_l/\rho_v)^{3/4}}{h_{fg}} \tag{6.4}$$

where

$$\Delta T_{sub} = T_{sat} - T_{pool}$$

The following list indicates some commonly encountered CHF values for the specified fluid–surface combinations:

- Water–copper: $q''_{max} = 0.62$ to $0.85\ \text{MW/m}^2$
- Water– steel: $q''_{max} = 1.29\ \text{MW/m}^2$
- Methanol–copper: $q''_{max} = 0.39\ \text{MW/m}^2$
- Methanol–steel: $q''_{max} = 0.39\ \text{MW/m}^2$

The *Leidenfrost point* refers to the minimum heat flux (see Figure 6.2). Zuber (1958) observed similarities in approaching the region of transition boiling, whether cooling to the minimum heat flux from film boiling (i.e., reducing the wall superheat) or heating to the CHF from nucleate boiling (i.e., increasing the temperature difference). When reducing the wall superheat below the minimum heat flux, the rate of vapor generation was not high enough to sustain a stable vapor film. In this case, Equation (6.3) is modified as follows:

$$q''_{min} = C h_{fg} \rho_v \left[\frac{\sigma g(\rho_l - \rho_v)}{(\rho_l + \rho_v)^2} \right]^{1/4} \tag{6.5}$$

where $C \approx 0.09$ (Berenson, 1961) was obtained from experimental data for horizontal surfaces. The result is accurate within approximately $\pm 50\%$ at moderate pressures for most fluids.

6.2.4 Film Pool Boiling

In film boiling, a thin vapor film separates the bulk liquid from the heating surface. Several methods for heat transfer predictions in film boiling have been proposed during the past several decades. Bromley (1950) extended previous theories of film condensation to film boiling on tubes and vertical plates with the following result:

$$\bar{h} = 0.62 \left[\frac{k_v^3 h_{fg} \rho_v (\rho_l - \rho_v) g}{D \mu_v \Delta T} \left(1 + \frac{0.4 c_p \Delta T}{h_{fg}} \right) \right]^{1/4} \tag{6.6}$$

where $\bar{h}$ and h_{fg} refer to heat transfer coefficient and latent heat of vaporization, respectively.

Alternatively,

$$\overline{Nu_D} = \frac{\overline{h}D}{k_v} = C\left[\frac{g(\rho_l - \rho_v)h'_{fg}D^3}{v_v k_v(T_w - T_{sat})}\right]^{1/4} \tag{6.7}$$

where $C \approx 0.62$ (horizontal cylinders) or 0.67 (spheres). In Equation (6.7), the subscripts v, l, w, and sat refer to vapor, liquid, wall, and saturation, respectively. The vapor properties are evaluated at the film temperature, i.e., $(T_w + T_{sat})/2$, and the liquid properties are evaluated at T_{sat}. Also, $h'_{fg} \approx h_{fg} + 0.8c_{p,v}(T_w - T_{sat})$ accounts for contributions arising from latent heat and sensible heat to maintain the temperature in the vapor film above the saturation temperature. Berenson (1961) and Duignam et al. (1991) applied the correlation in Equation (6.7) to film boiling on horizontal surfaces, based on a modification of D. Also, a detailed similarity solution was given by Nishikawa et al. (1976) for film boiling along vertical plates. Their results demonstrated that the role of temperature-dependent properties was significant for the case of density and specific heat in the vapor phase, but less important for viscosity and thermal conductivity.

In film boiling, the radiative heat transfer contribution often becomes significant. In cases with a large-wall superheat level, the effective heat transfer coefficient (including convection and radiation) can be expressed as

$$h^{4/3} = h_{conv}^{4/3} + h_{rad}h^{1/3} \tag{6.8}$$

For the case of $h_{rad} < h_{conv}$, the heat transfer coefficient can be approximated through

$$h \approx h_{conv} + 0.75h_{rad} \tag{6.9}$$

$$h_{rad} = \sigma\epsilon_s\left(\frac{T_w^4 - T_{sat}^4}{T_w - T_{sat}}\right) \tag{6.10}$$

where ϵ_s refers to the surface emissivity and σ is the Stefan-Boltzmann constant.

The treatment of radiative heat transfer in film boiling with Equations (6.8) to (6.10) is based on the following assumptions: (i) the phase interface and wall are flat and parallel, and (ii) the interfacial radiative transport occurs as a blackbody. These correlations have been mainly applied to basic geometrical configurations, including horizontal and vertical plane surfaces. In the following case study, boiling heat transfer along inclined surfaces is considered.

6.2.5 Case Study: Pool Boiling on Inclined Surfaces

Surface inclination can have a significant impact on boiling heat transfer (Hendradjit et al., 1997; Naterer et al., 1998). The heat transfer processes may be classified into two periods: liquid and vapor. During the liquid period, heat is extracted by transient conduction from the wall to the liquid. In the vapor period, bubbles expand from the surface cavities and the thin microlayer beneath each bubble evaporates. In the following analysis, both periods of heat transfer will be considered.

6.2.5.1 Nucleate Boiling (Liquid Period)

Heat transfer during the liquid period is assumed to be governed by the following transient heat conduction equation:

$$\frac{\partial T}{\partial t} = \alpha \frac{\partial^2 T}{\partial y^2} \tag{6.11}$$

subject to the boundary conditions of $T(y, 0) = T_{sat}$ and $T(0, t) = T_w$. Based on a similarity solution, the following temperature profiles and wall heat flux are obtained during the liquid period:

$$\frac{T - T_w}{T_{sat} - T_w} = erf\left(\frac{y}{2\sqrt{\alpha t}}\right) \tag{6.12}$$

$$q_w = \frac{k(T_w - T_{sat})}{\sqrt{\pi \alpha t}} \tag{6.13}$$

where $erf(w)$ is the Gaussian error function. The slope at the end of the liquid period (at $t = t_l$), based on Equation (6.13), is used for a linear approximation of the initial temperature profile in the vapor period.

6.2.5.2 Nucleate Boiling (Vapor Period)

In this period, the vapor bubble covers the heated surface. The function $\delta(t)$ will be used to refer to the diameter of the growing bubble. Transient conduction in the vapor phase is assumed to be governed by the following heat conduction equation:

$$\frac{\partial T}{\partial t} = \alpha \frac{\partial^2 T}{\partial y^2} \tag{6.14}$$

A conformal mapping between a two-dimensional region surrounding the bubble and a one-dimensional planar region with a thickness of $\delta(t)$ is used to represent the heat transfer process (Naterer et al., 1998). Under this change of variables, heat transfer through the microlayer (beneath the bubble) is approximated by one-dimensional heat conduction across the planar layer. The thickness of the microlayer decreases when liquid in the microlayer is evaporated.

In this way, the boundary condition at $y = \delta$ involves a moving boundary problem. The initial and boundary conditions in the vapor period become

$$T(y, t_l) = T_w - \lambda(T_w - T_{sat})\frac{y}{\delta}; \quad T(0, t) = T_w; \quad T(\delta, t) = T_{sat} \tag{6.15}$$

where $\lambda = \delta/\sqrt{\pi \alpha t_l}$. An exact solution of this equation is given by

$$\frac{T_w - T}{T_w - T_{sat}} = (1 - \lambda)\sum_{n=0}^{\infty}\left\{erf\left[\frac{\delta(2n+1) + y}{2\sqrt{\alpha(t - t_l)}}\right] - erf\left[\frac{\delta(2n+1) - y}{2\sqrt{\alpha(t - t_l)}}\right]\right\} + \lambda\left(\frac{y}{\delta}\right) \tag{6.16}$$

Also, the wall heat flux can be obtained by Fourier's law and Equation (6.16) with the following result:

$$q_w = \frac{k(1 - \lambda)\Delta T}{\sqrt{\pi\alpha(t - t_l)}}\sum_{n=0}^{\infty}\left\{exp\left[\frac{-(2n\delta + \delta + y)^2}{4\alpha(t - t_l)}\right] + exp\left[\frac{-(2n\delta + \delta - y)^2}{4\alpha(t - t_l)}\right]\right\} + \frac{k\lambda\Delta T}{\delta} \tag{6.17}$$

Furthermore, the average heat flux during a cycle throughout a liquid–vapor period is obtained by integration of the previous heat flux expressions over the appropriate time periods, i.e.,

$$\overline{q}_w = \frac{1}{t_v}\int_0^{t_l} q_l(t)dt + \frac{1}{t_v}\int_{t_l}^{t_v} q_v(t)dt \tag{6.18}$$

where t_l and t_v refer to the liquid and entire period durations, respectively.

The interfacial constraint involves the following modes of heat conduction and the latent heat of vaporization at the phase interface:

$$-k_l\left.\frac{\partial T}{\partial y}\right|_{y=\delta_-} - \rho L\frac{d\delta}{dt} = -k_v\left.\frac{\partial T}{\partial y}\right|_{y=\delta_+} \tag{6.19}$$

where the latter term is negligible within the bubble under the lumped capacitance approximation. Using the result from Equation (6.16) and rearranging terms,

$$\frac{d\delta}{dt} = \frac{k(1-\lambda)\Delta T}{\rho L\sqrt{\pi\alpha(t-t_l)}} \sum_{n=0}^{\infty} \left\{ exp\left[\frac{-\delta^2(n+1)^2}{\alpha(t-t_l)}\right] + exp\left[\frac{-\delta^2 n^2}{\alpha(t-t_l)}\right] \right\} + \frac{k\lambda\Delta T}{\rho L\delta} \tag{6.20}$$

In this result, higher order terms become negligible in the summation.

During late stages of bubble growth ($\lambda \sim 1$), the wall heat flux in Equation (6.17) is simplified as

$$q_w = 2\frac{(\sqrt{\pi\alpha t_l} - \delta)k\Delta T}{\pi\alpha\sqrt{t_l(t-t_l)}} exp\left[\frac{-\delta^2}{4\alpha(t-t_l)}\right] + \frac{k\Delta T}{\sqrt{\pi\alpha t_l}} \tag{6.21}$$

where the interface position is obtained from the solution of Equation (6.20), with the following result:

$$\delta(t) = \frac{\sqrt{\pi\alpha t_l}(\rho L\pi\alpha - 2k\Delta T)}{2k\Delta T} \left\{ exp\left[-2\frac{k\Delta T}{\rho L\pi\alpha}\sqrt{\frac{t-t_l}{t_l}}\right] - 1 \right\} + \sqrt{\pi\alpha(t-t_l)} \tag{6.22}$$

Due to the complexity of the boiling process on inclined surfaces, the above models remain empirical in nature by virtue of their dependence on correlations involving bubble diameter and liquid period duration. Despite these limitations, the current analysis can provide a physical basis from which various boiling trends can be understood.

6.2.5.3 Film Boiling Conditions

In the case of film boiling conditions on the inclined surface (see Figure 6.3), the thin film approximation leads to the following reduced forms of the momentum and energy equations in the vapor phase:

$$\mu_v \frac{\partial^2 u}{\partial y^2} = -g_x(\rho_l - \rho_v) \tag{6.23}$$

$$k_v \frac{\partial^2 T}{\partial y^2} = 0 \tag{6.24}$$

At the wall, velocity and temperature fields are known. Also, compatibility conditions for mass, momentum, and energy, respectively, are applied at the vapor–liquid interface, with the result that

$$u(0) = v(0) = 0; \quad T(0) = T_w \tag{6.25}$$

$$-\frac{\dot{m}_l}{A} = \rho_v u_v \frac{d\delta}{dx}\bigg|_\delta = \rho_l u_l \frac{d\delta}{dx}\bigg|_\delta \tag{6.26}$$

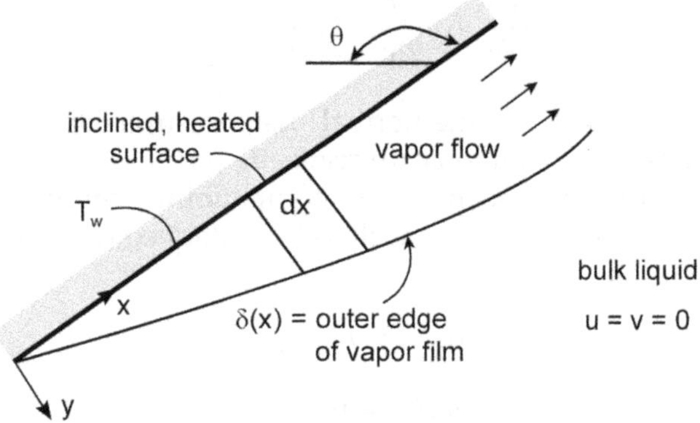

FIGURE 6.3
Film boiling schematic for inclined surface.

$$\mu_v \frac{\partial u_v}{\partial y}\bigg|_\delta = \mu_l \frac{\partial u_l}{\partial y}\bigg|_\delta \tag{6.27}$$

$$-k_v \frac{\partial T}{\partial y}\bigg|_\delta = -k_l \frac{\partial T}{\partial y}\bigg|_\delta + \frac{\dot{m}_l}{A} h_{fg} \tag{6.28}$$

In the vapor region, conservation of mass requires that the rate of evaporation in the vapor film must balance the supply rate from the liquid region, i.e.,

$$\frac{1}{W}\frac{d\dot{m}_v}{dx} = \frac{d\Gamma}{dx} = \frac{\dot{m}_l}{A} \tag{6.29}$$

where $\Gamma = \dot{m}_v/A$ is the mass flow rate per unit area of plate and W refers to the plate width. Integrating the energy equation, Equation (6.24), across the vapor film and substituting the resulting heat flux into the interfacial balance, Equation (6.28), we obtain

$$\frac{d\Gamma}{dx} = \frac{k_v(T_w - T_{sat})}{\delta h_{fg}} \tag{6.30}$$

In the liquid region, the governing equations including free convection are written as

$$\mu_l \frac{\partial^2 u}{\partial y^2} = -g_x(\rho_{l\infty} - \rho_l) \tag{6.31}$$

$$u\frac{\partial T}{\partial x}+v\frac{\partial T}{\partial y}=k_l\frac{\partial^2 T}{\partial y^2} \tag{6.32}$$

We will define $y=M$ as the normal distance from the wall where the undisturbed ambient conditions are reached.

Then, in addition to the previous compatibility conditions, matching conditions of velocity at $y=\delta$, as well as boundary conditions at $y=M$, are applied as follows:

$$u_v(\delta)=u_l(\delta)=U_i \tag{6.33}$$

$$u_l(M)=0 \tag{6.34}$$

$$T_l(M)=T_\infty \tag{6.35}$$

where U_i represents the unknown interface velocity. Then the liquid momentum equation may be solved for the liquid velocity, subject to specified interface and ambient velocity conditions, so that

$$u_l(y)=\left(\frac{\Delta\rho_l g_x}{2\mu_l}\right)(y-\delta)(M-y)+U_i\left(\frac{M-y}{M-\delta}\right) \tag{6.36}$$

where $\Delta\rho_l=\rho_{l,\infty}-\rho_l$.

The momentum equation is solved in a similar manner in the vapor film. However, in this case, the no-slip boundary condition at the wall and the compatibility constraints at the phase interface are used to determine the unknown constants of integration. As a result,

$$u_v=\left(\frac{\Delta\rho_v g_x}{2\mu_v}\right)(2\delta y+M\lambda y-y^2)+\frac{\mu_l}{\mu_v}U_i\left(\frac{y}{\delta-M}\right) \tag{6.37}$$

In Equation (6.37), $\Delta\rho_v=\rho_l-\rho_v$ and $\lambda=\Delta\rho_l/\Delta\rho_v$. Substituting $y=\delta$ into Equation (6.37),

$$U_i=\frac{(\delta-M)\Delta\rho_v g_x(\delta+M\lambda)\delta}{2(\delta\mu_v-M\mu_v-\delta\mu_l)} \tag{6.38}$$

We can now substitute this expression back into Equations (6.36) and (6.37) in order to find explicit expressions for the vapor and liquid velocity profiles. Also, the mass flow rate of vapor, Γ, may then be obtained by integration of the vapor velocity profile from $y=0\rightarrow\delta$.

This mass flow rate may be expressed in terms of the ratio of film thickness, δ, to liquid penetration distance, M (i.e., distance to an undisturbed position in the bulk liquid), where $B=\delta/M$, as follows:

$$\Gamma = \int_0^{\delta} \rho_v u_v dy = \left(\frac{\rho_v \Delta\rho_v g_x}{4\mu_v}\right)\left(\frac{4}{3} + \frac{\lambda}{B} - \frac{\mu_l(B+\lambda)}{(1-B)\mu_v + B\mu_l}\right)\delta^3 \tag{6.39}$$

After combining this expression with the interfacial energy balance in Equation (6.30), we have

$$\Gamma^{1/3}\frac{d\Gamma}{dx} = \frac{k_v \Delta T}{h_{fg}}\left(\frac{\rho_v \Delta\rho_v g}{4\mu_v}\right)^{1/3}\left(\frac{4}{3} + \frac{\lambda}{B} - \frac{u_l(B+\lambda)}{(1-B)\mu_v + B\mu_l}\right)^{1/3}\left(\frac{g_x}{g}\right)^{1/3} \tag{6.40}$$

where $g_x/g = sin(\theta)$ for a plane, noncurved surface that is inclined at an angle of θ with respect to the horizontal axis.

Integrating from $x = 0$, where $\Gamma = 0$, to x,

$$\Gamma = \left(\frac{4k\Delta T}{3h_{fg}}\right)^{3/4}\left(\frac{\rho_v \Delta\rho_v g}{4\mu_v}\right)^{1/4}\left(\frac{4}{3} + \frac{\lambda}{B} - \frac{\mu_l(B+\lambda)}{(1-B)\mu_v + B\mu_l}\right)^{1/4}\left[\int_0^x \left(\frac{g_x}{g}\right)^{1/3} dx\right]^{3/4} \tag{6.41}$$

Substituting this result into Equation (6.39) and rearranging,

$$\delta = \left[\frac{3h_{fg}\rho_v \Delta\rho_v g}{16k\Delta T\mu_v}\left(\frac{4}{3} + \frac{\lambda}{B} - \frac{\mu_l(B+\lambda)}{(1-B)\mu_v + B\mu_l} + \right)\right]^{-1/4}\left[\int_0^x \left(\frac{g_x}{g}\right)^{1/3} dx\right]^{1/4} \times \left(\frac{g}{g_x}\right)^{1/3} \tag{6.42}$$

Then we can evaluate the average heat transfer rate as

$$\bar{h}\Delta T = \frac{1}{S}\int_0^S \left(\frac{q}{A}\right) dx = \frac{1}{S}\int_0^S \left(\frac{k_v \Delta T}{\delta}\right) dx \tag{6.43}$$

Substituting our expression for δ and integrating over the surface,

$$\bar{h} = 0.88\left[\frac{k_v^3 h_{fg}\rho_v \Delta\rho g}{S\mu_v \Delta T}\left(\frac{4}{3} + \frac{\lambda}{B} - \frac{\mu_l(B+\lambda)}{(1-B)\mu_v + B\mu_l}\right)\right]^{1/4} \cdot \left[\frac{1}{S}\int_0^S \left(\frac{g_x}{g}\right)^{1/3} dx\right]^{3/4} \tag{6.44}$$

where $B = \delta/M$ is assumed to be independent of x. The value of B can be estimated by matching the earlier solution to the previously reported results in Equation (6.6) at the common angle of $\theta = 90°$. Comparisons between the currently predicted results and the experimental data have shown close agreement for both nucleate and film boiling problems (Naterer et al., 1998). The previous analysis indicates how analytical methods developed in earlier chapters for single phase flows (such as integral methods) can be applied to problems involving multiphase heat transfer.

6.2.6 Boiling with Forced Convection

Various complex flow patterns may arise in two-phase flows involving forced convection and boiling. In this section, these patterns will be described for two-phase flows in certain geometrical configurations, such as vertically and horizontally aligned tubes.

6.2.6.1 Inside a Vertical Tube

The flow regimes for vertical flow with boiling of water in a pipe will be considered. Liquid enters the pipe, and due to heat transfer, boiling occurs when the temperature of the liquid reaches the saturation temperature. Initially, bubbles form along the walls and a two-phase flow occurs. Its structure and characteristics largely depend on the flow Reynolds number. With sufficiently large heat input, the fluid exits as vapor once evaporation of the liquid is completed.

Various flow regimes between the inlet and the outlet are observed in this two-phase flow (see Figure 6.4). In particular, *bubble flow* refers to conditions involving a continuous liquid flow with a dispersion of bubbles within the liquid. A *slug flow* or *plug flow* occurs whenever the bubbles coalesce to make large groups of bubbles with a combined size approaching the pipe diameter. Then a *churn flow* arises if the slug flow bubbles continue to break down to give an oscillating gas–liquid distribution. Furthermore, an *annular flow* occurs whenever the liquid mainly flows as a film along the walls, with some internal liquid entrainment in the core, while the vapor flows through the center of the pipe. Finally, a *wispy annular flow* refers to the case where the concentration of drops in the core increases to form large

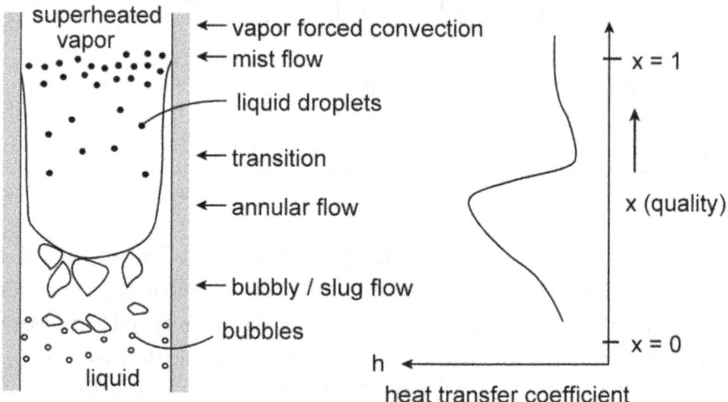

FIGURE 6.4
Two-phase flow regimes in vertical pipe flow.

lumps or streaks (or wisps) of liquid. This type of flow occurs as the liquid flow rate is increased. Other flow regimes appear in problems involving different geometries (i.e., horizontal pipe, rather than a vertical pipe) or boundary conditions.

In pool boiling, the vapor flow is largely buoyancy driven. In contrast, forced convection boiling involves bulk motion of the liquid and buoyancy effects. In Figure 6.4, the vertical tube is exposed to a uniform heat flux and supplied with a subcooled liquid upflow. The liquid evaporates as it flows up through the tube. Despite recent advances in the detailed understanding of these two-phase flows, generalized theories are not available due to various flow complexities, i.e., bubble growth, separation and coalescence, effects of flow hydrodynamics and velocities, and variations in the patterns of the two-phase flow.

In two-phase vertical flows, separate correlations will be considered for each of the following three regions: (i) single phase (subcooled liquid or superheated vapor) regions, (ii) slug or bubbly flow, and (iii) annular or annular mist. The classification of two-phase flows into various flow regimes is often illustrated by two-phase flow maps (see Figure 6.5). In the subcooled or superheated regions, the single phase convection correlations may be used. Furthermore, for slug or bubbly flows, the total heat flux may be expressed as the sum of heat flux contributions from boiling (based on pool boiling results) and convection (based on single phase convection correlations).

Correlations for annular boiling flows have been presented by Chen (1963). In the annular or annular mist regions, the correlation for total heat transfer coefficient, h_{tot}, will be obtained in terms of convective and boiling components, h_c and h_b, respectively. For the boiling coefficient, h_b, the phase-averaged velocity, G, and density-weighted phase quality (or

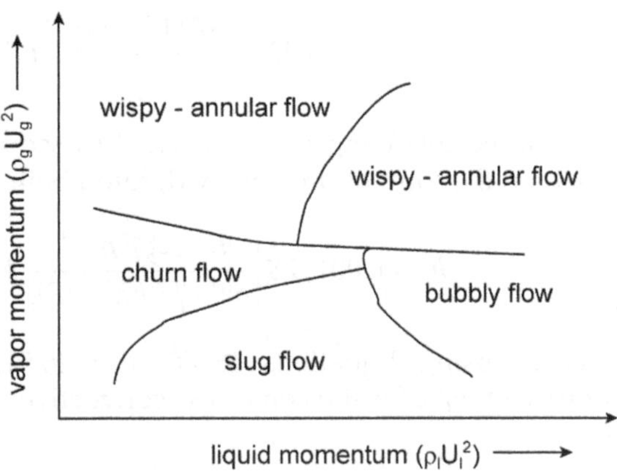

FIGURE 6.5
Two-phase flow map.

Martinelli parameter), X_{tt}, are required as follows:

$$G = \frac{\dot{m}}{\pi D^2/4} \tag{6.45}$$

$$X_{tt} = \left(\frac{1-x}{x}\right)^{0.9}\left(\frac{\rho_v}{\rho_l}\right)^{0.5}\left(\frac{\mu_l}{\mu_v}\right)^{0.1} \tag{6.46}$$

where x is the quality (vapor fraction) at the point of interest. The Martinelli parameter is a numerical approximation of the square root of the liquid pressure gradient divided by the vapor pressure gradient in the axial (flow) direction of the two-phase flow.

Also, the dynamic transition factors, F and S, are defined as follows:

$$F = \begin{cases} 2.35(0.213 + 1/X_{tt})^{0.736} & \text{if } 1/X_{tt} > 0.1 \\ 1 & \text{if } 1/X_{tt} \le 0.1 \end{cases} \tag{6.47}$$

$$S = \begin{cases} 0.1 & \text{if } Re_T > 70 \\ 1/(1 + 0.42 Re_T^{0.78}) & \text{if } 32.5 < Re_T < 70 \\ 1/(1 + 0.12 Re_T^{1.14}) & \text{if } Re_T < 32.5 \end{cases} \tag{6.48}$$

where

$$Re_T = \frac{GD(1-x)}{\mu_l} F^{1.25} \times 10^{-4} \tag{6.49}$$

The single phase heat transfer coefficient, h_c, is based on previously outlined convection correlations (Chapter 3) using saturated liquid properties and Re_T in Equation (6.49), i.e.,

$$h_c = 0.023F\left[\frac{GD(1-x)}{\mu_l}\right]^{0.8} Pr^{0.4}\left(\frac{k_l}{D}\right) \tag{6.50}$$

Also, the following result is widely used for the boiling coefficient, h_b, based on the form of previously derived boiling correlations in this chapter:

$$h_b = 0.00122S\left[\frac{k_l^{0.79} c_{p,l}^{0.45} \rho_l^{0.49}}{\sigma^{0.5} \mu_l^{0.29} h_{fg}^{0.24} \rho_v^{0.24}}\right]\Delta T^{0.24} \Delta p_{sat}^{0.75} \tag{6.51}$$

Then, using Equations (6.50) and (6.51), the overall heat transfer coefficient, h_{tot}, for the combined convective–boiling flow is approximated by

$$h_{tot} = Fh_c + Sh_b \tag{6.52}$$

which includes the contributions of the convection coefficient, h_c, and the boiling heat transfer coefficient, h_b.

Although two-phase flow maps are available for distinguishing flow regions within two-phase flows, such as slug and bubbly flows in pipes (see Figure 6.5), it should be noted that the subject of flow transition between these regimes is an area of active research. Transition between various flow regimes often involves flow instabilities. For example, in vertical flows, irregular coalescence of bubbles is often observed to initiate a bubble–slug transition. Gradual bubble growth to large bubbles occupying the entire cross-sectional area of the pipe generates slugs in the flow. However, as the fluid velocity increases, chaotic eddy motion in the turbulent flow acts to break up these bubbles, while offsetting the previously mentioned coalescence process. An alternative explanation may be available through the formation of "void waves" in the flow. The transport of waves in the streamwise direction could promote packing of bubbles; coalescence arises as a result of this induced packing.

6.2.6.2 Inside a Horizontal Tube

For horizontal two-phase flows in a pipe, the role of gravity is apparent in its effects on the transition between the flow regimes. The buoyancy of bubbles and weight of liquid have a smaller role in the processes of bubble production and movement. In horizontal flow with low flow velocities, a stratified flow regime occurs whereby liquid settles under its own weight along the bottom of the pipe. However, selective liquid settling along the lower section in annular flow was not observed in the previous vertical two-phase flow configuration. Thus, the orientation of the gravity vector with respect to the flow direction affects numerous aspects of the phase change heat transfer. These aspects include the force balances leading to bubble departure, as well as production and expansion of vapor in the cavities during initial bubble growth.

In the bubble flow regime, Rohsenow and Griffith (1955) recommended the following method for calculating the heat flux for forced convection with boiling:

$$\left.\frac{q}{A}\right|_{total} = \left.\frac{q}{A}\right|_{forced\ \ convection} + \left.\frac{q}{A}\right|_{nucleate\ \ boiling} \tag{6.53}$$

The heat flux corresponding to nucleate pool boiling, i.e., the last term in Equation (6.53), is effectively separated from the forced convection contribution. It can be computed from previously developed correlations, such as Equation (6.2). It should be noted that Equation (6.53) is applicable to local forced convection boiling, where the near-wall fluid is at or above the saturation temperature, but the bulk portion of the remaining liquid temperature is subcooled.

Over a broader range of flow conditions, Altman et al. (1960) developed the following correlation for refrigerants flowing in horizontal tubes:

$$\overline{Nu}_D = \frac{\bar{h}D}{k_f} = c[Re_D^2 F]^b \tag{6.54}$$

in the range of $10^9 < Re_D^2 F < 0.7 \times 10^{12}$. For incomplete evaporation, $c = 0.0009$ and $b = 0.5$, whereas for complete evaporation, $c = 0.0082$ and $b = 0.4$. Also, $Re_D = GD/\mu_l$ and the dimensionless load factor, F, is defined by

$$F = 102\frac{h_{fg}\Delta x}{L} \tag{6.55}$$

where L is the tube length (m), h_{fg} is the enthalpy of evaporation (kJ/kg), and Δx refers to change in the flow quality along the pipe length.

6.2.6.3 *Inside Vertical or Horizontal Tubes*

Kandlikar (1990) recommended a single general correlation for two-phase flows with boiling for both vertical and horizontal tube orientations. This correlation was based on experimental data obtained from 24 sources involving various working fluids (such as refrigerants and water). The correlation utilizes a convection number, C_0, Froude number, Fr_l, and boiling number, B_0, as follows:

$$C_0 = \left(\frac{1-x}{x}\right)^{0.8}\left(\frac{\rho_v}{\rho_l}\right)^{0.5} \tag{6.56}$$

$$Fr_l = \frac{G^2}{\rho_l^2 gD} \tag{6.57}$$

$$B_0 = \frac{q}{AGh_{fg}} \tag{6.58}$$

Then the ratio of the desired two-phase heat transfer coefficient, h_{tp}, to the corresponding single-phase flow coefficient, h_l, is

$$\frac{h_{tp}}{h_l} = C_1 C_0^{C_2}(25Fr_l)^{C_5} + C_3 B_0^{C_4} F_{fl} \tag{6.59}$$

where h_l is based on the Dittus–Boelter correlation (Chapter 3), i.e.,

$$\overline{Nu_D} = \frac{h_l D}{k_l} = 0.023 Re_l^{0.8} Pr_l^{0.4} \tag{6.60}$$

Empirical coefficients for Equation (6.59) are summarized in Table 6.1. It should be noted that $C_5 = 0$ for horizontal tubes (and vertical tubes with $Fr_l > 0.04$).

The coefficient F_{fl} in Equation (6.59) is dependent on the type of working fluid, i.e., $F_{fl} = 1$ (water), $F_{fl} = 4.7$ (nitrogen), $F_{fl} = 1.24$ (R-114), $F_{fl} = 1.1$ (R-113, R-152a), and $F_{fl} = 1.63$ (R-134a). Also, it should be noted that Kandlikar's correlation in Equation (6.59) is applicable over the range of $0.001 < x < 0.95$.

The peak heat flux for two-phase flows with boiling cannot be generally predicted based on previously described correlations involving pool boiling, such as Equation (6.3). The onset of transition boiling from nucleate boiling is more complicated than pool boiling, due to the combined effects of inertia and gravity on the bubble dynamics. Gambill (1962) recommended that the maximum heat flux in forced convection with boiling can be predicted based on a superposition (addition) of the CHF from pool boiling and the heat flux arising from forced convection. This approach is an approximation that does not imply that linear superposition can be systematically applied over a wide range of conditions to the nonlinear processes of forced convection with boiling.

6.2.6.4 *Outside a Horizontal Tube*

For boiling heat transfer outside of horizontal tubes, nucleation usually initiates at the base of the tube and bubbles move along the surface and depart from the top of the tube. As a result, a bubble layer is created around the tube, which leads to an angular variation of the heat transfer coefficient around the circumference of the tube (see Figure 6.6).

TABLE 6.1

Coefficients for Kandlikar Correlation

	$C_0 < 0.65$ **(Convective Region)**	$C_0 > 0.65$ **(Boiling Region)**
$C_1 =$	1.1360	0.6683
$C_2 =$	−0.9	−0.2
$C_3 =$	667.2	1058.0
$C_4 =$	0.7	0.7
$C_5 =$	= 0.3	0.3

In Figure 6.6, α and U refer to the heat transfer coefficient and vertical liquid velocity, respectively. Thus, $U = 0$ corresponds to pool boiling over a horizontal tube. The test results were obtained for a 27-mm-diameter tube and various incoming liquid velocities, U (m/sec), with boiling of R-113 at a pressure of 1 atm. The value of the heat transfer coefficient increases from the base due to the increasing bubble layer velocity. The vapor layer processes lead to these observed trends in Figure 6.6, in contrast to different variations of the heat transfer coefficient for single-phase flows.

6.3 Condensation Heat Transfer

Condensation occurs when vapor comes into contact with a cooled surface below the saturation temperature. For quiescent vapor in contact with a cooled surface, two fundamental modes of condensation are observed, namely, dropwise and film condensation. In dropwise condensation, the liquid does not entirely wet the surface. This is often desirable, since a large heat transfer rate occurs when the vapor at the saturation temperature, T_{sat}, comes into direct contact with the wall at T_w. Droplets form along the

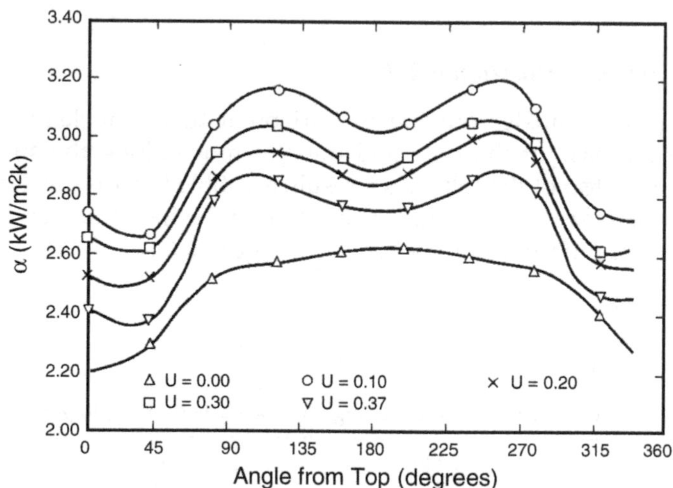

FIGURE 6.6
Heat transfer coefficient for boiling with forced convection at $q'' = 25$ kW/m^2. (From Hewitt, G.F., Shires, G.L., and Polezhaev, Y.V., Eds., *International Encyclopedia of Heat and Mass Transfer*, Copyright 1997, CRC Press, Boca Raton, FL. Reprinted with permission.)

surface. As their diameters grow in size, they flow downward along the surface under the influence of gravity. Although some experimental data involving condensation heat transfer is available (Takeyama and Shimizu, 1974), there are generally few comprehensive models of dropwise condensation processes.

In film condensation, a liquid film covers the entire surface and a lower rate of heat transfer occurs between the vapor and the wall. It is helpful to compare the approximate ranges of condensation heat transfer coefficients under various conditions. As mentioned in chapter 5, for condensation of steam on surfaces about 3 to 20°C below the saturation temperature, the average heat transfer coefficient, h, ranges between 11,000 and 23,000 W/m^{2}K for horizontal tubes (25- to 75-mm outer diameter) and between 5,700 and 11,000 W/m^{2}K for vertical surfaces (3 m high). In the following sections, it will be shown that the condensation heat transfer coefficient depends on various thermophysical properties, as well as geometrical configuration, thereby explaining the differences between reported values of heat transfer in different problems.

The rate of heat transfer during film condensation is usually at least an order of magnitude smaller than that during dropwise condensation, particularly at low values of the temperature excess (i.e., difference between the wall and saturation temperatures). The proximity increases between the two modes as the temperature excess increases, since eventually the condensed droplets form a continuous liquid film. In the following sections, we will mainly consider film condensation, since dropwise condensation is not usually sustained over long periods of time. We will adopt a scaling analysis, as well as an integral formulation, in analyzing this mode of heat transfer.

6.3.1 Laminar Film Condensation on Axisymmetric Bodies

Consider a vapor at T_{sat} in contact with a plate (length of L) aligned in the vertical direction at a uniform temperature of T_w (see Figure 6.7). This arrangement could be observed on a window inside a humid room exposed to cold outdoor conditions. Sample measurements could be taken to indicate an approximate 0.2-mm film thickness (δ_l) for $L \approx 5$ m with $\Delta T = T_w - T_{sat} = 15$°C. Together with measurements of the associated condensate runoff mass flow rate from the plate, these sample values suggest an approximate average film velocity of 0.08 m/sec (streamwise direction). Additional characteristic scales can be estimated based on these measurements. For example, a characteristic time scale and v velocity component (cross-stream direction) can be estimated by $t \approx L/u_l = 63$ sec and $v_l \approx \delta/t = 3 \times 10^{-6}$ m/sec, respectively.

Based on these measurements, we can now consider discrete scaling of the governing equations in the liquid film (i.e., Navier–Stokes equations, energy equation) in order to determine the dominant terms. Performing this scaling analysis, we find that viscous drag and the body force (buoyancy) are the two main terms in the momentum balance, whereas conduction across the film layer is much larger than other terms in the energy equation, i.e.,

$$\mu_l \frac{\partial^2 u_l}{\partial y^2} = -g_x(\rho_l - \rho_v) \tag{6.61}$$

$$\frac{\partial^2 T}{\partial y^2} = 0 \tag{6.62}$$

where g_x refers to the component of gravitational acceleration in the direction along the surface. For example, $g_x = g$ for a vertical plate, whereas $g_x = 0$ for a horizontal plate.

We must also include the following mass conservation equation and boundary conditions for closure of the formulation in the liquid region:

$$\frac{\partial u_l}{\partial x} + \frac{\partial v_l}{\partial y} = 0 \tag{6.63}$$

$$y = 0: \quad u_l = 0 = v_l, \quad T = T_w \tag{6.64}$$

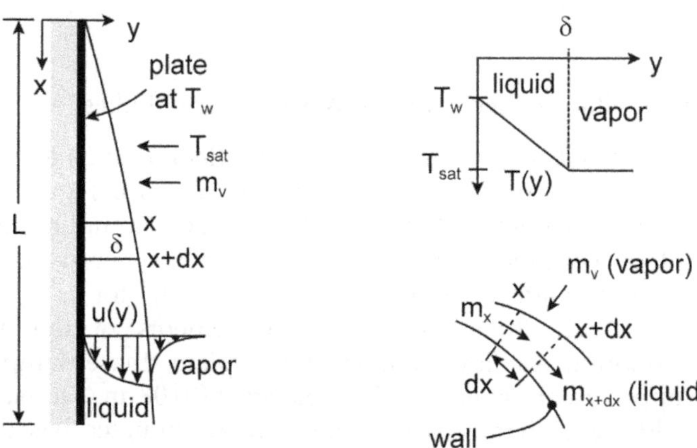

FIGURE 6.7
Condensation schematic with typical velocity and temperature profiles.

$$y=\delta: \quad \mu\frac{\partial u_l}{\partial y}\approx 0, \quad T=T_{sat}, \quad k_l\frac{\partial T}{\partial y}\approx \dot{m}_v''h_{fg} \tag{6.65}$$

Solving Equation (6.61) subject to boundary conditions,

$$u_l(y)=\left(\frac{\rho_l-\rho_v}{\mu_l}\right)g_x\left(-\frac{1}{2}y^2+\delta y\right) \tag{6.66}$$

Also, Equation (6.62) suggests that the conduction heat flux in the y direction into a differential control volume within the film balances the flux out of the control volume. The observation that the scaled magnitude of $k_l\partial^2 T/\partial y^2$ is large (relative to the convection and dissipation terms) remains consistent with Equation (6.62). The net heat flux into the differential control volume is zero (i.e., inflow term balances the outflow term) at the steady state.

From a similar development in the vapor region, we have $T=T_{sat}$ as the solution of the energy equation (assuming pure vapor with no noncondensable gases such as air). Without temperature gradients, forced flow, or buoyancy in the vapor, all vapor motion is generated by shear stresses at the phase interface.

Solving Equation (6.62), subject to the boundary conditions,

$$T(y)=T_w=\left(\frac{T_{sat}-T_w}{\delta}\right)y \tag{6.67}$$

Also, we will define the rate of liquid mass flow along the surface (per width W) as $\Gamma=\dot{m}_l/W$. From a mass balance within the liquid film, we find that the rate of change of this mass flow in the x direction (streamwise direction) must balance the vapor supply rate. Adopting this definition of mass flow, and using Equation (6.67) together with the interfacial heat balance, we have

$$\frac{d\Gamma}{dx}=\dot{m}_v''=\frac{k_l(T_{sat}-T_w)}{\delta h_{fg}} \tag{6.68}$$

Thus, the film growth rate and heat transfer are controlled by the thermal resistance of the film, which depends on the local thickness of the liquid layer.

In addition to the previous differential equation for the mass flow rate, Γ, we can estimate the film mass flow by integration of the velocity profile, Equation (6.66), across the film (from $y=0\rightarrow\delta$), i.e.,

$$\Gamma = \int_0^{\delta} \rho_l u_l dy = \left[\frac{\rho_l(\rho_l - \rho_v)g_x}{3\mu_l}\right]\delta^3 \tag{6.69}$$

After combining and solving Equations (6.68) and (6.69), we obtain

$$\Gamma = \left[\frac{4k_l(T_{sat} - T_w)}{3h_{fg}}\right]^{3/4}\left[\frac{g\rho_l(\rho_l - \rho_v)}{3\mu_l}\right]^{1/4}\left[\int_0^x \left(\frac{g_x}{g}\right)^{1/3} dx\right]^{3/4} \tag{6.70}$$

The film thickness can be determined from Equation (6.69), based on the result in Equation (6.70). An improvement of Equation (6.70) modifies the latent heat of vaporization according to

$$h'_{fg} = h_{fg} + 0.68c_{p,l}(T_{sat} - T_w) \tag{6.71}$$

which includes thermal advection effects (Rohsenow, 1956).

Furthermore, the local heat transfer is given by the Fourier heat flux. Since the temperature profile in the liquid remains linear, i.e., zero second derivative in Equation (6.62), we find that the heat flux is proportional to ΔT ($= T_w - T_{sat}$) divided by the film thickness, δ. Then, the local Nusselt number can be obtained and integrated along the surface, thereby giving the following average Nusselt number:

$$\overline{Nu}_s = \frac{\overline{h}S}{k_l} = \frac{(\overline{q}_w/\Delta T)S}{k_l} = \frac{4}{3}\left[\frac{g\rho_l(\rho_l - \rho_v)h'_{fg}S^3}{4k_l(T_{sat} - T_w)\mu_l}\right]^{1/4} \times \left[\frac{1}{S}\int_0^S \left(\frac{g_x}{g}\right)^{1/3} dx\right]^{3/4} \tag{6.72}$$

Good agreement is achieved between this modeling and experimental data for problems involving laminar film condensation. However, discrepancies may arise due to factors such as wall roughness, turbulence (to be discussed shortly), or noncondensable gases in the vapor. Noncondensable gases have important effects on the boundary layer dynamics during condensation (Chin, Ormiston, and Soliman, 1998).

6.3.2 Wavy Laminar and Turbulent Condensation

During film condensation along a vertical surface, the liquid film initially flows as a laminar film, but as its distance downstream increases, transition to turbulence may occur. A Reynolds number, based on the condensate flow rate, $\dot{m}$, is defined by

$$Re_{\delta} = \frac{4\dot{m}}{\mu_l P} \tag{6.73}$$

where properties are evaluated at the film temperature. Also, P is the wetted perimeter. It refers to the plate width for a vertical plate, whereas

$P = \pi D$ for a vertical tube and P is twice the tube length for a horizontal tube. During condensation, the following three regions can be identified along the surface: (i) laminar (wave-free) for $Re_\delta < 30$, (ii) wavy–laminar (transition) for $30 \le Re_\delta \le 1800$, and (iii) turbulent ($Re_\delta > 1800$). We will consider each region separately in the following set of heat transfer correlations.

In the laminar region, the previous film condensation results can be written in the following form:

$$\frac{h_x(\nu_l^2/g)^{1/3}}{k_l} = \left(\frac{3}{4}Re_\delta\right)^{-1/3} \equiv Nu_x^* \quad (6.74)$$

This result gives a modified Nu number with a characteristic length of $(\nu_l^2/g)^{1/3}$, based on the fluid viscosity (rather than the plate length or width).

In the wavy–laminar and turbulent regions, experimental correlations are generally required. For the wavy laminar region (Kutateladze, 1963),

$$\overline{Nu}^* = \frac{\overline{h}_L(\nu_l^2/g)^{1/3}}{k_l} = \frac{Re_\delta}{1.08Re_\delta^{1.22} - 5.2} \quad (6.75)$$

Furthermore, for heat transfer in turbulent condensate flow (Labuntsov, 1957),

$$\overline{Nu}^* = \frac{\overline{h}_L(\nu_l^2/g)^{1/3}}{k_l} = \frac{Re_\delta}{8750 + 58Pr^{-0.5}(Re_\delta^{0.75} - 253)} \quad (6.76)$$

Once the heat transfer coefficient is computed, the heat transfer from the surface can be obtained by multiplying this heat transfer coefficient by the exposed surface area and temperature difference (between the surface and ambient temperatures).

6.3.3 Other Configurations

6.3.3.1 Outside a Single Horizontal Tube

Film condensation outside horizontal tubes arises in many engineering applications. For example, shell-and-tube condensers are widely used in power plants and processing industries. Integrating our earlier condensation result around the circumference of a cylinder, we obtain the result that

$$\bar{h} \approx 0.725 \left[\frac{g\rho_l(\rho_l - \rho_v)h'_{fg}k_l^3}{\mu_l D(T_{sat} - T_w)} \right]^{1/4} \tag{6.77}$$

6.3.3.2 *Outside Several Horizontal Tubes*

For N tubes, we can replace $D \rightarrow ND$ in Equation (6.77). For example, there is a reduction in the coefficient from 0.725 to 0.56 for the case of $N = 10$. Alternatively, for a horizontal tube bank with N tubes placed directly above one another in the vertical direction, the convection coefficient for the N tube arrangement, h_N, can be approximated in terms of the single tube coefficient, h, as follows (Kern, 1958):

$$\bar{h}_N = N^{-1/6}\bar{h} \tag{6.78}$$

Also, for surfaces (such as plates or cylinders) inclined at an angle of θ with respect to the horizontal direction, the resulting heat transfer can be approximated by replacing the gravitational acceleration, g, by $g\sin(\theta)$ to reflect the modified component of gravity in the flow direction of condensate along the surface.

6.3.3.3 *Outside a Sphere*

A similar result is obtained for a film condensation around a sphere (diameter D),

$$\bar{h} \approx 0.815 \left[\frac{g\rho_l(\rho_l - \rho_v)h'_{fg}D^3}{\mu_l k_l(T_{sat} - T_w)} \right]^{1/4} \tag{6.79}$$

It can be observed that the heat transfer coefficient decreases with higher values of μ_l and temperature excess $(T_{sat} - T_w)$. For a higher liquid viscosity or temperature excess, a thicker liquid film is formed, which creates a higher thermal resistance to heat transfer between the surface and surrounding saturated vapor.

6.3.4 Condensation with Forced Convection

6.3.4.1 *Inside a Vertical Tube*

In many practical applications, an external vapor flow within or past the condensation surface alters the rate of heat transfer at the surface. In particular, for the case of film condensation inside a vertical tube with forced convection, the following Carpenter–Colburn correlation can be used to find $\bar{h}$ for pure vapors of steam and hydrocarbons up to $G \approx 150$ m/

sec,

$$\bar{h}Pr^{1/2} \approx 0.046 c_{p,l} G_m \sqrt{\frac{\rho_l}{\rho_v}} f \tag{6.80}$$

where

$$G_m = \sqrt{\frac{1}{3}(G_1^2 + G_1 G_2 + G_2^2)} \tag{6.81}$$

$$f = \frac{0.046}{Re_v^{1/5}} \tag{6.82}$$

and $Re_v = G_m D/\mu_v$. The values G_1 and G_2 refer to the mass velocity of the inlet and exit vapor flows, respectively, i.e., $G_1 = x_1 \dot{m}/A$, where A is the tube cross-sectional area.

6.3.4.2 Inside a Horizontal Tube

Another important configuration is condensation with forced convection inside horizontal tubes. These two-phase flows occur widely in practical applications, such as condensation in refrigeration and air conditioning systems. Due to the complexity arising from transition between several types of flow regimes (i.e., annular, stratified, etc.), two-phase flow maps are frequently used to identify the flow patterns. For example, Travis and Rohsenow (1973) presented a flow mapping for freon-12. In this configuration, the following superficial liquid (subscript *sl*) and gas (subscript *sg*) velocities are defined as usual by

$$V_{sl} = \frac{(1 - x)G}{\rho_l} \tag{6.83}$$

$$V_{sg} = \frac{xG}{\rho_v} \tag{6.84}$$

where $G = 4\dot{m}/(\pi D^2)$ is the mass velocity.

Stratified flow inside horizontal tubes occurs when the vapor velocity is low enough (i.e., $Re_v < 35{,}000$ at the inlet) that a condensate film forms along the upper, inner surface of the tube and flows downward under gravity to collect in the lower section of the tube. The terminology of stratified reflects the process where the laminar film flows downward and collects as a 'stratified' layer in the lower section of the tube. In stratified flow, it is anticipated that the resulting heat transfer coefficient varies with vapor fraction (or quality), x, in the flow field, since this parameter affects the thickness of the liquid film, and thus thermal resistance to heat transfer to the surface. This coefficient also varies along the periphery of the tube, while having its largest value at the bottom. This dependence was identified by Zivi (1964) in terms of a void fraction, α, where

$$\alpha = \frac{1}{1 + (\rho_v/\rho_l)^{2/3}(1 - x)/x} \tag{6.85}$$

Based on this void fraction and a modification of the condensation result obtained earlier in Equation (6.72), Jaster and Kosky (1976) recommended the following stratified flow correlation for the average heat transfer coefficient, h:

$$\bar{h} = 0.725\alpha^{3/4}\left[\frac{g\rho_l(\rho_l - \rho_v)h_{fg}k_l^3}{D(T_{sat} - T_w)\mu_l}\right]^{1/4} \tag{6.86}$$

Alternatively, based on a similar modification of Equation (6.72), but performed on an average (over the range of phase fractions in the stratified regime), another commonly adopted result in the stratified regime is given by

$$\bar{h} = 0.555\left[\frac{g(\rho_l - \rho_v)k_l^3 h'_{fg}}{v_l D(T_{sat} - T_w)}\right]^{1/4} \tag{6.87}$$

where the modified latent heat is

$$h'_{fg} = h_{fg} + 0.375c_{p,l}(T_{sat} - T_w) \tag{6.88}$$

Another flow regime arising with condensation inside horizontal tubes is annular flow. In annular flow, a condensate film grows along the walls while vapor occupies the inner core of the tube. This situation resembles forced convection with vapor inside a horizontal tube, with a liquid thermal resistance due to the condensate film along the wall. Jaster and Kosky (1976) reported various criteria for transition between stratified and annular flow based on the following transition factor, F:

$$F = \frac{f_l u_l^2}{2g\delta} \tag{6.89}$$

where u_l refers to the condensate velocity and δ refers to the condensate film thickness. The liquid friction factor, f_l, is calculated as though liquid alone is flowing through the pipe.

Also, the condensate velocity can be approximated by

$$u_l = \frac{V_l}{1-\alpha} \tag{6.90}$$

where V_l and α refer to the liquid velocity in the pipe (based on the same mass flow rate, but in the absence of vapor) and fraction of vapor flow area divided by total area, based on Equation (6.85), respectively. Based on these definitions, it was observed that $F \geq 29$ exhibited annular flow, $F \leq 5$ yielded stratified flow, and $5 < F < 29$ is a transition regime between stratified and annular flow.

In the annular flow regime, the following heat transfer correlation is widely used:

$$h_x \approx h_{lo}\sqrt{\frac{\rho_l}{\rho_m}} \tag{6.91}$$

$$\rho_m = \left[\frac{x}{\rho_v} + \frac{(1-x)}{\rho_l}\right]^{-1} \tag{6.92}$$

where h_{lo} refers to the heat transfer coefficient for liquid flowing at the same total flow rate.

Other regimes in condensation with forced convection include the wavy and slug flow regimes. In these two-phase flow regimes, a wavy interface and condensate slugs appear as the vapor velocity increases beyond the limit of stratified flow. In addition to the correlations described herein, other two-phase flow maps (Mandhane, 1974) are available for further analysis of two-phase flows under conditions involving gas and fluid phases.

6.3.4.3 Outside a Single Horizontal Tube

For external flow of vapor past a cool surface, the shear stress exerted by the vapor on the surface film affects the rate of condensation. For example, a downward flow of vapor past the surface reduces the thickness (and thus thermal resistance) of the liquid film, which leads to an enhancement of the heat transfer rate. The shear stress, τ_δ, at the vapor–liquid interface ($y = \delta$) is computed as follows:

$$\tau_\delta = \mu_l \left.\frac{\partial u}{\partial y}\right|_\delta \tag{6.93}$$

Based on the profile of vapor velocity at the phase interface, the shear stress can be obtained and subsequently used to find the film thickness and heat transfer to the surface.

For external flow of high-velocity saturated vapor past a single horizontal tube, Shekraladze and Gomelauri (1966) recommended the following correlation for the heat transfer coefficient:

$$\overline{h} = \left[\frac{1}{2}\overline{h}_s + \left(\frac{1}{4}\overline{h}_s^4 + \overline{h}_g^4\right)^{1/2}\right]^{1/2} \tag{6.94}$$

where $\overline{h}_g$ refers to the coefficient arising from gravity alone, i.e., Equation (6.77). Also, $\overline{h}_s$ due to shear stresses is given by

$$\overline{Nu}_d = \frac{\overline{h}_s d}{k_l} = CRe_d^{1/2} \tag{6.95}$$

where d is the tube diameter. Also, $Re_d = (\rho_l u_v d/\mu_l)$ and $C = 0.9$ for $Re_d < 10^6$, whereas $C = 0.59$ is recommended when $Re_d > 10^6$.

6.3.4.4 Outside Tube Bundles

The previous results for a single tube can be extended to the case of external flow of saturated vapor past tube bundles. In particular, the vapor velocity, u_v in Equation (6.95), is replaced by

$$u_v \rightarrow \frac{u_{vo}}{\alpha} \tag{6.96}$$

where α refers to the void fraction of the tube bundle (ratio of the free volume to the total volume) and u_{vo} represents the vapor velocity that would be obtained without the presence of the tubes. The gravity component of the heat transfer coefficient would be based on an earlier correlation obtained for N tubes in a vertical alignment, i.e., Equation (6.78).

6.3.4.5 Finned Tubes

Another frequently encountered configuration is condensation on horizontally finned tubes. Correlations for the heat transfer coefficient, $\overline{h}$, in this configuration have been presented by Beatty and Katz (1948) and Katz and Geist (1948); i.e.,

$$\overline{h} = 0.725[g\rho_l(\rho_l - \rho_v)h'_{fg}k_l^3\mu_l(T_w - T_{sat})]^{1/4}\left(\frac{A_b}{AD_i^{1/4}} + \frac{1.3\eta A_f}{AL^{1/4}}\right) \tag{6.97}$$

The coefficient 0.725 is based on the heat transfer coefficient obtained for horizontal tubes, Equation (6.77). However, based on experimental data, Beatty and Katz (1948) recommended that the leading coefficient should be 0.689, rather than 0.725, to yield closer accuracy with measured data. The variable L refers to the average fin height over the circumference of the tube, i.e.,

$$L = \frac{\pi}{4}\left(\frac{D_o^2 - D_i^2}{D_o}\right) \tag{6.98}$$

where D_o and D_i are the outer and inner diameters of the finned tube, respectively. The terms A, A_b, and A_f refer to the total area and areas of the base and fin, respectively. Also, in Equation (6.97), η refers to the fin efficiency, which is related to the geometrical quantities as follows:

$$A = A_b + \eta A_f \tag{6.99}$$

The fin efficiency can be calculated from conventional fin relations (as discussed in Chapter 2).

In experimental data reported by Beatty and Katz (1948), the fin efficiency of copper tubes with short fin heights (less than 1.6 mm) is generally larger than 0.96. Equation (6.97) is based on the assumption that the temperature difference (between the wall and surrounding vapor) is the same along the base (root portion) of the fin as it is throughout the fin. However, it is anticipated that the temperature difference between the vapor and wall will be larger along the base surface than farther down the fin, particularly as the fin efficiency decreases. In this context, Young and Ward (1957) recommended that the fin efficiency in the last expression in Equation (6.97) should be raised to the three-fourths power, rather than the first power, to accommodate this effect.

It should be noted that even low-finned tubing provides an appreciably higher heat transfer coefficient than an unfinned tube of equal diameter. The fins lead to a smaller average thickness of the condensate film, thereby reducing the thermal resistance in comparison to a plain tube. This effect can be observed through the last parenthetical term in Equation (6.97), which can be interpreted as an equivalent diameter raised to the power of $-\frac{1}{4}$. This equation assigns a smaller equivalent diameter to a finned tube than to a plain tube (diameter of D), thereby leading to a larger heat transfer coefficient. Various types of extended surfaces on tubes are actually used in condenser equipment, including fins of sawtooth shape, as well as wires loosely attached to the tubes.

6.4 Devices with Vapor–Liquid Phase Change

6.4.1 Thermosyphon

Closed, gravity-assisted, two-phase thermosyphons (GATPTs) arise in various energy and industrial applications, including heat exchangers, solar energy devices, thermal control of food storage, and other applications. In a typical thermosyphon, the working fluid (such as water) in the evaporator section absorbs sensible heat (temperature change) and latent heat (phase change) from the heater (see Figure 6.8). Then, vapor rises upward due to buoyancy until it reaches the condenser section, where it releases its latent heat when it condenses to liquid. The condensed film then flows down along the wall under the effect of gravity back to the evaporator section, where the cycle continues.

6.4.2 Heat Pipe

A heat pipe is similar to a thermosyphon, but with some important exceptions. It is a closed device containing a liquid that transfers heat under isothermal conditions as follows: vaporization of liquid in the evaporator, transport and condensation of vapor, and liquid return flow by capillary action (in a wick structure) back to the evaporator (see Figure 6.9). The adiabatic section is designed to fit the external geometrical requirements (i.e., spacing limitations) of the heat pipe. At the evaporator section, thermal energy from the external source is transferred to the working fluid in the

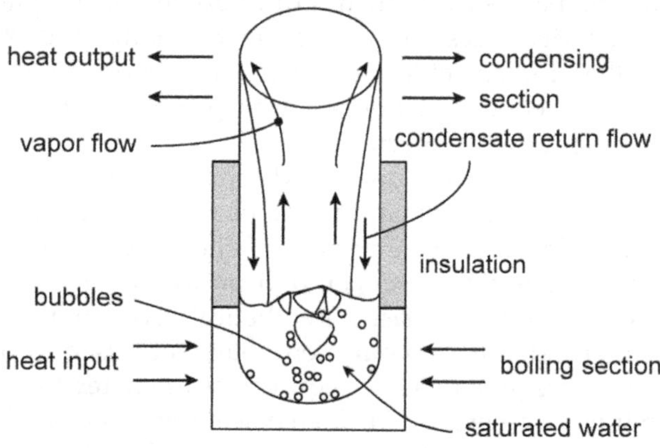

FIGURE 6.8
Schematic of a thermosyphon.

heat pipe. A buffer volume may be constructed at the end of the heat pipe to enclose a noncondensable gas (such as helium or argon) for controlling the operating temperature, by controlling the pressure of the inert gas. Vapor flow is transported through the core interior region of the heat pipe at high velocities to the condensing section (i.e., up to 500 mi/h in some cases).

A porous wick material with small, random interconnected channels is constructed along the inner wall of the container of the heat pipe. The pores in the wick act as a capillary "pump"; the word "pump" is used because of the wick's analogous role to regular pumping action on fluids in pipes by pumps. It provides an effective means of transporting liquid back to the evaporator through surface tension forces acting within the wick. Also, the wick serves as an effective separator between vapor and liquid phases, thereby allowing more heat to be carried over greater distances than other pipe arrangements.

Heat pipes are used in various applications, such as heating, ventilating, and air conditioning (HVAC) heat recovery systems, microelectronics cooling, and spacecraft thermal control. For example, a series of heat pipes in an air-to-air HVAC heat recovery system allows effective storage of thermal energy (i.e., storage of thermal energy contained in exiting combustion gases). Heat transfer through heat pipes offers key advantages over conventional techniques, including low maintenance (no moving parts), long life, and cost savings. Another example arises in the important technology of microelectronics cooling. Heat pipes can be up to 1000 times more conductive than metals (at the same weight). Laptop and other computers, as well as telecommunications equipment, have adopted heat pipes with success in their thermal designs. Furthermore, heat pipes appear

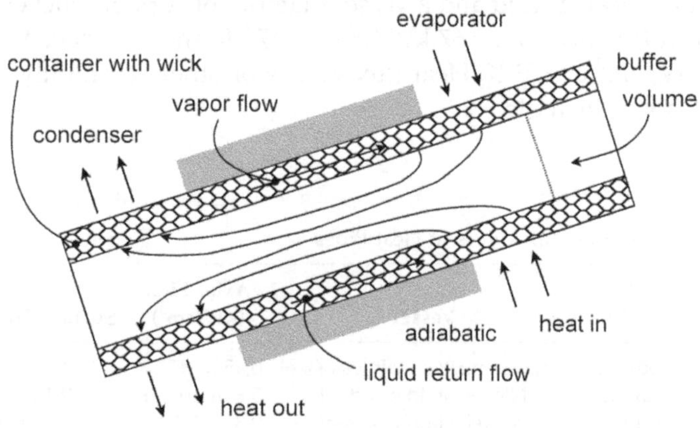

FIGURE 6.9
Schematic of a heat pipe.

in several spacecraft thermal control applications. For example, heat pipes are used in satellites to transfer heat generated by electronic equipment to radiation panels that dissipate heat into space. Heat pipes have also been adopted in the tubing of satellites. They provide the effective control of temperatures required for reliable performance of electrical components on the satellite.

Several heat transfer processes occur during heat pipe operation. At the evaporator section, heat is transferred by conduction from the energy source through the container wall and wick–liquid matrix to the vapor–liquid interface. Liquid is then evaporated at the vapor–liquid interface. Then heat is transferred by convection of vapor (laminar or turbulent) from the evaporator to the condenser. The vapor temperature is approximately the average between the source and sink temperatures at the ends of the heat pipe. Condensation of vapor then occurs at the liquid–vapor interface in the condenser. Heat transfer by conduction occurs through the wick–liquid and container wall to the heat sink. Finally, condensate returns to the evaporator through the wick structure (generally laminar) return flow.

Several aspects must be carefully considered in the design and construction of a heat pipe, including the working fluid and wick type. Desirable characteristics of the working fluid include a high latent heat of evaporation, high thermal conductivity, high surface tension, low dynamic viscosity, and suitable saturation temperature. Also, the working fluid should effectively wet the wick material.

Examples of typical working fluids include water or ammonia for operation at moderate temperatures, or liquid metals, such as sodium, lithium, or potassium, at high temperatures (above 600°C). A heat pipe with water as the working fluid and a vessel material of copper–nickel provides an axial heat flux of about 0.67 kW/cm^2 at 473 K and a surface heat flux of about 146 W/cm^2 at 473 K. Heat flux values of other commonly used heat pipes are shown in Table 6.2.

TABLE 6.2

Typical Measured Heat Flux Values of Heat Pipes

Range (K)	Fluid	Vessel	Axial Flux (kW/cm^2)	Surface Flux (W/cm^2)
230–400	Methanol	Copper, nickel, stainless steel	0.45 at 373 K	75.5 at 373 K
673–1073	Potassium	Nickel, stainless steel	5.6 at 1023 K	181 at 1023 K
773–1173	Sodium	Nickel, stainless steel	9.3 at 1123 K	224 at 1033 K

Example 6.4.2.1

Comparison between Heat Pipe and Solid Material.

Compare the axial heat flux in a heat pipe, using water as the working fluid at 200°C, with the heat flux in a copper bar (10 cm long) experiencing a maximum temperature difference of 80°C.

For the copper bar at 200°C, the heat flux per unit area can be estimated by Fourier's law as follows:

$$q'' \approx -k\frac{\Delta T}{\Delta x} = -374\left(\frac{-80}{0.1}\right) = 299.2[kW/m^2] \approx 0.03[kW/cm^2] \tag{6.100}$$

On the other hand, the axial heat flux for a water heat pipe under the given conditions is about 0.67 kW/cm^2. Thus, the heat pipe transfers heat at a rate more than 20 times higher than that of the copper rod, even under a substantial temperature gradient in the rod. This example demonstrates an important reason why heat pipes are used in many applications.

An important design requirement involves the circulation criterion for the heat pipe. Proper liquid circulation is maintained within the heat pipe as long as the driving pressure (capillary forces) within the wick exceeds the sum of frictional pressure drops (liquid and vapor) and the potential (gravity) head of the liquid in the wick structure. Let us consider each of these physical mechanisms separately.

First, capillary action in the wick arises from surface tension forces at the liquid interface. Consider a representative case of the capillary rise in a tube. For example, a tube is partially submerged beneath a liquid surface, and the internal rise of the liquid level (due to surface tension, σ) within the tube is observed. It can be shown that the pressure difference in the liquid due to this capillary rise can be obtained by a force balance on the fluid, yielding

$$\Delta P_\sigma = \frac{2\sigma cos(\theta)}{r_p} \tag{6.101}$$

where r_p and θ refer to the radius of the capillary tube and the angle subtended by the liquid rise along the capillary wall, respectively. A value of $\theta = 0$ yields the maximum capillary pressure, and it corresponds to an assumption of perfect wetting within the tube (or wick). In the actual wick structure, a liquid flow is generated by the capillary action due to liquid entrainment within the wick structure. The spatial differences in local capillary pressure differences (due to different curvatures of liquid menisci) induce the capillary flow.

Second, a liquid pressure drop occurs when liquid flows (generally laminar) through grooves in the wick from the condenser back to the evaporator. From Darcy's law for laminar flow in a porous medium, and an analogy to laminar pipe flow, we have

$$\Delta P_L \approx \frac{\mu_l l_{eff} \dot{m}}{\rho_l K_w A_w} \approx \frac{64}{Re_{D_h}} \frac{l_{eff}}{D_H} \frac{\rho_l V_l^2}{2} \tag{6.102}$$

where K_w, A_w, and D_h refer to wick permeability (wick factor), wick cross-sectional area, and hydraulic diameter (four times liquid flow area divided by the wetted perimeter), respectively. Typical values of K_w are 3.8×10^{-9} m^2 (nickel foam) and 4.15×10^{-10} m^2 (monel beads). Also, l_{eff} refers to the effective length between the condenser and evaporator sections (evaluated from the midpoint of the condenser to the midpoint of the evaporator). Typical wick structures might involve a wrapped screen along the inner wall of the heat pipe or screen-covered grooves. Wick parameters for other commonly encountered heat pipes are shown in Table 6.3.

Third, a vapor pressure drop arises since vapor drag in the core region may impede liquid flow in the grooves of the wick at high vapor velocities. From Moody's chart (in analogy to laminar, fully developed pipe flow), an expression for the vapor pressure drop can be established. However, in practice, it is often a small contribution to the overall force balance, since the vapor density is much smaller than the liquid density. Also, vapor drag is often reduced by covering the grooves in the wick structure with a screen.

Finally, the gravity head can be positive (gravity assisted) or negative, but the latter case defeats the purpose of the wick. A positive head implies that the evaporator is above the condenser. The pressure term arising from this factor is written as

$$\Delta P_G = \rho_l g l_{eff} sin(\phi) \tag{6.103}$$

Upon assembling the previously mentioned four pressure contributions, while neglecting the vapor pressure drop, we obtain the following equilibrium design condition:

$$\frac{2\sigma cos(\theta)}{r_p} = \frac{\mu_l l_{eff} \dot{m}}{\rho_l K_w A_w} + \rho_l g l_{eff} sin\phi \tag{6.104}$$

This condition is significant because operating beyond the maximum capillary pressure value (left side of this equation with $\theta = 0$) can dry out the wick (i.e., "burnout" condition).

TABLE 6.3

Typical Wick Parameters of Heat Pipes

Material	Pore Radius (cm)	Permeability (m^3)
Copper foam	0.021	1.9×10^{-9}
Copper powder (45 μm)	0.0009	1.74×10^{-12}
Copper powder (100 μm)	0.0021	1.74×10^{-12}
Felt metal	0.004	1.55×10^{-10}
Nickel felt	0.017	6.0×10^{-10}
Nickel fiber	0.001	0.015×10^{-11}
Nickel powder (200 μm)	0.038	0.027×10^{-10}
Nickel powder (500 μm)	0.004	0.081×10^{-11}
Nickel 50	0.0005	6.635×10^{-10}
Nickel 100	0.0131	1.523×10^{-10}
Nickel 200	0.004	0.62×10^{-10}
Phosphorus/bronze	0.0021	0.296×10^{-10}

Example 6.4.2.2

Heat Transfer Capability of a Water Heat Pipe.

A water heat pipe is inclined at 12°C (evaporator above condenser). Its length and inner diameter are 40 and 2.4 cm, respectively. The wick consists of the following characteristics: five layers of wire screen, $r_p = 1 \times 10^{-5}$ m, and $K_w = 0.1 \times 10^{-10}$ m^2 (wire diameter = 0.01 mm). If the heat pipe operates at 100°C at atmospheric pressure, find the maximum heat transfer capability and the liquid flow rate.

Properties of water at the saturation temperature (100°C) are approximated as follows: $h_{fg} = 2.26 \times 10^6$ J/kg, $\rho_l = 958$ kg/m^3, $\mu_l = 279 \times 10^{-6}$ Ns/m^2, and $\sigma_l = 58.9 \times 10^{-3}$ N/m. With five layers of wire in the wick, the resulting wick area is obtained as

$$A_w \approx 2\pi Rt = 2\pi(0.012)5(1 \times 10^{-5}) = 3.77 \times 10^{-6} \text{ m}^2 \tag{6.105}$$

Also, after neglecting the vapor pressure drop through the core region and assuming perfect wetting ($\theta = 0$), we can use the equilibrium design condition to estimate the mass flow rate. Multiplying this result by the latent heat of evaporation, h_{fg}, gives the following expression for the maximum heat flux:

$$q_{max} = \dot{m}_{max} h_{fg} = \left(\frac{\sigma \rho_l h_{fg}}{\mu_l}\right)\left(\frac{A_w K_w}{l_{eff}}\right)\left(\frac{2}{r_p} - \frac{\rho_l g l_{eff} sin\phi}{\sigma}\right) \tag{6.106}$$

The maximum mass flow rate can be determined from this equation, based on known parameters and thermophysical properties. As a result, we obtain a maximum flow rate of $\dot{m}_{max} = 7.11 \times 10^{-6}$ kg/sec. Furthermore, after multiplying this value by the latent heat of vaporization (as indicated by the previous equation), we obtain $q_{max} \sim 16$ W. The corresponding axial heat flux can be obtained after dividing this value by the heat pipe area ($\pi \times 0.012^2$ m^2), with the result that $q''_{max} \sim 3.55$ W/cm^2. From this analysis, it can be observed that the heat transport capability of the heat pipe can be increased significantly by adding extra layers of mesh screen within the wick material.

In addition to the design condition in Equation (6.104), the following limitation affects the performance of a heat pipe: (i) wicking, (ii) entrainment, (iii) sonic, and (iv) boiling limitation.

6.4.2.1 Wicking Limitation

This limitation on the axial heat flux arises if we consider the maximum flow rate through the wick at the maximum capillary pressure rise. In this case, the equilibrium design condition is rearranged to give

$$q_{max} = \dot{m}_{max} h_{fg} = M \left(\frac{A_w K_w}{l_{eff}} \right) \left(\frac{2}{r_p} - \frac{\rho_l g l_{eff} \sin\phi}{\sigma} \right); \quad M = \frac{\sigma \rho_l h_{fg}}{\mu_l} \tag{6.107}$$

where M refers to the Figure of Merit (see Figure 6.10). In the intermediate range of operating temperatures (i.e., 400 to 700 K), water is the most suitable and widely used working fluid in heat pipes.

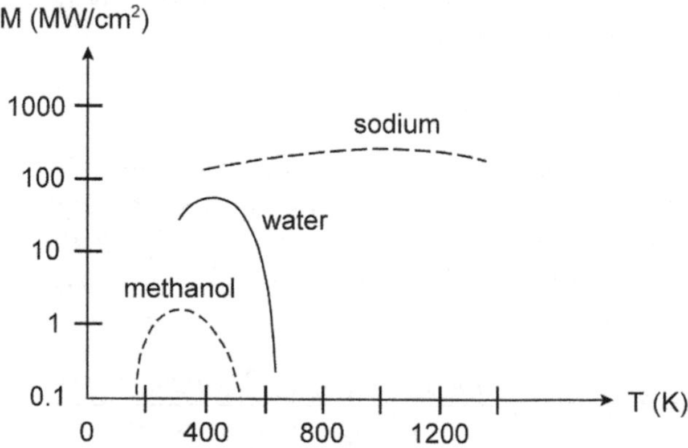

FIGURE 6.10
Figure of Merit.

In addition to M, the *heat transfer factor* is defined as follows:

$$heat\ transfer\ factor = q_{max} l_{eff} \mid_{\phi=0} = \frac{2MA_w K_w}{r_p} \tag{6.108}$$

The heat transfer factor in heat pipes can be improved by selecting wicks with large K_w values or small r_p values.

6.4.2.2 Entrainment Limitation

The vapor velocity may become sufficiently high to produce shear force effects on the liquid return flow from the condenser to the evaporator. In this case, waves can be generated on the liquid surface and droplets may be entrained by the vapor flow, since there would be inadequate restraining forces of liquid surface tension in the wick. The relevant condition at the onset of entrainment is expressed in terms of the Weber number (We). This dimensionless parameter is defined as the ratio of inertial effects in the vapor to surface tension forces in the wick. Then the onset of entrainment occurs approximately when

$$We \equiv \frac{\rho_v V_v^2 L}{\sigma} \approx 1 \tag{6.109}$$

where L refers to the hydraulic diameter of the wick surface pores. The actual vapor velocity should remain lower than the value outlined in the previously defined limit. Otherwise, entrainment of droplets into the vapor flow may cause starvation of liquid return flow from the condenser (called *dryout*).

6.4.2.3 Sonic Limitation

During start-up from near-ambient conditions, a low vapor pressure within the heat pipe can lead to a high resulting vapor velocity. In addition to the previous entrainment limit, if the vapor velocity becomes choked (i.e., sonic limit), this condition limits the axial heat flux in the heat pipe. This axial heat flux is defined in the following manner:

$$q''(axial) = \frac{q}{A} = \frac{\dot{m} h_{fg}}{A_v} = \frac{(\rho_v V_v A_v) h_{fg}}{A_v} = \rho_v V_v h_{fg} \tag{6.110}$$

The sonic limit, as well as the other limits, is often shown graphically in terms of the fluid operating temperature (see Figure 6.11; boiling limit to be discussed shortly).

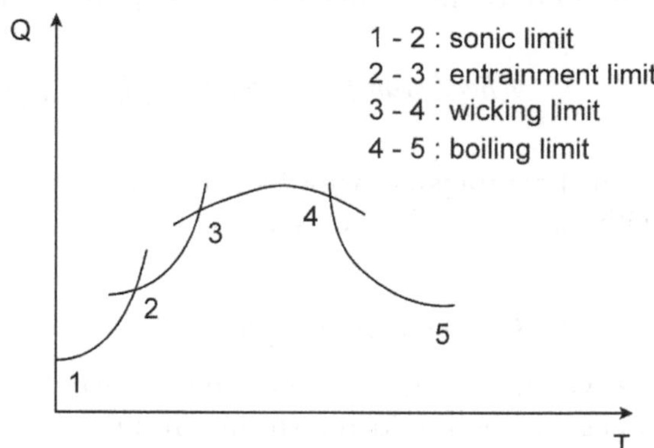

FIGURE 6.11
Heat pipe limitations.

The heat flux limits generally increase with evaporator exit temperature due to the effect of temperature on the speed of sound in the vapor. For example, for sodium, the heat flux limit increases from 0.6 kW/cm^2 at 500°C to 94.2 kW/cm^2 at 900°C. For a working fluid of potassium, the heat flux limit is 0.5 kW/cm^2 at 400°C (evaporator exit temperature), and it increases to 36.6 kW/cm^2 at 700°C. For high-temperature applications, lithium can be used since its heat flux limit ranges between 1.0 kW/cm^2 at 800°C and 143.8 kW/cm^2 at 1300°C.

6.4.2.4 Boiling Limitation

The previous three cases involved limitations on the axial heat flux in the direction of the vapor flow in the heat pipe. In this fourth case, the boiling limitation involves the radial heat flux through the container wall and wick. In particular, the onset of boiling within the wick interferes with and obstructs the liquid return flow from the condenser. Boiling within the wick may cause a *burnout* condition by drying out the evaporator containment. As a result, this situation places an additional limit on the design of the heat pipe. The four limitations are summarized graphically in Figure 6.11. Recent advances in heat pipe technology are providing innovative techniques for dealing with these thermal limitations and enhancing the overall capabilities of heat pipes.

This section has outlined certain overall features and aspects of heat pipe design. Additional sources of information, including a more detailed analysis of heat pipes, are provided in the following references: Kreith and Bohn (2001), Dunn and Reay (1982), and Chi (1976). For more detailed coverage of two-phase (vapor–liquid) flows, refer to material presented by Durst et al. (1979), Tong (1965), Carey (1992), or Hewitt et al. (1997).

References

M. Altman, R.H. Norris, and F.W. Staub. 1960. Local and average heat transfer and pressure drop for refrigerants evaporating in horizontal tubes, *J. Heat Transfer*, 189–196.

K.O. Beatty and D.L. Katz. 1948. Condensation of vapors on the outside of fined tubes, *Chem. Eng. Prog.*, 55–70.

P.J. Berenson. 1961. Film boiling heat transfer from a horizontal surface, *ASME J. Heat Transfer*, 83: 351–358.

L.A. Bromley. 1950. Heat transfer in stable film boiling, *Chem. Eng. Prog.*, 46: 221–227.

V.P. Carey. 1992. *Liquid-Vapor Phase Change Phenomena*, London: Taylor and Francis.

J.C. Chen. A Correlation for Boiling Heat Transfer to Saturated Fluids in Convective Flow, ASME Paper 63-HT-34, paper presented at 6th ASME-AIChe Heat Transfer Conference, Boston, August, 1963.

S.W. Chi. 1976. *Heat Pipe Theory and Practice*, Washington, D.C.: Hemisphere.

Y.S. Chin, S.J. Ormiston, and H.M. Soliman. 1998. A two-phase boundary-layer model for laminar mixed-convection condensation with a noncondensable gas on inclined plates, *Heat Mass Transfer*, 34(4): 271–277.

M.R. Duignam, G. Greene, and T. Irvine. 1991. Film boiling heat transfer to large superheats from a horizontal flat surface, *J. Heat Transfer*, 113: 266–268.

P.D. Dunn and D.A. Reay. 1982. *Heat Pipes*, 3rd ed., New York: Pergamon.

F. Durst, G.V. Tsiklauri, and N.H. Afgan, eds. 1979. *Two-Phase Momentum, Heat and Mass Transfer in Chemical Process and Energy Engineering Systems*, vols. 1 and 2, Washington, D.C.: Hemisphere Publishing Corp.

W.R. Gambill. 1962. Generalized Prediction of Burnout Heat Flux for Flowing, Subcooled, Wetting Liquids, Report to 5th National Heat Transfer Conference, AIChE Report 17, Houston.

W. Hendradjit, et al. 1997. Steady state boiling heat transfer from inclined surfaces to methanol, in *Experimental Heat Transfer, Fluid Mechanics and Thermodynamics*, vol. 4, M. Giot, F. Mayinger, and G.P. Celata, eds., Brussels, Belgium, pp. 2227–2232.

International Encyclopedia of Heat and Mass Transfer, G.F. Hewitt, G.L. Shires, and Y.V. Polezhaev, eds., Boca Raton, FL: CRC Press, pp. 509–512, 1997.

H. Jaster and P.G. Kosky. 1976. Condensation heat transfer in a mixed flow region, *Int. J. Heat Mass Transfer*, 19: 95–99.

S.G. Kandlikar. 1990. A general correlation for saturated two-phase flow boiling heat transfer inside horizontal and vertical tubes, *ASME J. Heat Transfer*, 112: 219–228.

D.L. Katz and J.M. Geist. 1948. Condensation on six finned tubes in a vertical row, *Trans. ASME*, 907–914.

D.Q. Kern. 1958. Mathematical development of tube loading in horizontal condensers, *J. Am. Inst. Chem. Eng.*, 4: 157–160.

F. Kreith and M.S. Bohn. 2001. *Principles of Heat Transfer*, 6th ed., Brooks/Cole Thomson Learning.

S.S. Kutateladze. 1948. On the transition to film boiling under natural convection, *Kotloturbostroenie*, 3.

S.S. Kutateladze. 1963. *Fundamentals of Heat Transfer*, New York: Academic Press.

D. Labunstov. 1957. Heat transfer in film condensation of pure steam on vertical surfaces and horizontal tubes, *Teploeneroetika*, 4, 72–80.

J.H. Lienhard, V.H. Dhir, and D.M. Riherd. 1973. Peak pool boiling heat flux measurements on horizontal finite horizontal flat plates, *ASME J. Heat Transfer*, 95: 477–482.

J.M. Mandhane, C.A. Gregory, K. Aziz. 1974. A flow pattern map for gas–liquid flow in horizontal pipes, *Int. J. Multiphase Flow*, 1(4): 537–554.

G.F. Naterer. Fluid Flow and Heat Transfer in Inclined Boundary Layers with Film Boiling, AIAA Paper 99-3449, paper presented at AIAA 33rd Thermophysics Conference, Norfolk, VA, June 28–July 1, 1999.

G.F. Naterer, W. Hendradjit, K. Ahn, and J.E.S. Venart. 1998. Near-wall microlayer evaporation analysis and experimental study of nucleate pool boiling on inclined surfaces, *ASME J. Heat Transfer*, 120: 641–653.

K. Nishikawa, T. Ito, and K. Matsumoto. 1976. Investigation of variable thermophysical property problem concerning pool film boiling from vertical plate with prescribed uniform temperature, *Int. J. Heat Mass Transfer*, 19: 1173–1182.

W.M. Rohsenow. 1952. A method for correlating heat transfer data for surface boiling of liquids, *Trans. ASME*, 74: 969–976.

W.M. Rohsenow. 1956. Heat transfer and temperature distribution in laminar film condensation, *Trans. ASME*, 78: 1645–1648.

W.M. Rohsenow and P. Griffith. 1955. Correlation of Maximum Heat Flux Data for Boiling of Saturated Liquids, paper presented at AIChE-ASME Heat Transfer Symposium, Louisville, KY.

I.G. Shekraladze and V.I. Gomelauri. 1966. Theoretical study of laminar film condensation of flowing vapor, *Int. J. Heat Mass Transfer*, 9: 581–591.

T. Takeyama and S. Shimizu. 1974. On the transition of dropwise-film condensation, in *5th Internal Heat Transfer Conference*, Vol. 3, Tokyo, p. 274.

L.S. Tong. 1965. *Boiling Heat Transfer and Two-Phase Flow*, New York: John Wiley & Sons, Inc.

D.P. Travis and W.M. Rohsenow. 1973. Flow regimes in horizontal two-phase flow with condensation, *ASHRAE Trans.*, 7: 31–34.

R. Vachon, G. Nix, and G. Tanger. 1968. Evaluation of constants for the Rohsenow pool-boiling correlation, *J. Heat Transfer*, 90: 239–247.

E.H. Young and D.J. Ward. 1957. How to design finned tube and shell-and-tube heat exchangers, *Refining Eng.*, 32–36.

S.M. Zivi. 1964. Estimation of steady state steam void fraction by means of the principle of minimum entropy production, *J. Heat Transfer Trans. ASME C*, 86: 247–252.

N. Zuber. 1958. On the stability of boiling heat transfer, *Trans. ASME*, 80: 711–720.

Problems

1. A liquid–vapor phase change process in microgravity conditions is carried out sufficiently slowly so that a one-dimensional linear profile of temperature is observed throughout the vapor. Liquid (initially at the saturation temperature, T_{sat}) is suddenly exposed to a heated wall kept at T_w (where $T_w > T_{sat}$). A planar vapor film is formed and it propagates outward over time. Find the variation of vapor temperature and phase interface position with time. Assume that the liquid ahead of the interface remains at T_{sat} and neglect convective motion.
2. Repeat the previous problem while using an integral method to account for time-dependent conduction in the vapor region (i.e., nonlinear temperature profile). Derive the governing differential equation for the change of interface position with time and outline how this equation can be used to find the wall heat flux. Assume a quadratic profile of temperature in the vapor for the integral solution.
3. After a bubble departs from a heated wall during boiling, heat conduction into the liquid yields a certain temperature profile, which is present at the onset of the next vapor period. Use the profile slope obtained from one-dimensional conduction into a semi-infinite liquid domain as the initial condition for subsequent heat conduction into the vapor (bubble). Solve the heat conduction equation in the vapor (subject to a fixed wall temperature, T_w) to find the dependence of vapor temperature on x (position), t (time), and thermophysical properties.
4. In a set of boiling experiments, a copper sphere with a diameter of 8 mm and an initial temperature of 500°C is immersed in a water bath at atmospheric pressure. The sphere cools along the boiling curve from point 1 (film boiling) to point 4 (nucleate boiling). These points are listed (temperature excess in K, heat flux in W/m^2) as follows: (1) 350, 10^5, (2) 120, 1.89×10^4 (minimum heat flux), (3) 30, 1.26×10^6 (critical

heat flux), and (4) 5, 10^4. Use the lumped capacitance method to estimate the time for the sphere to reach a temperature of 200°C in the transition boiling regime. Assume that the boiling curve may be approximated by a set of piecewise curves with a heat flux of $c\Delta T^n$ in each region (i.e., nucleate, transition, and film boiling regions), where c and n are constants (estimated from a curve fit). Use properties of copper at standard temperature/pressure (STP).

5. Brass tubes will be designed to boil saturated water at atmospheric pressure, and the tubes are operated at 80% of the critical heat flux. The diameter and length of each tube are 19 mm and 0.5 m, respectively. How many tubes and what surface temperature (each tube) are required to provide a vapor production rate of 10 kg/min?
6. Pool boiling occurs along the outer surface of a copper pipe (2-cm outer diameter) submerged in saturated water at atmospheric pressure. Up to what temperature excess, ΔT (i.e., difference between surface temperature and saturation temperature), will the heat transfer remain within the nucleate boiling regime? Find the heat transfer coefficient when $\Delta T = 12°C$.
7. Heat transfer to water at atmospheric pressure occurs by nucleate boiling along a heated steel surface. Estimate the amount of subcooling of the stationary liquid pool to yield a maximum heat flux that is 20% greater than the maximum heat flux for a saturated liquid pool. What heat flux is obtained under the subcooled conditions?
8. At copper bar at 700°C is immersed in a water bath at 60°C. The heat transfer from the bar is measured based on transient thermocouple measurements within the bar. It may be assumed that the Biot number is small (i.e., $Bi < 0.1$) and bath and liquid free surfaces within the container do not appreciably affect the boiling heat transfer along the surface of the bar.
   (a) Explain how the surface cooling rate varies with temperature during this quenching process. Discuss the physical mechanisms that distinguish the different regimes of heat transfer and the dominant modes of heat transfer in each regime.
   (b) What effects would different initial water temperatures and different surface roughness have on the results in part (a)?
   (c) How would agitation (fluid circulation) within the bath affect the maximum cooling rate? Explain your response.
9. An electrical current passes through a polished copper 2.5-cm-diameter conductor rod immersed in saturated water at 2.64 MPa.
   (a) Find the maximum rod temperature and heat transfer coefficient if the heat input must not exceed 70% of the critical heat flux during boiling.

(b) Determine the rate of evaporation over 0.8 m of conductor length.

10. After heat treatment in a furnace, a carbon steel bar is quenched in a water bath to achieve specified surface hardness requirements. The cylindrical bar (height of 16 cm) is removed from the furnace at 500°C and then submerged in a large water tank at atmospheric pressure. If the surface emissivity of the bar is 0.85, what bar diameter is required to produce a desired total heat transfer rate of 5 kW initially during film boiling?
11. Forced convective flow with boiling heat transfer occurs in an 8-cm heated vertical pipe. The water temperature and mass flow rate are 430 K and 360 kg/h, respectively. If the quality (vapor fraction) is 40% at a position where the wall temperature is 440 K, find the rate of heat transfer to the two-phase flow at that location.
12. Laminar film condensation occurs within an industrial device consisting of small square ducts located above one another with baffles separating each duct. The sides of the square duct (side length of L) are inclined at 45° with respect to the horizontal plane. A baffle is a plate that connects the base of a duct with the top corner of the duct below it. The condensate flows downward from the top duct to lower ducts. The wall temperature of the condensing surface is 60°C, and the space adjacent to the wall is occupied by stagnant saturated steam at atmospheric pressure.
    (a) Find the average Nusselt number in this geometry for N sets of ducts and baffles. What Nusselt number is obtained for the case of $N = 10$ with $L = 1$ cm?
    (b) Find an expression for the ratio of convective heat transfer coefficients, h_N/h_1, where h_N and h_1 refer to coefficients for N sets and one duct/baffle set, respectively. Explain its functional dependence on N. In other words, explain why this ratio increases (or decreases) with N.
13. A liquid film (representing condensate) flows steadily down along a flat plate inclined at an angle of θ with respect to the horizontal direction. It is assumed that the film thickness, δ, remains nearly uniform. Also, the film, wall, and air temperatures are T_o everywhere upstream of $x = 0$ (note: x measured along the direction of the plate). Then the surface temperature increases abruptly to T_n downstream of $x = 0$, and a thermal boundary layer, δ_T, develops and grows in thickness within the film.
    (a) Start with the relevant film momentum equation and derive the following result for the velocity distribution in the film:

$$\frac{u}{U} = 2\left(\frac{y}{\delta}\right) - \left(\frac{y}{\delta}\right)^2$$

where U refers to the velocity at the film–air interface. Assume that the terminal film velocity is attained (i.e., zero velocity change in x direction along the plate).

(b) Assume that the temperature profile within the thermal boundary layer can be approximated linearly, i.e., $T = a + by$, where the coefficients a and b are determined from appropriate boundary conditions. Then perform an integral analysis by integrating the relevant energy equation to obtain δ_T in terms of x, δ, U, and α. Estimate the location where δ_T reaches δ.

(c) Explain how the Nusselt number can be obtained from the results in part (b) (without finding the explicit closed-form solution for Nu).

14. A laminar film of condensate covers the outer surface of a horizontal tube (radius of r_o) in a heat exchanger. Suddenly, the wall temperature of the tube is slightly lowered to T_w. This temperature change is assumed to occur without generating any additional condensate. Use the similarity method with the similarity variable of $\eta = r^2/(4\alpha t)$ to solve the following heat conduction equation for the temperature in the condensate layer ($r > r_o$):

$$\frac{\partial^2 T}{\partial r^2} + \frac{1}{r}\frac{\partial T}{\partial r} = \frac{1}{\alpha}\frac{\partial T}{\partial t}$$

The boundary and initial conditions are $T = T_w$ at $r = r_o$ and $T = T_i$ at $t = 0$. This analysis can be used to evaluate the effects of wall temperature perturbations on the heat transfer.

15. A vertical plate with a height of 3 m is maintained at 40°C and exposed to saturated steam at atmospheric pressure.

(a) Determine the heat transfer and condensation flow rate per unit width of the plate.

(b) Would it be more desirable to use a shorter and wider plate instead of the current high and narrow plate? Explain your response in terms of total heat transfer enhancement. Assume that an equal surface area is maintained in both cases.

16. A heat exchange apparatus contains a 2-m-high vertical pipe exposed to saturated steam at atmospheric pressure. Cooling water flows internally within the pipe to maintain a pipe surface temperature of 50°C.

(a) What pipe diameter is required to produce 0.1 kg/sec of condensate on its outer surface? What heat transfer coefficient is obtained at this diameter?

(b) What effects on the heat transfer coefficient and heat transfer rate would be observed if the operating pressure increases? Explain your response.

17. A vertical arrangement of ten tubes (each of length 2 m and diameter of 4 cm) is used within a section of an industrial heat exchanger to condense saturated steam at atmospheric pressure.
    (a) What tube surface temperature is required to produce a condensate mass flow rate of 700 kg/h from the ten-tube set?
    (b) How would the presence of a noncondensable gas in the vapor (such as air) affect the results in part (a)?
18. The outer surface temperature of a metal sphere (diameter of 4 cm) is maintained at 90°C. Find the average heat transfer coefficient when saturated steam at atmospheric pressure is condensing on the outer surface of the sphere.
19. A cylindrical thermosyphon is oriented vertically with the heater surface at 120°C along the bottom boundary. The copper thermosyphon of 6 cm in height is equally divided into a lower insulated section (3 cm) and upper condensing section (3 cm). Saturated water is boiled at atmospheric pressure above the heater. The top surface (above the condensation section) is well insulated.
    (a) What is the heat flux during nucleate boiling?
    (b) Find the thermosyphon diameter required to produce a total condensation flow rate of 0.03 g/sec at steady state.
20. A suitable heat pipe is required to transport 20 kW over a distance of 24 cm in a high temperature application. For these specifications, find and compare the required pipe diameters for the following two types of heat pipes: (i) potassium–nickel and (ii) sodium–stainless steel.
21. A water heat pipe is tested in a ground-based facility before it is applied to spacecraft thermal control. The ground-based heat pipe produces a maximum liquid flow rate of 0.02 kg/h at an inclination angle of 5°. The internal wick consists of 200 μm of nickel powder. The heat pipe operates at 100°C at atmospheric pressure, and its inner diameter and length are 2 and 10 cm, respectively.
    (a) Is the same heat pipe capable of producing a maximum heat transfer rate of 60 kW/m^2 under microgravity conditions (where $g \approx 0$ is assumed)?
    (b) How can the heat transfer capability of the heat pipe be enhanced to further exceed its present capacity?
22. An ammonia heat pipe is constructed from a stainless steel tube with an effective length of 1.6 m and an outer diameter of 0.075 in. The

aluminum fibrous slab wick has a permeability of 16×10^{-10} m^2. A set of performance tests was carried out, whereby the burnout heat load was determined as a function of heat pipe inclination (note: adiabatic section maintained at 22 ± 1°C). The results for burnout heat load and heat pipe angle are listed as follows: q_{max} (W) at ϕ^o (inclination angle), 83 W at 0°, 69 W at 0.25°, 50 W at 0.5°, 31 W at 0.75°, and 12 W at 1.0°. Estimate the effective pore radius and cross-sectional area of the wick.

23. Compare the axial heat flux that can be achieved by a heat pipe (diameter of 1.2 cm) inclined at 10° with the heat flux through a copper rod. In both cases, the length is 0.3 m and the temperature difference for the rod (end to end) is 60°C. Also, the heat pipe operates with water at atmospheric pressure. Its wick consists of the following characteristics: five layers of wire screen, $r_p = 1 \times 10^{-5}$ m, and $K_w = 0.1 \times 10^{-9}$ m^2 (wire diameter = 0.01 mm). Estimate the vapor velocity for the heat pipe in this example.
24. A heat pipe uses water as the working fluid and copper–nickel as the vessel material. The 35-cm-long heat pipe must have an energy transport capability of at least 130 W at 200°C. If the measured axial heat flux at this temperature is 0.67 kW/cm^2, estimate the required cross-sectional area. How would you determine the axial heat flux at higher temperatures?
25. A water heat pipe provides a maximum heat load of 60 W in the horizontal orientation. The copper foam wick has a cross-sectional area of 2×10^{-6} m^2.
    (a) What is the effective heat pipe length?
    (b) What heat load is expected at a heat pipe inclination of 10°?
26. Component temperatures in a satellite are passively controlled by heat pipes of 28 cm in length and 1.1 cm in diameter using water as the working fluid. The wick material is nickel felt with a characteristic pore size of 0.34 mm and a thickness of 1 mm. What minimum number of heat pipes is required to ensure that the component temperatures do not exceed 110°C?

7

Gas–Solid (Particle) Systems

7.1 Introduction

Gas–solid flows arise in many engineering technologies, i.e., control of particulate pollution, combustion of pulverized coal, fluidized bed mixing, plasma-arc coatings, and other applications. The dynamics and heat transfer in gas–solid flows provide important information in the design of rocket propulsion systems with metallized fuels. Also, applications arise in the design of advanced spray techniques in materials processing, manufacturing, and pharmaceutical products. These two-phase systems involve the transport of suspended solid particles in a gas flow stream, and in many cases, heat transfer by convection, radiation, or phase change. In many cases, the particles are small enough to be considered isothermal, i.e., negligible spatial gradients of temperature within each particle. In this chapter, we will consider various aspects of heat and momentum transport in the analysis of gas–solid flows.

7.2 Classification of Gas–Solid Flows

In a gas–solid flow involving solid particles in a moving gas stream, the number of particles in the flow affect its characteristics. In a *dilute flow*, the particle motion is largely controlled by drag and lift forces on the particle, since there is a sufficiently large distance between individual particles to avoid significant particle–particle interactions. The particle–gas interactions are the dominant transport processes in this type of flow. This category of flow is commonly encountered in applications such as combustion of solid fuel particles and gas–particle flows in rocket nozzles.

On the other hand, particle–particle collisions are the primary mechanisms controlling particle motion in *dense flows*. Dense flows can be further classified as *collision-dominated flows* or *contact-dominated flows*. In the collision-dominated case, the particles collide in pairs and then move to the next collision. This process is usually inherent in the formation of clusters of particle clouds, such as particle formations in fluidized beds and other applications. The detailed modeling of these collisions is not necessary to resolve certain macroscopic features of the flow. However, in contact-dominated flows, details of the collisions must be included to predict the main features of the flow patterns. In contact-dominated flows, the particle comes into contact with many other particles simultaneously. This type of flow arises in applications such as flows in horizontal pipes with dense packing of particles.

In addition, the coupling between the particle and gas flow can be classified in terms of *one-way coupling* or *two-way coupling*. In one-way coupling, the gas stream affects the temperature and motion of the particles, but a particle does not affect the velocity or temperature of the gas. On the other hand, a dependent interaction between the gas and solid phases arises in a two-way coupling. Although all flows are two-way coupled to some extent, the effects on the gas phase become negligible as the particle concentration becomes sufficiently low. This critical concentration value indicating a transition between one-way and two-way coupling may depend on various factors, including particle size, turbulence levels in the gas stream, and other factors. These transport processes and various related applications are discussed in a summarized fashion by Hewitt et al. (1997).

An important example involving gas–solid flow is *pneumatic transport*. This type of transport refers to gas–solid flows in industrial applications involving the transport of metal particles, grains, coal, and other particles or materials through pipes or ducts. In vertical pneumatic transport, the gas velocity must be sufficiently larger than the settling velocity of the solid particles to retain particle transport. On the other hand, in horizontal pneumatic transport, the following four flow patterns may be observed: (i) *homogeneous flow*, (ii) *dune flow*, (iii) *slug flow*, and (iv) *packed-bed flow*. The transition between these types of flows depends on the gas velocity and particle concentration within the pipe (see Figure 7.1).

In homogeneous flow, the gas velocity is high enough to maintain the solid particles in suspension. As the gas velocity decreases or the particle concentration increases, some particles begin to settle along the bottom of a pipe (similar to sand dunes). This flow is called dune flow, and the velocity when particles begin to settle is called the *saltation velocity*. Slug flows arise with further reduction in the gas velocity since the flow pattern then resembles alternating regions between suspended particles and slugs (similar to slugs arising in vapor–liquid flows with boiling). Finally, a packed bed arises at low gas velocities; in this case, the particles behave similarly to a porous medium and the gas moves through interstitial regions between the particles. For example, gas transport through the packed bed

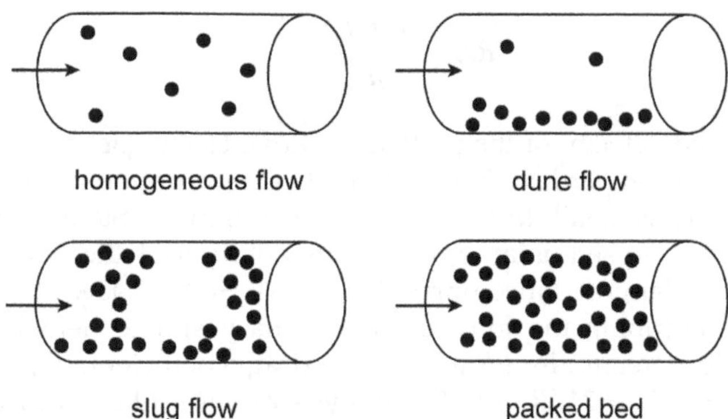

FIGURE 7.1
Gas–solid flow regimes.

becomes somewhat analogous to liquid flow through pockets or channels in a solid matrix during solidification of metal alloys. In addition to the porous flow behavior, the gas flow may produce a slow net movement of the packed bed along the bottom of the pipe.

7.3 Dynamics of Gas–Solid Flows

The dynamics of gas–solid flows includes the interactions between both phases and the role of these interactions in the transport of momentum and energy in the flow. In this section, we will examine these physical processes. In the range of conditions typically associated with many industrial processes involving gas–solid flows, departures from ideal gas behavior are often negligible. In the analysis of flows with fine particles, an important property is the mean free path of the gas molecules, λ. In particular, it will be a parameter in the calculations involving drag forces, and other quantities, in the gas and solid (particulate) phases. Based on the kinetic theory, it can be shown that this mean free path is typically proportional to the gas temperature and inversely proportional to pressure.

7.3.1 Gas–Particle Interactions

The particle movement relative to the gas is important because it largely influences the drag force occurring between the particles and the gas stream. The particle Reynolds number is defined in the following manner:

$$Re_p = \frac{\rho_g u_p d_p}{\mu} \tag{7.1}$$

where u_p is the velocity of the particle (relative to the gas) and d_p refers to the diameter of the particle. In many practical applications involving gas–solid flows, Re_p is small. In those cases, a solution for Stokes flow (where $Re_p < 0.1$) can be used as an approximate solution of the Navier–Stokes equations for laminar, incompressible flow. Stokes' law refers to the expression describing the force on a stationary sphere held in a slowly moving fluid stream. In terms of the drag coefficient, C_d, it can be represented by $C_d = 24/Re_p$. In the analysis of gas–solid flows, the Stokes flow result is considered to be equivalent to a spherical particle moving at a relative velocity with respect to the gas stream.

In the case of very small particles, a correction factor, C, is required to modify the Stokes flow result since the particle diameter may be comparable to the mean free path of the molecule. This proximity of scales leads to "slip effects," which refer to a microscopic momentum exchange due to interaction between the molecular and particle motions. As a result, the drag force is modified through the correction factor in the following manner:

$$F_d = \frac{3\pi\mu u_p d_p}{C} \tag{7.2}$$

In this result, C is the slip correction factor (or *Cunningham factor*). It is defined as the ratio between the drag force in continuum flow at the same Reynolds number to the drag force on the particle in the presence of slip effects. This factor is also called the *Cunningham coefficient*. Based on approximately spherical particles (Davies' correlation, 1945), C has been empirically correlated with the Knudsen number (Kn) as follows:

$$C = 1 + Kn\left[2.514 + 0.8exp\left(-\frac{0.55}{Kn}\right)\right] \tag{7.3}$$

where $Kn = \lambda/d_p$. Also, λ is the mean free path of the gas molecules. Extensions of this correlation to account for nonspherical particles have been developed (Clift et al., 1978).

At higher values of Re_p, the approximation of Stokes flow becomes invalid and alternative correlations for the drag force are required. From a dimensional analysis, if slip effects are neglected at high Re_p values in incompressible flow, we anticipate that the average drag coefficient varies predominantly with Re_p. The following Schiller–Nauman correlation can be used in the range of $Re_p < 800$:

$$F_d = \frac{\pi}{8} \rho_g u_p^2 d_p^2 C_d \tag{7.4}$$

$$c_d = \frac{24}{Re_p}(1 + 0.15Re_p^{0.687}) \tag{7.5}$$

In this correlation, c_d decreases with Re_p. However, the drag force is proportional to the square of velocity and diameter, suggesting that the drag force increases with $u_p d_p$ (as expected).

Thermal effects on the particle motion are also important considerations in industrial applications involving gas–solid flows. Gas viscosity and drag forces between the gas and particles usually increase with temperature. As a result, the removal of particles from the gas stream is more difficult at higher temperatures. The role of operating pressure arises through its influence on gas acceleration, as well as slip effects in the particulate phase. Increasing pressure is generally expected to make particle removal more difficult. In upcoming sections, the interactions between particle motion and heat transfer in the gas and particles will be examined.

7.3.2 Particle–Particle Interactions

In some applications, such as electrostatic precipitators, cohesive forces between particles (i.e., electrostatic and van der Waals forces) have important influences on the dynamics of the flow stream. Particles may exhibit an attractive positive–negative charge combination for grouping and collection. Electrostatic forces between particles arise from a surplus or deficit of electrons, thereby leading to a charged particle. For example, fly ash particles from coal combustion processes carry these types of charges. As a result, the particles are highly cohesive. Van der Waals forces arise from the attraction between dipoles at the atomic and molecular levels. These forces act over much shorter ranges (at atomic and molecular scales) than electrostatic forces. Electrostatic forces are more effective in separating particles from gases since they act across a larger spatial range than van der Waals forces. The Hamaker constant represents a characteristic property of the particles that describes the strength of the van der Waals cohesion.

If a liquid film resides on the surface of the particle in the gas stream, then capillary forces arise (typically several orders of magnitude larger than the van der Waals forces). In a manner analogous to that of capillary forces in heat pipes (as discussed in the previous chapter), surface tension in a film between two coalesced particles creates an attractive force. For a simplified case of zero particle separation and a fully wetting liquid, the cohesive force between the particles can be represented by (Hewitt et al., 1997)

$$F_c = \frac{\pi d_p \sigma}{[1 + tan(\phi/2)]} \tag{7.6}$$

where ϕ refers to the angle subtended by the line joining the particle midpoints and the edge of the film. Also, σ refers to the surface tension at the gas–solid interface. It can be observed that the cohesive force increases to a maximum value of $\pi d_p \sigma$ when ϕ approaches 0°.

7.3.3 Mass and Momentum Transport Processes

Consider the motion of a solid particle of diameter D suspended in a moving gas stream. The equation of motion (i.e., Newton's law) for the particle involves a balance between the particle's acceleration and the net force acting on it. If the gas becomes motionless, then only drag (friction) forces and gravity would be acting on the initially moving particle. In this case, the drag forces are dependent on the particle velocity and surface area. This dependence appears through the drag coefficient, c_d, of the particle. In the presence of a moving gas stream, the net drag force becomes proportional to the relative velocity between the gas and particle.

For a single particle, the equation of motion becomes

$$m\frac{du_p}{dt} = \frac{1}{2}\rho_g c_d \frac{\pi D^2}{4}(u_g - u_p)|u_g - u_p| + mg \tag{7.7}$$

where u_g and u_p refer to the gas and particle velocities, respectively. An assumed spherical particle mass can be written as $\rho_p \pi D^3/6$, so that the equation of motion can be written as

$$\frac{du_p}{dt} = \frac{C_d}{24/Re_p}\left(\frac{18\mu}{\rho_p D^2}\right)(u_g - u_p) + g \tag{7.8}$$

where Re_p is the particle Reynolds number based on the relative velocity $(u_g - u_p)$, particle diameter, and gas properties. The factor $24/Re_p$ is called the *Stokes drag* since it arises from the result of Stokes flow (i.e., laminar flow around a sphere when $Re_p < 1$).

It may be observed that the reciprocal of the second factor in Equation (7.8), i.e., $\rho_p D^2/(18\mu)$ has dimensions of time. As a result, it is often interpreted as a *particle velocity response time*, τ_v, or a measure of the time of response of a particle to a change in the gas velocity. The interpretation of particle velocity response time is analogous to that of the thermal response time, D^2/α, arising if the particle is suddenly subjected to an environment at a different temperature (discussed in chapter 2). At times much less than the response time, the body remains largely near its initial temperature, whereas when the actual time far exceeds the characteristic response time, the particle approaches the temperature of the environment. In between these limits, the particle's temperature is in transition between its initial temperature and the surrounding environment temperature. Response times, such as the above velocity and thermal response times, are effective ways of characterizing and understanding the transport mechanisms associated with discrete particles in a surrounding gas stream.

Local pressure losses are another important characteristic of gas–solid flows. The pressure distribution depends on the gas momentum relative to the particle packing. The following example demonstrates how pressure can be correlated with the particle density in horizontal pneumatic transport:

Example 7.3.3.1

Pressure Drop in Horizontal Pneumatic Transport.

In this example, a gas–solid flow in a horizontal pipe is considered (Hewitt et al., 1997). How does the pressure loss, ΔP, through the pipe depend on the particle loading within the pipe? Explain your response by developing an expression for ΔP in terms of the pipe length, relative velocity (between particles and gas), response time, particle packing density, and thermophysical properties.

The pressure drop between two locations in the pipe (at high gas velocities in the homogeneous regime) varies approximately as the square of velocity (in analogy with Bernoulli's equation for single-phase flow). However, the pressure drop increases with particle packing because more particles lose momentum upon impact with the wall. The net force in the gas flow by particle drag is given by

$$F_p \approx 3\mu n f \pi D L A (u_g - u_p) \tag{7.9}$$

where f, n, A, and L refer to the friction factor, number of solid particles per unit volume, cross-sectional area, and pipe length, respectively.

The increased pressure loss due to the particle packing is directly related to the applied drag force. If we consider an integral momentum balance on the duct, the forces arising from pressure and particle drag terms balance each another for a constant area duct. Performing the balance of forces, it can be shown that

$$(\Delta P)A \approx AL \frac{f \rho_{p,e}}{\tau_v}(u_g - u_p) \tag{7.10}$$

where $\rho_{p,e}$ is the effective or bulk particle density (i.e., mass of particles per unit volume). It is established as a result of the nonuniform distribution of particles in the control volume encompassing the pipe. In cases involving a variable cross-sectional area, additional differences between momentum fluxes at the inlet and outlet sections would need to be considered in an integral momentum balance. A simplification in this integral analysis arises with a constant cross-sectional area because these momentum fluxes mutually cancel each another. In this example, the gas velocity should be sufficiently high to prevent particle deposition on the walls (possibly leading to plugging of the main flow).

The previous example involved a momentum exchange for internal gas–solid flows within a pipe. A similar approach can be adopted for external flows, whereby an integral balance may be adopted over an appropriate flow region of interest comprising the gas stream and particles.

7.3.4 Convective Heat Transfer

Heat transfer in gas–solid flows with particles is largely influenced by momentum transport processes, as described in the previous section. In many cases, the thermal and velocity response times are often within the same order of magnitude. In Chapter 2, the Biot number (Bi) was described in terms of a dimensionless parameter that characterizes the extent of uniformity of temperature within an object (or particle).

Under conditions where the Biot number is small (less than 0.1), it may be assumed that spatial temperature gradients within the particles are negligible. Furthermore, neglecting phase change of the droplet or particle, we obtain the following energy balance for a spherically shaped particle in terms of the Nusselt number, Nu_D:

$$mc_p \frac{dT_p}{dt} = \left(\frac{Nu_D k}{D}\right) \pi D^2 (T_g - T_p) \tag{7.11}$$

where

$$Nu_D = \frac{hD}{k} \tag{7.12}$$

In Equation (7.11), c_p is the specific heat of the particle, and in Equation (7.12), h is the convective heat transfer coefficient. Correlations for the convection coefficient, h, are typically approximated by standard convection correlations involving forced convection in external flow past a sphere, i.e., Equations (3.129) and (3.130).

Although the gas velocity does not explicitly appear in Equation (7.11), an interequation coupling between velocity and temperature arises in the correlation involving the Nusselt number. The convective heat transfer coefficient depends on the particle velocity (relative to the gas stream), and the physical properties in Re_p are temperature dependent. An iterative procedure may be adopted to resolve this interequation dependence between velocity and temperature. Alternatively, the temperature-dependent property variations arising in Re_p may be neglected in cases involving a one-way coupling between the gas and solid phases. The variations may be neglected under conditions involving sufficiently small temperature intervals or small time steps in a numerical formulation (to be discussed in Chapter 11).

Packed-bed flow was illustrated in Figure 7.1. In this flow regime, the dense packing of solid particles appreciably affects the convective heat transfer between both the gas flow and particles. Correlations for convective heat transfer are typically based on a superficial fluid velocity, U_s, which would exist if the packed bed was empty (i.e., without particles). Also, an equivalent diameter of the packed particles, D_p, is typically defined based on the particle volume and surface area. Furthermore, a void fraction, f_v, is defined as the fraction of the packed bed that is empty (i.e., occupied by gas). Under these definitions, Whitaker (1972) recommends the following correlation for convective heat transfer in packed-bed flows:

$$\overline{Nu}_{D_p} = \frac{\bar{h}D_p}{k} = \frac{1-f_v}{f_v}(0.5Re_{D_p}^{1/2} + 0.2Re_{D_p}^{2/3})Pr^{1/3} \tag{7.13}$$

in the range of $20 < Re_{D_p} < 10^4$ and $0.34 < f_v < 0.78$, where

$$Re_{D_p} = \frac{D_p U_s}{\nu(1-f_v)} \tag{7.14}$$

In Equation (7.13), the variable D_p is defined as six times the volume of the particle divided by the particle surface area. For spherical particles in the packed bed, this relationship for D_p reduces to the diameter of the particle. The correlation in Equation (7.13) was reported to be accurate within $\pm 25\%$ for the indicated range of flow parameters.

7.3.5 Radiative Absorption and Emission

A simplified approach for modeling of radiative transport in gas flows with particles involves a linearization of Stefan–Boltzmann's law. A linearized coefficient, based on Stefan–Boltzmann's law, is added to the heat transfer coefficient in Equation (7.11). In this way, both convective and radiative components can be included in the energy balance, i.e.,

$$h = h_{conv} \pm h_{rad} \tag{7.15}$$

where

$$h_{rad} = \epsilon\sigma(T_g + T_p)(T_g^2 + T_p^2) \tag{7.16}$$

Also, $\sigma = 5.67 \times 10^{-8}$ W/m^{2}K^4 and ϵ refer to Stefan–Boltzmann's constant and the particle surface emissivity, respectively. The radiative component is typically significant only in cases of high-temperature conditions, such as conditions arising in combustion chambers or boilers.

It has been assumed that the gaseous medium between the particles, as well as the particles themselves, is not directly participating in the radiation exchange. Although this may be a fair approximation at low temperatures and in gases such as O_2, N_2, and H_2, it may become invalid for CO_2, SO_2, NH_3, CO, and hydrocarbon vapors, which emit and absorb radiation over wide ranges of wavelengths and temperatures. For example, water vapor in the gas stream absorbs thermal energy transported by radiation between high-temperature particles or surfaces enclosing the gas–solid flow. This absorption depends on a variety on factors, including gas temperature, T_g, particle (or surface) temperature, T_s, partial pressure, P_w, and mean beam length, L_e. The mean beam length may be interpreted as the radius of a hemispherical gas mass exhibiting the same effective emissivity as the region enclosing the radiative transport in the geometry of interest. In addition, correction factors, such as c_w, are typically required and based on results from relevant experimental data.

For water vapor (subscript w), carbon dioxide (subscript c), and a mixture containing both gases (subscript m), the following correlations are often adopted:

$$\alpha_w = c_w\left(\frac{T_g}{T_s}\right)^{0.45} \times \epsilon_w\left(T_s,\ P_w L_e \frac{T_s}{T_g}\right) \tag{7.17}$$

$$\alpha_c = c_c\left(\frac{T_g}{T_s}\right)^{0.65} \times \epsilon_c\left(T_s,\ P_c L_e \frac{T_s}{T_g}\right) \tag{7.18}$$

The correction factors, c_w and c_c, are required at pressures other than standard atmospheric conditions (1 atm).

For the gas mixture (subscript g),

$$\epsilon_g = \epsilon_w(H_2O) + \epsilon_c(CO_2) - \Delta\epsilon \tag{7.19}$$

$$\alpha_g = \alpha_w(H_2O) + \alpha_c(CO_2) - \Delta\alpha \tag{7.20}$$

Values of c_w, c_c, ϵ_w, ϵ_c, $\Delta\epsilon$, and $\Delta\alpha$, as well as correlations for other gases, have been determined and tabulated for a variety of conditions (Incropera and DeWitt, 1996). It should be noted that ϵ_g and α_g are functions of the shape of the gas mass as a result of their dependence on the container through the partial pressures and concentration of each gas constituent. In most practical cases involving radiative transport in high-temperature gas–solid flows, the gray gas assumption (i.e., $\epsilon_g = \alpha_g$) is invalid. As a result, the previously described equations, such as Equations (7.19) and (7.20), are needed for the radiation analysis.

Example 7.3.5.1

Radiative Exchange in a Closed Chamber.

Consider radiative heat exchange between a participating gas with particles at T_g and surfaces at a temperature T_s (see Figure 7.2). Find the net rate of radiative heat transfer between the gas and walls.

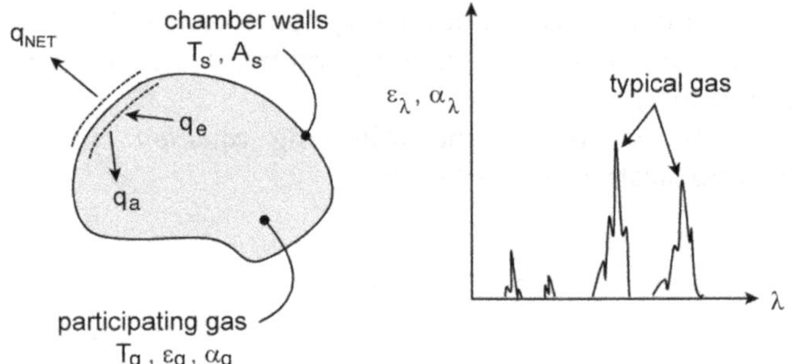

FIGURE 7.2
Radiation exchange in a closed chamber.

The radiative emission from the gas, q_e, and the heat absorbed by the gas, q_a, are given by

$$q_e = A_s \epsilon_g \sigma T_g^4 \tag{7.21}$$

$$q_a = A_s \alpha_g \sigma T_s^4 \tag{7.22}$$

As a result, the net radiative exchange is given by the following difference between emitted and absorbed radiation:

$$q_{net} = A_s \sigma (\epsilon_g T_g^4 - \alpha_g T_s^4) \tag{7.23}$$

It should be noted that the radiative emission involves gas emissivity, whereas the absorption involves surface temperature since the walls emit radiation that is absorbed by the gas.

When heat is transferred by radiation through a gas layer, an *equation of transfer* describes the transport in a way analogous to those of Fourier's law and Newton's law of cooling, which are used for conduction and convective heat transfer, respectively. Consider radiative heat transfer from a surface at $x = 0$ through an absorbing gas layer to another surface at $x = L$ (one-dimensional approximation; see Figure 7.3). The intensity of radiation, I_λ, decays as a result of absorption within the gas–particle layer. This intensity varies with wavelength (i.e., subscript λ). As a first approximation, it is assumed that the change of radiative intensity across a layer of thickness dx is proportional to the magnitude of this intensity, i.e.,

$$dI_\lambda = -\kappa_\lambda I_\lambda dx \tag{7.24}$$

where the proportionality constant (absorption coefficient), κ_λ, is tabulated as a function of wavelength, λ, gas temperature, T_g, partial pressure, and total pressure of the gas.

Dividing by I_λ, we obtain the following *equation of transfer* for a homogeneous, nonscattering gas layer:

$$\frac{dI_\lambda}{I_\lambda} = -\kappa_\lambda dx \tag{7.25}$$

Integrating the above equation from x = 0, where $I = I_{\lambda 0}$, to $x = L$, where $I = I_{\lambda L}$, we then obtain the following result (called *Beer's law*):

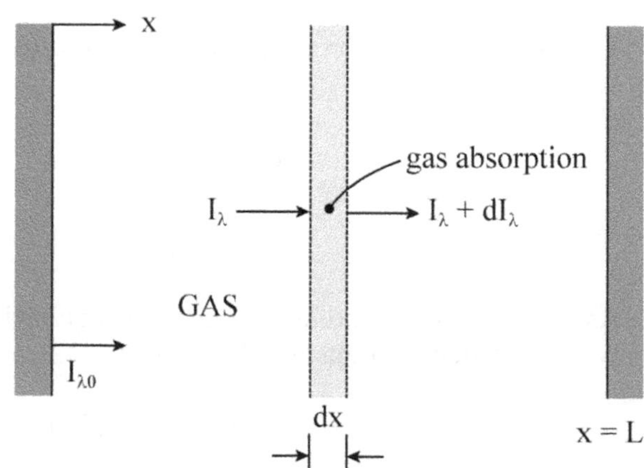

FIGURE 7.3
Radiative absorption in a planar gas layer.

$$I_{\lambda L} = I_{\lambda 0} exp(-\bar{\kappa}_\lambda L) \quad (7.26)$$

This result indicates that the intensity of radiation decays exponentially due to absorption of the radiation by particles and constituents in the gas.

Furthermore, the spectral absorptivity of the gas can be determined from the differences in radiative intensities at the different locations, i.e.,

$$\alpha_\lambda = \frac{I_{\lambda 0} - I_{\lambda L}}{I_{\lambda 0}} = 1 - exp(-\bar{\kappa}_\lambda L) \quad (7.27)$$

Based on Kirchkoff's law for a diffuse gas ($\alpha_\lambda = \epsilon_\lambda$),

$$\epsilon_g \equiv \frac{\int_0^\infty E_{b,\lambda}\epsilon_\lambda d\lambda}{\int_0^\infty E_{b,\infty} d\lambda} = \frac{1}{\sigma T_g^4}\int_0^\infty E_{b,\lambda}(1 - exp(-\bar{\kappa}_\lambda L))d\lambda \quad (7.28)$$

These radiative properties can be adopted in the net radiative exchange to predict the heat transfer rates. An example that outlines this radiative exchange for a case with soot particles is given in the following problem.

Example 7.3.5.2

Radiation Exchange Involving Soot Particles.

Consider an example of radiative heat transfer involving combustion of fuel in air within a combustion chamber. During the combustion process, soot particles forming in the flow actively participate in the radiative exchange between the gas flow and the reacting mixture. Find the effective emissivity of the soot–gas mixture for evaluation of the radiative heat transfer through the participating gas layer.

As a first approximation (assumed cylindrical isothermal cavity, i.e., blackbody enclosure), the net radiative exchange between the gas and walls of the enclosure can be approximated as

$$q_{rad} = A\sigma(\epsilon_g T_g^4 - \alpha_g T_s^4) \quad (7.29)$$

When soot particles are not present, these emissivity and absorptivity coefficients are readily available.

As a second approximation, consider the gas–particle mixture within gray walls (instead of a blackbody enclosure), while still assuming no soot is present. The incident radiation on the surface (or irradiation, G_λ) and radiosity (reflected and emitted radiation from the surface, J_λ) may be written as

$$G = \tau_g J + \epsilon_g E_{bg} \quad (7.30)$$

$$J = \rho_s G + \epsilon_s E_{bs} = \frac{\rho_s \epsilon_g E_{bg} + \alpha_g \epsilon_s E_{bs}}{1 - \rho_s \tau_g} \tag{7.31}$$

where the latter equality is obtained through substitution of Equation (7.30) into the first equality of Equation (7.31) with rearrangement. Also, it is assumed that $\epsilon_s = \alpha_s$ (gray wall) and $\alpha_s = 1 - \rho_s$.

From an energy balance on the surface of the enclosure, the net radiation exchange is given by

$$q''_{rad} = G - J = \tau_g J + \epsilon_g E_{bg} - J \tag{7.32}$$

Then, combining Equations (7.31) and (7.32),

$$q''_{rad} = \frac{\epsilon_g E_{bg} - \rho_s \tau_g \epsilon_g E_{bg} - \alpha_g \rho_s \epsilon_g E_{bg} - \alpha_g \epsilon_s E_{bs}}{1 - \rho_s \tau_g} \tag{7.33}$$

Simplifying the expressions in Equation (7.33), while invoking the gray wall assumption,

$$q_{rad} = \frac{\epsilon_s A \sigma}{1 - \rho_s \tau_g} (\epsilon_g T_g^4 - \alpha_g T_s^4) \tag{7.34}$$

where $\rho_s = 1 - \epsilon_s$ and $\tau_g = 1 - \alpha_g$.

The previous result in Equation (7.34) considered gray walls, but still without soot particles in the gas mixture. Another correction is required to account for the soot particles. For example, soot particles in the reacting flow are active parts of the radiative exchange in combustion processes. Unlike H_2O and CO_2, soot particles emit continuous radiation. Difficulties of sooty gas emissivity calculations arise due to their dependence on operating pressure, quality of fuel atomization, particle distribution within the chamber, air/fuel ratio, and other factors. In this analysis, Beer's law is used and the absorption coefficient can be approximated through experimental data, so that

$$\kappa_\lambda \approx \frac{\beta f_s}{\lambda} \tag{7.35}$$

where f_s and β refer to the soot particle volume fraction and an empirical factor accounting for the soot composition (usually between four and ten).

Integrating the spectral distribution of soot emission with Beer's law would yield a difficult expression for computational purposes. Instead, the emissivity of the soot–gas mixture can be approximated directly through Beer's law as follows:

$$(\epsilon_\lambda)_{mix} = 1 - exp(-\kappa_g L - \kappa_s L) \tag{7.36}$$

where K_g and K_s refer to coefficients for the gas and soot, respectively. Then,

$$(\epsilon_\lambda)_{mix} = (1 - exp(-\kappa_g L)) + (1 - exp(-\kappa_s L)) - (1 - exp(-\kappa_g L)) \times (1 - exp(-\kappa_s L)) \tag{7.37}$$

$$(\epsilon_\lambda)_{mix} = (\epsilon_\lambda)_g + (\epsilon_\lambda)_s - (\epsilon_\lambda)_g \times (\epsilon_\lambda)_s \tag{7.38}$$

Furthermore, treating the soot as a gray substance (i.e., $\epsilon_s = \alpha_s$),

$$\epsilon_{mix} \approx \epsilon_g + \epsilon_s - \epsilon_g \epsilon_s \tag{7.39}$$

Based on this result, we can now use our earlier energy balance to account for soot particles in the radiative exchange during gas–solid flows with combustion.

The previous example demonstrates that solid particles affect the temperature of the gas–solid flow through radiative exchange involving the particles. Careful analysis is required in determining the radiative heat transfer coefficients. Then appropriate energy balances involving the particle phase or surfaces enclosing the flow are written to obtain the radiative heat transfer throughout the flow.

Further details regarding gas–liquid flows and heat transfer are provided by Klinzing (1981), Kreith and Bohn (2001). In the following sections, additional examples of gas–solid flows in industrial applications will be considered. In many practical applications, mixing and dispersion of the solid particles in the gas stream is important for efficient operation of the system. Various modes of heat transfer, including convection and radiation, have important roles in improving the performance of the system. The upcoming sections will examine the following three important applications involving gas–solid flows: fluidized beds, particle separation technologies, and spray drying.

7.4 Fluidized Beds

A fluidized bed refers to a two-phase mixture (i.e., solid material and fluid) utilized in a variety of technologies involving physical or chemical processing of materials such as coal, organic waste, and others. *Homogeneous fluidization* refers to a bed fluidized with liquids, whereas *hetero-*

geneous fluidization refers to a bed fluidized with gas. In this section, we will mainly consider heterogeneous fluidization and the resulting gas–solid flow. However, similar correlations and methods of analysis are observed for homogeneous fluidization.

7.4.1 Flow Regimes

A fluidized bed usually consists of a vertical cylinder or tank loaded with solid particles and supplied with a gas through a perforated distributor plate at the bottom of the cylinder (see Figure 7.4). A variety of engineering applications, including coal gasification, polyethylene manufacturing, granulation, combustion of fuels, disposal of organic, toxic, and biological wastes, and particulate processing, utilize fluidized beds (Howard, 1983). The performance and operation of fluidized beds are predominantly influenced by thermofluid processes. In particular, the understanding of gas–solid flows is a key component of their effective design.

A fluidized bed is often encountered in applications involving chemical reactions because surface reactions and heat transfer are enhanced (in comparison to other arrangements) because the gas is exposed to a large effective solid surface area. Scaling and dynamic similarity between experimental and actual conditions are often needed for understanding the flow complexities in fluidized bed mixing. These complexities include the effects of radiative heat transfer, turbulence, chemical reactions, and phase change.

Similar to gas–solid flows in horizontal pipes, a wide range of flow conditions arise in gas fluidization (Geldart, 1973; Hetsroni, 1982). These conditions vary from dilute to dense particle packing. At conditions with

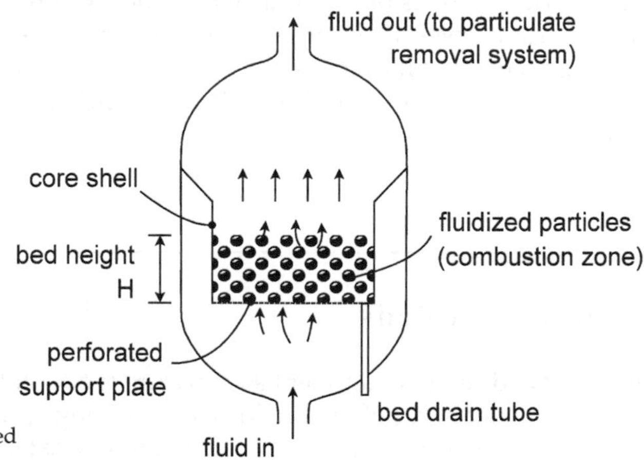

FIGURE 7.4
Schematic of a fluidized bed.

low incoming gas velocities, no significant particle motion occurs and the flow resembles flow through a porous medium (i.e., packed bed). As the incoming gas velocity increases, a condition is reached whereby the particle weight balances the upward momentum of the gas (called *particle fluidization*). Further increases in the gas velocity lead to bubble formation in the gas phase due to regions of low particle concentration. These bubbles move upward and enhance the mixing processes. They begin to grow and fill the tube in a slug flow as the gas velocity increases. Subsequently, the gas slugs further separate the solid particles into clusters with periodic and nonuniform mixing. At very high gas velocities, sufficient momentum may drive the particle clusters up and out of the tank, with some resulting backflow down along the walls. This case is called *fast fluidization*. As a result, more particles must be injected into the cylinder to maintain steady-state operation.

7.4.2 Velocity and Pressure in Fluidized Beds

The characteristics and behavior of fluidization processes depend on many factors, including the geometrical and physical properties of the solid particulate materials. The particle shapes and other characteristic dimensions affect the aerodynamic properties of the solid particles in the two-phase transport processes. The terminal velocity of a group of particles is related to the size distribution of particles in the fluidized bed and the relative interactions between gravitational and drag forces on the solid phase. The solid particles may be floating, motionless, or actively moving in the fluidized bed. Particle movement can be free and not obstructed within the central core of the bed, while mechanical devices, such as perforated plates, may obstruct this motion along outer regions. As a result, a large number of complex physical processes influence the velocity and pressure fields in the bed.

In a fixed bed, the immobile solid particles are packed in layers on top of each another. They exert gravitational forces (involving individual particles and the entire bed weight) at contact points joining each solid material in a layer. These forces spread in all directions from these contact points. The fluidized state arises when a sufficiently high velocity of a fluid stream penetrates through the bed to separate parts of the bed. When the fluid stream exceeds a certain critical velocity (called the *minimum fluidization velocity*), the solid particles begin moving and interacting with each other. The mean distance between particles increases since the particles move to occupy a larger portion of the volume enclosing the fluidized bed. The pressure drop remains approximately constant (i.e., bed weight divided by an appropriate surface area) within the fluidized bed.

The main parameters describing the macroscopic behavior of a fluidized bed include the minimum fluidization velocity, change of height of the fluidized bed, and the pressure drop across the bed. The transition between a fixed bed and a fluidized bed can be detected based on measurements of the pressure drop at various fluid velocities. In a fixed bed, the pressure drop increases in a nearly linear fashion, as the flow resembles gas flow through a porous medium. However, as the fluid velocity approaches and exceeds the minimum fluidization velocity, the pressure drop becomes constant when the gas stream transports solid particles by convection.

The pressure drop across the bed approaches a value associated with the overall particle weight (Hewitt et al., 1997),

$$\Delta P_b = (1 - \epsilon_{mf})(\rho_p - \rho_g) g H_b \tag{7.40}$$

where ϵ_{mf} and H_b refer to the fluidized bed porosity and the bed height, respectively. Based on measurements of the pressure drop at various fluidization velocities, ΔP_b becomes nearly constant beyond the minimum fluidization velocity, v_{mf}. This minimum fluidization velocity can be determined from the Wen–Yu relation as follows:

$$Re_{mf} = \frac{d_p v_{mf} \rho_g}{\mu_g} = (33.7 + 0.0408 A_r)^{1/2} - 33.7 \tag{7.41}$$

where

$$A_r = \frac{g d_p^2 \rho_g (\rho_p - \rho_g)}{\mu_g^2} \tag{7.42}$$

The Reynolds number at the minimum fluidization velocity corresponds to the condition where all particles in the fixed bed begin floating.

A *splash zone* is generated above the free surface in the fluidized bed as bubbles rise, grow, and burst across the free surface. Particles are ejected into the space above the free surface, and so the free surface is often wavy and chaotic rather than strictly horizontal. A large particle concentration exists below the free surface, and this concentration drops abruptly across this surface. Large particles with sufficient weight may drop back into the bed, and the remaining lighter particles are carried away from the bed by the gas stream.

The rise of the fluidized bed height is closely related to the volume fraction occupied by the bubbles and the fluidized bed porosity. A basic model for predicting the bed height, H, in terms of the minimum fluidization height, H_{mf}, involves an assumption that the increase in the bed height is related to the volume fraction occupied by the bubbles, f_b, as follows:

$$\frac{H}{H_{mf}} = \frac{1}{1 - f_b} = \frac{1 - \epsilon_{mf}}{1 - \epsilon} \tag{7.43}$$

A widely used correlation for the bed porosity is Todes' equation, i.e.,

$$\epsilon = \epsilon_{mf} \left(\frac{Re_p + 0.02Re_p^2}{Re_{mf} + 0.02Re_{mf}^2} \right)^{0.21} \tag{7.44}$$

Since the gas velocity at the free surface may considerably exceed the mean fluidization velocity, the larger particles may be ejected far from the free surface. Single particles and clusters of particles may be ejected from the free surface during this process. These particle clusters may break apart into smaller individual particles that may descend back down to the fluidized bed surface. Only the small particles will ascend out of the bed, but the specific size and dynamic factors affecting this ascent include the particle weight relative to the gas velocity and other relevant forces.

Fewer particles would be observed farther from the surface because many particles may have already fallen back to the free surface below those locations. It is anticipated that the particle cluster density decreases with height. There exists a critical or maximum height in the fluidized bed whereby only particles with a terminal velocity less than the gas velocity can be observed. Above this height (called the transport disengaging height (TDH)), the particle concentration becomes zero. The following Geldart equation is often adopted for an estimate of the TDH:

$$TDH = 1200 H_{mf} \left(\frac{Re_p^{1.55}}{A_r^{1.1}} \right) \tag{7.45}$$

This correlation is generally applied for particles with a diameter, d_p, between 0.075 and 2 mm.

Bubble formation and growth in heterogeneous fluidization is analogous to bubble dynamics in pool boiling. As bubbles depart from the surface, the presence of other bubbles causes lateral motion and coalescence of bubbles. Various forces acting on a large bubble may break it up prior to its arrival at the free surface. Bubbles are usually faster in fluidized beds with small particles than in beds with larger particles. Fluidized beds with slower bubble velocities are often more desirable than fast flows because the entire gas flow can mix fully during the chemical reaction. In the case of fast bubbles, the gas may pass through the fluidized bed without reacting with any particles.

The bubble rise velocity can be estimated analogously to bubbles ascending in a liquid container during pool boiling. In particular, the bubble rise velocity, v_b, can be approximated in terms of the bubble diameter, D_b, as follows (Hewitt et al., 1997):

$$v_b = 0.71(gD_b)^{1/2} \tag{7.46}$$

Unlike bubbles rising during pool boiling, upward gas flow in a fluidized bed keeps the particles located primarily along the upper section of the bubble. Bubbles interact with the gas flow, creating a wake behind the bubble. Overall mass is conserved so that upward particle movement in one section is accompanied by downward flow in another section. Circulation within these gas–solid flows provides enhancement of the heat transfer, mixing, and chemical reactions within the bed.

Difficulties arise in experimental simulations of fluidized bed flows due to a relative lack of scaling laws. Achieving dynamic similarity is challenging due to the assortment of complex flow processes. For example, the intensity and shape of recirculating cells within the fluidized bed depend on turbulence and radiative heat transfer within the chamber. Downward motion of particles, together with turbulent mixing and molecular gas diffusion, can disturb the upward gas flow and possibly create local flow reversals in the gas stream. The particle mixing process is generally more intensive in the streamwise (vertical) direction. It is typically 10 to 100 times more intensive than mixing in the cross-stream direction.

7.4.3 Heat and Mass Transfer in Fluidized Beds

Several simultaneous modes of heat transfer occur during heterogeneous fluidization. Heat is transferred by convection and radiation between the gas and particle phases. Also, heat is transferred between different locations within the bed and between fluidized stationary particles and larger particles moving through the bed. Combustion within the fluidized bed involves chemical reactions and radiative heat transfer between the particles and gas stream. As a result of these transient, three-dimensional, and turbulent thermal processes, the detailed analysis of heat transfer in a fluidized bed is a challenging task. Although heat transfer coefficients between individual particles and the gas stream may be relatively small, the overall heat transfer is generally large due to the large surface area of particles within the entire fluidized bed. Typical conditions within a fluidized bed may include a mean bed temperature of 900 K, with coefficients for heat transfer to the particles in the bed of about 20 W/m^2K.

From a dimensional analysis of heat transfer in a fluidized bed, the main variables include the heat transfer coefficient(s), characteristic geometric dimensions, thermophysical properties of the particle and gas stream, and the fluidized bed porosity, ϵ. Based on a dimensional reduction of variables through the Buckingham–Pi theorem, the relevant dimensionless groups are Nu, Re_p/ϵ, and Pr_g, where the subscripts p and g refer to particle and gas, respectively. Coefficients relating these dimensionless groups have been inferred through experimental data over a range of conditions in heterogeneous fluidization. Those results can be written as the following Gelperin and Einstein equations:

$$Nu_p = 0.016\left(\frac{Re_p}{\epsilon}\right)^{1.3} Pr^{0.33}; \qquad \frac{Re_p}{\epsilon} < 200 \tag{7.47}$$

$$Nu_p = 0.4\left(\frac{Re_p}{\epsilon}\right)^{2/3} Pr^{0.33}; \qquad \frac{Re_p}{\epsilon} \geq 200 \tag{7.48}$$

It can be observed that the gas-to-particle heat transfer coefficient increases with Re_p/ϵ (as expected) since the intensity of mixing is increased.

When chemical reactions occur between active particles and the fluidizing gas, the heat transfer becomes strongly coupled with mass transfer between the particles and the bed. Evaporation occurs during formation of reactants during the chemical reactions between the particles and gas stream (called combustion of char material). Mass transfer occurs when the reacting gas diffuses toward the surface of the active particle. This diffusion process largely affects the rate of combustion. Both diffusive and convective processes contribute to the mass transfer process. A suitable correlation for estimating the mass transfer rate is given by the following La Nuaze–Jang equation:

$$Sh = 2\epsilon_{mf} + 0.69\left(\frac{Re_p}{\epsilon}\right)^{1/2} Sc^{0.33} \tag{7.49}$$

where Sh and Sc refer to the Sherwood and Schmidt numbers (based on the particle diameter) for mass transfer, respectively. This correlation incorporates the movement of clusters of particles carrying fresh gas from the fixed bed toward the active particle.

The previous correlations and results have provided important tools for the analysis of fluidized beds. However, their limitations should not be overlooked, particularly considering the overall complexity of all processes during the heterogeneous fluidization. These processes include complicated interactions involving turbulence, combustion, radiation, phase change, and other processes. As a result, experimental data are widely scattered and the accuracy of the previous heat transfer formulations generally lies within a broad range of ±50%. In addition to the aforementioned flow complexities, this wide scatter arises because the formulations do not include many parameters contributing to the heat transfer (i.e., bubble sizes, tube arrangements, and other parameters). Further detailed treatment of transport processes in fluidized particle systems is given by Doraiswamy and Majumdar (1989).

7.5 Gas–Solid Separation

Another important application involving gas–solid flows is the separation of solid particulates from a gas stream. For example, the removal of particulates from a gas stream prior to emission from an industrial smokestack is an important consideration in pollution control. Two common devices for gas–solid separation are the *cyclone separator* and the *electrostatic precipitator*. Centrifugal acceleration separates solid particles from a gas stream in a cyclone separator. In electrostatic precipitators, particles are charged by Coulomb forces and later removed from the gas stream by a collecting surface attracting the charged particles. Although the cyclone separator is inexpensive and often highly effective, the electrostatic precipitator is often utilized because of its capabilities in removing smaller particles from a gas stream and its higher efficiency. In particular, furnaces often use electrostatic precipitators for removing dust, dirt, and other particles from the air prior to ventilation through a house or building. In addition, electrostatic precipitators are often utilized in coal-fired power plants in removing fly ash from exiting gas emissions from the plant.

The effective design of these systems requires a detailed understanding of both thermal and fluid flow processes associated with the separation process in order to avoid problems such as particle accumulation and blockage in certain entrance regions. For example, in the electrostatic precipitator, the gas–solid flow passes through a region enclosed by an array of vertically suspended and charged metal plates. Also, high-voltage wires between the plates produce an electric field between the walls and wires. The resulting magnetic effects on each particle act as forces driving particles toward the collector plates. This particle deflection in the gas stream depends on the local gas velocity and temperature. In particular, structural ribs are often placed between the collector plates for structural rigidity, but these plates must be carefully designed to prevent particle buildup in corners due to pressure losses of the gas flowing around the ribs.

The deflection velocity, or drift velocity, of particles toward the wall, u_d, can be written in terms of the electric field intensity, E, and charge to the mass ratio on the particle (q/m) in the following way:

$$u_d = \frac{q}{m}\frac{\tau_v E}{f} \tag{7.50}$$

where τ_v is the particle velocity response time (discussed following Equation (7.8)). Also, f represents a drag factor (Hewitt et al., 1997).

In the case of electrostatic precipitators, it is desirable to collect most of the entering particles on the collector plates. Thus, we define the collector's efficiency, η, as the ratio of particles entering to particles collected. The following Deutsch–Anderson equation predicts an exponential decay of efficiency with deflection velocity:

$$\eta = 1 - exp\left[-\left(\frac{u_d A}{u_g}\right)^k\right] \tag{7.51}$$

where A and k refer to the plate surface area and an empirical factor, respectively.

Despite recent advances in precipitator technologies, the detailed effects of the charged particles on the surrounding electric field are largely unknown. In addition, the effects of turbulence on particle dispersion are important in terms of the collector efficiency. A coupling between Maxwell's equations and the Navier–Stokes equations is required for an accurate understanding of the role of magnetic effects on the gas flow through the collector plates. Due to its relevance in other applications, such as aeronautics and astronautics, this coupling is receiving considerable attention through the merging of computational electromagnetics (CEM) and computational fluid dynamics (CFD).

7.6 Spray Drying

Another important example involving gas–solid flows is the production of spray-dried products, such as pharmaceutical powders and laundry detergents. In many production processes, drying products are highly sensitive to temperature, and thus accurate understanding of heat transfer is important in view of quality control of these products. In a typical spray dryer, slurry droplets are sprayed downward from the top of a mixing chamber and then dried by a hot gas stream entering from below the chamber in an opposite direction to the droplets. The gases flow into the chamber with a tangential velocity component, thereby introducing a swirling motion, as the gases pass upward and out through the top of the mixing chamber. The dried particles fall through the bottom of the chamber. They are later collected, processed, and packaged as a final powdered product.

Proper thermal control of the drying process is a main design consideration since the particle rates of temperature change are closely related to final product quality. Cooling of the drying gases by the downward particle flow reduces the drying effectiveness, so the inlet gas temperature(s) and velocity are key design criteria. The gas temperature cannot be too high because associated burning of the powder can adversely affect the taste of the powdered food product. Also, sufficient drying of particles must be achieved prior to their departure through the bottom of the drying chamber to prevent accumulation on the walls. Particle accumulation along the walls may become dangerous in view of initiating a fire in the chamber.

Rapid advances in predictive methods for two-phase flows (such as gas–solid flows) have been achieved in the past few decades (Crowe, 1999). Numerical models can provide valuable design tools in the production of powdered products, such as powdered foods. Two main techniques have been developed for these types of gas–solid flows: particle trajectory and two-fluid approaches. In the first approach, particle trajectory equations, based on force and heat balances, are solved for both velocity and temperature of the particles in the gas stream. Also, heat and momentum equations are solved for the gas field, and two-way coupling is established with the particle trajectories. Although the trajectory approach is effective for dilute flows, the two-fluid approach is often more suitable for dense flows. In the two-fluid method, the particulate phase is treated as a second fluid with mass-weighted thermophysical properties. Cross-phase interactions in the momentum and energy equations are included for coupling between the phases. It is anticipated that further advances in simulation technology can lead to better design tools for problems involving gas–solid systems, such as spray drying.

References

C. Crowe. Modeling Fluid-Particle Flows: Current Status and Future Directions, AIAA Paper 99-3690, paper presented at 30th AIAA Fluid Dynamics Conference, Norfolk, VA, 1999.

R. Cliff, J.R. Grace, and M.E. Weber. 1978. *Bubbles, Drops and Particles*, Academic Press, New York.

L.K. Doraiswamy and A.S. Majumdar. 1989. *Transport in Fluidized Particle Systems*, Amsterdam: Elsevier Science Publishers.

D. Geldart. 1973. "Types of gas fluidization," *Powder Technol.*, 7: 285–292.

Handbook of Multiphase Systems, G. Hetsroni, ed., Washington, DC: Hemisphere Publishing Corp 1982.

International Encyclopedia of Heat and Mass Transfer, G.F. Hewitt, G.L. Shires, and Y.V. Polezhaev, eds., Boca Raton, FL: CRC Press 1997.

Fluidized Beds Combustion and Applications, J.R. Howard, ed., London: Applied Science Publishers 1983.

F.P. Incropera and D.P. DeWitt. 1996. *Fundamentals of Heat and Mass Transfer*, 4th ed., New York: John Wiley & Sons.

G.E. Klinzing. 1981. *Gas-Solid Transport*, New York: McGraw-Hill.

F. Kreith and M.S. Bohn. 2001. *Principles of Heat Transfer*, 6th ed., Pacific Grove, CA: Brooks/Cole Thomson Learning.

S. Whitaker. 1972. "Forced convection heat transfer correlations for flow in pipes, past flat plates, single cylinders, single spheres, and for flow in packed beds and tube bundles," *AIChE J.*, 18: 361–371.

Problems

1. Spherical particles at a temperature of 30°C are injected into a cross-flow of air within a furnace. The problem parameters are listed as follows:

   - 5 m/sec relative velocity between the airstream and particles
   - Processed sandstone particles with properties of $\rho = 2200$ kg/m^3, $c_p = 920$ J/kgK, and $k = 1.7$ W/mK and an emissivity of $\epsilon = 0.9$
   - Mean particle diameter of 3 mm
   - Freestream air temperature of 100°C

   What is the initial rate of heating (K/sec) of the particles by the wall and airstream as they enter the furnace? It may be assumed that the particles do not appreciably participate in the radiation exchange with the furnace wall at 600°C.

2. An experimental apparatus has been proposed for the evaluation of the effective thermal conductivity of a packed bed at elevated temperatures. The apparatus consists of a furnace enclosure with internal heating elements (above the packed bed), a packed bed of polished alumina spheres (packed at the bottom of the enclosure), a series of thermocouples (inside the packed bed), and a lower heat exchanger (below the packed bed) with a coolant (water) flow. The heat exchanger directly contacts the packed bed with a surface area of 0.3 × 0.3 m. It extracts the heat from the radiant heaters by circulating cold water through its milled channels. Measurements of the water temperature rise between the exchanger inlet and outlet and the mass flow rate indicate that $\Delta T \sim 1.3$°C and $\dot{m} \approx 0.05$ [kg/sec], respectively. Also, the following thermocouple measurements at vertical locations (within the bed) were recorded: 300 K at the wall, 376 K at 2 cm, 444 K at 4 cm, 517 K at 6 cm, and 590 K at 8 cm.

   (a) Estimate the effective thermal conductivity of the packed bed. State the assumptions in your analysis.
   (b) Discuss the factors that may lead to experimental errors and suggest approaches to minimize these errors.

3. In a particulate removal process, solid particles migrate at 4 m/sec relative to an airstream at 500 K. Find the drag force of the gas on an individual particle when the particle diameter is 0.2 mm.
4. Spherically shaped particles of pulverized coal are injected into a cross-flow of air in a furnace. The initial temperature difference (at the point of injection) between the air and particles is ΔT_0. The particles are heated by convection and radiation (characterized by heat transfer coefficient h) during their trajectory over a specified distance, L. If the particle velocity, V, is assumed to remain constant throughout the trajectory, then find the rise of temperature of a particle in terms of ΔT_0, L, V, h, D (particle diameter), and thermophysical properties.
5. A packed bed with a 60% void fraction consists of spherically shaped particles (0.4 mm in diameter) flowing through air at 600 K. Estimate the convective heat transfer coefficient when the superficial fluid velocity is 12 m/sec.
6. A catalytic reactor contains an airflow that passes through a packed bed of spherical nickel pellets in a hydrogenation process for petroleum refining. Consider a typical packed-bed flow that operates under the following conditions: void fraction of 0.4, superficial gas velocity of 5

m/sec, pellet diameter of 4 mm, and an incoming gas (air) temperature of 350°C. If the initial temperature of pellets in the packed bed is 30°C, then what temperature will these pellets reach after the air flows through the packed bed for 10 sec?

7. A horizontal airstream (carrier phase, subscript c) carrying small particles flows past an inclined surface. Under certain conditions (i.e., particle temperature and velocities match airstream values), the thermal boundary layer equation may be written as

$$\beta_c \rho c_p u \frac{\partial T}{\partial x} + \beta_c \rho c_p v \frac{\partial T}{\partial y} = k_{eff} \frac{\partial^2 T}{\partial y^2} \tag{7.52}$$

where β_c refers to the volume fraction of the air phase (assumed constant) and k_{eff} is the weighted average of particle and air conductivities. The velocity field is predicted by

$$u = \frac{\partial \psi}{\partial y}; \quad v = -\frac{\partial \psi}{\partial x} \tag{7.53}$$

where

$$\psi = (U_\infty \nu x)^{1/2} f(\eta); \quad \eta = y\left(\frac{U_\infty}{\nu x}\right)^{1/2} \tag{7.54}$$

Also, the freestream velocity is $U_\infty = Cx^m$, where the exponent m is related to the surface inclination angle. Use the similarity solution method to find the temperature profile within the boundary layer, and explain how this solution can be used to find the local Nusselt number.

8. Spherically shaped particles of diameter D are dispersed uniformly in a gas flow and heated primarily by radiative exchange with furnace walls at a temperature of T_w. Find the time taken for the particles to reach a specified temperature, T_{spec}, when the particle emissivity is ϵ. Express your result in terms of ϵ, thermophysical properties, D, T_w, and T_0 (initial particle temperature).

9. Radiative heat exchange involves a participating gas at 1600 K containing finely dispersed particles. The mixture has emission bands between 1 and 3 μm (where $\epsilon = 0.7$) and between 6 and 9 μm (where $\epsilon = 0.5$). Find the total emissivity and emissive power (W/m^2) of the gas–particle mixture.

10. A gaseous mixture containing pulverized coal particles at 1200°C has a radiative absorption coefficient of $\kappa_\lambda \approx 4 - exp(-\lambda/6)$/m where λ refers to wavelength (units of μm). A monochromatic beam of radiation at $\lambda = 5$ μm enters a 20-cm thick layer of the mixture with an intensity of 9

kW/m^2 μm · sr. What beam intensity emerges from the layer? The solid particles emit and absorb radiation as they pass through the layer.

11. A monochromatic beam of radiation at a wavelength of 4 m and intensity of 8 kW/m^2 μm · sr enters a sooty gas layer (10 cm thick) containing finely dispersed particles. The particles absorb and emit radiation so that the beam intensity is reduced by 20% as it leaves the layer. What is the absorption coefficient of the gas–particle layer at the given wavelength of radiation?
12. How can the absorption coefficient be used to characterize the optical thickness or degree of transparency of a sooty gas with particles?
13. How can Beer's law be generalized to an equation of transfer for a participating gas–particle mixture, including emission and absorption of radiation by the particles? Assume that the radiant intensity becomes independent of position as the gas layer thickness becomes sufficiently large. Does your result approach the correct asymptotic trends in free space (i.e., without emission or absorption)?
14. Generalize the result from the previous question to a vector form of the equation of transfer for a participating medium.
15. For radiative absorption by solid particles in a sooty gas layer, the equation of transfer can be used to derive how the intensity of radiation is reduced due to the absorption. How can the radiative heat flux be determined at a particular location, r, when the intensity of the beam of radiation is known at that position? Provide an integral expression that shows the appropriate range of solid angles.
16. Extend the result from the previous question to derive the governing equation for combined convective–radiative transport through the gas–particle mixture. Show how the thermal energy equation can be written in terms of the radiative heat flux vector, $\mathbf{q}_r$, and alternatively, in terms of the intensity of radiation, I_λ.
17. What properties of the solid particles affect the characteristics and parameters of fluidized beds?
18. In a fluidized bed, how can the transition from the fixed-bed regime to the fluidized bed regime be detected? Explain your response by referring to specific measurements that can be taken in the fluidized bed.
19. What similarities exist between particle movement in fluidized beds and two-phase flows of liquid–gas or solid–liquid mixtures?
20. In uniform fluidization, the temperature difference between points in the fluidized bed often does not exceed 5°C. Also, a small temperature difference is encountered between the particles and gases leaving the fluidized bed. Furthermore, if heat transfer coefficients to particles in the bed are relatively small (generally up to about 25 W/m^2K), then

what mechanism(s) are primarily responsible for the intense heat transfer between particles and the fluidizing gas?

21. Convection of a gas stream with dispersed particles is encountered in a channel of height H and length L. The wall temperature, T_w, is constant. As a first approximation, heat released by particle combustion is represented by a constant heat source, $\dot{q}$, in the energy equation. Also, the wall heat flux is $q_w(1+x/L)$, which increases with distance, x, along the channel.
    (a) Explain the assumptions adopted to give the following governing equation:

$$\rho c_p \frac{\partial(uT)}{\partial x} + \rho c_p \frac{\partial(vT)}{\partial y} = k\frac{\partial^2 T}{\partial y^2} + \dot{q}$$

    The thermophysical properties are represented by a mass fraction-weighted average of solid (particle) and gas properties.
    (b) Use the integral method to find the variation of centerline temperature with axial position (x) throughout the channel. At a fixed position, x, assume that the cross-stream temperature profile can be approximated quadratically, i.e., $T = A + By + Cy^2$, where A, B, and C must be determined. In the spatial integrations, use a constant mean velocity, u_m, across the channel. Perform the integrations over a control volume of thickness dx and a height of H.

8

Liquid–Solid Systems

8.1 Introduction

Thermofluid systems involving liquid–solid flows and phase change arise in many engineering applications, such as materials processing (i.e., casting, injection molding, extrusion), aircraft de-icing, and phase change materials (PCMs) in thermal energy storage systems. In addition to technologies involving solidification and melting, solid–liquid flows arise in applications dealing with the transport of solid particles by a liquid phase. These applications require an understanding of both the mixture flow and relative flow between the liquid and the solid phases.

The importance of solid–liquid phase change in our everyday lives is evident in many ways. For example, consider water and its density difference between solid and liquid phases. Unlike many materials, its density is higher in the liquid phase. This property arises from the unique angular arrangement of hydrogen and oxygen atoms in a water molecule. As a result, ice freezes and floats on the top of bodies of water, such as lakes and oceans. Otherwise, if water solidification occurred with a higher density in the solid phase (like most metals), freezing ice would descend. In that case, it is speculated that oceans, lakes, and rivers could eventually completely freeze, so that life on Earth likely would not exist. The purpose of this example is to indicate that solidification and melting are important parts of common everyday experiences.

In Chapter 5, recall that the heat transfer, q, associated with phase change between solid and liquid phases can be expressed by

$$q = \dot{m}L \tag{8.1}$$

where $\dot{m}$ and L (or h_{sf}) refer to the rate of mass change due to the phase transition and the latent heat of fusion, respectively. In this chapter, processes leading to this phase change will be examined.

8.2 One-Dimensional Solidification and Melting

8.2.1 Stefan Problem

Problems involving heat conduction with solid–liquid phase change are often called *Stefan problems*. They are named after the 19th-century Slovenian physicist Jozef Stefan (see Figure 8.1), particularly in recognition of the original work of Stefan (1891) in analyzing the formation of ice in the Polar Seas. In this section, Stefan problems and one-dimensional problems with solid–liquid phase change will be analyzed through a variety of analytical techniques. In particular, phase change involving a planar interface will be considered hereafter.

A fundamental problem that often serves as a basis for phase change analysis is solidification of a liquid (initially at a temperature of T_i) that is cooled by a wall at a temperature of T_w (see Figure 8.2). Consider one-dimensional transient heat conduction in a semi-infinite domain. If the phase change temperature is denoted by T_f, the governing equations in the solid and liquid phases, respectively, are given by

$$\rho c_p \frac{\partial T_s}{\partial t} = k_s \frac{\partial^2 T_s}{\partial x^2} \tag{8.2}$$

FIGURE 8.1
Jozef Stefan (1835–1893). (From Hewitt, G.F., Shires, G.L., and Polezhaev, Y.V., Eds., *International Encyclopedia of Heat and Mass Transfer*, Copyright 1997, CRC Press, Boca Raton, FL. Reprinted with permission.)

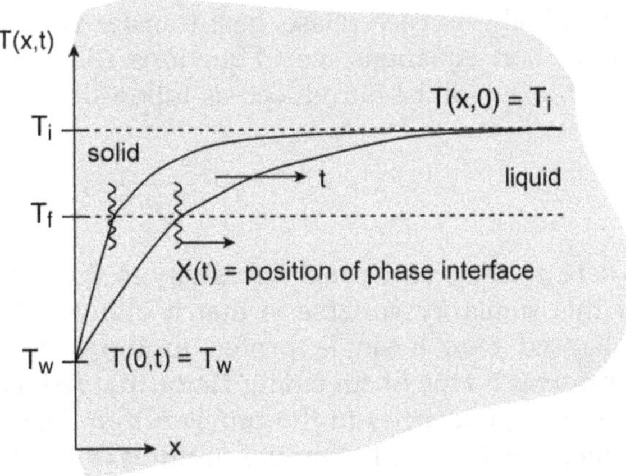

FIGURE 8.2
Schematic of phase change in a semi-infinite domain.

$$\rho c_p \frac{\partial T_l}{\partial t} = k_l \frac{\partial^2 T_l}{\partial x^2} \tag{8.3}$$

where the subscripts s and l refer to solid and liquid phases, respectively.

Also, the boundary/interface conditions are given by

$$T_s(0,\ t) = T_w \tag{8.4}$$

$$T_s(X,\ t) = T_f = T_l(X,\ t) \tag{8.5}$$

where X refers to the position of the solid–liquid interface. Although the forms of Equations (8.2) and (8.3) have been considered extensively earlier (Chapter 2), the main difficulty here is predicting the unknown position of the phase interface, which is required in Equation (8.5). At the phase interface, the heat balance requires that the difference between the solid–liquid conduction heat flows must balance the latent heat released by the liquid as it solidifies, i.e., at $x = X$,

$$-k_l \frac{\partial T_l}{\partial x} + k_s \frac{\partial T_s}{\partial x} = \rho L \frac{dX}{dt} \tag{8.6}$$

The phase change problem becomes nonlinear due to the latent heat term in the interfacial heat balance of Equation (8.6).

In earlier chapters, particularly in the similarity solutions leading to Equation (2.111) in Chapter 2 and Equation (3.95) in Chapter 3, it was observed that similarity variables could be used to solve problems involving transient heat conduction. In the present context, once the interface position, X, is established, it appears that this similarity method can also be used to obtain the temperatures in the liquid and solid phases individually. In each phase, heat transfer is governed by the transient heat conduction equations, i.e., Equations (8.2) and (8.3). Thus, a similarity variable, η, will be introduced as follows:

$$\eta = \frac{x}{2\sqrt{\alpha_s t}} \tag{8.7}$$

where α_s refers to thermal diffusivity in the solid phase. The physical basis of this similarity variable is that it allows all temperature profiles to be collapsed onto a single profile in terms of that similarity variable. It represents a type of stretching factor that reduces the combined transient–spatial dependencies in the problem to a dependence on a single variable alone, thereby simplifying the solution of the problem.

It is assumed that the temperature becomes a function of η alone. Also, the derivatives in terms of x and t in Equation (8.2) are transformed to derivatives in terms of η through the chain rule, thereby yielding

$$\frac{d^2 T_s}{d\eta} + 2\eta \frac{dT_s}{d\eta} = 0 \tag{8.8}$$

Solving this differential equation subject to $T = T_w$ at $x = 0$ (i.e., $\eta = 0$),

$$T_s = T_w + C_1 erf(\eta) = T_w + C_1 erf\left(\frac{x}{2\sqrt{\alpha_s t}}\right) \tag{8.9}$$

where C_1 is a constant of integration and the error function, $erf(\eta)$, is given by

$$erf(\eta) = \frac{2}{\sqrt{\pi}} \int_0^{\eta} e^{-\zeta^2} d\zeta \tag{8.10}$$

Based on a similar procedure in the liquid region, Equation (8.3) is solved to yield

$$T_l = T_i - C_2 \left[1 - erf\left(\frac{x}{2\sqrt{\alpha_l t}}\right)\right] \tag{8.11}$$

At the solid–liquid interface ($x = X$), the solid temperature in Equation (8.11) must balance the liquid temperature obtained from Equation (8.9), i.e.,

$$T_w + C_1 erf\left(\frac{x}{2\sqrt{\alpha_s t}}\right) = T_f = T_i - C_2 \left[1 - erf\left(\frac{x}{2\sqrt{\alpha_l t}}\right)\right] \tag{8.12}$$

It can be observed that equating both temperatures to each other gives expressions involving time in Equation (8.12). Since those temperatures must match each other at all values of time, the numerator in the error function arguments in Equation (8.12) must be proportional to $\sqrt{t}$ to eliminate the time dependence. This suggests that X becomes proportional to the square root of time. Thus, based on dimensional homogeneity, the following expression is adopted:

$$X = 2\beta\sqrt{\alpha_s t} \tag{8.13}$$

In Equation (8.13), the constant β is selected to allow simplification of the resulting expressions in view of canceling a similar term in the denominator of the error function arguments in Equation (8.12). Performing this substitution for $x = X$ in Equation (8.12),

$$T_w + C_1 erf(\beta) = T_f = T_i - C_2\left[1 - erf\left(\beta\sqrt{\frac{\alpha_s}{\alpha_l}}\right)\right] \tag{8.14}$$

Also, the derivatives of temperature (with respect to x) can be evaluated in terms of the constants, C_1 and C_2, based on Equations (8.9) to (8.11). Then, once these derivatives are substituted into Equation (8.6) and combined with Equation (8.14), it can be observed that three equations are obtained for the three unknown coefficients, C_1, C_2, and β. These three equations can be solved simultaneously for C_1, C_2, and β and then substituted into Equations (8.9) to (8.11) to give

$$T_s = T_w + \frac{T_f - T_w}{erf(\beta)} erf\left(\frac{x}{2\sqrt{\alpha_s t}}\right) \tag{8.15}$$

$$T_l = T_i - \frac{T_i - T_f}{1 - erf(\beta\sqrt{\alpha_s/\alpha_l})}\left[1 - \left(\frac{x}{2\sqrt{\alpha_l t}}\right)\right] \tag{8.16}$$

The value of β is described by the following implicit expression:

$$\frac{L\beta\sqrt{\pi}}{c_s(T_f - T_w)} - \frac{exp(-\beta^2)}{erf(\beta)} + \frac{k_l\sqrt{\alpha_s/\alpha_l}(T_i - T_f)exp(-\alpha_s\beta^2/\alpha_l)}{k_s(T_f - T_w)(1 - erf(\beta\sqrt{\alpha_s/\alpha_l}))} = 0 \tag{8.17}$$

Numerical solutions of β in Equation (8.17) are shown in Table 8.1 for water (Carslaw and Jaeger, 1959).

The Stefan solution (1891) is a special case of the previously described similarity analysis. This special case is obtained if the initial temperature of the liquid is equal to the phase change temperature, T_f. In this case, the third term on the right side of Equation (8.17) becomes zero, thereby

TABLE 8.1

Values of β for Water

$(T_f - T_w)$°C	$\Delta T = 0$°C	$\Delta T = 1$°C	$\Delta T = 2$°C	$\Delta T = 3$°C	$\Delta T = 4$°C
1	0.056	0.054	0.053	0.051	0.050
2	0.079	0.077	0.076	0.074	0.073
3	0.097	0.095	0.093	0.091	0.090
4	0.111	0.110	0.108	0.106	0.104
5	0.124	0.123	0.121	0.119	0.117

Note: $\Delta T = T_i - T_f$

simplifying the nonlinear equation for the parameter β. Approximating the error function in this reduced equation by a Taylor series expansion and retaining only the leading term of the series expansion if β is small,

$$erf(\beta) \approx \frac{2}{\sqrt{\pi}}\beta \tag{8.18}$$

As a result, Equation (8.17) becomes

$$\beta = \sqrt{\frac{c_s(T_f - T_w)}{2L}} \tag{8.19}$$

Then, based on Equation (8.13),

$$X = \sqrt{\frac{2k_s(T_f - T_w)t}{\rho_s L}} = \sqrt{2(Ste)\alpha t} \tag{8.20}$$

where the Stefan number is defined by

$$Ste = \frac{c_s(T_f - T_w)}{L} \tag{8.21}$$

The results in Equations (8.19) and (8.20) were originally presented by Stefan (1891). These approximate results are based on the assumption of sufficiently small values of β (i.e., neglecting higher order terms in the Taylor series expansion of $erf(\beta)$). For water and soil systems, such as problems involving freezing and thawing of the ground, the values of β are small, i.e.,

- $T_f - T_w = 0$°C → $\beta = 0.054$
- $T_f - T_w = 1$°C → $\beta = 0.076$

- $T_f - T_w = 2°C \rightarrow \beta = 0.094$
- $T_f - T_w = 3°C \rightarrow \beta = 0.108$

Thus, for freezing or thawing in the natural environment, the approximate solution in Equation (8.20) yields good accuracy. However, for metals, $\beta \approx 1$ and appreciably less accuracy is achieved with the Stefan solution. However, the solution approach of Stefan can be readily modified to include an additional term(s) in the series expansion approximation of Equation (8.18), thereby extending its applicability to materials other than water and soil systems.

8.2.2 Integral Solutions

Another general technique for analyzing phase change problems is based on integral solutions. This approach was described in earlier chapters (particularly Chapter 3). For example, in a manner similar to that in the momentum integral analysis of Chapter 3, the governing equations can be satisfied in an average sense over the selected control volume, rather than at each point. In boundary layer problems, the governing equations were integrated across the thickness of the boundary layer since parameters were well established (or known) at the edge of the boundary layer. In phase change heat transfer problems, an analogous range of integration is the penetration depth, $\delta(t)$. This depth refers to the distance that a temperature disturbance propagates into the material. Beyond this point, the temperature remains at or very close to its initial temperature.

Example 8.2.2.1

Integral Solution of Solidification in a One-Dimensional Domain.

Use an integral solution technique to find the rate of interface advance of a solidifying liquid in a one-dimensional domain (outlined in Figure 8.2). The spatial distribution of temperature in the solid can be assumed to vary linearly with x (note: x refers to the distance from the wall).

This problem was analyzed with a similarity solution method in the previous section (based on $T_i > T_f$), whereas an integral solution technique is used in this example. Equations (8.2) and (8.3) can be integrated from the wall to the interface (solid), as well as to the edge of the penetration depth, δ, in the liquid, i.e.,

$$\int_0^X \frac{\partial T_s}{\partial t} = \alpha_s \int_0^X \frac{\partial^2 T_s}{\partial x^2}\, dx \tag{8.22}$$

$$\int_X^\delta \frac{\partial T_l}{\partial t} = \alpha_l \int_X^\delta \frac{\partial^2 T_l}{\partial x^2}\, dx \tag{8.23}$$

Using the Leibnitz rule of calculus, based on Equation (3.116), these equations become

$$\frac{d}{dt}\int_0^x T_s(x,\ t)dx - T_f\frac{dX}{dt} - \alpha_s\frac{\partial T_s(X,\ t)}{\partial x} + \alpha_s\frac{\partial T_s(0,\ t)}{\partial x} = 0 \tag{8.24}$$

$$\frac{d}{dt}\int_X^\delta T_l(x,\ t)dx - T_i\frac{d\delta}{dt} + T_f\frac{dX}{dt} - \alpha_l\frac{\partial T_l(\delta,\ t)}{\partial x} + \alpha_s\frac{\partial T_l(X,\ t)}{\partial x} = 0 \tag{8.25}$$

where the fourth term in Equation (8.25) becomes zero since the temperature is unchanged at $x = \delta$. Thus, its temperature gradient is approximately zero at that location.

Based on similar arguments leading to Equation (8.13), the following forms of solutions are adopted for the interface position and the penetration depth:

$$X = 2\beta\sqrt{\alpha_s t} \tag{8.26}$$

$$\delta = 2\zeta\sqrt{\alpha_l t} \tag{8.27}$$

The solution of the governing equations, Equations (8.2) and (8.3), in an integral or average sense allows the multivariable problem to be reduced to single-variable problems involving ordinary differential equations (rather than partial differential equations).

However, this reduction of problem variables comes at the expense of needing to provide additional information that effectively compensates for the information lost in the averaging process. This additional information is provided in the form of the following assumed forms of the temperature profiles in the solid and liquid regions, respectively:

$$T_s = a_1 + a_2 x \tag{8.28}$$

$$T_l = b_1 + b_2 x + b_3 x^2 \tag{8.29}$$

A linear profile is adopted in the solid. At least a second order interpolation is required in the liquid to match the required slope at the interface, $x = X$, as well as a zero temperature slope in the liquid at $x = \delta$.

The unknown coefficients in Equations (8.28) and (8.29), from a_1 to b_3, can be determined by imposing the required interfacial and boundary conditions. These conditions include $T_s = T_w$ at $x = 0$, $T_s = T_f = T_l$ at $x = X$, and

$T_l = T_i$ at $x = \delta$. After finding the unknown coefficients and substituting them into Equations (8.28) and (8.29),

$$T_s = T_w + (T_f - T_w)\frac{x}{X} \tag{8.30}$$

$$T_l = T_f + 2\left(\frac{x - X}{\delta - X}\right)(T_i - T_f) - \left[\frac{x - X}{\delta - X}\right]^2 (T_i - T_f) \tag{8.31}$$

It will be useful to write Equation (8.31) in the following form:

$$T_l = T_f + 2\left(\frac{x - X}{X}\right)\frac{T_i - T_f}{p - 1} - \left(\frac{(x - X)^2}{X^2}\right)\frac{T_i - T_f}{(p - 1)^2} \tag{8.32}$$

where $p = \delta/X$ is a function of the Stefan number defined in Equation (8.21), temperatures (such as T_f and T_i), and thermophysical property values, but not time.

Evaluating the derivative and integrals of the temperature in Equation (8.32) and substituting those results into Equation (8.25),

$$\left(\frac{T_f}{3} + \frac{2T_i}{3}\right)(p - 1)\frac{dX}{dt} - T_i p\frac{dX}{dt} + \frac{2\alpha_l(T_i - T_f)}{(p - 1)X} + T_f\frac{dX}{dt} = 0 \tag{8.33}$$

Based on the functional form for the interface position, Equation (8.26), the derivative of X can be substituted into Equation (8.33) to obtain the following explicit expression for p in terms of β :

$$p = \sqrt{2.25 + \frac{3\alpha_l}{\beta^2\alpha_s}} - \frac{1}{2} \tag{8.34}$$

Then, the following result is obtained:

$$\left(\frac{1}{Ste} + \frac{1}{2}\right)\frac{dX}{dt} = \frac{\alpha_s}{X}\left[1 - \frac{2k_l(T_i - T_f)}{k_s(T_f - T_w)(p - 1)}\right] \tag{8.35}$$

Solving Equation (8.35) subject to $X = 0$ at $t = 0$ yields the initially adopted form in Equation (8.26), i.e.,

$$X = 2\beta\sqrt{\alpha_s t} \tag{8.36}$$

where the parameter β is given by

$$\beta^2 = -\frac{p_2}{2p_1} - \frac{\sqrt{p_2^2 - 4p_1(Ste)^2}}{2p_1} \tag{8.37}$$

and

$$p_1 = (2 + Ste)\left[2 + \frac{2k_l\alpha_s(T_i - T_f)Ste}{k_s\alpha_l(T_f - T_w)} + Ste\right] \tag{8.38}$$

$$p_2 = -\frac{4k_l^2\alpha_s(T_i - T_f)^2 Ste^2}{3k_s^2\alpha_l(T_f - T_w)^2} - 2Ste\left(2 + \frac{k_l\alpha_s(T_i - T_f)Ste}{k_s\alpha_l(T_f - T_w)} + Ste\right) \tag{8.39}$$

This result in Equation (8.36) gives the change of interface position with time.

The approximate solution in Equations (8.36) and (8.37) agrees closely with the exact solution using the β parameter in Equation (8.17). Based on comparisons documented by Lunardini (1988), the exact and approximate β values differ by less than 1% for Stefan numbers of less than about 0.1, with somewhat larger errors at higher Stefan numbers. Further favorable comparisons between the approximate and exact solutions were presented by Sparrow et al. (1978). As a result, the heat balance integral technique provides a good approximate method of analyzing phase change problems over a range of Stefan numbers and temperature conditions. Further accuracy in the integral solution can be obtained by using higher order interpolations in Equations (8.28) and (8.29), but this approach also entails further complexity in the resulting differential equations corresponding to Equation (8.33).

8.2.3 Phase Change Temperature Range

For pure materials, phase change occurs at a single discrete temperature. However, for multicomponent mixtures such as metal alloys, phase change occurs over a range of temperatures. In this section, modifications of the previously discussed exact (similarity) and approximate (integral) solutions will be described. For an initially liquid binary alloy, solidification begins at the liquidus temperature, T_{liq}, and it is completed at the solidus temperature, T_{sol} (see Figure 1.6). Thus, in addition to solid and liquid regions, a two-phase (or mushy) region is formed where solid coexists with liquid in equilibrium. For example, pockets or channels of liquid are interspersed within the solid dendritic matrix. The solution procedures described herein can be extended to other multiconstituent mixtures with more than two components in the material, such as tertiary alloys.

A solution based on the previously described similarity analysis was presented by Cho and Sunderland (1969). In this analysis, it is assumed that variations of thermophysical properties with temperature can be neglected. Also, the solid fraction, f_s, is assumed to vary linearly between the solidus interface ($x = X_{sol}$) and liquidus interface ($x = X_{liq}$), i.e.,

$$f_s = f_e\left(1 - \frac{x - X_{sol}}{X_{liq} - X_{sol}}\right) \tag{8.40}$$

where $X_{sol} < x < X_{liq}$ and f_e refers to the solid fraction at the eutectic composition. The liquid fraction, f_l, is 100% minus the solid fraction from Equation (8.40).

The approximation in Equation (8.40) effectively states that the liquid fraction increases linearly with position between the solidus and liquidus interfaces. This type of linear relation within the two-phase region is generally applicable to materials such as metal alloys. However, other materials may require different functional forms of the equation of state. For example, soil mixtures exhibit a range of temperatures over which phase change occurs. In this case, an exponential change of solid fraction through the two-phase range more closely approximates soil–water data. The range of phase change encompasses the values between the solidus and liquidus temperatures at a given mixture composition. These values are obtained from the binary phase equilibrium (i.e., Figure 1.6). It should be noted that a binary mixture problem is similar to the previously described pure material problems, except that two interfaces (solidus and liquidus) are tracked for binary systems, rather than the single interface in the pure material phase change.

In addition to the same governing heat equations, Equations (8.2) and (8.3) in the solid and liquid phases, respectively, the following equation is applicable in the two-phase (mushy) region:

$$\rho_{sl} c_{sl} \frac{\partial T_{sl}}{\partial t} = k_{sl} \frac{\partial^2 T_{sl}}{\partial x^2} + \rho_{sl} L \frac{df_s}{dt} \tag{8.41}$$

where the subscript sl refers to the two-phase region (i.e., solid and liquid simultaneously). Equation (8.41) is solved subject to the conditions of $T_s = T_{sol} = T_{sl}$ at $x = X_{sol}$ and $T_{sl} = T_{liq} = T_l$ at $x = X_{liq}$ (see Figure 8.3).

Also, the interfacial balance, Equation (8.6), is modified to include the following balances at the solidus and liquidus interfaces, respectively:

$$-k_s \frac{\partial T_s}{\partial x} + \rho_s L(1 - f_e) \frac{dX_{sol}}{dt} = -k_{sl} \frac{\partial T_{sl}}{\partial x} \tag{8.42}$$

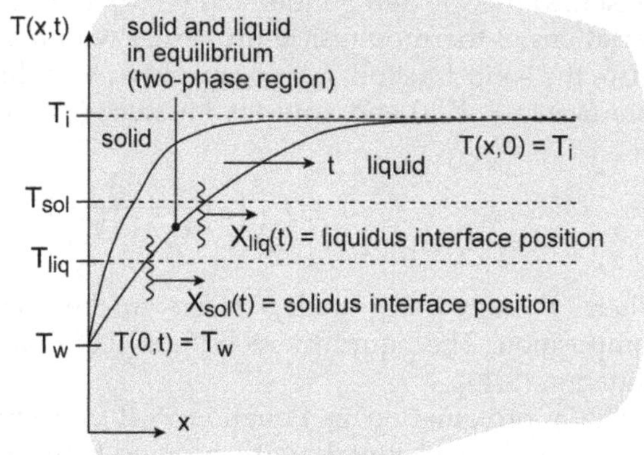

FIGURE 8.3
Schematic of solidification with a two-phase region.

$$-k_{sl}\frac{\partial T_{sl}}{\partial x} = -k_l\frac{\partial T_l}{\partial x} \tag{8.43}$$

The latent heat absorbed or released at the liquidus interface is modeled within the two-phase region by Equation (8.41), and thus does not appear in the interfacial heat balance in Equation (8.43).

Unlike Equation (8.26), where a single interface is tracked, the current binary alloy problem requires tracking of two interfaces, namely, the solidus and liquidus interfaces. Thus, extending Equation (8.26) to accommodate two interfaces,

$$X_{sol} = 2\beta_{sol}\sqrt{\alpha_s t} \tag{8.44}$$

$$X_{liq} = 2\beta_{liq}\sqrt{\alpha_s t} \tag{8.45}$$

In a manner similar to the analysis leading to Equation (8.17), the following results are obtained in the solid, two-phase (mushy), and liquid regions (Lunardini, 1988; Cho and Sunderland, 1969), respectively:

$$\frac{T_s - T_w}{T_{sol} - T_w} = \frac{erf\left(\dfrac{x}{2\sqrt{\alpha_s t}}\right)}{erf(\beta_{sol})} \tag{8.46}$$

$$\frac{T_{sl}-T_w}{T_{sol}-T_w}=1-\frac{Lf_e\left(-\beta_{sol}+\dfrac{x}{2\sqrt{\alpha_s t}}\right)}{c_l(T_{sol}-T_w)(\beta_{liq}-\beta_{sol})}+\frac{T_{liq}-T_{sol}+\dfrac{Lf_e}{c_l}}{T_{sol}-T_w}$$
$$\times\left[\frac{erf\left(\dfrac{x}{2\sqrt{\alpha_l t}}\right)-erf\left(\beta_{sol}\sqrt{\dfrac{\alpha_s}{\alpha_l}}\right)}{erf\left(\beta_{liq}\sqrt{\dfrac{\alpha_s}{\alpha_l}}\right)-erf\left(\beta_{sol}\sqrt{\dfrac{\alpha_s}{\alpha_l}}\right)}\right] \tag{8.47}$$

$$\frac{T_l-T_i}{T_{liq}-T_i}=\frac{1-erf\left(\dfrac{x}{2\sqrt{\alpha_l t}}\right)}{1-erf\left(\beta_{liq}\sqrt{\dfrac{\alpha_s}{\alpha_l}}\right)} \tag{8.48}$$

The parameters β_{sol} and β_{liq} in Equations (8.46) to (8.48) are given implicitly by

$$-\frac{k_l\alpha_s\left(T_{liq}-T_{sol}+\dfrac{Lf_e}{c_l}\right)exp\left(-\beta_{sol}^2\dfrac{\alpha_s}{\alpha_l}\right)}{k_s\alpha_l(T_{liq}-T_w)\left(erf\left(\beta_{liq}\sqrt{\dfrac{\alpha_s}{\alpha_l}}\right)-erf\left(\beta_{sol}\sqrt{\dfrac{\alpha_s}{\alpha_l}}\right)\right)}+\frac{exp(-\beta_{sol}^2)}{erf(\beta_{sol})}$$
$$+\frac{\sqrt{\pi}f_eLk_l}{2c_l(T_{sol}-T_w)(\beta_{liq}-\beta_{sol})k_s}=\frac{\sqrt{\pi}(1-f_e)L\beta_{sol}}{c_s(T_{sol}-T_w)} \tag{8.49}$$

$$(T_{liq}-T_i)\frac{k_l\sqrt{\dfrac{\alpha_m}{\alpha_l}}exp\left(-\beta_{liq}^2\dfrac{\alpha_s}{\alpha_l}\right)}{k_m\left(1-erf\left(\beta_{liq}\sqrt{\dfrac{\alpha_s}{\alpha_l}}\right)\right)}+\frac{T_{liq}-T_{sol}+\dfrac{Lf_e}{c_m}\,exp\left(-\beta_{liq}^2\dfrac{\alpha_m}{\alpha_s}\right)}{erf\left(\beta_{liq}\sqrt{\dfrac{\alpha_s}{\alpha_l}}\right)-erf\left(\beta_{sol}\sqrt{\dfrac{\alpha_s}{\alpha_l}}\right)}$$
$$=\frac{Lf_e\sqrt{\pi\dfrac{\alpha_m}{\alpha_l}}}{2c_m(\beta_{liq}-\beta_{sol})} \tag{8.50}$$

An approximate solution of this same problem, based on the heat balance integral method, is documented by Tien and Geiger (1967). The exact solution in Equation (8.46) is adopted in the solid region, but Equation (8.41) is integrated across the two-phase region (between X_{sol} and X_{liq}). Then, based on an assumed quadratic temperature profile in the two-phase region and substitution of this profile into the integrated two-phase equation, the resulting differential equation is solved for the interface positions, as well as the associated parameters β_{sol} and β_{liq}. A main benefit of using the integral-based approach is its flexibility in adapting the method to other, more complicated geometrical configurations or problem conditions.

8.3 Phase Change with Convection

8.3.1 Solidification of Flowing Liquid (Perturbation Solution)

If a flowing liquid comes into contact with a chilled surface at a temperature below T_f (phase change temperature), then solidification occurs at the surface and proceeds outward. Heat transfer occurs by convection from the flowing liquid to the solidified layer, and subsequently by heat conduction through the solid to the chilled boundary. In this section, a perturbation solution will be described for the solution of this type of phase change problem.

The fundamental basis of perturbation solutions is that difficult problems (i.e., nonlinear equations or complicated geometrical configurations) can be solved by perturbing (successively modifying) solutions of related simpler problems. In other words, a known solution can be continuously changed so that comparable problems can be solved by gradual modifications of the original problem and its solution.

For example, Laplace's equation is given by

$$\frac{\partial^2 T}{\partial x^2} + \frac{\partial^2 T}{\partial y^2} = 0 \tag{8.51}$$

whose solution will be designated by $T_0(x, y)$. Equation (8.51) can be modified in the following manner:

$$\frac{\partial^2 T}{\partial x^2} + \frac{\partial^2 T}{\partial y^2} + \epsilon f(T) = 0 \tag{8.52}$$

where $f(T)$ is a nonlinear function of temperature and $0 \leq \epsilon \leq 1$. Then it can be shown that solutions of Equation (8.52) may be obtained by adding small perturbations to the solution of Equation (8.51), i.e., using a power series expansion:

$$T = T_0 + \epsilon T_1 + \epsilon^2 T_2^2 + \epsilon^3 T_3^3 + \ldots \tag{8.53}$$

Thus, by modifying the solution of Equation (8.51) using Equation (8.52), an approximate solution of the following nonlinear equation when $\epsilon = 1$ can be obtained:

$$\frac{\partial^2 T}{\partial x^2} + \frac{\partial^2 T}{\partial y^2} + f(T) = 0 \tag{8.54}$$

In this way, Equation (8.53) provided a link between the solutions to Equations (8.51) and (8.54).

In addition to modified differential equations, the perturbation method can also be used to solve problems involving irregular domains. For example, the solution of Equation (8.51) inside a deformed square domain can be obtained by perturbing the solution of Equation (8.51) inside a regular square. If a side of the regular square is given by $x = 1$, then the boundary condition along the deformed square boundary, i.e., $x = 1 + cx^2$ (where c is a constant), can be considered as an extreme case of a boundary condition described along $x = 1 + \epsilon cx^2$, where $0 \le \epsilon \le 1$. The solution method is carried out by perturbations of the boundary conditions rather than by the governing equation.

Furthermore, perturbation methods can be used to solve initial value problems, as well as nonlinear differential equations and problems involving irregular geometries (Farlow, 1982). The following example will show how the individual functions in the power series, i.e., T_0, T_1, etc. in Equation (8.53), can be determined.

Example 8.3.1.1

Wall Cooling and Freezing of a Flowing Liquid.

In this example, solidification of a flowing liquid occurs when the liquid comes into contact with a chilled surface at a specified temperature below the phase change temperature, T_f (see Figure 8.4). The purpose of this example is to apply the perturbation method to find the temperature distribution and movement of the phase interface over time.

The heat conduction equation is solved subject to specified temperature boundary conditions at the chilled surface ($T = T_w$ at $x = 0$), as well as convection boundary conditions at the phase interface ($x = X$). The governing equation is given as follows:

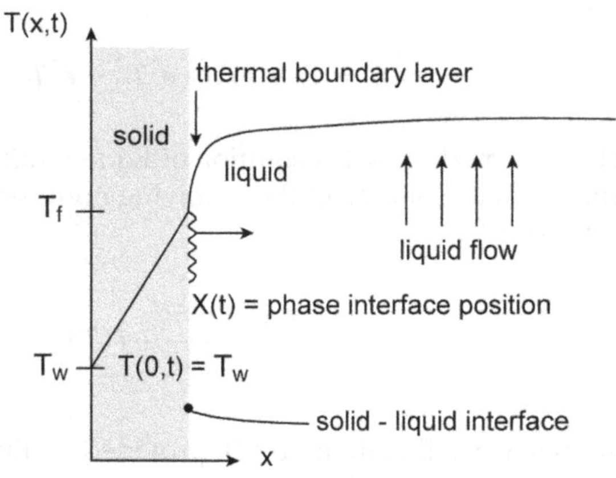

FIGURE 8.4
Schematic of freezing with flowing liquid.

$$\frac{\partial T}{\partial t} = \frac{1}{\alpha} \frac{\partial^2 T}{\partial x^2} \tag{8.55}$$

subject to the following interface and boundary conditions:

$$k \left. \frac{\partial T}{\partial x} \right|_X - h(T_\infty - T_f) = \rho L \frac{dX}{dt} \tag{8.56}$$

$$T(0,\ t) = T_w \tag{8.57}$$

$$T(X,\ t) = T_f \tag{8.58}$$

where T_∞ refers to the freestream liquid temperature.

The steady-state solution can be obtained by removing the transient term in Equation (8.55) and solving in terms of the unknown interface position, X, using the boundary conditions in Equations (8.57) and (8.58). Also, solving for the steady-state interface position, X_s, based on Equation (8.56), using the derivative of the linear temperature profile in the solid,

$$X_s = \frac{k}{h} \left(\frac{T_f - T_w}{T_\infty - T_f} \right) \tag{8.59}$$

Furthermore, the following dimensionless variables are defined:

$$T^* = \frac{T - T_w}{T_f - T_w} \tag{8.60}$$

$$t^* = \frac{h(T_\infty - T_f)}{\rho L X_s} t \tag{8.61}$$

$$x^* = \frac{x}{X_s} \tag{8.62}$$

$$X^* = \frac{X}{X_s} \tag{8.63}$$

Then the dimensionless variables transform Equations (8.55) to (8.58) into the following governing equation:

$$\frac{\partial^2 T^*}{\partial x^{*2}} = Ste \frac{\partial T^*}{\partial X^*} \frac{dX^*}{dt^*} \tag{8.64}$$

where the Stefan number, *Ste*, is defined by

$$Ste = \frac{c_p(T_f - T_w)}{L} \tag{8.65}$$

Also, the boundary conditions, Equations (8.56) to (8.58), become

$$\frac{dX^*}{dt^*} = \left.\frac{\partial T^*}{\partial x^*}\right|_{X^*} - 1 \tag{8.66}$$

$$T^*(0,\ t^*) = 0 \tag{8.67}$$

$$T^*(X^*,\ t^*) = 1 \tag{8.68}$$

Using the perturbation solution method, the temperature and interface velocity are expanded in the form of a power series about the perturbation parameter, *Ste*, as follows:

$$T^* = T_0^* + (Ste)T_1^* + (Ste^2)T_2^* + \ldots \tag{8.69}$$

$$\begin{aligned} \frac{dX^*}{dt^*} &= \frac{dX_0^*}{dt^*} + Ste\frac{dX_1^*}{dt^*} + Ste^2\frac{dX_2^*}{dt^*} + \ldots \\ &= \left(\frac{\partial T_0^*}{\partial x^*} + Ste\frac{\partial T_1^*}{\partial x^*} + Ste^2\frac{\partial T_2^*}{\partial x^*} + \ldots\right) - 1 \end{aligned} \tag{8.70}$$

where the latter equality is based on Equation (8.66).

Substituting the differentiated Equation (8.69), as well as Equation (8.70), into the governing equation, Equation (8.64),

$$\left(\frac{\partial^2 T_0^*}{\partial x^{*2}} + Ste\frac{\partial^2 T_1^*}{\partial x^{*2}} + Ste^2\frac{\partial^2 T^*}{\partial x^{*2}} + \dots\right)$$
$$= Ste\left(\frac{\partial T_0^*}{\partial X^*} + Ste\frac{\partial T_1^*}{\partial X^*} + Ste^2\frac{\partial T_2^*}{\partial X^*} + \dots\right)$$
$$\times\left(-1 + \frac{\partial T_0^*}{\partial x^*} + Ste\frac{\partial T_1^*}{\partial x^*} + Ste^2\frac{\partial T_2^*}{\partial x^*} + \dots\right) \quad (8.71)$$

Setting the terms with the same powers of *Ste* equal to each other, the following sequence of problems is obtained:

$$\frac{\partial^2 T_0^*}{\partial x^{*2}} = 0 \quad (8.72)$$

$$\frac{\partial^2 T_1^*}{\partial x^{*2}} = \frac{\partial T_0^*}{\partial X^*}\left(\frac{\partial T_0^*}{dx^*} - 1\right) \quad (8.73)$$

$$\frac{\partial^2 T_2^*}{\partial x^{*2}} = \frac{\partial T_1^*}{\partial X^*}\left(\frac{\partial T_0^*}{dx^*} - 1\right) + \frac{\partial T_0^*}{\partial X^*}\frac{\partial T_1^*}{\partial x^*} \quad (8.74)$$

and similarly for higher order equations (corresponding to higher order terms involving *Ste*).

It can be observed that each problem can be solved successively, based on the solution from the previous problem. For example, Equation (8.72) and its boundary conditions are solved to give T_0. Then, substituting the differentiated T_0 into Equation (8.73), the next solution of T_1 is obtained. The differentiated T_0 and T_1 expressions are then substituted into Equation (8.74) to yield the solution for T_2, and so on.

Furthermore, Equation (8.70) shows how individual terms in the power series involving temperature are related to corresponding terms in the power series of interface position terms, i.e.,

$$\frac{dX_0^*}{dt^*} = \left.\frac{\partial T_0^*}{\partial x^*}\right|_{X^*} - 1 \quad (8.75)$$

$$\frac{dX_1^*}{dt^*} = \left.\frac{\partial T_1^*}{\partial x^*}\right|_{X^*} - 1 \quad (8.76)$$

and similarly for higher order terms.

Thus, once the temperature solution for each T_i^* is obtained at step i, where $i = 0, 1, 2$, and so on, then the interface position component, X_i^*, can be obtained by integrating Equation (8.75), Equation (8.76), or the appropriate equation involving X_i^*. The final expression for interface position is then obtained by assembling all X_i^* components together by a power series expansion similar to that in Equation (8.69), but involving X_i^* instead. Due to the term-by-term integration, the final solution will describe the interface position implicitly in terms of time.

Based on the previously described procedure, the solutions for the first few cases are obtained as follows (Lunardini, 1988):

$$T_0^* = \frac{x^*}{X^*} \tag{8.77}$$

$$T_1^* = -\frac{(1 - X^*)}{6X^{*3}}(x^{*2} - X^{*2})x^* \tag{8.78}$$

$$T_2^* = \frac{(1 - X^*)}{12X^{*3}}\left[\left(\frac{3 - 2X^*}{X^{*2}}\right)\frac{x^{*5}}{10} + \frac{x^{*3}}{3} - \frac{X^{*2}}{5}\left(\frac{19}{6} - X^*\right)x^*\right] \tag{8.79}$$

In addition, the variation of interface position with time is outlined implicitly by

$$t^* = t_0^* + (Ste)t_1^* + (Ste^2)t_2^* + \dots \tag{8.80}$$

The following results are obtained for the first few cases:

$$t_0^* = -X^* - ln(1 - X^*) \tag{8.81}$$

$$t_1^* = \frac{1}{3}\ t_0^* \tag{8.82}$$

$$t_2^* = -\frac{1}{90}[3X^{*2} + 2X^* + 2ln(1 - X^*)] \tag{8.83}$$

Once individual terms are substituted and assembled into Equation (8.80), a final resulting equation is obtained that implicitly describes the variation of interface position, X^*, with dimensionless time, t^*. This result completes the perturbation solution of this problem.

The perturbation solution of the previous phase change and convection problem was first presented by Seeniraj and Bose (1982).

8.3.2 Convective Cooling by Ambient Fluid (Quasi-Stationary Solution)

Another commonly encountered problem involves solidification due to convective cooling from one side of the liquid. For example, freezing of a lake typically starts along the top surface of the lake and proceeds downward into the water. Two thermal resistances are evident in this problem: (i) conduction resistance in the solid (increasing with time due to the growing ice thickness) and (ii) convective resistance (constant, unless the convection coefficient, h, or T_∞ vary with time). As time elapses, the relative significance of the convective resistance diminishes relative to conduction through the solid.

A useful approach for analyzing this type of problem is based on a *quasi-stationary approximation*. This approach assumes that heat conducts throughout the solid rapidly in comparison to the time scale associated with movement of the phase interface. As a result, a transient term appears in the heat balance at the phase interface, but it may be neglected in comparison to the spatial heat conduction term(s) in the governing equation within the solid.

Based on the quasi-stationary approximation, the following governing, boundary, and interfacial equations are obtained (see Figure 8.5):

$$\frac{\partial^2 T}{\partial x^2} = 0 \tag{8.84}$$

subject to

$$T(X,\ t) = T_f = T(x,\ 0) \tag{8.85}$$

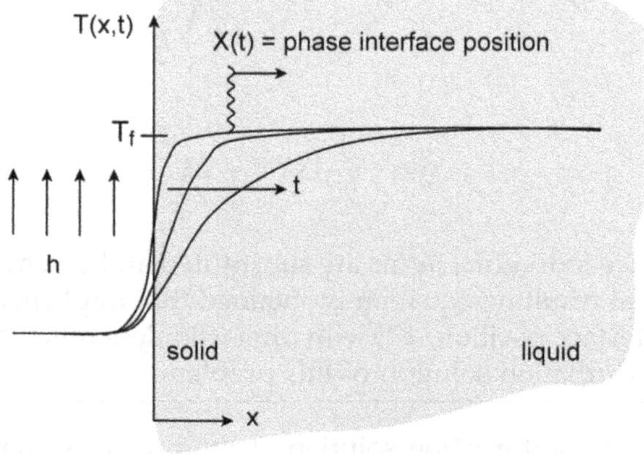

FIGURE 8.5
Schematic of solidified layer with convective cooling.

$$k \left. \frac{\partial T}{\partial x} \right|_X = \rho L \frac{dX}{dt} \tag{8.86}$$

$$k \left. \frac{\partial T}{\partial x} \right|_0 = h(T(0, t) - T_\infty) \tag{8.87}$$

Equations (8.84) to (8.87) indicate that the transient effects are included in the interfacial heat balance, Equation (8.86), but not the heat equation in the solid, Equation (8.84).

Solving Equation (8.84) subject to Equations (8.85) and (8.87),

$$T = T_\infty + \left(\frac{x + k/h}{X + k/h} \right)(T_f - T_\infty) \tag{8.88}$$

Differentiating Equation (8.88) and substituting into Equation (8.86),

$$\frac{dX}{dt} = \frac{k}{\rho L} \left(\frac{T_f - T_\infty}{X + k/h} \right) \tag{8.89}$$

which can be integrated and rearranged to yield the following differential equation for interface position, X:

$$\frac{d}{dt} \left(X + \frac{k}{h} \right)^2 = \frac{2k}{\rho L}(T_f - T_\infty) \tag{8.90}$$

Although this analysis assumes that the fluid temperature, T_∞, and convection coefficient, h, remain constant with time, the integrated results can be readily modified to permit their variations with time.

Assuming that $X = 0$ at $t = 0$, Equation (8.90) is integrated to give

$$X(t) = \sqrt{\frac{2k}{\rho L}(T_f - T_\infty)t + \left(\frac{k}{h} \right)^2} - \frac{k}{h} \tag{8.91}$$

The temperature in the solid follows from substitution of Equation (8.91) into Equation (8.88). This analysis of solidification with convective cooling by a fluid was considered by London and Seban (1943). These results provide good accuracy for small Stefan numbers. However, the resulting accuracy of the quasi-stationary approximation is reduced at higher Stefan numbers. The lower magnitude of latent heat (relative to sensible cooling) increases the interface velocity, thereby allowing less time for thermal equilibrium in the solid (compared to cases with low Stefan numbers).

8.3.3 Constant Heat Flux at the Phase Interface

Another type of problem condition occurs when the heat flux in the material undergoing phase change is assumed to be approximately constant (see Figure 8.6). A solution of this problem has been documented by Foss and Fan (1972) and Lunardini (1988), based on the quasi-stationary approximation. Convective cooling occurs from the surrounding gas (air) that is initially at the freezing temperature, T_f, but then it is suddenly dropped below T_f to initiate solidification. It is assumed that the interface moves sufficiently slowly so that the quasi-stationary approximation can be used.

The governing, boundary, initial, and interfacial conditions are listed as follows:

$$\frac{\partial^2 T_s}{\partial x^2} = 0 \tag{8.92}$$

$$T_s(0,\ 0) = T_f = T_\infty(0) = T_s(X,\ t) \tag{8.93}$$

$$-k_s \left.\frac{\partial T_s}{\partial x}\right|_0 = h(T_s(0,\ t) - T_\infty) \tag{8.94}$$

$$k_s \left.\frac{\partial T_s}{\partial x}\right|_X - k_l \left.\frac{\partial T_l}{\partial x}\right|_X = \rho L \frac{dX}{dt} \tag{8.95}$$

where the subscripts s and l refer to solid and liquid regions, respectively. In Equation (8.95), the second term is assumed to be constant, i.e.,

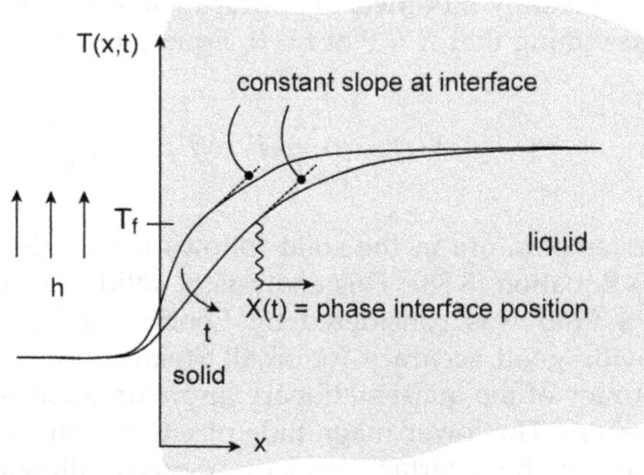

FIGURE 8.6
Schematic of phase change with constant interfacial heat flux.

$$k_l \frac{\partial T_l}{\partial x}\bigg|_X = q_w \tag{8.96}$$

which implies that the heat flux from the liquid at the solid–liquid interface is constant.

Solving Equation (8.92) subject to Equations (8.93) and (8.94),

$$T_s = \left(\frac{T_f + hT_\infty X/k_s}{1 + hX/k_s}\right)\left(1 + \frac{hx}{k_s}\right) - \frac{h}{k_s} T_\infty x \tag{8.97}$$

Differentiating this expression with respect to x and substituting that result with Equation (8.96) into Equation (8.95),

$$\frac{dX}{dt} = \frac{h(T_f - T_\infty)}{\rho L(1 + hX/k_s)} - \frac{q_w}{\rho L} \tag{8.98}$$

Furthermore, integrating Equation (8.98) for the case of a constant ambient fluid temperature (Lunardini, 1988),

$$\left(\frac{hq_w}{k_s \rho L}\right) t = \frac{h(T_f - T_\infty)}{q_w}\, ln\left[\frac{h(T_f - T_\infty) - q_w}{h(T_f - T_\infty) - q_w(1 + hX/k_s)}\right] - \frac{hX}{k_s} \tag{8.99}$$

which gives an implicit closed-form solution for the movement of the phase interface, X, with time.

Once the interface position is known, the temperature distribution can be obtained from Equation (8.97). Also, setting the left side of Equation (8.98) to zero and solving for X on the right side of the same equation yields the following steady-state position of the solid–liquid interface:

$$X_{max} = k_s\left(\frac{T_f - T_\infty}{q_w}\right) - \frac{k_s}{h} \tag{8.100}$$

As expected, this steady-state position is larger at lower values of the ambient fluid temperature, T_∞, since more convective cooling occurs from the boundary of the ambient fluid.

8.4 Phase Change with Coupled Heat and Mass Transfer

The solubility limit in a multicomponent system refers to the maximum concentration of solute that can dissolve within the solvent. The solubilities of various components within the material change under phase transition between liquid and solid phases. For example, in a binary mixture, the concentration of solute (tin) in the solid remains less than the original mean solute composition in the liquid. As a result, the balance of solute from that original concentration remains in the liquid after phase transition. These combined heat and species concentration processes will be discussed in this section.

As discussed in Chapter 5, this transfer of solute at the phase interface creates a solute diffusion layer that reduces the equilibrium liquidus (freezing point) temperature. In multicomponent systems, convective flows can arise from thermal or solutal buoyancy, which enhances the species transport at the phase interface. Also, there is no fixed single freezing point. Solutal buoyancy is induced by variations of fluid density with concentration in the mixture. Thermal and solutal buoyancy can act independently. For example, buoyancy may be induced by an unstable concentration gradient, even in the presence of a stable thermal gradient.

During cooling below the eutectic temperature in a binary system, the simultaneous growth of two or more phases from the liquid phase is observed. In addition to phase change between liquid and solid phases, solid state transformations may occur. A particular solid phase refers to a specific type of lattice structure in the solid phase. In a diffusion-dependent solid transformation of a multicomponent system, a change in the number and composition of phases is encountered. The phase equilibrium diagram and material composition describe the type of the resulting solid phase. In a diffusionless transformation, a meta-stable (nonequilibrium) phase is formed, such as a martensite transformation in some steel alloys. For example, this type of transformation may occur under rapid solidification conditions when a large number of atoms experience a cooperative movement in the solid phase of the lattice structure.

Eutectics are composed of more than one solid phase, and these phases can exhibit a wide variety of geometrical arrangements. Consider an example involving two particular solid phases, called α and β phases. During phase transition, separate solute diffusion layers are formed ahead of the solid (α and β) phases. Due to these resulting differences in solute concentrations, a lateral concentration gradient leads to mass transfer laterally. Thus, a *diffusion coupling* is created, which refers to the simultaneous solute rejection from the solid phase at the advancing interface, together with lateral diffusion between regions ahead of the α and β solid phases. The result of this diffusion coupling is that interfacial concentration differences are reduced due to the lateral diffusion.

The combined lateral diffusion of solute and axial diffusion in the direction of the advancing phase interface affects the temperature gradient in the liquid. Also, it affects λ, which refers to the spacing between a set of α

and β phases (representing the effective wavelength of the interface). A change in the wavelength affects the forces, such as the interphase tension, acting on the solidified phases. For example, consider mechanical equilibrium at a point joining the α (solid), β (solid), and liquid phases with angles of θ_1 and θ_2 formed at the interface (see Figure 8.7).

Then mechanical equilibrium at this point suggests that

$$\gamma_{\alpha\beta} = \gamma_{\alpha 1} cos(\theta_1) + \gamma_{\alpha 2} cos(\theta_2) \tag{8.101}$$

$$\gamma_{\alpha 1} sin(\theta_1) = \gamma_{\alpha 2} sin(\theta_2) \tag{8.102}$$

where γ is the surface tension (units of N/m). This system of two equations can be solved simultaneously to find the two interphase angles, θ_1 and θ_2, respectively, and subsequently λ. A small value of λ leads to more lateral diffusion of solute ahead of the phase interface, which reduces the liquidus temperature, thereby increasing the interface velocity since the freezing temperature is lowered.

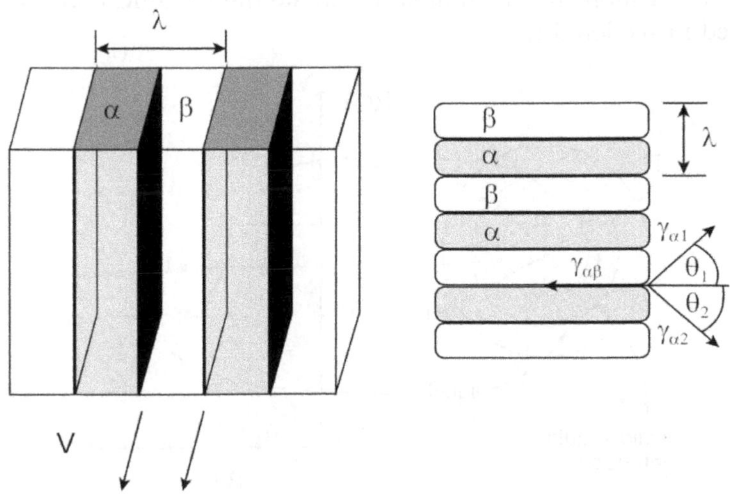

FIGURE 8.7
Equilibrium in diffusion-coupled solidification.

8.5 Problems in Other Geometries

The previous problems have been formulated in Cartesian coordinates, but many engineering problems are better suited to cylindrical or other coordinates. Solidification and melting problems in cylindrical systems occur in various types of applications. For example, thawing around a hot pipe buried in the frozen ground involves a cylindrical geometry. In this section, various cylindrical and spherical configurations will be considered, together with appropriate methods of analysis for these types of problems.

8.5.1 Cylindrical Outward Phase Change

One-dimensional solidification of an initially subcooled single-phase liquid in a semi-infinite cylindrical domain (see Figure 8.8) is formulated in terms of radial coordinates as follows:

$$\frac{1}{\alpha}\frac{\partial T}{\partial t} = \frac{\partial^2 T}{\partial r^2} + \frac{1}{r}\frac{\partial T}{\partial r} \tag{8.103}$$

where r refers to radial position. The liquid is initially at a temperature of T_i, which is below the phase change temperature, T_f. The onset of solidification occurs once liquid solidifies in contact with a nucleation site along the wall. The liquid can be initially below T_f without freezing since the wall nucleation site is required to initiate the solidification process (as discussed in section 5.2.3).

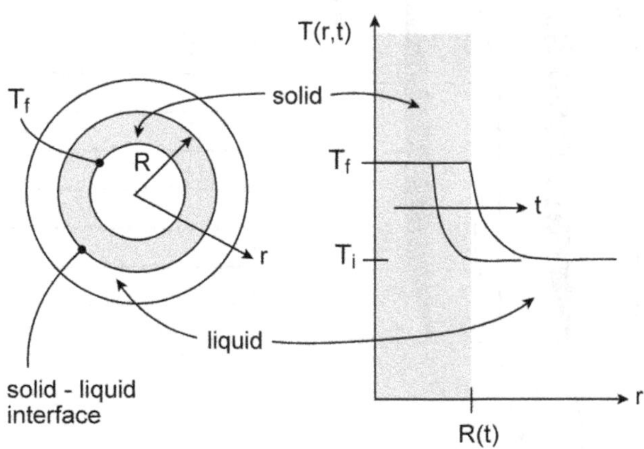

FIGURE 8.8
Schematic of outward cylindrical phase change.

The boundary, initial, and interfacial conditions are given as follows:

$$T(R,\ t) = T_f = T(r < R,\ t) \tag{8.104}$$

$$T(r,\ 0) = T_i \tag{8.105}$$

$$-k\left.\frac{\partial T}{\partial r}\right|_R = \rho L\frac{dR}{dt} \tag{8.106}$$

In Equation (8.104), it is assumed that the solidified region ($r < R$) remains at the fusion temperature. This situation would typically arise if the external boundary was heated and maintained at or near the fusion temperature. For example, although icing of overhead power transmission lines does not represent a semi-infinite domain, it nevertheless represents a case closely resembling Equation (8.104), when internal heat generation within the cable yields an ice temperature of nearly 0°C.

The following change of variables,

$$\zeta = \frac{r^2}{4\alpha t} \tag{8.107}$$

is used to transform Equation (8.103) into

$$\zeta\frac{d^2T}{d\zeta^2} + (1+\zeta)\frac{dT}{d\zeta} = 0 \tag{8.108}$$

Based on the transformed variable, ζ, the initial condition in Equation (8.105) becomes $T(\zeta \to \infty) = T_i$.

The solution of this problem was reported by Carslaw and Jaeger (1959) and Lunardini (1988). Solving Equation (8.108) by utilizing the transformed initial condition,

$$T = T_i - CE_i(\zeta) \tag{8.109}$$

where C refers to a constant of integration and the exponential integral, $E_i(\zeta)$, is defined by

$$E_i(\zeta) = \int_{\zeta}^{\infty}\frac{exp(-v)dv}{v} \tag{8.110}$$

Based on the anticipated rate of interface advance as obtained in Equation (8.13),

$$R = 2\beta\sqrt{\alpha t} \tag{8.111}$$

The remaining conditions in Equation (8.104), particularly at $r = 0$ and $r = R$, as well as Equation (8.106), are used to find the remaining constant in Equation (8.109).

Differentiating Equation (8.111) with respect to t (transforming to ζ), combining with the differentiated temperature from Equation (8.109), and substituting into Equation (8.106),

$$\beta^2 LE_i(-\beta^2)exp(\beta^2) + c_p(T_f - T_i) = 0 \tag{8.112}$$

Once β is obtained, the temperature distribution can be determined and rewritten in terms of the original coordinates. In this problem, the latent heat released by the solidifying liquid is transferred to the liquid at the phase interface, rather than into the solid, thereby raising the temperature of the subcooled liquid to T_f. This heat transfer into the subcooled liquid sustains the movement of the phase interface into the remainder of the subcooled liquid.

8.5.2 Constant Phase Change Rate

Without initial subcooling in the liquid, an analytical solution may also be obtained when the liquid temperature is assumed to be initially equal to the phase change temperature, T_f. Furthermore, it is assumed that the temperature gradient in the solid is approximately constant. The problem is solved based on the following governing equation:

$$\frac{1}{\alpha}\frac{\partial T}{\partial t} = \frac{1}{r}\frac{\partial}{\partial r}\left(r\frac{\partial T}{\partial r}\right) \tag{8.113}$$

This transient heat equation and its solution refer to the solid phase since $T = T_f$ in the liquid. The initial, boundary, and interfacial conditions are given as follows:

$$T(r > r_o,\ 0) = T_f \tag{8.114}$$

$$T(r_o,\ t) = T_w \tag{8.115}$$

$$T(R,\ t) = T_f \tag{8.116}$$

$$k\left.\frac{\partial T}{\partial r}\right|_R = \rho L\frac{dR}{dt} \tag{8.117}$$

Since the interfacial temperature gradient is assumed to be constant, integration of Equation (8.117) and substitution of $R = r_o$ at $t = 0$ give the following result:

$$R(t) = \left(\frac{k}{\rho L} \frac{\partial T}{\partial r} \bigg|_{R} \right) t + r_o \tag{8.118}$$

This result indicates that the interface advances linearly in time. If the pipe wall temperature, T_w, is assumed to vary with time, then a series solution for temperature can be obtained (Kreith and Romie, 1955).

A special case is obtained when a constant pipe wall temperature, T_w, is assumed in Equation (8.115). Using the heat balance integral approach, Equation (8.113) is integrated from the wall, $r = r_o$, to the phase interface, $r = R$. Performing this integration and rearranging based on Leibnitz's rule,

$$\frac{d}{dt} \int_{r_o}^{R} rTdr - RT_f \frac{dR}{dt} = \alpha \left(R \frac{\partial T}{\partial r} \bigg|_{R} - r_o \frac{\partial T}{\partial r} \bigg|_{r_o} \right) \tag{8.119}$$

Rather than a polynomial approximation for temperature, a logarithmic profile is assumed since the area in the path of heat transfer is increasing with r. A logarithmic profile that satisfies Equations (8.115) to (8.117) is

$$\frac{T - T_w}{T_f - T_w} = \frac{ln(r/r_o)}{ln(R/r_o)} \tag{8.120}$$

Substituting this temperature profile into Equation (8.119) and solving the resulting differential equation for R (Lunardini, 1988),

$$\frac{Ste}{4} \left(R^{*2} - 1 - \sum_{n=1}^{\infty} \frac{2^n (lnR^*)^n}{nn!} \right) + \frac{1}{2} R^{*2} lnR^* - \frac{1}{4} R^{*2} + \frac{1}{4} = t^* \tag{8.121}$$

where $Ste = c_p(T_f - T_w)$. Also, $R^* = R/r_o$ and $t^* = \alpha(Ste)t/r_o^2$ refer to the dimensionless interface position and time, respectively.

8.5.3 Effects of Superheating or Subcooling

During melting of a cylindrical region at an initial temperature of T_i, a thermal diffusion layer develops in the liquid region ahead of the phase interface. Heat is transferred by conduction from the wall (at a temperature of T_w) through the melted region to the phase interface (at a temperature of T_f). The extent of the diffusion layer is determined by the region between the phase interface and some location sufficiently far from the interface where the temperature disturbance of the advancing interface is not experienced. The temperature penetration distance, denoted by δ, is analogous to the momentum penetration distance utilized in the integral analysis of boundary layer problems (see Chapter 3).

Based on the quasi-stationary approximation, the one-dimensional governing equation and boundary conditions are given by

$$\frac{d}{dr}\left(r\frac{dT}{dr}\right)=0 \tag{8.122}$$

$$T_s=T_f=T_l; \qquad r=R \tag{8.123}$$

$$T_s=T_w; \qquad r=r_o \tag{8.124}$$

$$T_l=T_i; \qquad r=\delta \tag{8.125}$$

Equation (8.122) will be solved separately within each phase. The subscripts s, l, f, w, and i refer to solid, liquid, fusion (phase change temperature), wall, and initial, respectively.

Solving Equation (8.122) in the solid phase subject to the boundary and interfacial conditions,

$$T_s=T_w+\frac{(T_f-T_w)}{ln(R/r_o)}\,ln\left(\frac{r}{r_o}\right) \tag{8.126}$$

Similarly in the liquid phase,

$$T_l=T_i+\frac{T_f}{ln(R/\delta)}\,ln\left(\frac{r}{\delta}\right) \tag{8.127}$$

The heat added from the wall is the sensible heat to raise the temperature of the solid between the phase interface and the outer extent of the thermal penetration distance, as well as the latent heat to melt the material between the wall and the phase interface, i.e.,

$$Q_{tot}=\rho c_s\int_{r_o}^{R}(T_s-T_f)2\pi r dr+\rho c_l\int_{R}^{\delta}(T_l-T_f)2\pi r dr+\pi(R^2-r_o^2)$$
$$\times\left[L+\rho c_l(T_f-T_i)\right] \tag{8.128}$$

Also, the rate at which this heat is absorbed by the melting material balances the rate of heat addition from the wall. This overall energy balance can be expressed as follows:

$$\frac{dQ_{tot}}{dt}=-k_s(2\pi r_o)\left.\frac{dT_s}{dr}\right|_{r_o} \tag{8.129}$$

The differentiated temperature from Equation (8.126) is substituted into

Equation (8.129). Also, Q_{tot} is integrated in Equation (8.128) and then differentiated in Equation (8.129). These steps lead to the following solution for interface growth over time (Lunardini, 1988):

$$\left(\frac{4\alpha_s Ste}{r_o^2}\right)t-(Ste)\left[\left(\frac{R}{r_o}\right)^2+\gamma-1+ln\left(ln\left(\frac{R}{r_o}\right)\right)-2E_i ln\left(\frac{R}{r_o}\right)\right]$$

$$-M\left[2\left(\frac{R}{r_o}\right)^2 ln\left(\frac{R}{r_o}\right)+1-\left(\frac{R}{r_o}\right)^2\right]=0 \qquad (8.130)$$

where

$$M=1+Ste\frac{c_l}{c_s}\left(\frac{T_f-T_i}{T_w-T_f}\right)\left[\frac{\delta^2-R^2-2R^2 ln(\delta/R)}{2R^2 ln(\delta/R)}+1\right] \qquad (8.131)$$

The Stefan number is given by $Ste=c_s\ (T_f-T_w)/L$. Also, $\gamma=0.577215\ldots$ is the Euler constant, and E_i refers to the exponential integral defined by Equation (8.110).

Once the interface position, R, is obtained from Equation (8.131), the resulting temperature distributions can be determined from Equations (8.126) and (8.127). The current solution was based on the quasi-steady approximation in Equation (8.122), thereby limiting the solution applicability to problems involving small Stefan numbers. Further accuracy at higher Stefan numbers can be obtained through a heat balance integral solution, but typically at the expense of considerably more labor and complexity in the solution analysis. In addition to phase change outside of a cylindrically heated (or cooled) wall, many problems involve heat transfer inside a cylindrical domain (i.e., freezing of water in pipelines). Inward phase change analysis is described in the following section.

8.5.4 Cylindrical Inward Phase Change

For inward solidification of a liquid initially at a temperature of T_f, the governing equations are given by Equations (8.113), (8.114), (8.116), and (8.117). For a constant rate of interface growth in the negative radial direction (inward), the temperature gradient in the corresponding equation, Equation (8.117), becomes a negative constant, rather than a positive constant (arising in the case of outward phase change). Furthermore, for inward phase change, the radial position of the interface, $r=R$, decreases with time (demonstrated in the following example).

Example 8.5.4.1

Solidification Time for Inward Phase Change.

Estimate the time required for complete solidification of a liquid initially at T_f that occupies a tube of radius r_o (see Figure 8.9). A constant rate of phase change may be assumed during the solidification process.

In a manner similar to that in the integration of Equation (8.117) leading to the result in Equation (8.132),

$$R = r_o - \left(\frac{k}{\rho L} \frac{\partial T}{\partial r}\bigg|_R \right) t \tag{8.132}$$

This result gives the growth of the phase interface (at $r = R$) over time, where the temperature gradient at the interface is assumed to be a known (or specified) constant.

The time required to achieve complete solidification can be readily estimated by setting the right side of Equation (8.132) to zero (i.e., interface reaches the center of pipe) and solving for the resulting solidification time in terms of the temperature gradient and thermophysical properties. The following result is obtained:

$$t_f = \frac{\rho L r_o}{k(dT/dr)} \tag{8.133}$$

As expected, the time required for complete solidification is higher with materials having a larger latent heat of fusion or a lower thermal conductivity.

For problems involving small Stefan numbers, the heat conduction equation in the solid can be solved effectively with the quasi-stationary approximation. Neglecting the transient term in Equation (8.113) and

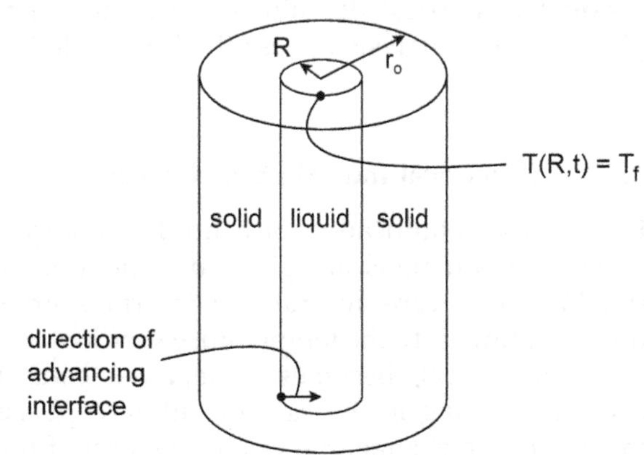

FIGURE 8.9
Schematic of inward cylindrical phase change.

solving the differential equation subject to the boundary conditions, Equations (8.115) and (8.116), yields

$$T_s = T_w + \frac{(T_f - T_w)}{ln(R/r_o)} ln\left(\frac{r}{r_o}\right) \tag{8.134}$$

Substituting this differentiated temperature profile into Equation (8.132) gives an explicit expression for the interface position in terms of the thermophysical properties. Then the time required for complete solidification can be computed in a manner similar to that described for Equation (8.133). The expression for the interface position is equated with $R = r_o$ to find the time required for the interface to reach the center of the cylindrical domain, yielding

$$t_f = \frac{\rho L r_o^2}{4k(T_f - T_w)} \tag{8.135}$$

which is the required time of solidification.

It can be observed that this result is similar to Equation (8.133), but a specific expression for the temperature gradient has been included.

8.5.5 Problems in Spherical Geometries

Another important geometrical configuration is a spherical system. For one-dimensional heat transfer in spherical (radial) coordinates,

$$\frac{1}{\alpha}\frac{\partial T}{\partial t} = \frac{1}{r^2}\frac{\partial}{\partial r}\left(r^2 \frac{\partial T}{\partial r}\right) \tag{8.136}$$

As outlined in Equation (8.107), a change of variables is useful, i.e.,

$$\zeta = \frac{r}{2\sqrt{\alpha t}} \tag{8.137}$$

Based on this change of variables, Equation (8.136) becomes

$$\frac{d^2 T}{d\zeta^2} + 2\left(\zeta + \frac{1}{\zeta}\right)\frac{dT}{d\zeta} = 0 \tag{8.138}$$

The solution of Equation (8.138) can be written as

$$T = A + BF(\zeta) \tag{8.139}$$

The values of A and B are constants of integration, and $F(\zeta)$ is the spherical function defined by

$$F(\zeta) = \frac{1}{2\zeta}\, exp(-\zeta^2) - \frac{\sqrt{\pi}}{2}(1 - erf(\zeta)) \tag{8.140}$$

Specific solutions of Equation (8.139) can be obtained once appropriate boundary conditions are applied. The following two cases will be examined in this section: (i) freezing of a subcooled liquid and (ii) freezing at a constant phase change rate.

8.5.5.1 Solidification of a Subcooled Liquid

Freezing of a subcooled liquid (initially at a temperature of T_i; see Figure 8.10) in a semi-infinite domain is determined by the following boundary, initial, and interfacial conditions:

$$T(R,\ t) = T_f \tag{8.141}$$

$$T(r,\ 0) = T_i \tag{8.142}$$

$$-k\left.\frac{\partial T}{\partial r}\right|_R = \rho L \frac{dR}{dt} \tag{8.143}$$

It is assumed that the temperature in the solid phase remains at T_f (phase change temperature). The thermophysical properties in Equation (8.143) are evaluated in the liquid phase.

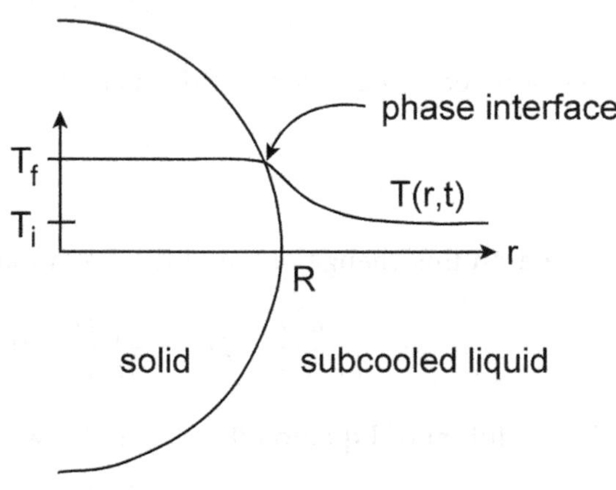

FIGURE 8.10
Schematic of spherical phase change.

Based on similar arguments leading to Equation (8.13),

$$R = 2\beta\sqrt{\alpha t} \tag{8.144}$$

Then Equations (8.141) and (8.142) can be used to find the constants of integration in Equation (8.139), thereby yielding the temperature profile. Furthermore, an expression for the coefficient β in Equation (8.144) is obtained after substituting the differentiated interface position, Equation (8.144), and the differentiated temperature into Equation (8.143). The following results (Lunardini, 1988) are obtained:

$$T = T_i + \frac{F(\zeta)}{F(\beta)}(T_f - T_i) \tag{8.145}$$

where

$$\beta^2 exp(\beta^2)[exp(-\beta^2) - \sqrt{\pi}\beta + \sqrt{\pi}\beta erf(\beta)] = \frac{1}{2}\ Ste \tag{8.146}$$

and $Ste = c_s\ (T_f - T_i)/L$ is the Stefan number.

8.5.5.2 Constant Rate of Phase Change

In the previous case, a negative temperature gradient in the liquid was obtained with a constant temperature of T_f in the solid. However, in this second case, the liquid is initially at a temperature of T_f (no subcooling), and a positive temperature gradient arises in the solid since heat is extracted through the solid to the wall. As a result, the negative sign in Equation (8.143) becomes positive, and thermophysical properties are evaluated in the solid (rather than the liquid).

When a constant temperature gradient at the phase interface is assumed, the interfacial condition in Equation (8.143) can be integrated directly, i.e.,

$$R(t) = \left(\frac{k}{\rho L}\left.\frac{\partial T}{\partial r}\right|_R\right)t + r_o \tag{8.147}$$

Thus, rather than increasing with the square root of time in Equation (8.144), the phase interface grows with inverse proportionality to the specified (constant) temperature gradient in this case.

These solution methods for planar, cylindrical, and spherical geometries have been based on one-dimensional heat transfer. However, practical applications often involve multidimensional variations of temperature. Phase change heat transfer in multidimensional problems will be examined in the following section.

8.6 Multidimensional Solidification and Melting

Multidimensional solidification and melting arise in many engineering and scientific problems, such as materials processing (i.e., casting solidification, extrusion, plastic injection molding), icing of aircraft, thermal energy storage, and other applications. For example, the production of advanced materials of high purity involves zone refinement with phase change (Jacobs et al., 1993). Semiconductors, dental materials, optical systems, and jewels are examples of materials with high purity requirements. Many materials have only recently become pure enough for true thermophysical property evaluation. These applications and others often require a detailed understanding of multidimensional solidification and melting.

8.6.1 Physical Processes and Governing Equations

In solidification of multicomponent mixtures, phase transition occurs over a range of temperatures (see Figure 1.6), whereas it occurs at a single temperature in pure materials. Consider phase change of a binary mixture. An initially liquid alloy begins to solidify at the liquidus temperature with the formation of a small amount of solid at a composition in accordance with the phase equilibrium diagram. The solid concentration remains less than the original mean solute composition, and the balance of the solute from that original concentration is transferred to the liquid. This solute rejection at the phase interface arises due to differences in solubilities of constituents within each phase. The solubility limit of a constituent refers to the maximum concentration of the constituent that can dissolve in solvent to form the solid solution.

In equilibrium phase transition, solidification is slow enough to allow extensive solid mass diffusion, so that the solid and liquid are homogeneous with compositions in accordance with the phase equilibrium diagram. However, during rapid phase change, the mean solute composition in the layer of solid near the phase interface is usually different than the corresponding solidus value. In this case, the reduced mass diffusion suggests that the successive layers of solid retain their original compositions. This state can be considered to be a nonequilibrium state in the sense that the mean composition of the solid remains less than the solid composition at the phase interface (i.e., composition associated with the phase equilibrium diagram).

In addition to thermal and mass diffusion in multidimensional problems, advection processes arise under the influence of pressure, buoyancy, and other forces. Convective transport due to local pressure differences in the flow field affects the phase transition temperatures and the interface characteristics. If the fluid ahead of the phase interface is in a supercooled state (i.e., temperature less than the equilibrium freezing temperature), then dendritic structures could appear. Also, if the binary mixture has a lower density in the liquid phase, it will contract on solidification. In this case, mass flow toward the interface accompanies the change of volume due to volumetric shrinkage. In addition to associated pressure gradients driving the shrinkage flow, a free liquid–gas surface may form within the enclosure, thereby altering the heat transfer from the external boundaries to the phase change material.

8.6.1.1 Mass, Momentum, and Energy Equations

The governing equations for solid–liquid phase transition include the Navier–Stokes equations in conjunction with the binary alloy phase diagram (i.e., Figure 1.6) and an equation of state. The mixture conservation equation for a scalar quantity in phase k (i.e., $k = 1$, solid; $k = 2$, liquid) was outlined in Chapter 5. In order to derive the mixture continuum equations, the individual phase equations are summed over all phases within a control volume. Then the variables are rewritten in terms of the resulting mixture variables. A mixture quantity, B, is defined here as the mass fraction-weighted sum of individual phase components of B, i.e.,

$$B = \sum_{k=1}^{2} f_k B_k \tag{8.148}$$

For example, $\mathbf{v} = f_l \mathbf{v}_l + f_s \mathbf{v}_s$ is the mixture velocity and $\rho = f_l \rho_l + f_s \rho_s$ is the mixture density.

In the following analysis of binary systems, phase transition is considered only above the eutectic temperature. Also, solid and liquid densities differ, but density variations with concentration or temperature in a particular phase are neglected. Furthermore, a continuous liquid–solid mixture is considered without internal gas voids, and the solid phase will be assumed to remain stationary during phase transition. Viscous dissipation in the thermal equation is assumed to be negligible. As a result, the following mixture equations for mass, momentum, and energy are obtained:

$$\frac{\partial \rho}{\partial t} + \nabla \cdot (\rho \mathbf{v}) = 0 \tag{8.149}$$

$$\frac{\partial (\rho \mathbf{v})}{\partial t} + \nabla \cdot (\rho v \mathbf{v}) = -\nabla p + \nabla \cdot \left(\frac{\rho_l}{\rho} \mu_l \nabla \mathbf{v} \right) + \rho \mathbf{g} \beta_T (T - T_0) + \rho \mathbf{g} \beta_C (C - C_0) + \mathbf{G} \tag{8.150}$$

$$\frac{(\partial \rho h)}{\partial t} + \nabla \cdot (\rho \mathbf{v} h) = \nabla \cdot (k \nabla T) + \nabla \cdot \rho (h - h_l) \mathbf{v} \tag{8.151}$$

where h and h_k refer to the mixture and phase enthalpy, respectively, and

$$h_k = h_{r,k} + \int_{T_0}^{T} c_{r,k}(\zeta) d\zeta \tag{8.152}$$

Also, k and ρ refer to the mixture conductivity and density, respectively, i.e., $k = f_l k_l + f_s k_s$. In Equation (8.150), **g**, **G**, β_T, and β_C represent the gravity vector, cross-phase interactions, and thermal and solutal expansion coefficients, respectively. In Equation (8.152), $c_{r,k}(T)$ refers to the reference specific heat of phase k. The subscript r refers to reference value.

8.6.1.2 Species Concentration Equations

For multicomponent mixtures, an additional species conservation equation is required. Diffusive mass transfer occurs in an manner analogous to that of the conduction of heat. Mass transfer by microscopic motion occurs when atoms migrate from lattice site to lattice site. *Fick's law* describes the process of species mass diffusion in the presence of a concentration gradient.

Fick's law can be understood by considering diffusion of interstitial atoms (component B) through a solvent (component A) across an imaginary plane over a microscopic distance of Δx. Let Γ_B, n, n_1, and n_2 refer to the average number of interstitial atom jumps per second in random directions, number of adjacent sites (i.e., $n = 6$ in three dimensions), number of B atoms per unit area in plane 1 (left side of the imaginary plane), and the number of B atoms per unit area in plane 2 (right side of the imaginary plane), respectively.

Then the net flux of interstitial atoms across the imaginary control surface is

$$J_B = J_{B,1\to 2} - J_{B,2\to 1} \tag{8.153}$$

or alternatively,

$$J_B = \frac{1}{n}\Gamma_B n_1 - \frac{1}{n}\Gamma_B n_2 \tag{8.154}$$

A positive expression for J_B suggests that there is a net species mass flux from the left side to the right side of the control surface.

The concentration of B atoms at plane 1, $C_{B,1}$, is

$$C_{B,1} = \frac{n_1}{\Delta x} \tag{8.155}$$

and similarly for $C_{B,2}$. Also, performing a Taylor series expansion,

$$C_{B,2} = C_{B,1} + \Delta x \frac{\partial C_B}{\partial x} \tag{8.156}$$

where higher order terms have been neglected.

Combining Equations (8.154) and (8.155),

$$J_B = -\left(\frac{\Gamma_B}{6}\Delta x^2\right)\frac{\partial C_B}{\partial x} \tag{8.157}$$

where the coefficient in parentheses is called the diffusion coefficient (denoted by D_{AB}) of constituents A and B. Equation (8.157) is called Fick's law and can be generalized to multidimensional diffusion as follows:

$$\mathbf{J} = -D_{AB}\nabla C \tag{8.158}$$

Fick's law describes the diffusion of mass due to a species concentration gradient in a way analogous to that of Fourier's law, which describes the diffusion of heat due to a temperature gradient. Typical values of diffusion coefficients are listed in the appendix.

Including both mass diffusion and convection, the mixture conservation equation for solute concentration becomes

$$\frac{\partial(\rho C)}{\partial t} + \nabla\cdot(\rho \mathbf{v} C) = \nabla^2(\rho_l f_l D_l C_l + \rho_s f_s D_s C_s) + \nabla\cdot\rho(C - C_l)\mathbf{v} \tag{8.159}$$

The problem of solid–liquid phase change for multicomponent systems is now governed by Equations (8.149) to (8.151) and (8.159), subject to the equation of state, Equation (8.152), and appropriate boundary and initial conditions.

8.6.2 Supplementary Relations

In the previous governing equations, it can be observed that additional relations are required for closure of the overall formulation. In particular, supplementary relations are required for the liquid fraction (in terms of the other solution variables) and momentum phase interactions (**G**). Furthermore, the Second Law of Thermodynamics provides an important complement to the analysis and understanding of these phase change processes (Naterer, 2000, 2001).

8.6.2.1 Interphase Liquid Fraction

In binary systems, the liquid fraction can be determined based on the solute concentrations. The sum of the phase fractions is unity, so that

$$f_s + f_l = 1 \tag{8.160}$$

and the mean concentration, C, is given by

$$C = f_s C_{sol} + f_l C_{liq} \tag{8.161}$$

The previous equations can be combined to give the following *lever rule*:

$$f_l = \frac{C - C_{sol}}{C_{liq} - C_{sol}} \tag{8.162}$$

Equation (8.162) represents an equilibrium relationship. However, nonequilibrium conditions can arise if the cooling rate is too rapid to permit sufficient mass diffusion in the solid phase, thereby meaning that the average solid composition remains less than the equilibrium solidus concentration (see Figure 1.6). This difference occurs since the separate solid layers retain their original interface compositions. Nonequilibrium conditions can be accommodated by time-dependent diffusion in the solid phase (Naterer and Schneider, 1995).

8.6.2.2 Momentum Phase Interactions

Another supplementary relation required for closure of the overall formulation involves the cross-phase interactions, **G**, in the mixture momentum equation, Equation (8.150). For solid–liquid phase change, this cross-phase transport is analogous to Darcy's law for flow through a porous medium. In that case, $\mathbf{G}K = \nu(\mathbf{v}_l - \mathbf{v}_s)$ can be adopted for the dependence of **G** on the medium permeability, K, and liquid fraction, f_l. Furthermore, in solid–liquid systems, the following *Blake–Kozeny equation* is often adopted for the solid permeability:

$$K = K_0 \left[\frac{f_l^3}{(1 - f_l)^2} \right] \tag{8.163}$$

This model is based on a physical analogy between interdendritic flow and Hagen–Poiseuille viscous flow through a noncircular tube with an equivalent hydraulic radius based on the local liquid fraction (see Figure 8.11).

The Blake–Kozeny model is based on an analogy to axial fluid motion within a duct, whereas interdendritic crossflow *normal* to a dendrite bank may produce higher pressure changes due to viscous interactions with the dendritic material. As a result, Naterer and Scheider (1995) incorporate both axial and cross-flow effects in the interdendritic modeling (see Figure 8.11). Since the primary dendrites grow in a direction normal to the local isotherm, the selection of an axial or cross-flow model is based on the interdendritic velocity field relative to the local temperature gradient.

Rather than Equation (8.163), the following result is obtained for analogies with cross-flows normal to protruding dendrites (Naterer and Schneider, 1995):

$$K = K_0\left(\frac{f_l}{\sqrt{1-f_l}}\right) \tag{8.164}$$

From this result, it can be observed that as $f_l \rightarrow 0$, then $K \rightarrow 0$ (i.e., solid has zero permeability). Since the effective viscosity becomes large and $\mathbf{G} \rightarrow \infty$, the fluid velocity is essentially damped to zero. Also, as $f_l \rightarrow 1$, then $K \rightarrow \infty$ and $\mathbf{G} \rightarrow 0$. In this case, the momentum equations have zero phase interaction forces and the single phase flow equations are returned.

A suitable combination of the previously described parallel and cross-flow-based permeabilities can provide an effective physically based approach. However, a weighting factor between both models is still required. When the primary dendrite arms grow in the direction of the local temperature gradient, a flow alignment weighting factor (called χ) is used to represent a scale factor between the axial and cross-flow directions, i.e.,

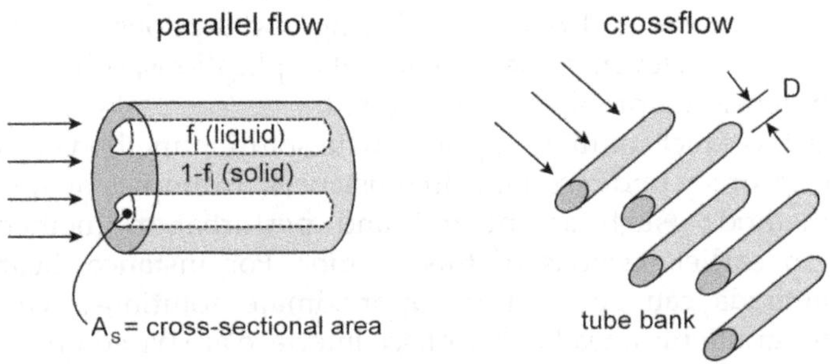

FIGURE 8.11
Parallel flow and cross-flow in the interdendritic region.

$$\chi = \frac{|\mathbf{v} \cdot \nabla T|}{|\mathbf{v}|\,|\nabla T|} \tag{8.165}$$

Then, the following *isotherm gradient model* incorporates a suitable weighting between the two relevant nonlinear functions of liquid fraction:

$$K = K_0 \chi F_A(f_l) + K_0(1-\chi)F_C(f_l) \tag{8.166}$$

where

$$F_A = \frac{f_l^3}{(1-f_l)^2}; \quad F_C = \frac{f_l}{\sqrt{1-f_l}} \tag{8.167}$$

In these correlations, F_A and F_C represent axial and cross-flow permeabilities, respectively. For example, $\chi = 0$ corresponds to a cross-flow condition and $\chi = 1$ represents an axial flow. In this way, the appropriate permeability factor is used in each limiting case. This cross-phase permeability model completes the closure of the momentum equations.

8.6.3 Solution Methods

In the analysis of solid–liquid phase change problems, a primary difficulty is predicting the unknown movement and position of the phase interface. In direct problems, the boundary conditions are given and the spatial variations of problem variables within the problem domain need to be determined. On the other hand, thermal control in many solidification and melting applications often requires inverse solutions, whereby the desired output is the unknown controlling boundary temperature (Xu and Naterer, 2001).

Phase change heat transfer, species concentration redistribution, and convective transport introduce strong nonlinearities in the governing equations. As a result, exact solutions are generally limited to one-dimensional domains and other simplifying assumptions, as observed in earlier sections. However, in many practical applications, solution methods for multidimensional problems are required.

In contrast to exact solutions, approximate solution methods can provide reasonable accuracy under certain circumstances. Examples of approximate solution methods (such as integral and perturbation methods) were discussed in earlier sections of this chapter. For instance, heat balance integral methods can be used for approximate solutions, whereby the temperature gradients ahead of the phase interface are based on polynomial interpolation. The heat equation is solved in the solid phase, and the region ahead of the liquidus interface is handled separately and later coupled to the position of the solidus interface. For conduction-dominated problems in a one-dimensional semi-infinite domain, the position of the phase interface and the temperature distribution(s) can be obtained in this manner. Furthermore, similarity solutions may be available for certain types of problems. Similarity transformations involving the solidus and liquidus positions could be used so that the boundary conditions yield algebraic relations for the positions of the solidus and liquidus interfaces.

Nevertheless, semianalytic solutions have inherent geometrical limitations, and thus more complicated configurations usually require fully discrete methods. Numerical models of solid–liquid phase change can be broadly categorized into techniques involving fixed-domain methods or interface tracking. In the former approach, the interface location is not tracked explicitly, but instead it is determined by the application of appropriate conservation principles. The latent heat of fusion is apportioned between the nearest nodal points in a chosen mesh. In these methods, the location of the interface can only be resolved with confidence to within one mesh spacing. In the latter approach (interface tracking), iterative procedures with interface constraints are required to track the unknown position of the phase interface. This approach often involves moving grids or coordinate transformations, which add to the complexity and execution time of the simulations. Based on these factors, as well as their limited capability of handling multiple interfaces, the former approach (fixed-domain methods) is often used for the discrete analysis of solid–liquid phase change.

A further classification of these numerical models includes the following two categories: source-based models and apparent heat capacity models. In source-based models, the enthalpy (or energy) discontinuity when latent heat is absorbed or released at the phase interface appears as a lagged source term in the discrete equations. After substituting an enthalpy equation of state into the transient term of the heat equation, the latent heat portion is transferred to the source term. Further interequation iterations are generally required to determine the magnitude of this source term and its liquid fraction dependence on temperature and composition. In apparent heat capacity models, the phase change and enthalpy phase discontinuity appear through the active transient term in the discrete equations. The apparent (or effective) heat capacity includes both sensible and latent heat components for temperature and phase fraction changes, respectively. In this approach, the numerical prediction of the unknown phase heat capacity requires phase iterations, specific heat adjustments, or lagged liquid fraction modifications for solution convergence. Both types of methods have been applied successfully to a wide range of multidimensional solid–liquid problems.

In numerical modeling of solid–liquid phase change, finite volume methods invoke discrete conservation balances of mass, momentum, species, and energy. In this way, a physical basis for analyzing the diffusive and convective flows is retained in the computer algorithm. In segregated solutions involving fluid flow, the mass and momentum equations are solved separately and interequation iterations are used to find the pressure field. Alternatively, an implicit and simultaneous closure of the conservation equations can be obtained through integration point mass and momentum equations providing a simultaneous velocity–pressure coupling (Naterer and Schneider, 1995). Simultaneous solutions allow the pressure field to appear directly in the continuity equation, so that a decoupling between problem variables will not occur. Further descriptions of these numerical methods are given in Chapter 11.

In the previous sections, direct problems involving solidification and melting have been examined. In these cases, heat transfer occurs largely through phase change when latent heat is absorbed or released at the phase interface. However, in many other applications, heat transfer and fluid flow arise in solid–liquid flows when phase change is not a predominant aspect of the problem. An example is liquid flow carrying solid particles in a pipe. The analysis of these types of flows is described in the following section.

8.7 Dynamics of Liquid–Solid Flows

In Newtonian flows of a solid–liquid mixture, shear stresses in the liquid are approximately linearly related to strain rates (or velocity gradients) in the fluid. In this section, it will be assumed that flows involving common liquids, such as water and liquid metals, are considered to be Newtonian flows. On the other hand, liquid polymers represent an example of non-Newtonian fluids (see Figure 3.3).

8.7.1 Flow Regimes

Flow patterns characterize the types of flow regimes in liquid–solid flows. For example, consider fluid transport with particle settling in a pipe. The degree of solid particle mixing, segregation, and deposition depends on a variety of factors, including fluid properties, operation parameters, flow direction, and solid properties. The fluid properties (particularly viscosity) affect the interfacial momentum exchange and forces on the particles, thereby influencing their motion relative to the liquid phase. Operation parameters, such as fluid velocity, lead to laminar or turbulent fluid motion. In this example, the pipe diameter largely influences the structure of the fluid boundary layer. Furthermore, solid properties, such as the particle shape and size, influence their momentum and heat exchange with the surrounding fluid. For an example involving crystal formation in casting solidification, the crystal shape often affects buoyant transport during crystal sedimentation processes. These processes are given as examples of factors contributing to the overall flow behavior and patterns in liquid–solid flows.

In vertical flows of liquid–solid mixtures, there is a low tendency for segregation of solid particles as a result of the magnitude and direction of forces acting on the solid particles. In general, a fairly uniform solid distribution is often observed in vertical transport of solid–liquid flows in the presence of gravity. However, in the case of horizontal transport, such as the transport of solid particles by a liquid flow in a pipe, asymmetrical forces arise. Segregation and collection of particles are often observed along the bottom section of a pipe. As a result, a solid concentration profile (describing the spatial distribution of particle density or concentration) becomes an important design consideration in horizontal slurry flows. These flows are called *settling slurries*.

In Newtonian slurries, *colloidal dispersion* refers to the transport of very fine particles (i.e., $Re_s < 10^{-6}$, where Re_s refers to the particle Reynolds number). These particles are maintained in suspension in the liquid flow. The *homogeneous flow regime* is observed for larger particles ($10^{-6} < Re_s < 0.1$). Low levels of turbulence are required to maintain the particles in a homogeneously suspended state. A *pseudohomogeneous flow regime* is obtained when the particle sizes increase to the range of $0.1 < Re_s < 2$. In the case of horizontal flow, a certain degree of segregation is obtained. Solid particles collect in a denser region analogous to settling slurries. In other configurations, sufficient flow rates and turbulence levels are adequate to maintain the particles in suspension. Finally, for larger particles with $Re_s > 2$, a *heterogeneous flow regime* is observed with a higher degree of segregation of solid particles throughout the mixture. At the *critical deposition velocity*, the solid particles begin to settle out of the mixture. The pressure drop through the mixture is reduced as a result of less drag and resistance of individual solid particles suspended in the mixture flow.

Other detailed flow characteristics, such as the boundary layer structure, are often summarized in the form of phase diagrams (Hewitt et al., 1997). These diagrams incorporate factors such as pipe diameter, friction velocity, particle size, and other factors. For example, a phase diagram for horizontal liquid–solid transport would typically illustrate the particle settling velocity, w_{so}, in terms of a friction velocity, u_o. The laminar sublayer near the wall of the pipe corresponds to a low value of u_o, and higher values occur toward the inner turbulent core of the pipe flow. At high u_o and low w_{so}, homogeneous flow arises, whereas transition to longitudinal standing particle waves and transverse waves (consisting of a grouping of solid particles) arises when u_o decreases and w_{so} increases. Also, limits involving the minimum transport for homogeneous flow with particles in suspension are often indicated in these graphs.

In the case of non-Newtonian slurries, the settling of particles is slow in comparison to slurries of Newtonian mixtures. Also, turbulence is not generally required to prevent settling out of particles in the flow. As a result, the particles generally remain suspended during transport over short distances. However, transport over longer distances may require turbulent conditions to prevent settling out of the particles.

8.7.2 Liquid–Solid Flows in Pipes

In vertical flows of liquid–solid mixtures, the solid particles can be transported upward when the fluid velocity, u_l, exceeds the terminal settling velocity, w_s. This terminal settling velocity depends on various factors, including solid concentration by volume, C, density differences, inertial effects and, grain size, d_s (Maude and Whitmore, 1958). It can be approximated in the following manner:

$$w_s = w_{so}\left(1 - \frac{V_s}{V_s + V_l}\right)^C \tag{8.168}$$

$$w_{so} = \sqrt{\frac{4d_s}{3c_d}\frac{(\rho_s - \rho_l)}{\rho_l}g} \tag{8.169}$$

where the latter fraction within parentheses in Equation (8.168) refers to the local solid concentration by volume. Also, the drag coefficient, c_d, can be correlated in terms of the Reynolds number as follows:

$$c_d = \frac{4}{\sqrt{Re_s}} + \frac{24}{Re_s} + 0.4 \tag{8.170}$$

where

$$Re_s = \frac{w_{so}d_s}{\nu_l} \tag{8.171}$$

In addition to empirical correlations involving c_d, the previous results can be derived based on discrete approximations of settling and inertial effects of the particles during transport by the liquid–solid mixture.

In the case of horizontal flows in pipes, the solid particles are transported horizontally without deposition when the fluid velocity, u_l, exceeds the critical deposition velocity, u_c, where

$$u_c = 3.525\left(\frac{\dot{V}_s}{\dot{V}_s + \dot{V}_l}\right)^{0.234}\left(\frac{d_s}{D}\right)^{1/6}\left[2Dg\left(\frac{\rho_s - \rho_l}{\rho_l}\right)\right]^{1/2} \tag{8.172}$$

based on experimental results of Wasp et al. (1977). In this equation, D refers to the pipe diameter. Also, the first fraction in parentheses (exponentiated by a factor of 0.234) denotes the delivered or transport concentration by volume.

Certain non-Newtonian and nonsettling slurries are classified as *Binghamian slurries*, based on their nonlinear dependence of shear stress on strain rates (velocity gradients, i.e., dv/dy) in the flow field (see Figure 3.3). In particular, these slurries obey the following type of constitutive relation:

$$\tau = \tau_o + \mu_b \frac{du}{dy} \tag{8.173}$$

where μ_b and τ_o refer to the Bingham viscosity and the slurry yield stress, respectively. A difficulty with applying this relation in the two-phase equations (in comparison to Newtonian, linear constitutive relations) is the resulting additional nonlinearities in the governing equations. As a result, some type of linearization is generally required for computational formulations involving non-Newtonian flows.

Economical transport of Binghamian slurries over short piping distances can be achieved with laminar flow. A sufficient flow rate and mixture velocity are required to prevent settling out of solids in the mixture. Additional frictional resistance with fluid turbulence means that more pumping power is typically required in the transport of turbulent liquid–solid flows. Similar to the transition to turbulence in single phase flows, the following Durand–Condolios correlation represents the critical Reynolds number, Re_{crit}, below which the flow remains laminar (Hewitt et al., 1997):

$$Re_{crit} = \frac{\rho_m u_{m,c} D}{\mu_B} = 1000\left[1 + \left(1 + \frac{\tau_o D^2 \rho_m}{3000\mu_B^2}\right)^{1/2}\right] \tag{8.174}$$

where $u_{m,c}$ refers to the critical (transition) velocity and D is the pipe diameter. The subscripts B and m refer to Binghamian and mixture, respectively.

At some particle size, the tendency of settling exceeds the tendency of the particle to remain suspended in the flow. If the particle's weight exceeds its net upward force arising from dynamic effects in the flow, settling out of solid particles will occur. Particles with a sufficiently small diameter (less than $d_{s,c}$) will retain stable laminar flow (Dedegil, 1986), where

$$d_{s,c} \approx \frac{3\pi\tau_o}{2(\rho_s - \rho_l)g} \tag{8.175}$$

If particles exceed this diameter, settling of solid particles is anticipated to occur.

The pressure drop of laminar Binghamian slurry flows can be estimated based on the following approximation of the *Buckingham equation* (Wasp, 1977):

$$\frac{\Delta P}{\Delta L} = \frac{16}{3}\frac{\tau_o}{D} + \mu_B\left(\frac{32u_m}{D^2}\right) \tag{8.176}$$

This pressure drop is an important quantity in the design of pumping systems for slurries and liquid–solid flows. In addition to this pressure drop, settling of particles is an important concern for economical transport of liquid–solid flows because settling particles generally entail additional pumping power, flow blockage through restricted sections, and other difficulties. If particles exceed the previous critical diameter of $d_{s,c}$, then increasing the flow rate (i.e., velocity) through the pipe to turbulent flow conditions may prevent settling out of particles. However, the mixing due to turbulence is increased. As a first approximation, this additional effect of turbulence can be approximated through an increased (modified) Binghamian viscosity, while retaining the regular friction and drag coefficient curves in laminar flow.

8.8 Applications

8.8.1 Materials Processing

Heat transfer, solidification, and melting arise in many materials processing technologies. For example, metallic structures with the highest strength, hardness, and wear resistance are created under well-controlled solidification processes. The final properties of a material depend on the amount, size, shape, distribution, and orientation of the solidified phases, which are all predominantly controlled by heat transfer during material solidification. A *grain boundary* is the interface that separates two adjoining crystals having different crystallographic directions. Controlled solidification permits the formation of grain boundaries in specified directions that can better resist conditions of high material stress, thereby improving the strength of the final solidified material.

An important example of materials processing involves material purification and refining. Zone refining and melting processes were originally introduced by Pfann (1958). The essential feature of zone refinement is the movement of a molten zone along the length of an initially solid material. This process transfers material impurities to one end of the original material. Consider a metal within a cylindrical container that is heated by a ring heater. As the heater moves upward, a finite molten zone forms along the bar and moves upward with the heater, and the lower section resolidifies. Due to different constituent solubilities in the solid and liquid phases, solute is transferred into the molten region and the solvent concentration increases beyond its initial value in the solid. Due to the solute's different molecular structure in the solid phase, the solute is unable to align itself into the crystal structure at the interface. As a result, it remains in the liquid phase. The net effect is a solute transfer from the original end to the final end of the material. This zone refinement process can be repeated for several iterations to purify the material.

Although zone refinement can purify the solidified material, the process involves certain limitations. For example, thermal or solutal convection, stratification in the molten zone, or contamination from the container walls may lead to material impurities. Recent advances have overcome certain difficulties by zone refinement in microgravity conditions, such as float zone furnaces (Jacobs et al., 1993). The molten zone can be supported between the ends of the bar by surface tension and electromagnetic forces. The following example gives an analytical solution relevant to problems involving the purification of solidified materials.

Example 8.8.1.1

Phase Change with a Uniform Interface Velocity.

In a Bridgman crystal growth process, the vertical axis of a mold moves through a furnace at a uniform rate, which permits a stationary interface between the solid and liquid phases. This problem can be represented by a long rod moving at a steady velocity, subject to fixed Dirichlet boundary conditions. The purpose of this example is to find the temperature profile in the solid and liquid regions, including the effects of phase transition at the solid–liquid interface. One-dimensional conditions are assumed in the analysis of this problem.

For one-dimensional steady-state heat transfer,

$$\frac{d^2T}{dx^2} - \frac{u}{\alpha}\frac{dT}{dx} = 0 \tag{8.177}$$

where α and u refer to the thermal diffusivity and bar velocity, respectively. In the solid region, α represents the solid thermal diffusivity. In the liquid phase, α represents the liquid thermal diffusivity. The solution of this first-order differential equation subject to Dirichlet conditions, $T(x_1) = T_1$ and $T(x_2) = T_2$, may be expressed in the following form in the solid:

$$T(x) = \frac{(T_M - T_1)[1 - exp(u(x - x_1)/\alpha)]}{[1 - exp(u(x_M - x_1)/\alpha)]} + T_1 \tag{8.178}$$

where $x_1 \leq x \leq x_M$. The subscript M denotes the position of the phase interface. In the liquid,

$$T(x) = \frac{(T_2 - T_M)[1 - exp(u(x - x_M)/\alpha)]}{[1 - exp(u(x_2 - x_M)/\alpha)]} + T_1 \tag{8.179}$$

where $x_M < x \leq x_2$. The coordinate of x_M represents the (unknown) position of the phase interface.

In order to determine the interface location, the following heat balance at the phase interface is applied:

$$-k \left.\frac{dT}{dx}\right|_{x_M^+} + k \left.\frac{dT}{dx}\right|_{x_M^-} = \rho u L \tag{8.180}$$

where the notation of x_M^+ and x_M^- refers to the liquid and solid sides of the interface, respectively. Also, L is the latent heat of fusion. Differentiating the previous expressions involving temperatures in Equations (8.178) and (8.179) and substituting those results into Equation (8.180),

$$\frac{(T_2 - T_M)}{1 - exp(u(x_2 - x_M)/\alpha)} + \frac{(T_M - T_1)}{1 - exp(-u(x_M - x_1)/\alpha)} = \frac{L}{c_P} \tag{8.181}$$

Proceeding to solve this algebraic solution for the unknown interface position, x_M, leads to the following quadratic equation:

$$\left[\frac{T_2 - T_1}{T_M - T_1} - \frac{1}{Ste}\right]\chi^2 + \left[\frac{1}{Ste}(1 + exp(Pe)) - exp(Pe) - \frac{T_2 - T_1}{T_M - T_1} + 1\right]\chi - \frac{exp(Pe)}{Ste} = 0 \tag{8.182}$$

where

$$\chi = exp\left[\frac{u(x_M - x_1)}{\alpha}\right] \tag{8.183}$$

Also, $Pe = u(x_2 - x_1)/\alpha$ and $Ste = c_P(T_M - T_1)/L$ refer to the Peclet and

Stefan numbers, respectively. The physical root of Equation (8.182) lies in the interval of $0 \leq \chi \leq 1$.

Comparisons between this analysis and finite element simulations have been carried out in the technical literature (Pardo and Weckman, 1990; Naterer and Schneider, 1995). From those results and Equation (8.180), a sharp difference in the temperature gradients across the phase interface is observed, particularly for low values of the Stefan number (i.e., high latent heat of fusion). The position of the phase interface is indicated by the sharp change in curvature of the temperature profile. In addition, as *Pe* increases, the interface position, x_M, increases because the additional energy transport into the domain decreases the effective rate of heat removal from the other end of the problem domain.

If the material is a binary alloy, then solute transfer across the phase interface can be examined in a similar manner. Based on a one-dimensional balance of solute concentration in a control volume that moves at the interface velocity u,

$$j_x + uC + \frac{d}{dx}(uC)dx = j_x + \frac{d}{dx}(j_x)dx + uC \tag{8.184}$$

where the two sides of this equation represent the inflow and outflow, respectively. Also, higher order terms in the Taylor series expansion have been neglected.

Using Fick's law with a constant mass diffusivity, D, Equation (8.184) becomes

$$\frac{d^2C}{dx^2} + \frac{u}{D}\frac{dC}{dx} = 0 \tag{8.185}$$

The solid and initial compositions are denoted by C_0. The solution of Equation (8.185) for the liquid composition ahead of the phase interface (i.e., $\hat{x}$ measured outward from the interface) is given by

$$C_L = \frac{C_0}{k}\, exp\left(-\frac{u}{D}\,\hat{x}\right) \tag{8.186}$$

The value of k represents the solute–solvent partition coefficient of the phase diagram (i.e., ratio of the solidus to liquidus slopes in Figure 1.6). This result in Equation (8.186) indicates that the liquid concentration decreases exponentially from the phase interface. The characteristic decay distance is D/u. In other words, D/u represents the distance in which the liquid concentration falls to $1/e$ of its interface value.

In the previous example, a binary alloy was considered. As a result, only one species concentration equation was required since the second equation automatically requires that the sum of concentrations of both constituents be 100%. In general, for multicomponent systems involving N constituents, a total of $N-1$ equations would be required to find each constituent concentration. The concentration of the Nth constituent would be obtained as 100% minus the sum of the other constituent concentrations.

During solidification of metals and alloys, some contraction on phase change is usually encountered due to the differences in density between solid and liquid phases. For example, in casting solidification of metal alloys, risers are added to permit a volume change and prevent void formation in the solidified material. The riser is often surrounded with a layer of exothermic material that burns at the melting point of the metal so that it does not add or extract heat from the metal. In this case, a balance of pressure below the layer of solid riser with the external atmosphere is required to prevent riser punctures at its surface. If shrinkage is not properly compensated by feeding mechanisms or risers, macroscopic voids may be created inside the casting, thereby leading to diminished mechanical properties of the final solidified material.

The occurrence of shrinkage voids in a solidified material can be characterized by the magnitude of the following shrinkage factor proposed by Niyama (Sahm and Hansen, 1984):

$$f = \frac{\sqrt{\partial T/\partial t}}{\partial T/\partial x} \tag{8.187}$$

The derivatives in the numerator and denominator represent temporal and spatial derivatives of temperature, respectively. If the shrinkage factor becomes sufficiently large, then solidification shrinkage and voids are expected to occur. Metals and alloys usually contract upon solidification, and shrinkage voids are created when interdendritic channels become closed to liquid inflow at the phase interface. Shrinkage voids are usually undesirable since they often lead to defects in the material's properties.

A related material defect occurs in *hot cracking* when interdendritic fluid forces pull newly solidified dendrites apart. A proper feeding mechanism is required to properly accommodate the solidification shrinkage. If the metal casting is not fed properly, liquid cannot flow between the separating dendrites to heal the tears. A criterion that can be used to predict hot cracking, proposed by Hansen (Sahm and Hansen, 1984), is based on

$$\phi = \frac{1}{V_{h,s}} \left(\frac{d(\Delta \rho)}{dt} \right) \tag{8.188}$$

where $V_{h,s}$ refers to the solidus interface speed. The latter fraction represents the contraction rate due to the change of material density with time (arising from the change of solid fraction in the two-phase region). If ϕ becomes sufficiently large, then hot cracking is expected to occur. The solute content in the material affects the change of phase fraction during solidification, thereby influencing the required strength to resist hot tearing near the liquidus interface. Based on this characterization, a numerical procedure could be developed to predict and prevent hot cracking in a material. For example, boundary conditions could be modified to alter the solidification trends near the anticipated hot cracking area. Extra chill boundaries could be added to eliminate the regions exhibiting hot cracking (i.e., regions of high ϕ).

The use of certain metals and alloys in manufacturing processes is largely influenced by their mechanical properties. Once solidified, a material usually undergoes certain applied forces and stresses. The term *elastic* deformation refers to a nonpermanent deformation that is totally recovered after the release of an applied stress. Elastic deformation is described by a linear relation between applied stress, σ, and strain, ϵ. The constant of proportionality is called E (modulus of elasticity), which outlines the resistance to separation of interatomic bonding forces. A *plastic* deformation is permanent and nonrecoverable. It involves the breaking of bonds to form new bonds. *Ductility* is a measure of the degree of plastic deformation when a material fractures. A *brittle* material exhibits little or no plastic deformation upon fracture. *Resilience* refers to the capacity of a material to absorb energy when it is deformed elastically. *Hardness* is a measure of the material's resilience to small plastic deformations. A *Vickers hardness test* involves a small pyramid diamond indenter that is forced into the surface of the test specimen. The resulting impression is observed under a microscope and measured to estimate the resulting surface hardness. The degree of difficulty of penetration (measured by the magnitude of the indentation) indicates the surface hardness of the material. Most or all of these properties of a manufactured material are closely related to its initial solidification process.

For example, the ultimate tensile strength (UTS) of a material is closely related to the solidification process in the material, i.e.,

$$UTS = c_1 (GAP)^{c_2} \tag{8.189}$$

where the gradient acceleration parameter (GAP) of Pillai (Sahm and Hansen, 1984) is given by

$$GAP = \frac{VG}{t_s} \tag{8.190}$$

The variables V, G, and t_s refer to solidification velocity, temperature gradient at the phase interface (i.e., dT/dx in units of K/mm), and solidification time (time to pass from the liquidus to the solidus temperature), respectively. Also, c_1 and c_2 are empirical constants. For example, $c_1 = 4.3$ and $c_2 = 0.12$ for an Al 7%Si alloy.

Another example relating material properties with the solidification process involves the yield stress, σ_y. It can be related to the standard reference yield strength, σ_o, as follows:

$$\sigma_y = \sigma_o + cV^{1/4} \tag{8.191}$$

where c is another empirical constant and V is the solidification velocity, estimated by Hall and Petch (Sahm and Hansen, 1984) as

$$V = \frac{1}{G}\left(\frac{\partial T}{\partial t}\right) \tag{8.192}$$

This solidification velocity is approximated as the ratio of the cooling rate, $\partial T/\partial t$, to the applied temperature gradient in the material, G.

The previous examples of property dependence on solidification processes show how effective thermal control of solidification can be used to optimize the final properties of the material. These optimized properties include microscopic homogeneity without segregation (i.e., compositional uniformity throughout the material, grain size and shape, and dendrite spacings). *Microsegregation* refers to compositional variations at the scale of the grain diameter–interdendritic distance. For example, the dendrite arm spacing, λ_d, is related to G (interfacial temperature gradient) and V (solidification velocity) by (Sahm and Hansen, 1984)

$$\lambda_d \approx CG^{-1/2}V^{-1/2} \tag{8.193}$$

where C is an empirical constant. On the other hand, *macrosegregation* refers to compositional variations on a scale of the cast part's dimensions.

The properties and costs of various materials and their alloys affect their use in industrial applications. For example, aluminum and its alloys and magnesium and its alloys are commonly used in aircraft parts due to their high strength per pound. Copper is frequently used for electrical wiring due to its high conductivity. Nickel is widely used in gas turbine parts in view of its high strength at elevated temperatures. Also, due to its high strength and relatively low cost, iron is widely used in steel parts for beams, pipes, automotive parts (i.e., doors, engine), and other commonly used industrial machines. Solidification processing of these materials has an important role in establishing these final mechanical properties of the material.

8.8.2 Manufacturing Processes

Solidification and melting are encountered in a variety of manufacturing processes, such as welding, casting solidification of metals and alloys, extrusion, and plastic injection molding. In these applications, heat transfer processes are important in view of the final material quality, production time, and other manufacturing measures. In this section, an overview of the importance of solidification and melting will be given in relation to various manufacturing processes.

8.8.2.1 Investment Casting

An *investment casting* refers to a type of precision casting, such as a casting for a die stamp component. This manufacturing process is generally comprised of the following steps:

- Inject the pattern material (i.e., liquid wax or plastic) through an ingate into the mold enclosing the inner shell of the pattern.
- Solidify the inner shell (casting) by heat rejection to a chilled boundary and heat-absorbing sand mold.
- Remove the pattern by opening the mold.
- Assemble the component as a part of a cluster tree that represents the actual full pattern.
- Dip (invest) the cluster tree into a ceramic slurry and allow it to solidify and dry.
- De-wax/heat the outer shell mold with high pressure steam to melt the interior pattern mold (formed in the first step).
- Heat the resulting hollow ceramic cluster tree in a furnace to burn out the remaining wax and preheat the mold.
- Pour molten metal into the hollow cluster tree with a ladle and solidify.
- Remove the outer ceramic shell using a vibrator to leave the final metal (inner) pattern.

Solidification and melting heat transfer have key roles in the previously described investment casting process. A low-strength pattern would be obtained with a randomly oriented grain structure during solidification, whereas a higher strength material could be achieved if the grain boundaries were formed through a carefully controlled solidification process. For example, a columnar structure forms and no grain boundaries are created in certain directions, so that ruptures would not be initiated at the grain boundaries. The grain boundaries can be carefully controlled through thermal processing during solidification by growing fewer grains or a single grain with a direction of highest strength oriented to resist conditions of high stress.

A practical example of the benefits of carefully controlled solidification technology appears in the development of cast node joints for offshore drill platforms in the North Sea. These joints require very sturdy constructions for an ocean depth of 200 to 300 m. The most critical stress conditions prevail in the node joint sections. In the actual platforms, both sizable weight and cost reductions were obtained by cast node joints, as compared with welded construction (Sahm and Hansen, 1984). For example, a welded construction yielded a joint weight of 22.2 tons, whereas the cast construction weighed 18.4 tons with a production cost that was 65% of that of the welded construction.

8.8.2.2 Wire Casting

Wire casting is commonly used in the production of wire-type products. Molten metal is supplied from a chamber as a thin liquid stream that is cooled and solidified in the form of a thin wire or similar product (see Figure 8.12). In this process, the critical design factors include the design of the supply chamber, the type and sizes of the cooling zones that cool the molten liquid stream, and the required microstructure of the wire product. Each of these factors is largely affected by the heat transfer, wire solidification, and supply of molten metal in the supply chamber.

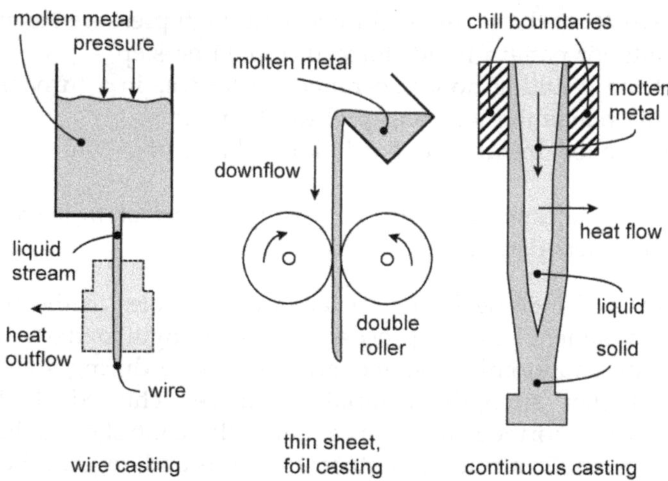

FIGURE 8.12
Wire, foil, and continuous casting.

8.8.2.3 Thin Sheet or Foil Casting

Medium- to high-thickness sheet castings of metal alloys are often produced in *thin sheet* or *foil casting* processes (Figure 8.12). In this case, the molten stream is supplied from a ladle and forced to flow through two counter-rotating rollers maintained at a specific temperature. The cooling rate of the solidifying sheet or foil is typically proportional to the contact angle tangent to the edge of the roller and the incoming liquid stream as well as and the temperature difference between the roller and the liquid stream. The final quality of the solidified sheet or foil is closely dependent on the roller surface characteristics, homogeneity of the supplied molten liquid, interactions with the surrounding gas pressure, and rate of nucleation in the solidifying material.

8.8.2.4 Continuous, Centrifugal, and Die Casting

In *continuous casting* solidification, the molten material is typically poured into water-cooled, open-ended molds (see Figure 8.12). It is widely used in the production of bars and slabs. This process differs from *centrifugal casting* (used primarily for pipes where molten material is poured into a rapidly rotating mold by a centrifuge) and *die casting* (molten material is forced under pressure into a die).

Molten metal is introduced into the casting through a gating system that usually consists of a channel carrying the liquid into the mold. The mold filling time, t_f, is an important design factor that is linked closely with the flow velocity, V_f. Transition to turbulence for internal liquid metal flows occurs at about a Reynolds number (based on the channel width) of 2000, with fully turbulent flow above an *Re* of 4000. The filling time may be roughly estimated from a solution of the Bernoulli and continuity equations for liquid flow through the gating system.

The following list summarizes some important design considerations examined by thermal engineers in casting solidification problems:

- Fixation of parts: accessibility of locations to be worked, direction of material stresses, spatial requirements
- Gating system: number of ingates, configuration of ingates
- Feeders: number, configuration, shape, size

Feeders provide the molten metal supply, and the gating system must be effectively constructed to transfer the molten material to various parts of the casting. For example, top gating supplies molten metal to the top of the casting, whereas bottom gating introduces the molten metal from the side or bottom of the casting. Step gating can be used to introduce the molten liquid at various heights throughout the casting. In ingot manufacturing processes, cast pieces are often further worked under heat treatment, including rolling and forging operations. *Inverse segregation* occurs as columnar dendrites thicken within the ingot, so that solute rich liquid flows back between the dendrites to compensate for solidification shrinkage. This raises the solute content of the outer parts of the ingot relative to the center of the ingot. The structure of an ingot usually includes a columnar zone that consists of columnar grains growing perpendicular from nucleation sites along the ingot walls. Equiaxed zones are often formed when dendrite arms and nucleating crystals are carried by convection in the center portions of the ingot. Nucleating crystals can be broken away from the walls due to natural convection currents in the molten liquid. A *negative segregation* can be created due to gravity effects when denser solid descends, carries less solute downward, and creates a nonuniform solute distribution in the solidified material. These processes all have important contributions to the final structure and quality of the solidified material.

8.8.2.5 Welding Processes

In gas metal arc (GMA) welds, liquid metal is deposited into the weld pool from an electrode (usually moving at a constant velocity, V; see Figure 8.13). Solidification of the weld pool is largely controlled by convection and radiative heat losses from the solidifying material. The weld geometry consists of the weld profile (shape) and the size of the newly formed volume, ΔV. For design purposes, this weld geometry could be estimated by an iterative procedure as follows.

- Estimate the weld temperature, phase, and weld pool depth.
- Solve the transient heat equation subject to appropriate convective and radiative boundary conditions.
- Adjust the weld geometry so that the heat balance on the newly solidified weld pool is satisfied.

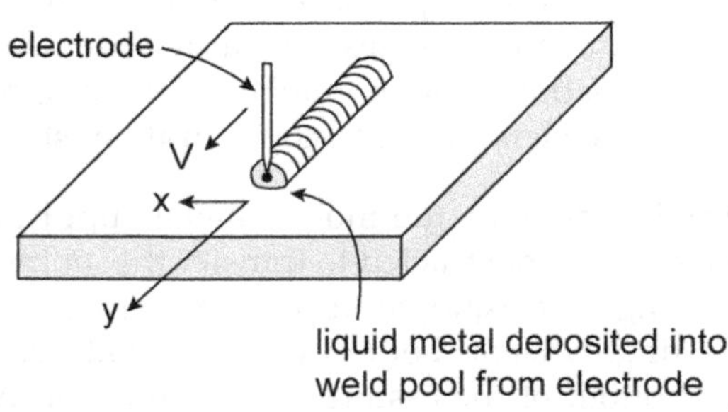

FIGURE 8.13
Schematic of gas metal arc welding.

- Return to the first step and repeat the steps until convergence between the heat losses and resulting weld geometry is achieved.

If the temperature of melted metal in the weld pool is much larger than the base metal temperature, a high temperature gradient is encountered in the material. This generally leads to cellular or columnar structures in the welded materials. Also, the welding speed, V, largely affects the melt pool and weld properties. For example, equiaxed dendrites and columnar structures are usually formed under high-velocity conditions.

8.8.2.6 Growth of Crystals

Czochralski crystal pulling involves a rotating rod with a seed crystal at an end of the rod that is lowered into a molten liquid and withdrawn for a continued crystallization to proceed. The crystal diameter is controlled by adjusting the pull rate and the heat input to the molten liquid in the crystal-forming enclosure. The heat input is important in view of the solidification process leading to the formation of the crystal structure. Crystals of high purity are required in a variety of industrial and other applications, such as semiconductors and biological applications.

8.8.3 Energy Storage with Phase Change Materials

In many applications, PCMs represent a less costly and less bulky method of storing thermal energy than other conventional techniques, such as heating of fluids in solar collectors. For example, it can be observed that the heat storage capacity of PCMs largely exceeds the capacity of sensible storage media, such as rock and water, for an air heating system with air at 20°C (see Figure 8.14). In Figure 8.14, the horizontal axis represents the temperature of the storage media. It can be observed that no energy is stored at 20°C since the temperature matches the air temperature. Also, the energy stored by the sensible storage media rises linearly with temperature according to the specific heat of the material. On the other hand, a step change of energy storage occurs with the PCM due to the heat absorbed during the change of phase (i.e., approximately 30°C for $CaCl_2 \cdot 6H_2O$). Above the phase change temperature, the energy stored by the PCM rises linearly with temperature in a way similar to that of the sensible storage of the rock and water (i.e., based on the specific heat of the liquid PCM).

In Figure 8.14, it can be observed that energy is stored or released while changing between solid and liquid phases in a PCM. Phase change materials are often used in applications such as solar receivers for storage of solar energy and spacecraft thermal control. Energy storage in the PCM occurs by melting during a cycle of high incoming solar radiation. The heat is released by solidification during periods of low incoming solar radiation,

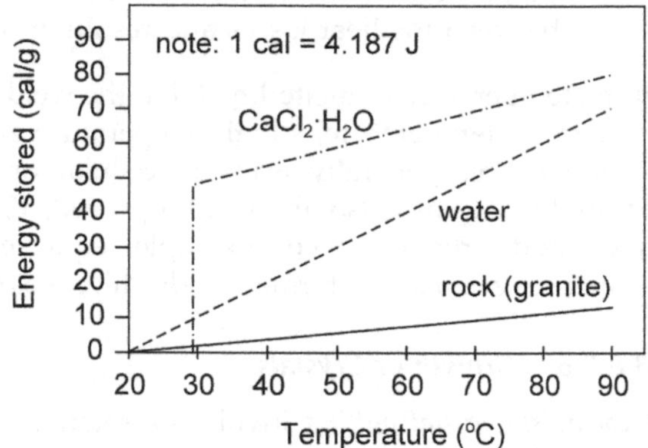

FIGURE 8.14
Comparison of energy storage capacities for 20°C air heating system. (From Lane, G.A., *Solar Heat Storage: Latent Heat Materials*, Copyright 1996, CRC Press, Boca Raton, FL. Reprinted with permission.)

while remaining isothermal during heat transfer to the working fluid in the receiver.

Phase change materials are usually classified as *congruent* or *noncongruent*. In a congruent system, the liquid and solid phases in equilibrium at the melting point have the same composition. For example, pure materials represent examples of congruent systems since there is a sharp, single melting point. Thermophysical properties of commonly used PCMs are shown in Figure 8.15 and Appendix A3.

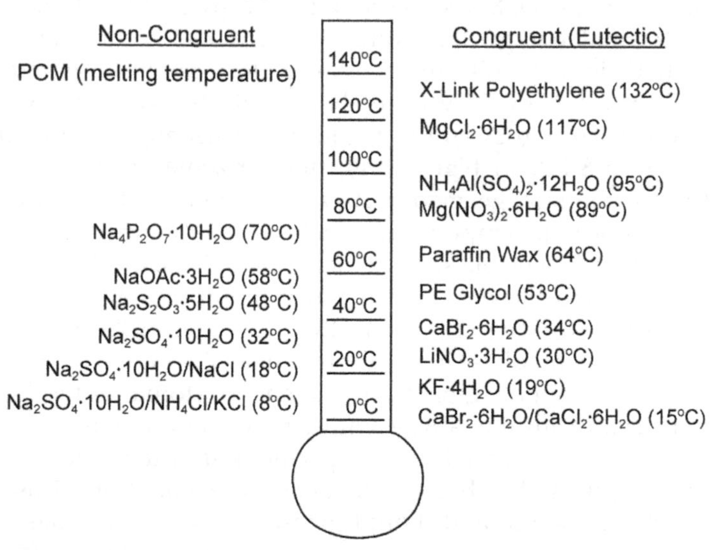

FIGURE 8.15
Melting temperatures of PCMs. (From Lane, G.A., *Solar Heat Storage: Latent Heat Materials*, Copyright 1996, CRC Press, Boca Raton, FL. Reprinted with permission.)

Also, minimization of entropy production is useful for the effective design of multiphase systems involving PCMs (Bejan, 1996). The following example demonstrates how the efficient utilization of PCMs can be assessed through this entropy-based technique.

Example 8.8.3.1

Solar Power Generation with a PCM.

A solar concentrator consists of several mirrors focusing solar radiation on a heat exchanger. It receives solar energy and transfers it to a PCM as a heat source for subsequent power generation (see Figure 8.16). Determine the optimal phase change temperature, T_m, based on entropy generation minimization (i.e., maximized power output) of the power station.

The external system refers to the energy absorbed and reflected from the solar receiver, whereas the internal system refers to the heat exchanger and heat engine delivering the power output. The energy balances for the external and internal systems, as well as the total entropy generated (external and internal irreversibilities), are

$$Q_e = Q^* - Q_H \qquad (external) \tag{8.194}$$

$$Q_H = UA(T_c - T_m) \qquad (internal) \tag{8.195}$$

$$\dot{S}_{gen} = \left(\frac{Q_H}{T_m} - \frac{Q_H}{T_c}\right) + \left(\frac{Q_e}{T_o} - \frac{Q_e}{T_c}\right) \tag{8.196}$$

where U, T_c, and T_o refer to the total thermal conductance in the heat

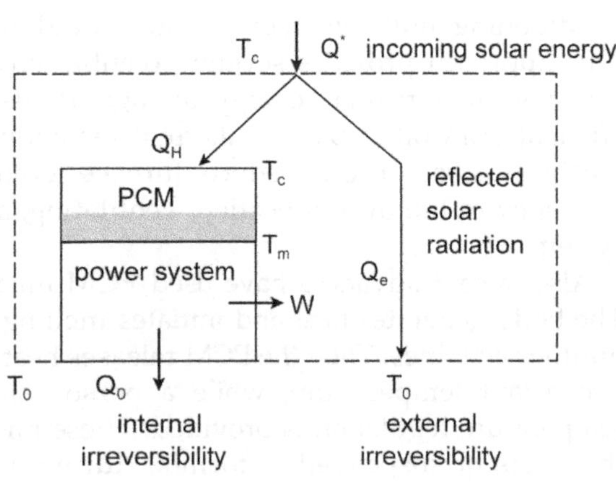

FIGURE 8.16
Schematic of a heat engine with PCMs.

exchanger(s), surface temperature of the solar collector, and temperature of the heat sink where heat is rejected from the power plant.

Combining the previous expressions,

$$\dot{S}_{gen} = UA(T_c - T_m)\left(\frac{1}{T_m} - \frac{1}{T_o}\right) + Q^*\left(\frac{1}{T_o} - \frac{1}{T_c}\right) \tag{8.197}$$

Then the optimal phase change temperature can be determined after setting the derivative of entropy generation (with respect to T_m) to zero, thereby yielding

$$T_{m,opt} = \sqrt{T_o T_c} \tag{8.198}$$

This result minimizes the entropy generation and irreversibilities, while maximizing the system efficiency and power output. The same result could have been obtained by maximizing the power output.

Phase change materials have been adopted in many other innovative applications. For example, PCM drywall consists of pellets of salt hydrates, paraffins, or fatty acids with melting points near room temperature. There is promising potential for PCM pellets to be added to drywall mixtures during manufacturing for subsequent use in the construction of walls and roofs in houses and buildings. This approach may provide an effective method for solar based heating in houses and buildings. The drywall absorbs heat in melting of the phase change material when the furnace or incoming solar radiation provides heat input during cold weather conditions. Then, it releases heat during freezing (i.e., heat released overnight in a house). A reverse moderating effect would occur during operation of an air conditioning unit, whereby cooling could be provided in hot climates. This approach provides several potential advantages: larger heat storage capacity than regular energy storage fluids and less costly and more efficient than other passive thermal techniques. The latter benefit occurs partly because it can reduce furnace cycling, thereby increasing the efficiency of the overall heating, ventilating, and air conditioning (HVAC) system.

Also, recent advances have used PCM microcapsules in clothing fibers. The body generates heat and initiates melting of the microcapsules during outdoor activities. Then the PCM releases heat when it freezes (maintaining a constant temperature), while a person sits and relaxes. In this way, temperature regulation is provided. These microcapsules are embedded in the fibers or suspended in foam(s) during manufacturing of the clothing materials.

PCMs have also been applied to de-icing of structures. For example, concrete slabs in bridges may use PCMs to prevent freezing on the bridge before it occurs on adjacent roads leading to the bridge. The phase change material is stored in a pellet form in the concrete such that it stores daytime heat and releases it when the temperature approaches 0°C. As a result, this delays icing on the bridge so that it coincides with the icing on the adjacent road. In addition to reducing potential accidents as a result of icing, this technique has other advantages: fewer potholes on the bridge and less cracking due to repeated freezing and thawing of the material.

Phase change materials are used in various other applications. These applications include heat-resistant coatings for aircraft, firefighter suits, cooling systems for computer and telecommunications equipment, boots, food delivery containers, blood storage systems, and others.

8.8.4 Freezing in Pipelines

The freezing of water in pipes can lead to substantial problems in water supply systems. Freezing of water in other constrained flows arises in various other applications, such as water flowing through turbines and outlet sections of hydroelectric power stations in cold weather conditions. In the following example, water flow in a pipe with phase change will be considered and the time required for freezing will be estimated.

Example 8.8.4.1

Freezing of Water in Pipes.

Water flows through a pipe cooled by surrounding ambient air at a subzero temperature. If the water temperature is initially T_o (above 0°C), then find

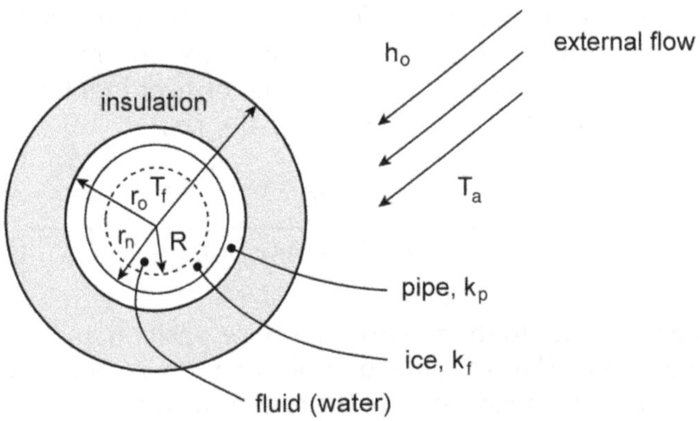

FIGURE 8.17
Schematic of thermal resistances in pipe flows.

the time elapsed before the water completely freezes within the pipe. One-dimensional heat transfer in the radial direction may be assumed in this problem.

From an energy balance, the rate of energy decline in the water balances the net rate of heat loss from the pipe. The rate of heat loss is given by the temperature difference, $T_w - T_a$, divided by the total resistance, composed of a sum of convection and conduction resistances,

$$\pi r_n^2 \rho_w c_w \frac{dT_w}{dt} = -\frac{2\pi(T_w - T_a)}{(r_n h_i)^{-1} + ln(r_o/r_n)/k_p + ln(r_i/r_o)/k_i + (r_i h_o)^{-1}} \tag{8.199}$$

where the subscripts p, w, a, i, and o refer to pipe, water, ambient, ice (or outer edge of insulation), and outer, respectively. The heat transfer coefficients for inner and outer convection are denoted by h_i and h_o, respectively.

This equation can be rewritten as a first-order differential equation with constant coefficients. Solving that equation, subject to $T(0) = T_0$, we find that

$$\frac{T_w - T_a}{T_o - T_a} = exp\left[\frac{-2t}{(r_n^2 \rho_w c_w)((r_n h_i)^{-1} + ln(r_o/r_n)/k_p + ln(r_i/r_o)/k_i + (r_i h_o)^{-1})}\right] \tag{8.200}$$

By setting $T_w = T_f$ in this equation, we can determine the time required for the water to cool to the freezing temperature, T_f.

Once the water reaches the phase change temperature, T_f, freezing will begin and the time required for freezing can be estimated. From the interfacial heat balance at the ice–water interface, the rate of latent heat release from the freezing ice balances the rate of heat loss through the ice, pipe, and insulation to the air (temperature T_a), i.e.,

$$LR\frac{dR}{dt} = \frac{T_a - T_f}{ln(r_n/R)/k_f + ln(r_o/r_i)/k_p + ln(r_i/r_o)/k_i + (r_i h_o)^{-1}} \tag{8.201}$$

where $R(t)$ and L refer to the moving interface position and latent heat of fusion, respectively. The left side of the equation is negative since the interface moves in the negative radial direction (inward).

Solving the previous equation and rearranging in terms of the elapsed time, t, we can obtain the following time required to form an ice layer of thickness R:

$$t = \frac{L}{k_f(T_f - T_a)} \times \left[\left(\frac{1}{2} + k_f\left(\frac{1}{k_p}\, ln\left(\frac{r_o}{r_n}\right) + \frac{1}{k_i}\, ln\left(\frac{r_i}{r_o}\right) + \frac{1}{r_n h_o}\right)\frac{r_n^2 - R^2}{2}\right) + R^2 ln\left(\frac{R}{r_n}\right)\right] \quad (8.202)$$

By setting $R = 0$ in this equation (i.e., ice interface moves to the center of the pipe), we can find the time required to completely freeze the water.

The previous analysis can be used for various practical problems in the maintenance of pipelines, such as the prevention of piping damage from ice expansion by freezing of water in the pipeline. Ice prevention is a widely encountered concern in various other engineering and scientific applications. In many instances, such as de-icing of aircraft surfaces or overhead power transmission lines, the heat transfer processes involve three-phase conditions (i.e., solid–ice, liquid–droplets, and air, simultaneously). Heat transfer in three-phase systems will be discussed in the following chapter.

References

A. Bejan. 1996. *Entropy Generation Minimization*, Boca Raton, FL: CRC Press.

H.W. Carslaw and J.C. Jaeger. 1959. *Conduction of Heat in Solids*, 2nd ed., Oxford: Clarendon Press.

S.H. Cho and J.E. Sunderland. 1969. "Heat conduction problems with melting or freezing," *J. Heat Transfer* 91: 421–426.

M.Y. Dedegil. Drag Coefficient and Settling Velocity of Particles, paper presented at International Symposium on Slurry Flows, ASME, FED, Anaheim, CA, December 7–12, 1986.

S.J. Farlow. 1982. *Partial Differential Equations for Scientists and Engineers*, New York: John Wiley & Sons, Inc.

S.D. Foss and S.S.T. Fan. 1972. "Approximate solution to the freezing of the ice-water system with constant heat flux in the water phase," *J. Water Resour. Res.* 8: 1083–1086.

International Encyclopedia of Heat and Mass Transfer, G.F. Hewitt, G.L. Shires, and Y.V. Polezhaev, eds., Boca Raton, FL: CRC Press, 1997.

J. Jacobs, et al. Design of a Float Zone Furnace for Microgravity Purification and Crystal Growth, AIAA Paper 93-0474, paper presented at AIAA 31st Aerospace Sciences Conference, 1993, Reno, NV.

F. Kreith and F.E. Romie. 1955. "A study of the thermal diffusion equation with boundary conditions corresponding to solidification or melting of materials initially at the fusion temperature," *Proc. Phys. Soc. B* 68: 277–291.

A.L. London and R.A. Seban. 1943. "Rate of ice formation," *Trans. Am. Soc. Mech. Eng.* 65: 771–779.

V.J. Lunardini. *Heat Conduction with Freezing or Thawing*, CRREL Monograph 88–1, U.S. Army Corps of Engineers, 1988.

A.D. Maude and R.L. Whitmore. A generalized theory of sedimentation, *Br. J. Appl. Phys.* 9, 1958.

G.F. Naterer. 2000. "Predictive entropy based correction of phase change computations with fluid flow: part 2: application problems," *Numerical Heat Transfer B* 37: 415–436.

G.F. Naterer. 2001. "Applying heat-entropy analogies with experimental study of interface tracking in phase change heat transfer," *Int. J. Heat Mass Transfer* 44: 2917–2932.

G.F. Naterer and G.E. Schneider. 1995. "PHASES model of binary constituent solid-liquid phase transition: part 2: applications," *Numerical Heat Transfer B* 28: 127–137.

E. Pardo and D.C. Weckman. 1990. "A fixed grid finite element technique for modeling phase change in steady-state conduction-advection problems," *Int. J. Numerical Methods Eng.* 29: 969–984.

W.G. Pfann. 1958. *Zone Melting*, New York: John Wiley & Sons.

P.R. Sahm and P. Hansen. *Numerical Simulation and Modelling of Casting and Solidification Processes for Foundry and Cast House*, CIATF, CH-8023, International Committee of Foundry Technical Associations, 1984, Zurich, Switzerland.

R.V. Seeniraj and T.K. Bose. 1982. "Planar solidification of a warm flowing liquid under different boundary conditions," *Warme Stoffubertragung* 16: 105–111.

E.M. Sparrow, S. Ramadhyani, and S.V. Patankar. 1978. "Effect of subcooling on cylindrical melting," *ASME J. Heat Transfer* 100: 395–402.

J. Stefan. 1891. "Uber die Theorie des Eisbildung, Insbesonder uber die Eisbildung im Polarmere," *Ann. Phys. Chem.* 42: 269–286.

R.H. Tien and G.E. Geiger. 1967. "A heat transfer analysis of the solidification of a binary eutectic system," *J. Heat Transfer* 89: 230–234.

E.J. Wasp, J.P. Kenne, and R. Gandhi. Solid liquid flow, slurry pipeline transportation, Trans Tech Publications, pp. 9–32, 1977, Zurich, Switzerland.

R. Xu and G.F. Naterer. 2001. "Inverse method with heat and entropy transport in solidification processing of materials," *J. Mater. Process. Technol.* 112: 98–108.

Problems

1. Liquid metal at an initial temperature of T_i in a large tank is suddenly exposed to a chilled surface at T_w (below the phase change temperature, T_f). Solidification begins at this surface and the solid–liquid interface propagates into the liquid over time. Assume that spatially linear profiles may be used to represent the temperature profiles in the solid and liquid, where the ratio of profile slopes is nearly constant over time. Derive an expression for the change of interface position with time in terms of thermophysical properties, T_f, T_w, and the ratio of slopes of temperatures in the liquid and solid.
2. A large tank contains a liquid at a temperature of T_i (above the phase change temperature, T_f). At $t = 0$, the surface temperature of the tank is dropped to T_w (where $T_w < T_f$) so that phase change occurs and the solid–liquid interface moves linearly in time. Can a similarity transformation, Equation (8.7), be used to find the rate of interface advance in terms of thermophysical properties and the given temperatures? Explain your response by reference to a one-dimensional transient solution of the problem.
3. A liquid alloy undergoes phase transition over a range of temperatures between the liquidus and solidus temperatures. During solidification, a layer of the material releases the latent heat of fusion under one-dimensional conditions. This release of latent heat may be characterized by a source of heat that varies linearly (i.e., equal to βx, where β is a constant) across the layer. A temperature of T_1 is imposed at the left edge ($x = 0$), while a specified heat flux, $q = q_w$, is required at the right edge of the layer ($x = L$).
   (a) Set up the heat conduction equation and boundary conditions for this problem.
   (b) Solve the governing equation to find the temperature variation across the layer of material. Express your answer in terms of β, L, q_w, and the effective conductivity of the material, k.
   (c) What values of β are required to maintain a unidirectional (rather than bidirectional, leftward and rightward) flow of heat at the left boundary?
4. Consider a region of liquid initially at the phase change temperature (T_f) that suddenly begins freezing outward from a wall due to cooling at a uniform rate through the wall at $x = 0$.
   (a) Using the quasi-stationary approximation for a one-dimensional semi-infinite domain, estimate the position within the solid

where the temperature reaches a specified fraction of T_f (based on Kelvin units) at a given time. Express your answer in terms of the wall cooling rate, given time, thermophysical properties, and T_f.

(b) For a wall cooling rate of 19 kW/m^2, estimate where the temperature of tin reaches $0.96 \times T_f$ after 100 min.

(c) What minimum cooling rate is required to ensure that the specified temperature in part (b) is attained anywhere in the domain?

5. A layer of solid with a given thickness, W, is initially at a temperature of T_f (fusion temperature). Suddenly, both outer edges of the solid are heated and held at a specified temperature of T_w (where $T_w > T_f$). In this way, the outer parts of the solid are melted over time. The melted solid is self-contained within the same region so that the boundary temperature of T_w is applied subsequently at the outer edges of the liquid.
   (a) Using the one-dimensional quasi-stationary approximation, find the temperature distribution in the liquid.
   (b) Estimate how much time is required to melt the entire solid layer.
6. Repeat the previous problem and compare the results obtained for the following three specific materials: aluminum, lead, and tin. Use an initial solid width of 10 cm and a wall temperature of 10°C above the fusion temperature for each material.
7. A long metal bar with a specified thickness, W, is initially at a temperature of T_f (change temperature), while it remains enclosed within a rigid container. The right boundary ($x = W$) is suddenly heated and maintained at a temperature of T_w (where $T_w > T_f$). Other surfaces (except the left boundary at $x = 0$) are well insulated such that a one-dimensional approximation may be adopted. What uniform heat flux, q_w, must be applied at the left boundary ($x = 0$) to melt the entire bar in a specified amount of time, t_s? Derive the expression for q_w in terms of thermophysical properties, specified time, and temperatures. Use a quasi-stationary analysis based on the assumption that the liquid is retained within the same rigid container after it melts (i.e., without flowing away).
8. A liquid is well mixed inside a container and held at a uniform temperature of T_i (above the phase change temperature, T_f). Initially, the surface temperature is T_f and solidification is initiated when the external boundary of the container is cooled at a uniform rate. The solidification process is carefully controlled to produce strengthened mechanical properties in the final solidified material. During this thermal control, a constant heat flux is provided from the liquid region at the phase interface. Using a one-dimensional, quasi-stationary

approximation, find expressions for the interface position over time and the spatial temperature distribution within the solid.

9. A layer of solid of thickness W is initially kept at a temperature of T_i (below the phase change temperature, T_f) in a rigid container. Suddenly, two external surfaces of the container are heated and held at a temperature of T_w (where $T_w > T_f$), while other surfaces are well insulated, so that the solid begins melting from both sides. The heat flux to the solid is controlled to approximately decline with inverse proportionality to the interface position, $X(t)$ (i.e., the heat flux decreases when X increases).
   (a) Estimate the time required to melt all of the solid. Use a one-dimensional, quasi-stationary approximation in your analysis.
   (b) What range of constants of proportionality are required to ensure that the solid is melted entirely?
10. A pipe is submerged in liquid at an initial temperature of T_i (above the phase change temperature, T_f). Then coolant is passed through the pipe and outward freezing of the liquid is observed over time. The wall temperature of the pipe and the temperature throughout the solid can be assumed to be uniform at T_f. Also, a constant rate of phase change is produced, whereby the heat flux from the liquid at the phase interface is nearly constant over time.
    (a) Use a one-dimensional, quasi-stationary approximation to find the temperature distribution in the liquid. Express your answer in terms of T_f, R, thermophysical properties, and the phase change rate.
    (b) At what position does the liquid temperature exceed T_f by 10°C?
11. A liquid surrounding a pipe of radius r_i is initially held at a temperature of T_i (above the phase change temperature, T_f). Then the wall of the pipe is cooled and its temperature is kept uniformly at T_w (where $T_w < T_f$). The liquid begins to freeze progressively over time and heat is conducted inward radially through the liquid. Use a one-dimensional, quasi-stationary analysis to obtain the rate of interface advance with time, $R(t)$, as well as the temperatures in the solid and liquid phases. It may be assumed that the heat flux from the liquid at the phase interface is nearly constant over time.
12. Inward phase change from a cylindrical surface occurs during freezing of water in an annular region between two pipes. The initial liquid temperature is T_i (above the phase change temperature, T_f). The outer pipe surface (radius of r_o) is cooled at a uniform rate to initiate solidification of the water between the pipes. The heat flux from the liquid at the phase interface is assumed to move with inverse proportionality to the interface position, $R(t)$.

(a) Derive expressions for the spatial variation of temperature in the solid. Use a one-dimensional, quasi-stationary approximation in your analysis.
(b) Find the change of interface position with time, $R(t)$.
(c) Find the highest cooling rate (from the outer pipe) below which freezing cannot occur under the specified conditions. Express your answer in terms of T_i, thermophysical properties, r_o, and the constant of proportionality.

13. Water inside a pipe is initially at a temperature of T_i; it begins to freeze inward when the outer wall temperature of the pipe is lowered to T_f (phase change temperature, T_f). Use a one-dimensional quasi-stationary analysis in responding to the following questions.
(a) Find the time required for the position of the phase interface to reach one half of the outer radius of the pipe. Express your answer in terms of thermophysical properties, pipe radius, and the constant phase change rate. Assume that the interface propagates inward at a constant phase change rate and the temperature throughout the solid is nearly uniform.
(b) Find the spatial temperature distribution in the liquid at the time calculated in part (a).

14. A solid material at a temperature of T_f (phase change temperature) in a spherical container is heated externally. As a result, the solid begins melting inward at a nearly constant phase change rate (i.e., constant heat flux from the liquid at the interface). It may be assumed that the temperature of the solid remains nearly uniform at T_f. Perform a one-dimensional, quasi-stationary analysis to answer the following questions.
(a) Find the time required to melt the spherically shaped solid entirely. Express your answer in terms of the constant interfacial heat flux, thermophysical properties, and the initial (unmelted) sphere radius.
(b) Determine the rate of change of wall temperature with time.

15. The wall of a solar energy collector consists of a phase change material (solid–liquid phases) of thickness L and height H. Temperatures at the left and right boundaries are T_H and T_M (phase change temperature). The top boundary is insulated, and a material with known properties (i.e., k_w, ρ_w, $c_{p,w}$) is located beneath the PCM. A buoyant recirculating flow arises in the liquid as a result of the differentially heated boundaries. Derive the dimensionless parameters affecting the Nusselt number for two-dimensional heat transfer across the PCM.

16. By imposing a fixed pressure gradient, a non-Newtonian slurry (liquid–sand mixture) is forced to flow through a channel of height $2H$ with a mass flow rate of $\dot{m}$. The fully developed laminar flow in the

x direction is unheated up to $x = a$, but then heated uniformly by a wall heat flux, q_w, between $a \leq x \leq b$. The shear stress of the slurry is $\tau = \tau_b + \mu_b(du/dy)$, where τ_b refers to the Bingham yield stress and μ_b refers to the Bingham viscosity. Write the governing equations and boundary conditions for this problem. Nondimensionalize these equations to derive the parameters affecting the temperature and Nusselt number distributions. It may be assumed that the thermophysical properties, such as c_p of the mixture, remain constant, and the inlet temperature is T_i.

17. Laser heating of a metal initiates melting at the surface of the metal. At time $t = 0$, the solid is initially at a temperature of T_o, and a uniform heat flux, q_w, is applied and held at the wall ($x = 0$) where the laser heat is experienced. The resulting temperature disturbance propagates into the metal, but at $x > \delta(t)$, where δ varies with time, the solid temperature is approximately T_o. The purpose of this problem is to use a one-dimensional integral analysis to estimate the time elapsed, t_m, before the surface reaches the melting temperature, T_m (note: $T_m > T_o$).
    (a) Set up the integrated energy equation for this problem.
    (b) Assume a temperature profile of the form

$$T = a_0 + a_1\left(\frac{x}{\delta}\right) + a_2\left(\frac{x}{\delta}\right)^2$$

    and find the constants using appropriate boundary constraints.
    (c) Find the expressions for δ_m and t_m when the surface at $x = 0$ reaches T_m.

18. The mechanical properties of an alloy are affected by the interdendritic flow of liquid metal during a casting solidification process. Consider a liquid metal undergoing one-dimensional solidification at a depth of L below the liquid free surface in the container. The pressure along the free surface is p_o, and a permeability of $K = -c_1/f_l{}^2$ is adopted, where c_1 is a constant and f_l refers to the liquid fraction. Using a velocity of $v = -c_2\, y$ (into the phase change interface, where c_2 is a constant) and Darcy's law, estimate the pressure distribution in the liquid based on gravitational effects only. Express your result of pressure in terms of y, p_o, c_1, c_2, L, and thermophysical properties.

19. The permeability of a solidifying material, K, with respect to interdendritic flow was given as $K \approx K_0 f_l^3/(1-f_l)^2$ (Blake–Kozeny equation), where f_l refers to the liquid fraction and K_0 is a constant. Derive this expression for K by treating the interdendritic flow as a viscous flow through a tube (i.e., Hagen–Poiseuille flow) of a hydraulic diameter equivalent to the solid matrix with a finite void

or liquid fraction. It may be assumed that a constant pressure gradient is experienced by the flow through the solid matrix and each fluid element travels an equal distance through the packed bed.

20. In a silver extraction process, spherically shaped silver particles 1 mm in diameter at 7°C are heated after injection into a water flow at 27°C. Find the distance traveled by the particles in the slurry flow before they reach within 0.1°C of the temperature of the water. Assume that a relative velocity of 0.1 m/sec between the water and particles is maintained over the distance of travel.
21. Compare the distances required under the same conditions in the previous problem for the following three particle materials: gold, titanium, and limestone. Explain the trend observed in your results.
22. A compartment of PCM is used for thermal control of a satellite. The satellite generates heat uniformly throughout its orbit from internal power. When the satellite is shaded from solar radiation by the Earth, latent heat is released by the PCM during solidification. On the bright side of its orbit around the Earth, heat is absorbed by the melting PCM, and the satellite is heated by an incident solar heat flux. Estimate the heat flux due to phase change of PCM that should be maintained so that the satellite temperature remains approximately uniform throughout its orbit. Express your answer in terms of the uniform incoming solar flux on the bright side of the orbit. List all other assumptions adopted in your analysis.

9

Gas–Liquid–Solid Systems

9.1 Introduction

Three-phase systems involving gas, liquid (droplet), and solid phases arise in various engineering and scientific applications. Some examples include droplets impinging on a wet iced aircraft surface, spray deposition of solidified droplets in manufacturing processes, and combustion of liquid droplets with particles in combustion chambers. In this chapter, heat transfer in three-phase systems will be examined.

A general formulation covering three-phase systems can be considered through multiphase flows with droplets interacting with a solid phase (i.e., near-wall region, solid particles, or solid–liquid phase change). The conservation equations for multiphase flow with droplets can be obtained through spatial averaging of the individual phase equations over a fixed control volume containing a mixture of phases (i.e., air, liquid droplets, and solid simultaneously). This Eulerian approach adopts a uniform spatial droplet distribution within each control volume. Rather than tracking of individual droplets, spatially averaged approximations can be used for interphase processes, such as momentum and thermal interactions between the droplets and main (air) flow. In this chapter, the two-dimensional conservation equations for these multiphase flows will be derived as special cases of the general conservation equation for a scalar quantity. These derivations can be readily extended to analogous forms of the three-dimensional equations.

Consider the two-dimensional transport of a general scalar quantity, ϕ_k, associated with phase k in a control volume containing a mixture of phases at a differential level (see Figure 9.1). Based on conservation of ϕ_k within the control volume,

$$\frac{\partial(\rho_k \phi_k)}{\partial t} + \frac{\partial(\rho_k \phi_k u_k)}{\partial x} + \frac{\partial(\rho_k \phi_k v_k)}{\partial y} + \frac{\partial(\mathbf{j}_{k,x})}{\partial x} + \frac{\partial(\mathbf{j}_{k,y})}{\partial y} = \hat{S}_k \tag{9.1}$$

In Equation (9.1), the terms can be interpreted as the transient accumulation of ϕ_k in the control volume occupied by phase k, net advection of ϕ across the phase k portion of the control surface, diffusion into phase k across the interfacial surface, and the source or sink of ϕ_k (i.e., contributions due to evaporation or freezing) in the phase k portion of the control volume, respectively. Although Equation (9.1) can be readily interpreted for a continuous phase (such as air), its meaning for dispersed phases (i.e., droplets or particles) is less clear. Careful attention with interphase modeling is required to properly describe the dispersed phase quantities when performing spatial averaging across a control volume.

In addition to the conservation equations, interfacial balances are applied at the interface between individual phases. In establishing these interfacial balances, the following symbols will be adopted (see Figure 9.1): a, surface area; $\mathbf{n}_x$, the outward normal (x direction); $\mathbf{j}_k$, the net diffusive flux of the conserved quantity; V_k, the volume of phase k; and $\mathbf{v}_i$, the interface velocity. Also, the following Leibnitz and Gauss rules will be required in the spatial averaging:

$$\frac{\partial}{\partial t}\int_{V_k} B dV = \int_{V_k} \frac{\partial B}{\partial t} dV + \int_a B(\mathbf{v}_i \cdot \hat{\mathbf{n}}_k) dS \quad (Leibnitz) \tag{9.2}$$

$$\int_{V_k} \nabla \cdot \mathbf{b} dV = \frac{\partial}{\partial x}\int_{V_k} \hat{\mathbf{n}}_x \cdot \mathbf{b} dV + \int_a (\hat{\mathbf{n}}_k \cdot \mathbf{b}) dS \quad (Gauss) \tag{9.3}$$

where B and $\mathbf{b}$ refer to arbitrary scalar and vector quantities, respectively.

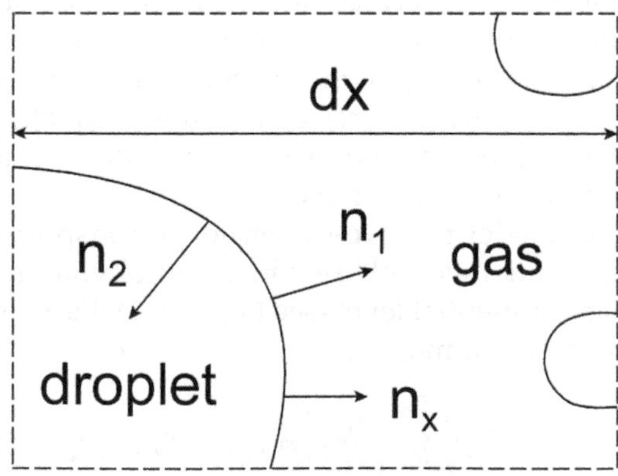

FIGURE 9.1
Multiphase control volume.

For brevity in the following derivations, only the x direction (one-dimensional) terms will be shown; however, the full multidimensional equations can be readily inferred by adding analogous terms in the other directions.

A volume-averaged quantity is defined by

$$\langle \beta_k \rangle = \frac{1}{V_k} \int_{V_k} \beta_k dV \tag{9.4}$$

Furthermore, the phase volume fraction, C_k, is given by

$$C_k = \frac{V_k}{V} \tag{9.5}$$

This volume fraction can be written in terms of a mass fraction through appropriate multiplication by density. In subsequent discussions, if C is written without the subscript, it refers to the volume fraction of the dispersed phase (i.e., liquid droplets) in the gas stream (i.e., air).

Based on volume averaging of Equation (9.1) by integration of the governing equation over V_k, together with Equations (9.2) and (9.3),

$$\frac{\partial}{\partial t} C_k \langle \rho_k \phi_k \rangle + \frac{\partial}{\partial x} C_k \langle \hat{\mathbf{n}}_x \cdot (\rho_k \phi_k \mathbf{v}_k + \mathbf{j}_k) \rangle + \frac{1}{V} \int_a (\dot{m}_k'' \phi_k + \mathbf{j}_k \cdot \hat{\mathbf{n}}_k) dS = C_k \langle \hat{S}_k \rangle \tag{9.6}$$

In Equation (9.6), the $\dot{m}_k''$ term refers to the interphase mass flux (units of mass per unit time and area). Also, it should be noted that the symbol $\int_a$ in Equation (9.6) refers to integration over the total area, including the interfacial area per unit volume (i.e., a_i along the boundary separating two distinct phases within the control volume), and any area of phase k in contact with the external walls, a_w, or boundaries of the system. The $\dot{m}_k''$ term is generally dropped upon evaluation along the walls due to the no-slip condition.

Equation (9.6) represents a general conservation equation, while specific conservation equations (such as mass or energy equations) follow as special cases of Equation (9.6). Predictions of multiphase flows with droplets (or particles) based on Equation (9.6) can be subdivided into three types of methods: (i) mixture modeling, (ii) tracking of individual droplet trajectories, and (iii) three-phase modeling.

In the first mixture approach, the dispersed and continuous phases are assumed to be uniformly distributed as a homogeneous mixture within a specified discrete (or infinitesimal) control volume. Based on spatial averaging of individual phase quantities, a single velocity is obtained that represents the mean velocity of the dispersed and continuous phases. This type of method is commonly used in other types of two-phase problems, such as problems involving casting solidification, since the velocity in one phase is well known (i.e., stationary solid phase often assumed), so the other phase velocity can be readily determined from the mixture velocity. However, in multiphase flows with droplets, this mixture approach encounters difficulties in view of properly subdividing the components of mixture velocity into their respective phases.

In the second approach, trajectories of droplets are tracked throughout the flow field. Based on force balances on these droplets or particles, together with appropriate empirical estimates of associated drag coefficients due to droplet interaction with the gas stream, the resulting droplet acceleration (and thus velocity through time integration) can be obtained. The resulting conservation equations for the droplet phase require appropriate relations for the effective thermophysical properties (such as thermal conductivity). Several limitations and difficulties are often encountered in these tracking methods, such as the complexity of tracking many droplets. Also, implementing Lagrangian-type models can be difficult to implement in conventional numerical methods and codes with a fixed grid in Eulerian-type formulations.

In the third approach, the two phases (or three phases, i.e., air, droplets, and solid phase in solidification problems) are spatially averaged within a control volume, and the resulting transport equations are solved for these individual phase equations. Cross-phase interactions are handled through appropriate phase interaction terms, such as spatially averaged force interactions between the air and droplet phases in the momentum equations. The approach has also been used for collision-dominated flows (dense, not dilute), where collisions between droplets (or particles) can be modeled based on sampling techniques derived in rarefied gas dynamics. Furthermore, statistical models that incorporate the physics of the dispersed phases can be used to give more detailed subscale models, such as particle rotation and friction during contact or coalescence of droplets. In this way, the advantages of the mixture model (reduced computational time) and the droplet tracking approach (capturing the detailed physics of interphase interactions) can be effectively retained, while providing a framework for implementation into existing Eulerian-type algorithms. This approach permits the droplet flow equations to be written in a standard scalar transport form in conjunction with gas (air), liquid (droplet), and solid phases (such as a moving ice interface) simultaneously.

9.2 Droplet Flows with Phase Change

9.2.1 Overview of Transport Processes

Droplet flows with phase change arise in many industrial and environmental flows. Some examples include icing of aircraft and structures, spray coating in manufacturing processes, droplet flows with evaporation in combustion systems, and others. In addition to the main (air) flow, the dynamics of droplet motion is an essential part of these processes. In the case of impinging droplets with solidification (i.e., icing problems, manufacturing processes such as spray deposition), splashback and runback of droplets along the surface impose unique challenges of accurate modeling. On the other hand, droplets with liquid–gas phase change (i.e., evaporating droplets, burning droplets, etc.) present their own challenges in terms of the interaction between the main flow and droplets.

Impinging droplets with solidification have common characteristics in their respective applications, including impinging droplets on iced structures, spray deposition in manufacturing processes, and others. For example, these problems involve droplet kinetic energy, convective heat transfer, and solid–liquid phase change. As a result of this common physical framework, we will examine the thermal and fluid flow phenomena as they relate to flows with impinging droplets and water phase change (i.e., freezing and icing processes). Although ice accretion is only one example of a three-phase system, it is considered in the following example in view of its similarities with other types of droplet flows in three-phase applications.

Example 9.2.1.1

Three-Phase Conditions in Air, Droplet, and Ice Processes.

The purpose of this example is to give a brief insight into the physical processes inherent in three-phase flows. In particular, the thermal processes leading to icing of structures (i.e., aircraft, overhead power lines) are considered. These processes include multiphase flow with droplets in the freestream, as well as solidification and melting along the moving ice boundary where incoming droplets adhere.

As incoming droplets come into contact with a solid surface, there is a mass conservation balance that essentially must relate the rate of solid (ice) accumulation to the rate of wind transport of droplets onto the surface. By neglecting splashback, runback, and evaporation of impinging droplets, an approximate expression for solid accumulation can be obtained analytically.

For example, the rate of ice accretion on a circular conductor can be accurately predicted by a mass balance when droplets solidify immediately upon impact (called *rime ice*) (Goodwin, 1982; Poots, 1996). Even in the absence of considering heat transfer for rime ice, a difficulty remains in the mass balance as a result of side effects. In particular, droplets passing by the side of the solid surface (i.e., parallel to the surface tangent) may adhere to the surface due to droplet breakup and/or surface tension effects. Due to the complexity of these microscopic processes, some researchers have used a variety of empirical factors (i.e., trigonometric shape factors) to accurately predict the changing surface shapes due to impinging and solidifying droplets. These factors have been correlated with various parameters, such as air temperature (Wakahama, 1977). In some atmospheric icing problems, precipitation occurs in the form of snowflakes rather than freezing droplets. The differences in mass, drag, and size distribution affect the resulting trajectories of snowflakes and their impact velocities on the solid surface (Skelton and Poots, 1991).

In certain problems, incoming droplets on multiple iced surfaces must be considered simultaneously. For example, rime ice accretion on twin conductor bundles was examined by Larcombe et al. (1988). In this type of problem, droplets near the downstream conductor are affected by the scattering and capturing of droplets by the upstream conductor. A viscous airflow solution is required to capture the flow separation on the upstream conductor and the interaction of the resulting wake region with the downstream conductor. In the case of power lines, stranding also occurs along the conductor. A perturbation solution for the attached potential flow may be used around an equivalent surface of a stranded conductor (Skelton and Poots, 1993). The impinging droplet velocities are altered by the modified potential function obtained from the perturbation expansion. In this way, appropriate correlations may be obtained for a circular conductor having an equivalent diameter to the stranded conductor. Analogous physical processes and solution techniques can be applied to the analysis of aircraft de-icing.

Glaze (wet) ice is another form of three-phase condition when incoming droplets partially solidify, while an unfrozen water layer forms along the ice surface. In some aircraft icing conditions, the ice growth may begin as rime ice, but latent heat is released and absorbed by the ice during freezing of incoming supercooled droplets, so that an unfrozen water layer is eventually generated. The release of latent heat may lead to a local temperature rise in the ice and transition to glaze ice along the outer edge of the ice surface. These three-phase conditions usually arise at air temperatures close to 0°C. The unfrozen film along the ice surface is typically very thin. This glaze ice growth is closely dependent on the impingement (or collection) efficiency of the surface, denoted by E. This efficiency refers to the ratio of mass flow of impinging droplets on the conductor on the upstream side to the mass flow that would contact the same surface if the droplets were not deflected by the airstream. Current techniques of reducing or preventing these icing problems include joule heating methods (i.e., resistance heating, or current loading of power lines), ice-retardant attachments and other more effective alternatives investigated by researchers.

The specific governing equations for multiphase flows involving three phases will be described in the following section.

9.2.2 Governing Equations

The governing equations for multiphase flow (potentially including three phases) under consideration are the following mass, momentum, and energy equations.

9.2.2.1 *Mass Equation*

For the continuity (conservation of mass) equation, set $\phi_k = 1$, $j_k = 0$, and $\hat{S}_k = 0$ in Equation (9.6) to obtain the following volume-averaged result:

$$\frac{\partial}{\partial t} C_k \langle \rho_k \rangle + \frac{\partial}{\partial x} C_k \langle \rho_k u_k \rangle + \frac{\partial}{\partial y} C_k \langle \rho_k v_k \rangle + \langle \dot{m}_k'' \rangle_i = 0 \tag{9.7}$$

where the subscript i refers to calculation along the interfacial area of the control volume. The interfacial mass flux, ($\langle \dot{m}_k'' \rangle_i$), involves phase change at the interface and accumulation or destruction of mass in one phase at the expense of another phase. For example, evaporation of droplets increases the volume fraction in the vapor phase at the expense of liquid droplet mass. In Equation (9.7), the subscript k refers to phase k, and in the case of the liquid phase, u_k refers to the instantaneous droplet(s) velocity in the x direction. Since the governing differential equations are spatially averaged over the control volume (Figure 9.1), this velocity may involve more than a single droplet.

Alternatively, Equation (9.7) can be written in terms of the mass of water (droplets) per unit volume of the air–water mixture, $\tilde{\rho}_l$, where

$$\tilde{\rho}_l = \frac{m_l}{V} = C_l \langle \rho_l \rangle \tag{9.8}$$

In certain applications (such as aircraft icing), $\tilde{\rho}_l$ is called the *liquid water content*. It is analogous to the concept of specific humidity, except that it includes discrete droplets rather than a continuous mixture of moisture in the air.

Combining Equations (9.7) and (9.8),

$$\frac{\partial \tilde{\rho}_l}{\partial t}+\frac{\partial(\tilde{\rho}_l u_l)}{\partial x}+\frac{\partial(\tilde{\rho}_l v_l)}{\partial y}+\langle \dot{m}_l'' \rangle_i=0 \tag{9.9}$$

Droplet coalescence and evaporation processes are represented by $\dot{m}_l''$ in Equation (9.9). It can be observed that Equation (9.9) is similar to the regular continuity or species concentration equation for single phase flow. As a result, it may be considered an analogous type of species transport equation, whereby the "concentration" of droplets (or phase fraction) is tracked throughout the flow field.

Once the solution of Equation (9.9) is obtained, the volume fraction, C, can be computed based on $\tilde{\rho}_l$. Alternatively, Equation (9.9) can be written and solved in terms of the volume fraction of droplets occupying the droplet–air control volume. Then an additional phase fraction, λ, can be introduced to distinguish liquid and solid phases when droplet freezing (or other solidification-type process) occurs along a solid boundary.

9.2.2.2 Momentum Equations

In this case, set $\phi_k=\mathbf{v}_k$, $\mathbf{j}_k=p_k\mathbf{I}-\tau_k$, and $\hat{S}_k=\mathbf{G}_k$, where $\mathbf{I}$, τ_k, and $\mathbf{G}_k$ refer to the identity matrix, stress tensor (molecular and turbulent), and body or interphase forces, respectively. Also, the component of the resulting Equation (9.6) in the x direction is obtained by taking the dot product of this equation with the unit x direction vector, $\hat{\mathbf{n}}_x$, yielding

$$\begin{aligned}&\frac{\partial}{\partial t}C_k\langle\rho_k u_k\rangle+\frac{\partial}{\partial x}[C_k\langle\rho_k u_k^2\rangle+\rho_k\langle\hat{\mathbf{n}}_x\cdot(\tau_k\cdot\hat{\mathbf{n}}_x)\rangle]+\frac{\partial}{\partial y}\\&\quad\times[C_k\langle\rho_k u_k v_k\rangle+\rho_k\langle\hat{\mathbf{n}}_x\cdot(\tau_k\cdot\hat{\mathbf{n}}_k)\rangle]\\&=-\frac{\partial}{\partial x}C_k\langle p_k\rangle-\frac{1}{V}\int_a\{\dot{m}_k u_k+\hat{\mathbf{n}}_x\cdot(\hat{\mathbf{n}}_k p_k)-\hat{\mathbf{n}}_x\cdot(\hat{\mathbf{n}}_k\cdot\tau_k)\}dS+C_k\langle G_{x,k}\rangle\end{aligned} \tag{9.10}$$

Similar equations for the y and z direction momentum equations can be obtained by taking the dot product with $\hat{\mathbf{n}}_y$ and $\hat{\mathbf{n}}_z$, respectively. Distribution coefficients express the ratio of a volume average of products in Equation (9.10), such as $\rho_k u_k u_k$, to the product of respective averaged quantities. In view of limited experimental data, these distribution coefficients are often taken to be approximately unity (Banerjee and Chan, 1980), so that volume averages of products can be approximated by a product of averaged quantities.

In Equation (9.10), the phase interaction forces are typically combined with other interfacial forces, such as the τ_k terms, using some form of resistance law for the droplets. These cross-phase interactions are often assumed to be proportional to the relative velocity difference between the air and droplets, i.e.,

$$G_{x,k} = \chi_k(\langle u_{a,i}\rangle - \langle u_{l,i}\rangle) \tag{9.11}$$

where χ_k is an empirical coefficient that depends on flow conditions, thermophysical properties (such as viscosity), and other parameters. An example of cross-phase interactions between the dispersed phase and the gas stream (Hewitt et al., 1997) is given by

$$G_{x,k} = \frac{18\mu_a}{d^2}\frac{c_d Re_l}{24}(u_a - u_l) \tag{9.12}$$

where $Re_l = \rho_g |u_a - u_l| d/\mu_a$ is the droplet Reynolds number (based on the relative velocity) and d is the mean droplet diameter in the flow.

Various modifications of the resistance force in Equation (9.12) have been reported. For example, the resistance force is approximately proportional to the velocity difference (between the air and droplets) raised to a certain exponent, rather than the difference alone (Tsuboi and Kimura, 1998). Assuming a spherically shaped droplet, the following curve fit may be adopted for the drag coefficient:

$$c_d = \frac{24}{Re_l} + \frac{4.73}{Re_l^{0.37}} + 0.00624 Re_l^{0.38} \tag{9.13}$$

which is suitable over the entire range of Reynolds numbers.

In many cases, the flow of droplets in the air (or other gas) stream is sustained mainly by forces imparted on each droplet from the main (air) flow, rather than by internal pressure gradients within the droplet. Furthermore, pressure and Eulerian-based spatial averaging are well defined for a continuous (air) phase, but less clear within the droplets when the spatial averaging is performed, since that phase is dispersed within the control volume. The resulting forces arising from the spatial averaging (including pressure and friction) can be empirically modeled by appropriate resistance laws (note: in addition to gravitational forces for the y direction equation). Under the relevant simplifications in the dispersed (droplet) phase, Equation (9.10) becomes

$$\frac{\partial(\tilde{\rho}_l u_l)}{\partial t} + \frac{\partial(\tilde{\rho}_l u_l u_l)}{\partial x} + \frac{\partial(\tilde{\rho}_l u_l v_l)}{\partial y} + \langle \dot{m}_l'' u_l\rangle = C_l\langle G_{x,l}\rangle \tag{9.14}$$

In the y direction momentum equation, an additional gravity term appears in the negative y direction, so that a droplet falls under its own weight. Once $\tilde{\rho}_l$ (or C_l) is obtained from a solution of Equation (9.9), Equation (9.14) and an analogous y direction momentum equation can be solved to give the components of droplet velocities, u_l and v_l, for two-dimensional flows.

In addition to the droplet (or solid particle) phase, substitution of the phase subscript, k, as the gas (subscript g) phase in Equation (9.10) yields the gas flow equations. In these gas (air) flow equations, a pressure gradient term appears in the usual fashion of the Navier–Stokes equations. Furthermore, the stress tensor in Equation (9.10) includes molecular diffusion for laminar flows, as well as the Reynolds stresses for turbulence. Extending Equation (9.10) with a similar procedure leading to Equation (9.14), the following volume-averaged momentum equations (x and y directions) are obtained:

$$\frac{\partial}{\partial t} C_g \langle \rho_g u_{i,g} \rangle + \frac{\partial}{\partial x_j} C_g \langle \rho_g u_{i,g} u_{j,g} \rangle$$

$$= \frac{\partial}{\partial x} C_g \langle p \rangle + C_g \frac{\partial \tau_{ij}}{\partial x_j} + C_g \langle G_{i,g} \rangle - \rho_g \frac{\partial}{\partial x_j} C_g \langle u_i' u_j' \rangle \tag{9.15}$$

where u_i' refers to the deviation from the volume-averaged mean velocity. The cross-phase interaction term, $G_{i,g}$, for the u_i velocity equation represents the force on the main (gas) flow due to interactions with the droplets (or particles). Although empirical expressions are often required for $G_{i,g}$, this term is generally a complex function of droplet size, shape, relative Reynolds number, and other factors (Crowe, 1999).

The last term in Equation (9.15) involves the Reynolds turbulent stresses, which can be roughly approximated in terms of a turbulent eddy viscosity through the Boussinesq approximation outlined in Equation (3.232). The presence of droplets (or particles) in the flow alters the length scales of turbulence, due to varying spacings between the droplets, and sizes and shapes of the droplets. However, conventional two-equation turbulence models, such as the models described in Chapter 3 for single phase flows, are often used as a basis for subsequent extensions to turbulent multiphase flows. For example, the turbulent kinetic energy and dissipation rate transport equations can be extended to include additional terms that account for the effects of droplets (or particles). In general, large droplets tend to enhance turbulence, while small droplets reduce turbulent mixing (Crowe, 1999). Multiphase turbulence is an active field of current research, due its fundamental importance in mixing processes and heat transfer in various industrial applications.

In the previous discussion, the droplet flow field was modeled by a discrete number of droplets that are transported by the main (air) flow through volume averaging of the conservation equations. In contrast to this approach, a two-fluid model treats the droplet (or particle) phase as a second separate continuum. A resulting equation similar to Equation (9.15) is obtained, but the turbulent stresses are summed over both phases individually, i.e., for the droplet (liquid, subscript l) phase,

$$\frac{\partial}{\partial t} C_l\langle \rho_l u_{i,l}\rangle + \frac{\partial}{\partial x_j} C_l\langle \rho_l u_{i,l} u_{j,l}\rangle$$

$$= \frac{\partial}{\partial x} C_l\langle p_l\rangle + C_l\frac{\partial \tau_{ij}}{\partial x_j} + C_l\langle G_{i,l}\rangle - \rho_l\frac{\partial}{\partial x_j}\sum_n C_{l,n} u'_{l,i,n} u'_{l,j,n} \tag{9.16}$$

In this equation, the pressure, droplet velocities, and shear stresses represent the mass-averaged quantities and the subscripts l, i, and n refer to liquid (droplet), coordinate direction, and droplet number, respectively. The dashed velocities (u') refer to the deviation from the mean velocity. The last term in Equation (9.16) refers to the turbulent Reynolds stresses due to each individual particle's interaction with the main (air) flow. These interactions involve the dispersed phase in dilute flow; otherwise, interactions with other droplets may also become significant. In a dilute flow, this droplet phase turbulent stress arises mainly from turbulence in the gas (carrier) phase. As a result, it is often written in terms of a turbulent eddy viscosity through the Boussinesq approximation outlined by Equation (3.232), but special near-wall formulations are required to provide the correct details of particle–wall contact.

An alternative to using the Boussinesq approximation for the effective viscosity is the application of kinetic theory to obtain the constitutive relations (Gidaspow, 1994). In this approach, an additional equation is formulated and solved to give the kinetic energy associated with the fluctuation of the droplet velocity about its mean value. This resulting *granular temperature* is related to the bulk viscosity of the main (gas) flow and the particle phase viscosity. In this approach, various processes are neglected, such as particle–particle and particle–wall interactions, thereby limiting its applicability to certain types of flows (i.e., vertical but not horizontal flows).

9.2.2.3 Energy Equation

A procedure similar to that outlined previously for the mass and momentum equations can be used to find the energy equation for multiphase flows. The relevant energy terms are substituted into the general scalar conservation equation, Equation (9.6). In particular, let $e_k = \phi_k = \hat{e}_k + \frac{1}{2}\mathbf{v}_k \cdot \mathbf{v}_k$ (total energy), $\mathbf{j}_k = \mathbf{q}_k + (p_k \mathbf{I} - \tau_k) \cdot \mathbf{v}_k$ (Fourier heat flux and work contributions arising from pressure and viscous forces), and $\hat{S}_k = \mathbf{F}_{b,k} \cdot \mathbf{v}_k + \hat{S}_{k,e}$ (work contributions arising from body forces, such as gravity, and heat generated per unit volume). The energy equation for multiphase flow with droplets is obtained once these individual parts are substituted into Equation (9.6).

In addition to its relevance in the freestream, the energy equation is focused more specifically on the near-wall region when three-phase systems are considered. For example, interactions with the third (solid) phase often occur through a heat balance at the moving solid boundary where impinging droplets are collected. At this moving boundary, the three-phase conditions include the following heat flows: (i) convective heat transfer to and from the surrounding gas stream, (ii) impinging droplets with an unfrozen liquid film, and (iii) conduction and release of latent heat into the solid phase. These three-phase heat balances are discussed in the following case study.

9.2.3 Case Study: Atmospheric Icing of Structures

An analysis of impinging droplets with icing of overhead power lines is considered. The purpose of this case study is to demonstrate the interactions between various thermal processes arising in three-phase conditions. In particular, the three phases include phase change and conduction in the solid, external airflow, and impinging droplets. Similar processes are observed in other three-phase applications, such as aircraft de-icing and spray deposition in manufacturing processes.

In the present case of iced power lines, a single, unstranded, nonrotating conductor is fixed in a uniform external cross-flow with a full collection efficiency (i.e., $E = 1$). Constant thermophysical properties are assumed for the water, air, and ice. Also, evaporation, radiation, sublimation, splash-back, and turbulence effects on the surface ice accretion are assumed to be negligible. An approximate analysis of three-phase heat transfer under icing conditions was considered by Naterer (2002a) with coworkers (1999).

9.2.3.1 Glaze Ice Growth

Impinging supercooled droplets extract heat from the ice upon impact on the ice surface (see Figure 9.2). Then latent heat is released by the freezing droplets, while heat is transferred by convection to the surrounding airstream, as well as conduction into the ice. Although rime (dry) ice may initially appear on the surface, as the ice temperature reaches $T_F = 0°C$, an unfrozen water layer forms along the ice surface (called glaze ice). During glaze ice growth, the droplet mass influx freezes only partially and releases

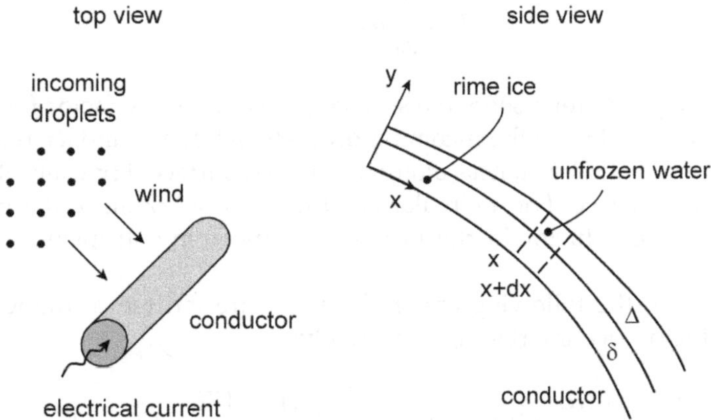

FIGURE 9.2
Ice accretion on a circular conductor.

latent heat. The remaining liquid flows in a thin film along the surface as unfrozen water.

Consider a mass balance for the glaze ice layer and unfrozen water film along the surface (see Figure 9.2). The origin of the (x, y) axis system is taken at the top of the conductor. In a small time step, dt, an incremental increase in the ice thickness, $d\delta$, occurs across dx, as a result of incoming droplets and the unfrozen liquid film along the ice surface, i.e.,

$$\rho_i d\delta(Wdx) = EVG(Wdx_p)dt + [\dot{m}_l]_x dt - [\dot{m}_l]_{x+dx} dt \tag{9.17}$$

where E, W, and G and the subscript p refer to collection efficiency, surface depth, liquid water content, and projected distance (normal to the direction of incoming droplets), respectively, Also, $\dot{m}_l / W = \Gamma$ is defined as the mass flow rate of unfrozen water per unit depth of surface.

Based on Equation (9.17), it can be observed that the amount of ice accumulation balances the water influx rate (into the control volume) minus the water outflow rate from the control volume at the downstream location, $x + dx$. In this case study, an integral formulation of the mass conservation equation will be adopted to estimate the ice growth rate. Using a Taylor series expansion and neglecting higher order terms in Equation 9.17,

$$\rho_i \frac{d\delta}{dt} = EVGN - \frac{d\Gamma}{dx} \tag{9.18}$$

where

$$\Gamma = \int_0^\delta \rho u dy \tag{9.19}$$

and $N = \int_c dx_p / \int_c dx$ represents the surface projection view factor for the side of the conductor facing the incoming droplets. Also, dx_p and dx refer to the projected and actual distances along the curved surface. For example, $N = 1$ for a planar surface. For a circular conductor, $N = 1/\pi$ since the projected width of the conductor in the presence of incoming droplets is $2r_o$ (not $2\pi r_o$).

In addition, the following energy balances are written at the ice–water and conductor–ice interfaces, respectively:

$$-k_i \frac{\partial T}{\partial y}\bigg|_i + \frac{VGL}{\pi} = h(T_F - T_a) + c_w \Delta T \frac{d\Gamma}{dx} + \frac{VG}{\pi} c_w (T_F - T_a) \tag{9.20}$$

$$-k_i \frac{\partial T}{\partial y}\bigg|_i = q_e \equiv \frac{I^2 R}{2\pi r_o} \tag{9.21}$$

where q_e, r_o, I, and R refer to the electrical heating per unit surface area, conductor (un-iced) radius, electrical current, and resistivity, respectively.

The last term in Equation (9.20) represents the heat transferred by the supercooled droplets as they extract heat from the interface to warm up to T_F in the liquid film. Furthermore, between x and $x + dx$ (see Figure 9.2) inside the unfrozen water layer,

$$\rho_i L \frac{d\delta}{dt} = h(T_F - T_a) + Vgc_w(T_F - T_a) + \dot{m}\hat{h}_F + \frac{\partial(\dot{m}\hat{h}_F)}{\partial x} dx$$

$$+ k_i \frac{\partial T}{\partial y}\bigg|_i - \dot{m}\hat{h}_F - \frac{\partial(\dot{m}\hat{h}_a)}{\partial x} dx \tag{9.22}$$

where $\hat{h}$ refers to the fluid enthalpy (in contrast to h, which refers to the convective heat transfer coefficient).

Combining the previous equations and solving Equation (9.22) subject to an initial condition of $\delta(0) = 0$,

$$\delta = \left[\frac{2VG/\pi + h/c_w}{\rho_i(1 + 1/Ste)}\right] t \tag{9.23}$$

where $Ste = c_w \Delta T / L$ is the Stefan number and zero joule heating, q_e, has been assumed at the conductor surface. In terms of a dimensionless time, t^*, and dimensionless ice thickness, δ^*,

$$\delta^* = \frac{1}{2}\left[1 - \left(\frac{\omega/Ste - 1}{1/Ste + 1}\right)\right]t^* \tag{9.24}$$

where

$$\omega = 1 - \frac{\pi h \Delta T}{VGL} \tag{9.25}$$

and

$$t^* = 2\frac{\rho_w}{\rho_i}\frac{Pt}{\pi r_o} \tag{9.26}$$

$$\delta^* = \frac{\delta}{r_o} \tag{9.27}$$

Also, $P = VG/\rho_w$ and r_o refer to the precipitation rate and conductor (un-iced) radius, respectively. In Equation (9.27), δ refers to the equivalent ice thickness, which represents the uniformly radial ice thickness yielding the same ice mass as the actual observed ice accretion.

9.2.3.2 Runback Water and Growth of the Unfrozen Water Layer

A quasi-steady approximation will be adopted for the unfrozen runback water in the glaze film. The variable $\hat{y} = y - \delta$ (see Figure 9.2) is defined, and the hat notation can be dropped for brevity in this analysis. From a mass balance in the unfrozen water film between x and $x + dx$, together with Leibnitz's rule along the moving film interface,

$$\frac{\partial}{\partial x}\int_0^{\Delta} u dy - u(\Delta)\frac{d\Delta}{dx} = VG \tag{9.28}$$

where $u(\Delta)$ refers to liquid velocity at the water–air free surface.

Based on the reduced form of the momentum balance in the thin liquid layer along the ice surface,

$$0 = \rho_w g_x + \mu \frac{d^2 u}{dy^2} \tag{9.29}$$

where μ_w and g_x are the liquid viscosity and gravitational component in the direction along the surface, respectively.

At the water–air interface, a slip boundary condition is required due to movement of the unfrozen water layer. Integrating Equation (9.29) and equating the resulting shear stress with the value obtained from the external flow solution at the edge of the glaze film (Naterer et al., 1999),

$$\mu_w \frac{du}{dy}\bigg|_{y=\Delta} = -\rho_w g y + constant = 2.1\left(\frac{\mu_a \rho_a V^3}{r_o}\right)^{1/2} \tag{9.30}$$

Also, the velocity distribution within the unfrozen water layer can be determined from Equation (9.29), subject to a no-slip boundary condition at the ice–water interface, i.e.,

$$u = -\left(\frac{\rho_w g}{2\mu_w}\right)y^2 + \left[\frac{\rho_w g}{\mu_w} + 2.1\left(\frac{\mu_a \rho_a V^3}{r_o \mu_w^2}\right)^{1/2}\right]y \tag{9.31}$$

Using this result in Equation (9.28),

$$\Delta^2 \frac{d\Delta}{dx} = 2\frac{\mu_w P}{\rho_w g} \tag{9.32}$$

Solving Equation (9.32) subject to a zero film thickness at the top of the conductor,

$$\Delta(x) = \left(\frac{6\mu_w P}{\rho_w g_x}\right)^{1/3} x^{1/3} \tag{9.33}$$

which indicates that the film thickness increases with precipitation rate and distance along the surface (as expected).

Furthermore, the film thickness is used to find the mass flow rate and its derivatives. Then Γ is substituted into the earlier energy balances to obtain the following result for the dimensionless ice growth rate:

$$\delta^* = \frac{1}{2}\left[1 - \left(\frac{2\mu_a}{3(1 + 1/Ste)\mu_w}\right)\left(\frac{\rho_w}{G}\right)^{1/3}\left(\frac{6\pi\mu_w V}{\rho_w g r_o^2}\right)^{2/3}\right]t^* \tag{9.34}$$

This formulation has been successfully compared with experimental data (Naterer et al., 1999). Despite the limitations of this approximate analysis, certain useful overall features and trends in the data can be inferred through the current modeling. In the following section, another case study involving three-phase heat transfer is considered.

9.2.4 Case Study: Melt Particularization

An important manufacturing process involving solidification or melting with droplets and airstream interaction is *melt particularization*. An example is powder production and droplet formation from a melted metal stream injected into a gas. Melt particularization is often adopted in applications such as ceiling sprays in drywall construction and manufacturing of parts that require close dimensional tolerances, such as gears. In this process, an injected molten stream is disintegrated by fluid interaction with a liquid or gas, thereby involving three-phase flow with heat transfer.

Initially, breakup of the injected molten stream into droplets is encountered, where the droplet diameter, d_o, is determined by a thermodynamic balance involving surface tension along the droplet and gravity. The disintegrated droplets come into contact with a cross-stream fluid to initiate secondary breakup into smaller droplets. In this secondary breakup, liquid evaporation and droplet coalescence are often realized. Finally, the droplets are solidified (see Figure 9.3).

The droplet and solidification dynamics are usually determined by the Weber number, We_d (ratio of droplet inertia to surface tension), and the Reynolds number, Re_d. At high Re_d and We_d numbers, the droplets are atomized (broken into distinct, separate droplets). De-stabilization and disintegration (stretched or wispy droplet streams) occur at lower values. In addition to this fluid dynamics, the rate of heat transfer from the solidifying droplets affects their structural properties following impact on the collection surface. The following Mehrotra correlation can be adopted for the convective heat transfer coefficient in melt particularization processes (Sahm and Hansen, 1984; Mehrotra, 1981):

$$h = \left(\frac{k_g}{d_o}\right)(2 + 0.6Re_d^{1/2}Pr^{1/3}) \tag{9.35}$$

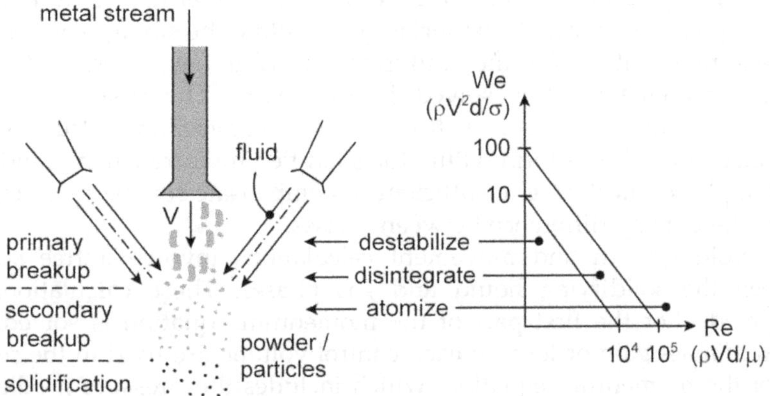

FIGURE 9.3
Schematic of melt particularization.

where k_g is the thermal conductivity of the gas stream. Convective heat transfer from the droplets is a predominant factor affecting the solidification process. Proper thermal control during melt particularization is essential in establishing the final desired quality of the solidified droplets.

In the previous cases studies, three-phase systems were examined in relation to droplet flows (air–liquid phases) with phase change (solid phase). In the following section, other examples of three-phase systems are examined, but without droplets in the freestream flow. Instead, gas flow is considered in the freestream, while solid–liquid phase change and heat transfer are encountered at other locations within the physical domain.

9.3 Gas Flows with Solidification and Melting

9.3.1 Case Study: Free Surfaces with Phase Change

The analysis of free surface flows is challenging, particularly in conjunction with phase change heat transfer. An example considered here is solidification shrinkage that initiates a free surface flow between a liquid and gas phase. This example arises in various applications, such as casting solidification of metals and alloys (Naterer, 1997), plastic injection molding, and thermal energy storage systems using phase change materials (Darling et al., 1993).

For instance, Darling et al. (1993) consider the void growth in thermal energy storage canisters for effective spacecraft thermal control, which involves phase change and free surface flow within the storage canister. The momentum equation for the multiphase system (i.e., solid and liquid phases simultaneously) is solved in two parts. The first part of the momentum equation involves the transient, convection, diffusion, and buoyancy terms. Voids form within the solidified mixture when liquid flow into the phase interface is insufficient to compensate for the mass change due to the density difference between phases.

The void growth and movement calculations involve a free surface between the solidifying liquid and gas phases. These calculations are performed after the first part of the momentum equation is solved. The values of mass gain or loss for each control volume are used in the second part of the momentum equation, which includes the pressure gradient to drive the transient changes of fluid momentum. If the solid density is larger than the liquid density, shrinkage flow is induced when the material contracts on solidification. The rate at which mass is drawn in (shrinkage) or expelled (expansion) is based on Equation (5.70). Once the solution of velocity is obtained from the mass and momentum equations, the resulting free surface displacement must be constrained appropriately by an overall conservation of mass.

In a numerical solution procedure, any mass adjustments in discrete control volumes are subject to two main constraints: (i) a control volume cannot have negative mass; and (ii) it cannot be overfilled. If shrinkage flow is induced into a solidifying control volume, each pocket of liquid drawn into the control volume solidifies and shrinks since the solid density is larger than the liquid density. As a result, even more liquid is required. A possible numerical solution is to completely change the phase of the control volume if its enthalpy is low enough to solidify any liquid entering the available portion of the control volume. Alternatively, the influence of mass exchange through solidification shrinkage may appear directly through the pressure gradient in the momentum equation and continuity (Xu and Li, 1991). In this approach, Xu and Li (1991) assumed that the pressure forces are always sufficient to drive the fluid from the riser (fluid entry section) into void locations subject to conservation of mass.

After a free surface forms due to the solidification shrinkage, a numerical method is required to accommodate the movement of the free surface in conjunction with the phase transition and surrounding gas stream. A widely used technique is called the volume of fluid (VOF) method (Hirt and Nichols, 1981). The application of this procedure in the context of a numerical method for multiphase flows is discussed in Chapter 11. During the numerical implementation, appropriate boundary conditions are required along the free surface (dividing solid–liquid and gas phases). These boundary conditions include matching conditions of shear stress, temperature, and velocity. In addition, surface tension forces act along the interface of the free surface. Once boundary conditions are established along the free surface with the gas flow, the problem can be reduced to single phase equations within each individual phase. Iterations are performed at the phase interface and free surface to resolve their respective locations.

9.3.2 Case Study: Ablation Shields

Another example involving three-phase heat transfer occurs in an *ablation shield*. During reentry into the atmosphere, the surface of a spacecraft encounters large heating due to frictional contact with the air and the resulting formation of a shock wave near the surface. An ablation shield is a heat shield in an outer layer of material along the surface of the spacecraft that melts to prevent overheating of the spacecraft. The melting material is selected to have a high latent heat of fusion and low thermal conductivity.

By reinforcing the shield with quartz fibers, the material is largely strengthened when it is molten. A solid part of the shield encloses the inner layer of the wall, outside of which resides the liquefied part of the shield, gaseous plasma (typically about 6000°C), shock wave, and surrounding atmospheric air.

Upon atmospheric reentry, the hypersonic reentry velocity generates a curved shock wave and intense friction in front of the spacecraft, which causes the ablation shield to become gaseous due to melting and evaporation. The interface between the liquid and gas at the outer layer of the shield reflects about 80% of the dissipative heat generated due to the shock wave back into the surrounding atmosphere by radiation. The proportion of heat that conducts through the wall and into the spacecraft is about 4% of the generated heat. As a result, ablation shields represent an effective method of thermal control of atmospheric reentry vehicle surfaces.

9.4 Chemically Reacting Systems

Chemically reacting systems represent another widely encountered example involving multiphase flows with gas, droplet, and solid (particle) phases. Chemically reacting mixtures with combustion involve heat transfer, species concentration equations, and multiphase flow equations. The thermofluid processes largely affect the rates and characteristics of the reactions. In this section, the fundamentals of chemical reactions with heat transfer will be considered. These reactions involve the compositions of reactants and products in the chemically reacting flows.

Combustion of pulverized coal particles is a common example of multiphase flow with solid and gas phases, whereas combustion of fuel droplets involves liquid and gas phases. In many instances, solid (such as soot particles), liquid (droplets), and gas (i.e., air, carbon dioxide) phases coexist simultaneously in combustion systems. Soot refers to black particles consisting mainly of carbon particles formed under the incomplete combustion of hydrocarbon fuels. Soot particles can be present in a combustion chamber, thereby creating a complex multiphase flow with particles contributing to radiative and convective heat transfer, while interacting with combustion of fuel droplets in the gas flow.

Consider a mixture involving N constituents. The following definitions will be used:

- Mass of mixture $= m = m_1 + m_2 + \ldots = \sum_1^N m_i$

- Mass fraction = $mf_i = m_i/m$
- Total number of moles of mixture $= n = n_1 + n_2 + \ldots = \sum_1^N n_i$
- Mole fraction = $y_i = n_i/n$

The mass of component i (given by m_i) and the number of moles of i (called n_i) are related by the molecular weight, M_i, as follows:

$$m_i = n_i M_i \tag{9.36}$$

which implies that

$$m = \sum_{i=1}^{N} n_i M_i \tag{9.37}$$

The units of molecular weight (or atomic weight; refer to Table A10 in the appendix) are kg/kmol.

The mixture molecular weight is given by

$$M = \frac{m}{n} = \sum_{i=1}^{N} y_i M_i \tag{9.38}$$

Also, the mass fraction and mole fraction are related by

$$\frac{mf_i}{y_i} = \frac{m_i/m}{n_i/n} = \left(\frac{m_i}{n_i}\right)\left(\frac{n}{m}\right) = \frac{M_i}{M} \tag{9.39}$$

A *molar analysis* specifies the number of moles of each component in a mixture, whereas a *gravimetric analysis* specifies the mass of each component.

For a mixture of ideal gases in a reacting flow, it can be shown that the mole fraction can be written in terms of the volume and pressure ratios as follows (see end-of-chapter problem):

$$y_i = \frac{V_i}{V} = \frac{P_i}{P} \tag{9.40}$$

where P_i and P refer to the partial pressure of component i and total pressure, respectively. The partial pressure refers to the pressure that the individual component would have if it occupied the entire volume of the mixture by itself. Also, V_i refers to the volume occupied by component i (i.e., the volume occupied if the individual component was isolated from the other components).

Mole fractions are frequently used in the analysis of reacting flows since the fractions of each reactant (by volume) are often known. Also, the properties of reacting mixtures are typically based on mole (or mass) fraction-weighted sums of individual component properties. For example, the specific heat of an air (subscript A)–fuel (subscript F) mixture is approximately

$$c_p = (mf)_A c_{p,A} + (mf)_F c_{p,F} \tag{9.41}$$

or alternatively, in terms of molar quantities,

$$\hat{c}_p = y_A \hat{c}_{p,A} + y_F \hat{c}_{p,F} \tag{9.42}$$

where y_A and y_F are the mole fractions of air and fuel, respectively. Also, mf_A and mf_B are the respective mass fractions. Once the mixture properties are evaluated in this manner, the remaining heat balance equations can be solved in the regular manner in terms of these mixture properties.

The analysis of reacting flows requires consideration of the chemical balance and rate equations that describe the chemical reaction(s). A commonly encountered reacting flow is the combustion of a hydrocarbon fuel (C_aH_b), such as methane (CH_4), in air. Air is approximately composed of 21% O_2 (oxygen) and 79% N_2 (nitrogen) by volume. In terms of the number of moles of air,

$$1\ Air = 0.21O_2 + 0.79N_2 \tag{9.43}$$

or

$$4.76\ Air = 1O_2 + 3.76N_2 \tag{9.44}$$

Thus, the chemical equation for the reaction of a hydrocarbon fuel (droplets or particles) in air can be modeled by

$$C_aH_b + 4.76\left(a + \frac{b}{4}\right)(0.21O_2 + 0.79N_2) \rightarrow aCO_2 + \frac{b}{2}\ H_2O + 3.76\left(a + \frac{b}{4}\right)N_2 \tag{9.45}$$

The leading coefficients of each term were obtained based on conservation of atoms for carbon, hydrogen, oxygen, and nitrogen on the left side of Equation (9.45). The second term on the left side of Equation (9.45) is the *stoichiometric air* (or *theoretical air*), which refers to the minimum amount of air required to provide complete combustion of the hydrocarbon fuel. The given by-products of combustion are formed under conditions of complete combustion when all C_aH_b is burned in air without any excess air remaining from the reaction. Otherwise, by-products such as carbon monoxide and others arise if not all of the C_aH_b is burned (called *incomplete combustion*). Incomplete combustion is usually undesirable for many reasons, e.g., the release of harmful pollutants into the atmosphere, reduced fuel efficiency. Thus, the amount of air supplied should be equal to or greater than the stoichiometric amount in Equation (9.45).

Example 9.4.1

Combustion of Methane in Air.

Write the chemical balance equation for the combustion of methane in air. Is the number of moles of reactants equal to the number of moles of products of combustion?

Using $a = 1$ and $b = 4$ in Equation (9.45) for methane, the chemical reaction for the combustion of methane in air is expressed by

$$CH_4 + 9.52(0.21O_2 + 0.79N_2) \rightarrow CO_2 + 2H_2O + 7.52N_2 \tag{9.46}$$

It can be observed that the sum of moles of reactants on the left side does not balance the sum of moles of products on the right side. However, the total mass of reactants must balance the total mass of products.

The mass of reactants or products, m, can be determined by the number of moles, n, multiplied by the molecular weight, M. Since air consists mainly of oxygen and nitrogen, the molecular weight is based on the mole fraction-weighted sum of individual molecular weights, i.e.,

$$M_{air} = \sum_{i=1}^{2} y_i M_i = y_{O_2} M_{O_2} + y_{N_2} M_{N_2} \tag{9.47}$$

where the subscript i refers to constituent i and y_i refers to the mole fraction of constituent i. Substituting the molecular weights of oxygen and nitrogen,

$$M_{air} = 0.21(32) + 0.79(28) = 28.84 \text{ kg/kmol} \tag{9.48}$$

The *air–fuel mass ratio* (AF) is defined by

$$AF = \frac{m_{air}}{m_{fuel}} = \frac{n_A M_A}{n_F M_F} = \frac{4.76(a + b/4)M_{air}}{aM_C + bM_H} \tag{9.49}$$

where $M_C = 12$ kg/kmol and $M_H = 1$ kg/kmol refer to the molecular weights of carbon and hydrogen, respectively. Also, *excess air* (EA) is defined as more air than the stoichiometric amount. For example, 25% excess air corresponds to 1.25 times the stoichiometric amount of air. Adapting Equation (9.45) to chemical reactions with excess air,

$$C_aH_b + (1 + EA)\left[4.76\left(a + \frac{b}{4}\right)(0.21O_2 + 0.79N_2)\right] \rightarrow aCO_2 + \frac{b}{2}\, H_2O$$

$$+(1+EA)4.76\left(a+\frac{b}{4}\right)(0.79\mathrm{N_2})+(EA)4.76\left(a+\frac{b}{4}\right)(0.21\mathrm{O_2}) \quad (9.50)$$

The last term in Equation (9.50) arises since the reaction is not stoichiometric. As a result, not all oxygen reacts. During a chemical reaction, the *heat of combustion* (or *enthalpy of combustion*) refers to the amount of heat release during the combustion process. The following example calculates these heating values during a combustion process.

Example 9.4.2

Heat Release in a Combustion Chamber.

Fuel and air enter a mixing chamber and react to form certain products of combustion. Find the amount of heat released by combustion while raising the products of combustion to an outlet temperature of T_P. Express the result in terms of the enthalpies of the reactants and products of combustion.

This problem is assumed to be a steady reacting flow through the combustion chamber. The amount of heat released by combustion is equivalent to the amount of heat required to be added to the combustion chamber in order to raise the reactants to a temperature of T_P. Defining a control volume that encompasses the combustion chamber, an energy balance for this control volume requires that

$$Q=\sum_{i=1}^{N_P}(\dot{m}h)_{i,P}-\sum_{j=1}^{M_R}(\dot{m}h)_{j,R} \quad (9.51)$$

where the subscripts P and R refer to products and reactants, and h refers to enthalpy (kJ/kg). Alternatively, in terms of the molar flow rate, $\dot{n}$, the required heat addition is

$$Q=\sum_{i=1}^{N_P}(\dot{n}\hat{h})_{i,P}-\sum_{j=1}^{M_R}(\dot{n}\hat{h})_{j,R} \quad (9.52)$$

where $\hat{h}$ refers to the molar enthalpy. These results indicate that the heat released by combustion is given by the difference of enthalpy between the products and reactants.

The mixture enthalpy can be evaluated in terms of mass fractions and individual enthalpies as follows:

$$mh=m\sum_{i=1}^{N_P}(mf)_i h_i=\sum_{i=1}^{N_P}m_i h_i \quad (9.53)$$

Alternatively, in terms of mole fractions,

$$n\hat{h} = n\sum_{i=1}^{N_P} y_i\hat{h}_i = \sum_{i=1}^{N_P} n_i\hat{h}_i \tag{9.54}$$

where the molar enthalpy, $\hat{h}$, has units of kJ/kmol.

The previous example involved enthalpy for energy balances based on the First Law of Thermodynamics. For the Second Law of Thermodynamics, the mixture entropy is required. It can be shown that the entropy of component i (subscript i) in the mixture can be written in terms of the absolute pressure and reference values as follows (see end-of-chapter problems):

$$s_i = s_{i,ref} + c_{p,i} ln\left(\frac{T}{T_{ref}}\right) - R_i ln\left(\frac{P}{P_{ref}}\right) - Rln(y_i) \tag{9.55}$$

If there is only one component in the mixture ($i = 1$ for a pure gas), then the last term in Equation (9.55) vanishes because $y_i = 1$ in this case. The subscript *ref* in Equation (9.55) refers to reference value. In general for reacting flows, reference values are typically taken at standard temperature/pressure (STP) (i.e., 298 K and 1 atm).

In the previous example, a negative Q in Equations (9.51) and (9.52) would indicate that heat is removed to lower the products to the temperature of T_P. A reaction is called *endothermic* if Q is positive (heat added) and *exothermic* if Q is negative (heat removed). The *heating value* of a fuel is defined as the heat of combustion for 1 mole of the fuel in stoichiometric air when the reactants and products are at STP. The *higher heating value* (HHV) refers to the heating value when H_2O in the products of combustion is in the liquid phase. Conversely, the *lower heating value* (LHV) is the heating value when H_2O in the products is in a gaseous phase.

The *adiabatic flame temperature*, T_{ad}, refers to the highest temperature that will be obtained by burning a fuel (in droplet or particle form) under a specified set of conditions. For example, this temperature is obtained when the reactants enter a well-insulated mixing chamber and the products of combustion leave the chamber at T_{ad}. In other words, all heat generated by the combustion reaction is transferred to raise the temperature of the combustion products. Typical values of adiabatic flame temperatures corresponding to STP for the reactants with stoichiometric air are readily available in tabular form for various fuels. For example, $T_{ad} \approx 2300$ K for certain hydrocarbon fuels. In Figure 9.4, it can be observed that T_{ad} varies when the air–fuel ratio, *AF*, is changed. In particular, T_{ad} decreases if more excess air is used since more energy is required to raise the temperature of

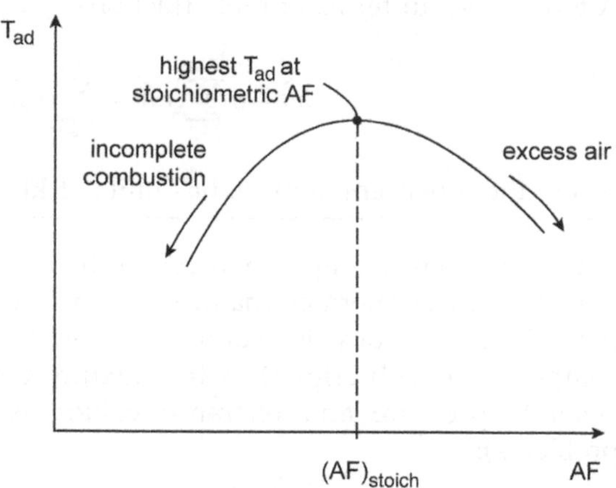

FIGURE 9.4
Adiabatic flame temperature.

the nonreacting air. Conversely, if less air than the stoichiometric amount is supplied, T_{ad} decreases due to the incomplete combustion. As a result, the maximum adiabatic flame temperature is reached at the air–fuel ratio corresponding to the stoichiometric amount of air.

For a nonreacting mixture, the fluid constituents that enter a control volume also come out. As a result, a reference enthalpy, h_{ref}, is often not required in these single phase problems since only enthalpy differences between the same constituents at the outlet and inlet are required. However, for a reacting mixture, the components or compounds entering the control volume do not all come out. In this case, h_{ref} becomes important. Reactions leading to new constituents or compounds are called *formation reactions* because a substance is formed from its elements in their respective *natural states* (i.e., gas, liquid, or solid) at STP.

For example, consider the heat released by the following formation reactions of multiphase systems:

- $\frac{1}{2}O_2(gas) + H_2(gas) \rightarrow H_2O(liquid)$ yields $\bar{q} \approx -286$ MJ of heat per kmol of H_2O
- $C(solid) + O_2(gas) \rightarrow CO_2(gas)$ yields $\bar{q} \approx -394$ MJ of heat per kmol of CO_2.

The overbar notation refers to the heat flow per kmol of product, which is represented by the total heat transfer in joules, Q, divided by the number of moles of products of combustion. The negative sign indicates that the products of combustion are very hot, so that heat must be rejected from the control volume (encompassing the reacting flow) to return the exit temperature down to STP. The previous $\bar{q}$ values are called the *standard heats of formation* (or *enthalpies of formation*). For a formation reaction, a notation of $\bar{h}_f^o = \bar{q}$ is used for the enthalpy of formation, where the superscript o denotes STP.

A fundamental part of analyzing reacting flows is the numerical computation of enthalpy. Consider the following reaction to calculate the enthalpy of formation of the product:

$$C(solid) + \frac{1}{2}\, O_2(gas) \rightarrow CO(gas) \tag{9.56}$$

In this example, $\bar{q} = \bar{h}^o_{f,CO}$ is difficult to measure because carbon (C) tends to burn to carbon dioxide (CO_2) rather than carbon monoxide (CO). However, the following *summation law of reactions* can be used:

$$C(solid) + O_2(gas) \rightarrow CO_2(gas) \tag{9.57}$$

For this reaction, $\bar{q} \approx -393$ MJ/kmol. Also,

$$CO(solid) + \frac{1}{2}\, O_2(gas) \rightarrow CO_2(gas) \tag{9.58}$$

with $\bar{q} \approx -283$ MJ/kmol. Subtracting Equation (9.58) from Equation (9.57), together with their respective enthalpies of formation,

$$C(solid) + \frac{1}{2}\, O_2(gas) \rightarrow CO(gas) \tag{9.59}$$

so that $\bar{q} = h^o_{f,CO} \approx -110$ MJ/kmol. It should be noted that CO is not an element in Equation (9.58), and thus $\bar{q} = -283$ MJ/kmol does not represent the enthalpy of formation of CO_2 on the right side of the equation.

A standard convention for enthalpy calculations is that the enthalpy of every element in its natural state at STP is set to zero. For example, $\bar{h}_f^o = 0$ for O_2 (gas), H_2 (gas), and C (solid). Also, an enthalpy scale is defined with reference to STP, i.e.,

$$\bar{h}_j(p,\ T) = \bar{h}^o_{f,j} + \int_{STP}^{T,p} d\bar{h}_j \tag{9.60}$$

where the first term represents chemical enthalpy and the second term is sensible enthalpy. The chemical enthalpy is set to zero for elements in their natural state. Evaluating the integral in Equation (9.60) for ideal gases,

$$\bar{h}_j(T) = \bar{h}_{f,j} + c_{p,j}(T - 298) \tag{9.61}$$

Variations of specific heat with temperature can be included in the evaluation of enthalpy in Equation (9.61). The following problem shows an example of how to use these enthalpy values to determine the rate of heat release during a combustion reaction.

Example 9.4.3

Multiphase Reacting Flow with Droplets and Excess Air.

Droplets of liquid propane (C_3H_8) are burned at atmospheric pressure with 150% excess air in a combustion chamber. Find the heat released within the combustion chamber if the products of combustion exit the chamber at T_P. The rate of fuel mass inflow is denoted by $\dot{m}$ for this steady flow. The fuel enters at STP, whereas the air inflow temperature is T_A. It may be assumed that the fuel burns completely with the excess air.

From Equation (9.50), with $EA = 0.5$, the chemical equation may be written as

$$C_3H_8 + 12.5O_2 + 47N_2 \rightarrow 3CO_2 + 4H_2O + 7.5O_2 + 47N_2 \tag{9.62}$$

Based on this chemical equation, the air–fuel ratio becomes

$$AF = \frac{m_A}{m_F} = \frac{n_{O_2}M_{O_2} + n_{N_2}M_{N_2}}{n_{C_3H_8}M_{C_3H_8}} \tag{9.63}$$

$$AF = \frac{12.5(32) + 47(28)}{1(44.1)} = 39\frac{kg(air)}{kg(fuel)} \tag{9.64}$$

Also, based on the conservation of mass,

$$\dot{m}_F + \dot{m}_A = \dot{m}_P \tag{9.65}$$

or alternatively,

$$\dot{n}_F M_F + \dot{n}_A M_A = \dot{n}_P M_P \tag{9.66}$$

Performing an energy balance over the control volume (combustion chamber),

$$Q = \sum_{i=1}^{3} n_{R,i}(\bar{h}_f^o + \bar{h} - \bar{h}^o)_{R,i} - \sum_{j=1}^{4} n_{P,j}(\bar{h}_f^o + \bar{h} - \bar{h}^o)_{P,j} \tag{9.67}$$

where $\bar{h}$ is a function of temperature alone for ideal gases. The values of enthalpy (reactants and products) are evaluated at the given temperatures. Once these values are retrieved from appropriate thermodynamic tables, they can be substituted into Equation (9.67) to find the heat released by combustion of the fuel droplets, i.e.,

$$\dot{Q} = \left(\frac{\dot{m}}{M}\right) Q \tag{9.68}$$

where M is the molecular weight of liquid propane (44 kg/kmol).

In this example, the outlet temperature was given so that the enthalpies could be readily computed as functions of temperature. However, in some cases, these temperatures are unknown. As a result, an iterative solution procedure is required. For example, if the heat transfer from the combustion chamber was given instead of the outlet temperature in the previous problem, the following sequence of steps could be used:

1. Estimate the outlet temperature and then calculate the enthalpy of the products of combustion based on this temperature.
2. Check and determine whether the energy equation, Equation (9.67), is satisfied.
3. Repeat steps 1 and 2 until the energy equation is satisfied.

In addition to the amount of heat released during the reaction, the amount of air used in the reaction is an important concern in combustion analysis. Furthermore, the state of the reactants and the degree of reaction completion affect the dynamics of combustion. In the following sections, further examples of chemically reacting flows in multiphase systems will be examined, particularly in three-phase systems, including gas, liquid (droplet), and solid (particle) phases.

9.5 Multiphase By-products of Reacting Flows

9.5.1 Gas–Gas Interactions

By-products of reacting flows can often arise in solid, liquid, or gas phases. The phases of combustion products can differ from the initial phases of the reactants. In this section, interactions between two gases will be considered as they react to form products in gaseous, liquid, or solid phases. Although the transport processes will be largely described in relation to a certain application (reaction of metallic vapor with oxygen), the fundamentals of the thermofluid processes are generally applicable to other types of gas–gas reactions yielding multiphase reactants, such as solid particles or liquid droplets.

Consider the combustion of droplets within a furnace or internal combustion engine, where volatilization of hydrocarbon vapor from the surface of a droplet interacts with diffusion and convection of oxygen toward the droplets. Volatilization refers to a process of changing readily to vapor (i.e., quickly evaporating). The gases react to form the products of combustion, which can include CO_2 (gas), CO (gas), H_2O (liquid or gas), soot particles (solid), and others. The black soot particles typically arise with carbon monoxide due to the incomplete combustion of the fuel. In this example, two gases have reacted to form a mixture of multiphase products.

Another common example involving reactions between gas streams to form multiphase products of combustion occurs in metallurgical applications. For instance, in fuming phenomena, oxygen reacts with evaporating metal droplets in steel-making processes to form a very fine layer of oxide particles in the boundary layer around the droplet (see Figure 9.5). Fuming arises in various metallurgical applications, such as the refining of nickel and zinc, as well as vacuum refining operations (Guthrie, 1993).

In fuming processes, the nucleation of particles in the gas boundary layer around the evaporating droplet leads to submicrometer-size dust particles that are difficult to filter out of the exhaust gases from the furnace. In some steel-making operations, up to about 1 tonne of iron oxide dust can be produced for every 100 tonnes of steel produced. This can lead to a considerable amount of environmental pollution. Also, this formation of solid particles from the reaction of two gas streams can cause potential

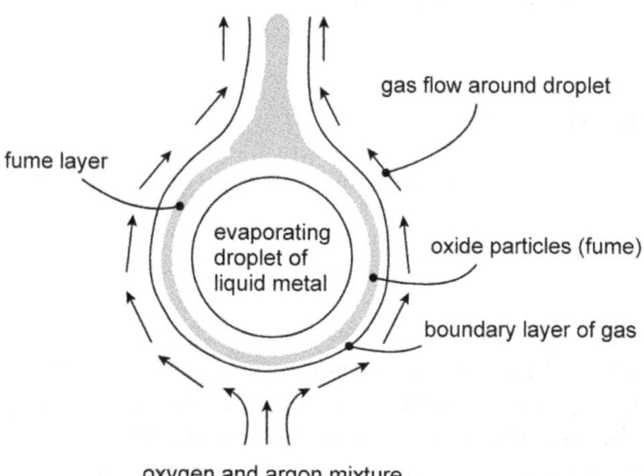

FIGURE 9.5
Fuming process in metallurgy.

injuries to personnel involved in these manufacturing processes. Thus, better control of the chemical and thermal processes is important in view of reducing and eliminating injuries and pollution to the environment.

The thermal and fluid dynamics of two interacting gas streams will be considered in reactions that produce products of combustion in other phases (i.e., solid oxide particles). During fuming around a liquid droplet, a concentration gradient of metallic vapor is established around the droplet. A diffusion layer of metallic vapor concentration is formed in the presence of the surrounding oxygen flow. If the partial pressure of oxygen in the gas flow is sufficiently high, then the two gas streams react to form an oxide fume, i.e.,

$$aF(gas) + \frac{b}{2}\,O_2 \rightarrow F_aO_b(solid) \tag{9.69}$$

where F refers to the element of the liquid metal droplet.

The formation of oxide particles, F_aO_b, in Equation (9.69) is subject to appropriate nucleation constraints for transformation to the solid phase (as discussed in Chapter 5). Also, the reactant represents the oxide of the metal under refinement or processing. For example, it represents FeO or ZnO (i.e., refining of zinc). The heat released by the reaction is transferred by convection to the surrounding gas flow and conduction to and from the droplet. Also, the nucleating oxide particles participate in the radiation exchange to reflect and absorb heat from the reacting layer. Furthermore, the formation of oxide particles affects the structure and composition of the gas boundary layer around the evaporating droplets. The solid particles extend the resistive effects of the droplets on the gas motion beyond the droplet surface to include the surrounding oxide particle layer. As discussed in Chapter 7, a pressure drop in the gas flow is anticipated due to the droplets and solid particles interspersed within the main (gas) flow.

During the chemical reaction in Equation (9.69), an outward diffusive flux of metallic vapor from a source of F (i.e., liquid or droplet surface) occurs simultaneously with an inward diffusive flux of oxygen from a source of O_2 in the freestream gas flow. The species concentration gradients lead to diffusion of F in the opposite direction to the diffusion of O_2 (called *equimolar counterdiffusion*). Under steady-state conditions, these species diffusion rates occur at equal rates in opposite directions if it is assumed that the total molar concentration is uniform when the pressure and temperature are uniform over time (Guthrie, 1993). On a per mole basis, this balance of diffusive flows based on Equation (9.69) requires that

$$\frac{\dot{N}''_F}{a} + \dot{N}''\frac{O_2}{b/2} = 0 \tag{9.70}$$

where $\dot{N}''$ refers to the diffusive flux of the constituent indicated by its subscript.

The thicknesses of the diffusion layers of both counterdiffusing gases (metallic vapor and oxygen) are denoted by δ_F and δ_{O_2}, respectively. Then, using the ideal gas law together with Equation (9.70) and Fick's law, Equation (5.37) or (8.158), the outward flux of metallic vapor becomes

$$\dot{N}''_F = -\left(\frac{2a}{b}\right)\frac{D}{(\delta_F - \delta_{O_2})}\left[\frac{P_{O_2} - 0}{RT}\right] \tag{9.71}$$

where D refers to the diffusion coefficient of gas A (metallic vapor) in gas B (oxygen). The numerator in the bracketed term in Equation (9.71) includes the partial pressure of oxygen outside the boundary layer (in the freestream). It is assumed that the oxygen concentration reaches zero at some location near the surface of the droplet inside the metallic vapor diffusion layer. This location reflects the reaction plane where vapor and oxygen react to form solid (oxide) particles.

It can be observed that the inward flux of oxygen increases with a higher partial pressure of reacting gas (oxygen) in the freestream. From Equation (9.70), this increased inward flux must be balanced by a higher outward flux of metallic vapor. This latter vapor flux is increased when the reaction plane moves closer to the surface of the metallic droplet since the concentration gradient steepens nearer the surface of the droplet. As a result, the formation of oxide particles in the boundary layer around the droplets, based on the reaction in Equation (9.69), can be controlled by the spatial distribution and partial pressure of oxygen in the surrounding gas flow.

The heat released due to evaporation from the surface of the droplet, as well as the reaction with the surrounding gas flow, affects the temperatures of the droplet and convective heat transfer to and from the gas flow. Latent heat is released in the chemical reaction described by Equation (9.69). During evaporation, liquid molecules with comparatively high energy overcome the cohesive forces that hold the molecules on the droplet surface. A cooling effect on the droplet occurs when these higher energy molecules leave the surface. The evaporative heat loss may be approximated similarly to Equation (5.40). Also, the heat released in forming the solid particles can be approximated by the difference of enthalpies of products and reactants, i.e., based on Equations (9.51) and (9.52). Previous results in Equation (5.41) have shown how the associated convection coefficients can be derived when the latent energy lost by the liquid droplet due to evaporation is supplied by heat transfer to the droplet from the surrounding gas flow. Furthermore, this analysis can be used to quantify the interaction between convective heat and mass transfer as it contributes to the reaction formation of solid particles.

In this section, the interaction of gases was considered in chemical reactions leading to multiphase by-products, namely, solid particles and other products of combustion. In the following section, further consideration is given to multiphase reactions involving gas and solid reactants.

9.5.2 Gas–Solid Interactions

In many engineering technologies, gases and solid particles react to form products in gas, liquid, or solid phases. A widely encountered example is the combustion of pulverized coal particles in thermal power plants. These combustion reactions involve hydrocarbon fuel, based on Equation (9.50), and the latent heats of transformation between solid and gas phases. The fluid dynamics and convective or radiative heat transfer of these multiphase flows are described by the transport processes outlined in previous chapters.

An important example of reactions between solid particles and gas flows to produce multiphase products involves the purification of metals using carbon-based gases. For instance, pellets of metal oxides (such as oxides of zinc, magnesium, and lead) in manufacturing processes are initially purified using carbon monoxide or carbon dioxide gases (Guthrie, 1993). Oxide ores are typically obtained from siliceous rocks by grinding, crushing, or flotation operations. The impurities in these ores can be removed by chemical reactions with gas mixtures such as CO or CO_2 (called *topochemical reduction*). The thermofluid processes within industrial furnaces during topochemical reductions are significant in view of the rate and quality of production of the final purified metals.

The rates of reaction are important factors that control the formation of products from the reactants. Chemical reactions that are characterized by *zeroth-order kinetics* can be described by

$$\dot{N}_A''' = k_0 \tag{9.72}$$

whereas *first-order reactions* are described by

$$\dot{N}_A''' = k_1 C_A \tag{9.73}$$

In Equations (9.72) and (9.73), $\dot{N}_A'''$, C_A, k_0, and k_1 refer to the volumetric species production rate of constituent A, concentration of constituent A, and constants, respectively. The units of k_0 and k_1 are kmol/sec m^3 and 1/sec, respectively. The chemical reaction may occur at a constant rate (i.e.,

Equation (9.72) is zero order) or a rate that is proportional to the local concentration (i.e., Equation (9.73) is first order). If $\dot{N}''_A$ is positive, then the reaction leads to a production of constituent A. On the other hand, if $\dot{N}''_A$ is negative, then the reaction is characterized by a consumption of constituent A. The following example shows how the kinetics of reactions can be used to estimate the time elapsed in a gas–solid reaction.

Example 9.5.2.1

Thermal Decomposition of a Metal Pellet.

In a limestone processing operation, individual pellets of calcium carbonate ($CaCO_3$) are heated at the base of a furnace (see Figure 9.6). The purpose of this problem is to estimate the required initial size of pellets for a thermal decomposition within a specified time interval. The pellets are decomposed by heat addition, followed by nucleation of CaO (solid) crystals and the formation of CO_2 (gas), which is transported away into the bulk gas flow by convection. Estimate the initial side length, L_0, required for a cube-shaped pellet that must be completely decomposed into the CaO crystals and CO_2 gas within a specified time of t_f.

In this example, the process is described by

$$CaO(solid) + CO_2(gas) \rightarrow CaCO_3(solid) \tag{9.74}$$

or

$$CaCO_3(solid) \rightarrow CaO(solid) + CO_2(gas) \tag{9.75}$$

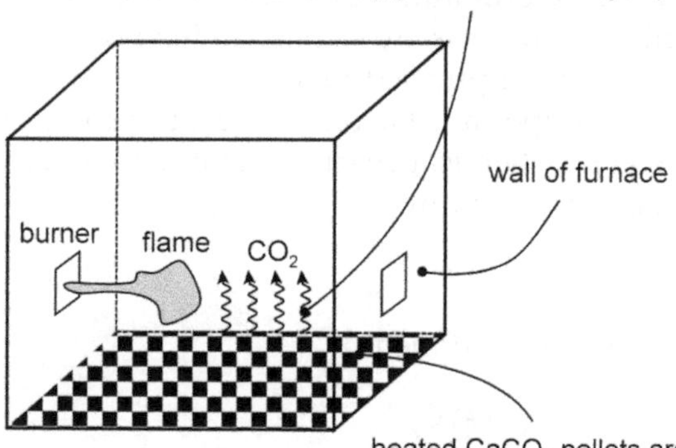

FIGURE 9.6
Thermal decomposition of limestone particles.

in conjunction with the release of heat through the chemical reaction. Assuming first-order kinetics in the chemical reaction,

$$\dot{N}'' = -k_1 C \tag{9.76}$$

where C and $\dot{N}''$ refer to the concentration of calcium carbonate and its rate of formation (per unit area and time), respectively.

Conservation of species concentration requires that

$$\frac{d}{dt}(CV) = -k_1 AC \tag{9.77}$$

where V and A refer to the volume and surface area of the pellet, respectively. Assuming that the concentration of the calcium carbonate remains constant over time during decomposition of the pellet,

$$C\frac{d}{dt}(L^3) = -k_1(6L^2)C \tag{9.78}$$

This equation can be rearranged to give

$$\frac{d(L)}{dt} = -2k_1 \tag{9.79}$$

which shows that the reaction front moves back into the pellet at a constant velocity of $2k_1$ during the thermal decomposition.

Solving Equation (9.79) subject to the conditions that $L = L_0$ at $t = 0$ (initial condition) and $L = 0$ at $t = t_f$ (final time), we obtain

$$L_0 = 2k_1 t_f \tag{9.80}$$

which represents the initial side length of the cube-shaped pellet. This length characterizes the size of pellet that is completely decomposed by the chemical reaction with the CO_2 gas in the required time of t_f.

Chemical reactions involving gas–solid interactions and multiphase products of combustion are often encountered in industrial blast furnaces. In the following case study, the operation of this type of furnace during metal purification is discussed.

9.5.3 Case Study: Blast Furnace

Blast furnaces are commonly used in the smelting of ores and refining of metals, including the production of iron ore and other metals, such as tin and lead. Hot and high-pressure gases are used to induce combustion

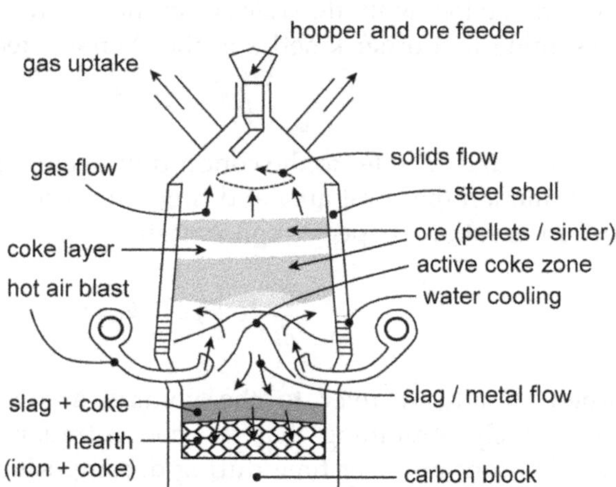

FIGURE 9.7
Schematic of internal processes within a blast furnace.

within the vertically oriented furnace. The combustion zone is called a *bosh*, and the region where the molten material is processed is called a *hearth* (see Figure 9.7). In modern production facilities, blast furnaces can exceed 30 m in height, 10 m in diameter, and 1700 tonnes of production per day.

During the production process in the furnace, layers of different materials in the form of pellets (such as limestone, coke, and iron oxide ore) are supplied at the top of the furnace. These materials slowly descend through the furnace and are heated by the hot, ascending gases. The ascending gases have a higher carbon monoxide content due to combustion that occurs lower in the furnace. The descending iron oxide pellets are transformed as a result of the chemical reaction and combined heat–mass transfer of carbon dioxide (and hydrogen) from the gas phase into the solid (pellet) phase. The first stage of the chemical transformation of solid is described by

$$3Fe_2O_3(solid) + CO(gas) \rightarrow 2Fe_3O_4(solid) + CO_2(gas) \tag{9.81}$$

The heat released by this reaction can be computed in a manner similar to that of the results derived earlier for chemically reacting flows, such as Equations (9.51) and (9.52).

The transformed iron oxide pellets react further with carbon monoxide in a second stage as follows:

$$Fe_3O_4(solid) + CO(gas) \rightarrow 3FeO(solid) + CO_2(gas) \tag{9.82}$$

The products of composition are wustite (FeO) and carbon dioxide (CO_2). As the wustite descends farther and mixes with carbon dioxide and carbon monoxide gases, the following final reactions are encountered:

$$FeO(solid) + CO(gas) \rightarrow Fe(solid) + CO_2(gas) \tag{9.83}$$

$$CO_2(gas) + C(solid) \rightarrow 2CO(gas) \tag{9.84}$$

$$FeO(solid) + C(solid) \rightarrow Fe(solid) + CO(gas) \tag{9.85}$$

It can be observed that the final products of combustion include the production of solid iron.

Thus, iron ore is transformed to iron by successive stages of combustion of the supply materials (including solid iron oxide pellets), in the presence of carbon monoxide gases. During these combustion reactions, a waste material (called *slag*) is formed as an upper, molten layer. This slag consists of oxides, ash, and impurities (such as silica, magnesia, and sulfur). It is important during the production process to remove certain substances from the material and protect the metal from oxidation. Solidified slag is used in various applications, such as phosphate fertilizers and aggregate material for road making. Slag is formed within the blast furnace when limestone decomposes and combines with ash and other impurities. Thus, impurities in the original metal pellets are removed from the base of the blast furnace in the form of melting and molten slag.

Stoves are used with blast furnace operations to provide preheated incoming air to the furnace. Incoming air into the stove is passed over hot, preheated brickwork, and the resulting heated air is then supplied to the blast furnace. The incoming air cools the brickwork. Thus, separate heating of the brickwork is required by periodically shutting off the cold air supply, opening a gas valve, and bringing in exit blast furnace gases to reheat the stove brickwork. Effective heat exchange in this process is essential since the recycling time for stoves is directly related to the metal production capacity and performance efficiency of the furnace.

9.5.4 Gas–Liquid Interactions

Another multiphase phenomenon of practical significance involves the chemical reaction of gas with liquid phases to form products in gas, liquid, or solid phases. In Chapter 6, the liquid–vapor phase change was described for nonreacting systems. In this section, the gas flow can react with a liquid phase to produce multiphase reactants. Examples involving processing of molten aluminum and liquid iron are considered. Also, a case study of steel production processes will be presented. This case study involves oxidation of impurities in metals by reactions between oxygen and liquid metal to form slag.

Consider the following example of reactions between argon–chlorine gas mixtures and liquid metals:

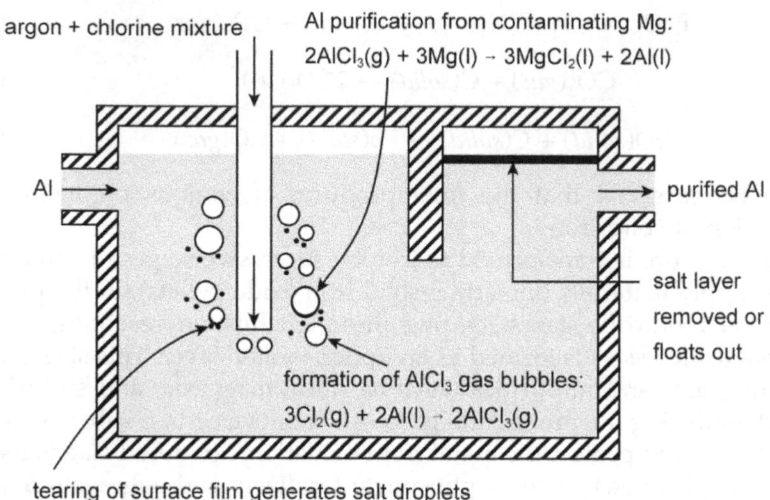

FIGURE 9.8
Refinement of molten aluminum.

$$3Cl_2(gas) + 2Al(liquid) \rightarrow 2AlCl_3(gas) \tag{9.86}$$

which is followed by

$$2AlCl_3(gas) + 3Mg(liquid) \rightarrow 3MgCl_2(liquid) + 2Al(liquid) \tag{9.87}$$

This example represents steps taken in the purification of liquid aluminum by removing magnesium impurities through chemical reactions with argon–chlorine gases (see Figure 9.8). In these reactions, Cl_2 gas bubbles react with liquid aluminum to form gas bubbles of $AlCl_3$, which subsequently react with Mg impurities in the liquid aluminum to form purified liquid aluminum and $MgCl_2$ droplets. In actual metal purification processes, the gas bubbles can further generate a fine distribution of salt particles in the diffusion layer around the bubbles. These particles are typically removed during final processing of the purified material. In the following example, another application is considered in view of the release and transfer of heat during the gas–liquid reaction.

Example 9.5.4.1
Heat Released by Desulfurization of a Liquid Iron Alloy.
Sulfur is separated out and removed from a liquid iron alloy during a magnesium injection operation. In this operation, magnesium gas bubbles react with sulfur in the liquid iron alloy (both at temperatures of T_R) to form particles of magnesium sulfide (at T_P), which rise to the surface of the

container. In this way, sulfur is removed from the mixture and the liquid iron is purified. The purpose of this example is to find the amount of heat released by this desulfurization process.

The chemical reaction that describes the desulfurization process is

$$\mathrm{Mg}(gas) + \mathrm{S}(liquid) \rightarrow \mathrm{MgS}(solid) \tag{9.88}$$

This reaction usually occurs in a sequence of steps. The Mg gas bubbles react with sulfur in the liquid iron alloy to form a film of MgS on the bubbles, which then precipitate on particles of magnesium sulfide. Thus, the particles are formed by the reaction between magnesium vapor bubbles and sulfur, which diffuses toward the gas–liquid interface at each bubble.

Assuming an insulated vessel containing only the cited reactants and product, the amount of heat released by the chemical reaction is equivalent to the difference of enthalpies of products and reactants. Defining a control volume that encompasses the reaction of Equation (9.88), an energy balance yields

$$Q = (n\hat{h})_{\mathrm{Mg}} + (n\hat{h})_{\mathrm{S}} - (n\hat{h})_{\mathrm{MgS}} \tag{9.89}$$

where h refers to the molar enthalpy (kJ/kmol). Also, the subscripts refer to the relevant products and reactants. In terms of the chemical enthalpy (enthalpy of formation) and the sensible enthalpy for the given reaction,

$$Q = \sum_{i=1}^{2} n_{R,i}(\hat{h}_f^o + \hat{h} - \hat{h}^o)_{R,i} - n_P(\hat{h}_f^o + \hat{h} - \hat{h}^o)_P \tag{9.90}$$

where the subscripts R and P refer to reactants and product (MgS), respectively. Also, the superscript o refers to evaluation at STP. The number of moles of reactants (per mole of product) is indicated by the leading coefficients in the chemical reaction equation, Equation (9.88).

For the reactants in Equation (9.90), enthalpies corresponding to the latent heats of phase change, $\hat{h}_{fg}$ and $\hat{h}_{sl}$, are included to accommodate the heats of transformation between liquid–gas phases and solid–liquid phases, respectively. Also, the enthalpies of formation of the elements (Mg and S) in their natural states at a temperature of 298 K are zero. The leading coefficients representing the number of moles of each reactant and product in Equation (9.90) are unity, based on the chemical reaction in Equation (9.88). Furthermore, the $\hat{h}$ values in Equation (9.90) are evaluated at the given temperatures of T_R and T_P for the reactants and products, respectively. The differences of sensible enthalpies can be approximated by the specific heat multiplied by the appropriate temperature difference. This expression would need to be integrated over the temperature range to accommodate any appreciable variation of specific heat with temperature.

The rate of sulfur removal can be calculated based on the number and size of MgS particles generated at the bubble interfaces. The species concentration, heat transfer, and fluid flow equations would need to be solved in the liquid alloy to yield the spatial distributions of Mg and S concentrations. These transport equations would describe how the magnesium and dissolved sulfur would codiffuse and react on particles of magesium sulfide throughout the mixture.

In the previous example, chemical reactions between gas and liquid phases were examined in an application involving purification of a liquid iron alloy. Another similar application arises in the production of steel. This application is discussed in the following case study.

9.5.5 Case Study: Steel Production

Steel was first widely produced in the mid-1800s, and its production is a vital part of many industries—automobile manufacturing, machinery production, building, and others. The raw materials in steel include reduced iron ore, scrap steel, and iron. Impurities in these materials are removed by oxidizing them with blasts of air or oxygen. In this way, most of the impurities (such as phosphorus, sulfur, and others) are converted to their respective oxides, which are combined with other waste material to produce slag. Once all impurities are removed, extra elements are added in careful proportions to modify or enhance the mechanical properties of the final solidified material. The molten steel is solidified in the shape of ingots or cast as continuous bars (called *strand casting*).

The process of impurity removal during steel production is often carried out in a basic oxygen furnace (BOF). High-velocity jets of oxygen are blown onto the molten iron in the BOF vessel (see Figure 9.9). In this way, carbon and other impurities in the molten iron react with the oxygen to form carbon monoxide (which escapes from the vessel) and other products of combustion that are removed as scrap slag. In particular, the chemical reactions for carbon and sulfur impurities in the iron are given by (Guthrie, 1993)

$$2\mathrm{C}(liquid) + \mathrm{O}_2(gas) \rightarrow 2\mathrm{CO}(gas) \tag{9.91}$$

$$\mathrm{Si}(liquid) + \mathrm{O}_2(gas) \rightarrow \mathrm{SiO}_2(liquid - solid) \tag{9.92}$$

In the former reaction, dissolved carbon within the liquid metal is transported to the gas (O_2)–liquid interface, and subsequently removed from the multiphase region as carbon monoxide through its chemical reaction to form CO with the impinging oxygen jet. In view of the high operating temperatures (typically around 1800 K), many complicated,

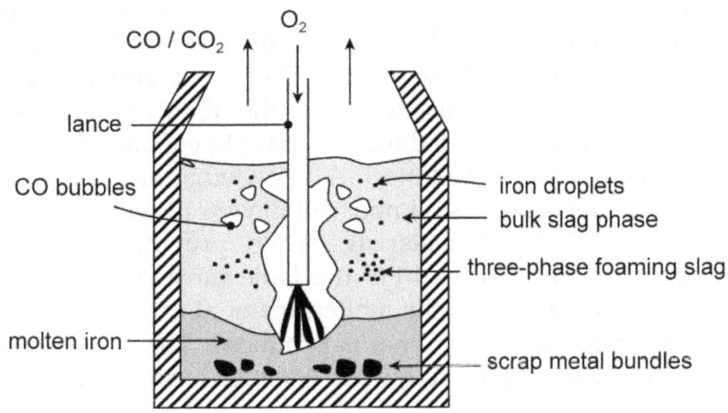

FIGURE 9.9
Schematic of BOF.

simultaneous modes of heat transfer arise in these operations, including phase change, turbulence, and radiative heat transfer.

During this impurity removal that produces carbon monoxide (CO) and slag (such as SiO_2 and others), a substantial amount of heat is transferred during the chemical reactions in Equations (9.91) and (9.92). In order to prevent overheating within the BOF, extra steel scrap (often up to 30% of the molten steel mass) is added as a coolant. Temperatures within the molten steel generally do not exceed about 1920 K. This required addition of scrap metal leads to other difficulties in the effective processing of the molten steel. For example, combined mass–heat transfer to the scrap metal can obstruct the transport mechanisms in the main portion of the molten steel.

Due to the difficulty of obtaining experimental data in these processes at high temperatures, the detailed interactions between fluid flow and heat transfer during impurity removal are not fully understood. Despite these limitations, many new innovations have appeared in steel-making technology. For example, in recent advances, oxygen is injected at the base of the steel-making furnace by a series of submerged jets, which allows more effective mixing of the molten steel and better slag–metal contact for impurity removal. As a result, reduced processing times can be achieved with a higher production rate of steel. In view of these recent advances, many BOF vessels have been upgraded to accommodate submerged gas-stirring facilities.

A ladle is used to move the molten metal to another location for subsequent processing and solidification. However, if the liquid steel is too highly oxidized (i.e., above 0.04% percentage by weight) when it is poured into the ladle, then appreciable defects (such as holes or voids) could be formed when the steel is solidified. De-oxidants (such as aluminum) are often added to bring down the dissolved oxygen content to levels that would prevent these defects. The de-oxidants would promote precipitation of condensed oxides to reduce the dissolved oxygen content. Also, other constituents are added (such as Mn) to enhance the mechanical properties of the final solidified steel. A knowledge of the resulting heat transfer and fluid flow patterns is important in view of reducing the costs of supplying these additional materials, while providing the necessary material properties. Heat losses from the ladle must be minimized since temperature differences within the molten material can significantly affect the quality of solidified material that is produced. These heat losses from the ladle include conduction through the walls and combined convection–radiation from the top surface.

The steel is solidified into slabs or billets. The rate of heat extraction through the walls affects the rate of solidification. This heat loss is typically controlled by suitable adjustments of the cooling rates with the use of sidewall insulators and other techniques. A gas–liquid interaction involving carbon monoxide largely affects the extent of the formation of defect voids in a solidified ingot. The oxygen content of the molten steel is carefully controlled to provide a sufficiently high partial pressure of gas ahead of the phase interface during solidification. In this way, the rate of formation of gas bubbles can be matched with the rate of volumetric shrinkage of the metal to reduce or eliminate the formation of large void defects in the solidified material. The voids are replaced with small dispersed voids, which can be readily removed by subsequent hot rolling and other operations.

In some cases, large amounts of time (up to 18 h for thick ingots) may be required for various constituents to be uniformly distributed throughout the steel. Recent thermal and fluids research has attempted to find more effective ways of establishing recirculating flow patterns in the molten steel to reduce slag entrainment and promote effective separation of the impurities. Also, research efforts are actively examining new ways to reduce the detrimental effects of microsegregation (nonuniform solute distribution) on the steel's properties.

Following solidification, the properties of the steel can be largely enhanced by subsequent heat treatment processes. For example, in *annealing*, the material is repetitively heated and cooled to remove stresses formed during the solidification process. In this way, the material becomes tougher. Also, *case hardening* can increase the steel's wear resistance and surface hardness by packing powdered charcoal around the steel and heating so the surface of steel absorbs carbon. The heated steel is then quickly cooled by quenching in a water bath to further enhance the final mechanical properties of the material. During the processing of steel, different phases in the solid (particularly ferrite, austenite, and cementite) may arise, which differ in microscopic structure and carbon content. The subject of steel metallurgy includes the study of how these and other solid phases affect the final properties of the steel.

9.5.6 Liquid–Liquid Interactions

Another important case of multiphase flows arises when two or more liquids come into contact with each other to react and form products in solid, liquid, or gaseous phases. Various examples of these types of reactions are described by Guthrie (1993).

In steel-making processes, carbon in droplets of iron can react with slag in the production furnace to produce carbon monoxide gas bubbles and bulk iron, i.e.,

$$C(liquid) + FeO(liquid) \rightarrow Fe(solid) + CO(gas) \quad (9.93)$$

When these two liquids react, the products of reaction include a natural element (Fe alone) and a new compound (CO). Thus, the release of heat from this reaction would include the chemical enthalpies (i.e., enthalpy of formation of CO), sensible enthalpy (due to the temperature of the products above STP), and the enthalpies of phase transformation (i.e., between liquid–solid and liquid–gas phases).

Another practical example of liquid–liquid reactions involving multiphase products of reaction is the aluminum de-oxidation of steel. In the processing of steel and its alloys, solid aluminum is melted and added to molten steel in the steel bath. During this mixing, liquid aluminum reacts with oxygen dissolved in the molten steel to form solid Al_2O_3 particles, i.e.,

$$Al(liquid) + O(liquid) \rightarrow Al_2O_3(solid) \quad (9.94)$$

This reaction is subject to the usual criteria for nucleation of solid crystals (as discussed in Chapter 5) and thermodynamic equilibrium conditions at the phase interface between liquid and solid phases.

References

S. Banerjee and A. Chan. 1980. "Separated flow models: I. Analysis of the time averaged and local instantaneous formulations," *Int. J. Multiphase Flow* 6: 1–24.

C. Crowe. Modeling Fluid-Particle Flows: Current Status and Future Directions, AIAA Paper 99–3690, paper presented at 30th AIAA Fluid Dynamics Conference, Norfolk, VA, 1999.

D. Darling, D. Namkoong, and R. Skarda. Modelling Void Growth and Movement with Phase Change in Thermal Energy Storage Canisters, AIAA Paper 93-2832, paper presented at AIAA 28th Thermophysics Conference, Orlando, FL, 1993.

D. Gidaspow. 1994. *Multiphase Flow and Fluidization*, London: Academic Press.

E.J. Goodwin, et al. Predicting Ice and Snow Loads for Transmission Lines, in *Proceedings of the 1st IWAIS*, 1982, pp. 267–273.

R.I.L. Guthrie. 1993. *Engineering in Process Metallurgy*, Oxford: Oxford University Press.

C.W. Hirt and B.D. Nichols. 1981. "Volume of fluid (VOF) method for the dynamics of free boundaries," *J. Comput. Phys.* 39: 201–225.

P.J. Larcombe et al. 1988. "Mathematical models for ice accretion on conductors using free streamline theory. Part I: single conductor," *IMA J. App. Math.* 41: 217–236.

G.F. Naterer. 1997. "Simultaneous pressure: velocity coupling in the two-phase zone for solidification shrinkage in an open cavity," *Modelling Simulation Mater. Sci. Eng.* 5: 595–613.

G.F. Naterer. 2002a. "Energy balances at the air/liquid and liquid/solid interfaces with incoming droplets at a moving ice boundary," *Int. Commn. Heat Mass Transfer* 29: 57–66.

G.F. Naterer. 2002b. "Multiphase flow with impinging droplets and airstream interaction at a moving gas/solid interface," *Int. J. Multiphase Flow* 28: 451–477.

G.F. Naterer, H. Deng, and N. Popplewell. 1999. "Predicting and reducing ice accretion on electric power lines with joule heating: theory and experiments," *CSME Trans.* 23: 51–70.

G. Poots. 1996. *Ice and Snow Accretion on Structures*, New York: John Wiley & Sons, Inc.

P.L.I. Skelton and G. Poots. 1991. "Snow accretion on overhead line conductors of finite torsional stiffness," *Cold Reg. Sci. Technol.* 19: 301–316.

P.L.I. Skelton and G. Poots. 1993. "Rime-ice accretion on fixed stranded conductors," *Mech. Res. Comm.* 20: 45–52.

K. Tsuboi and S. Kimura. Numerical Study of the Effect of Droplet Distribution in Incompressible Droplet Flows, AIAA Paper 98–2561, paper presented at AIAA 29th Fluid Dynamics Conference, Albuquerque, NM, 1998.

G. Wakahama, D. Kuroiwa, and D. Goto. 1977. "Snow accretion on electric wires and its prevention," *J. Glaciol.* 19: 479–487.

D. Xu and Q. Li. 1991. "Gravity and solidification induced shrinkage in a horizontally solidified ingot," *Numerical Heat Transfer* 20: 203–221.

Problems

1. Impinging supercooled droplets on an overhead power transmission line freeze immediately upon impact on the surface. The wind speed is 8 m/sec, and the liquid water content of droplets in the air is 0.9 g/sec. Estimate the equivalent ice thickness covering the cable after 1 h of precipitation.
2. During spray cooling of an industrial component, forced convection occurs with a thermal boundary layer and impinging droplets on a flat plate. The plate length is L, while the temperatures of the wall and ambient air are T_w and T_∞, respectively (i.e., $\Delta T = T_w - T_\infty$). The freestream air velocity is U_∞. In the freestream, the droplet and air temperatures are equal to each other.
   (a) Explain the physical meaning of each term in the following representation of the wall heat flux:

$$q_w = q_1 + q_2 + q_3$$

   where

$$q_1 = -k \left.\frac{\partial T}{\partial y}\right|_0; \qquad q_2 = VGc_w(T_w - T_\infty); \qquad q_3 = -\frac{1}{2}\, GV^3$$

   and V, G, and c_w refer to the impacting droplet velocity, liquid water content (i.e., mass of droplets per unit of volume of air–droplet mixture), and droplet specific heat, respectively. Evaporation and formation of a thin liquid film along the wall, due to the droplet influx, are not considered in this approximation of the wall heat flux.
   (b) Perform a scaling analysis of the flat plate boundary layer equations (single phase, air) to estimate the thermal boundary layer thickness, δ_t, at the end of a plate of length L. Neglect viscous dissipation and assume steady-state conditions.
   (c) Find the average Nusselt number, $\overline{Nu}_L$, based on q_w in part (a) and δ_t in part (b). Use this result to determine the required ranges of G and ΔT, so that only q_1 and q_2 contribute appreciably to the wall cooling.
3. The quality of solidified layers obtained in a plasma spray deposition process is largely affected by solidification of the

droplets impinging on a cold surface. Initially, the droplets are solidified upon impact on the surface, but after a certain amount of time has elapsed, latent heat released by the solidified droplets generates a thin liquid film above the solid layer. It is postulated that this liquid layer, $b(t)$, grows over time according to

$$\frac{db}{dt} = a_1 - \frac{a_2}{B(a_3 + b)}$$

where a_1, a_2, and a_3 are constants and $B(t)$ refers to the thickness of the solidified layer. Perform a one-dimensional analysis to estimate the time when the liquid film first appears above the solid layer. Express your answer in terms of the aforementioned constants, V (velocity of incoming droplets), G (liquid content of droplets in the air), and thermophysical properties.

4. Supercooled droplets are sprayed onto a plate (length of L, thickness of W) tilted at an angle of θ with respect to the horizontal plane. The droplets are partially solidified, and the remaining impacting liquid accumulates as a liquid film along the solidified layer. The rate of film growth with distance along the plate, x, is assumed to be constant. The incoming droplet velocity is V, and the liquid content in the air (kg/m^3) is G.
   (a) Find the rate of growth of the solidified layer over time, in terms of L, θ, V, and G.
   (b) How does the solid formation change if the film growth varies nonlinearly with distance along the plate? Explain how the film thickness can be computed under these conditions.
5. Consider the previous problem, but without the surface liquid film (i.e., droplets solidify immediately upon impact on the plate). Also, the thin plate is heated internally by electrical resistive elements so that the wall temperature (beneath the solidified layer) on the plate varies with time. The ambient temperature and convective heat transfer coefficient of the surrounding airflow (above the solidified layer) are T_a and h, respectively. Find the variation of plate temperature with time in terms of thermophysical properties, h, T_a, plate dimensions, and the electrical parameters (current, I, and resistance, R).
6. Metal powders and components with fine dimensional tolerances, such as gears, are produced in a multistage melt particularization process. Atomization of the liquid alloy through an impinging cool air jet separates individual liquid droplets, and the subsequent solidification of droplets creates a temperature fluctuation, ΔT_u. In

other words, the droplet temperature continually decreases as it is cooled until solidification, when a slight temperature rise is observed. During this small time interval of thermal fluctuation, assume radiative heat transfer is negligible.

(a) What physical process generates the temporary thermal fluctuation?

(b) Consider a liquid metal stream with a density of ρ, melt temperature of T_m, latent heat of fusion of L, and a final spherical particle diameter of D. The phase transformation occurs at the rate of $\rho L(df_l/dt)$ per unit volume, where f_l refers to the liquid fraction. Find an expression for the minimum convective cooling coefficient, h_{min}, required to ensure that the thermal fluctuation is avoided. Give your result in terms of ρ, L, air–melt temperature difference, and the solidification rate, df_l/dt.

7. Spray deposition (or spray casting) is a manufacturing process where liquid metal droplets are sprayed and deposited on a substrate. Consider a spherical droplet of liquid metal with an initially uniform temperature, which is suddenly subjected to convective and radiative cooling on its outer surface. The extent of its partial solidification prior to impact on the substrate is an important factor in assessing the properties of the final spray-formed material. Find the rate of solidification of the droplet prior to impact on the substrate in terms of the convection coefficient, h, droplet diameter, ambient gas temperature, T_∞, and surface temperature of the droplet, T_s.
8. In the previous question, the rate of solidification of droplets was expressed in terms of the convection coefficient for an accelerating or decelerating droplet. In this question, a closed form (analytic) solution is required for the transient temperature distribution in the droplet when the terminal velocity of the droplet is attained. Express your answer for droplet temperature in terms of the drag coefficient (spherical droplet), c_d, droplet diameter, droplet solidification rate (assumed constant), $\dot{m}$, ambient gas temperature, and thermophysical properties.
9. During a melt particularization process, a molten metal stream is injected into nitrogen gas at 120°C ($We = 3$). If the metal stream disintegrates at 1200°C when the fluid velocity is 22 m/sec, estimate the rate of heat loss from the disintegrated metal stream (at the transition to secondary breakup).
10. Derive Equation (9.40) by using the ideal gas to show that the mole fraction is equal to the volume fraction (*Amagat–Leduc law of*

additive volumes), as well as the pressure fraction (*Dalton's law of additive pressures*).

11. For an ideal gas mixture, show that $R = \Sigma\, mf_i R_i$. Also, show that $y_i/mf_i = R_i/R$, where R_i and R refer to the gas constants of component i and the entire mixture, respectively.
12. For an ideal gas mixture, derive Equation (9.55) to show how the specific entropy of component i can be computed in terms of the mole fraction of component i and mixture quantities (i.e., mixture temperature and total pressure).
13. A mixture of 60% nitrogen and 40% propane (by volume) flows through a well-insulated compressor. The inlet conditions of the gas mixture are 30°C and 0.4 MPa. Find the input power (per unit mass of mixture) required to compress the steady, nonreacting flow to outlet conditions of 100°C and 1.2 MPa (note: $c_p = 1.68$ kJ/kgK and $M = 44.1$ kg/kmol for propane; $c_p = 1.04$ kJ/kgK and $M = 28$ kg/kmol for nitrogen).
14. For the previous question, find the specific entropy change of each gas component separately (i.e., propane and nitrogen separately) between the inlet and outlet of the compressor. Which gas component in the mixture undergoes a larger specific entropy change? Give a physical interpretation to explain this difference of computed specific entropy values.
15. A rigid tank is initially divided by a partition into two sides. The left side contains methane (component A, mass of m_A with n_A moles) and the right side contains nitrogen (component B, mass of m_B with n_B moles). Both sides are initially at a temperature of T and pressure of P. Then the partition is removed and both sides of the insulated tank are completely mixed. Find the total entropy change of the gases due to the mixing process. Verify that this entropy change complies with the Second Law of Thermodynamics.
16. Explain how the adiabatic flame temperature, T_{ad}, can be calculated for the combustion of methane (CH_4) at STP in air. Show the relevant chemical reaction and energy balance equations, and give an outline of the steps required to calculate T_{ad}.
17. In practice, considerable variations can be observed between adiabatic flame temperatures obtained from actual experiments, tabulated values, and calculations based on the analysis considered in the previous question. Explain why these variations might occur and how more accurate calculations could be carried out in the previous question.
18. An expression for a controlled pellet size was obtained in Equation (9.80) for the thermal decomposition of calcium carbonate ($CaCO_3$)

into CaO (solid phase) and CO_2 (gas phase). Alternatively, the rate of thermal decomposition can be controlled by the ease with which CO_2 gas can diffuse through the gas boundary layer surrounding the pellet. Use Fick's law to estimate the required species concentration of CO_2 in the bulk gas phase (called C_∞) to provide a specified steady-state diffusive flux, $\dot{N}_s$, at the surface of a spherical pellet (at $r = R$).

19. During thermal decomposition of spherical limestone pellets ($CaCO_3$), solid crystals of CaO and CO_2 gas are formed. Using first-order kinetics of the chemical reaction with a proportionality constant of 0.02 mm/sec, estimate the time taken for a pellet to be decomposed to one half of its initial radius of 1 cm.
20. In a dehydration process, water is removed from a substance. Under what conditions of pressure and temperature can water be separated into ice and water vapor? Give an example of a basic procedure to remove water from a substance by phase transformations to solid and gas phases.

10

Heat Exchangers

10.1 Introduction

Heat exchangers are engineering devices that transfer thermal energy between fluid streams at different temperatures. They arise in many applications, including power generation, energy storage, air conditioning systems, materials processing, and various others. Commonly encountered heat exchanger configurations include *concentric tube*, *cross-flow*, and *shell-and-tube* heat exchangers. In this chapter, the design and analysis of these types of heat exchangers will be investigated. Both single phase and multiphase systems will be considered.

A concentric tube heat exchanger consists of two fluid streams, whereby an internal fluid flows through the inner tube and an external flow passes through the annular region between the inner and outer tubes. If the outer fluid flows in the same direction as the inner flow, then the configuration is called a *parallel flow*, and if the outer fluid flows in a direction opposite to the inner flow, then it is called a *counterflow* (see Figure 10.1). Empirical and numerical techniques are usually required for the detailed analysis of these heat exchangers. In some cases, semianalytic techniques can be used to determine the thermal effectiveness of heat exchange devices. For example, integral methods described in Chapter 3 can be applied for predictions of heat exchange in gas-fired water heaters (Naterer et al., 1996).

The second type of heat exchanger (cross-flow) typically consists of an outer flow passing across tubes carrying fluid that flows in a direction perpendicular to the cross-flow. In many cases, the tubes are covered with fins or other annular attachments to enhance the rate of heat transfer between the different fluid streams. If the cross-flow streams are separated from one another (i.e., fins separating fluid streams), then the configuration is *unmixed*, whereas a *mixed* configuration permits complete mixing of the fluid streams in the external cross-flow (see Figure 10.1).

The third main type of heat exchanger is called a shell-and-tube heat exchanger. This configuration consists of an outer shell pipe where fluid enters through one end, passes across internal tubes carrying a fluid at a

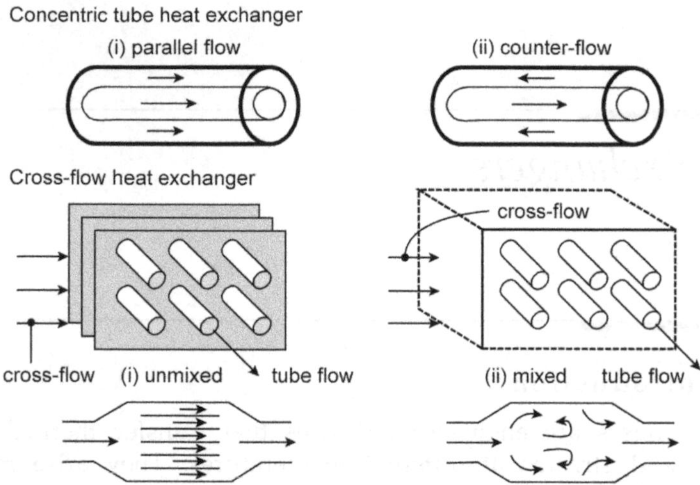

FIGURE 10.1
Concentric tube and cross-flow heat exchangers.

different temperature, and exits through the other end (see Figure 10.2). Baffles are usually placed perpendicular to the inner tubes to enhance mixing and turbulence of the outer fluid stream. Baffles are essentially perforated plates that obstruct some region of the outer flow while directing the flow around the remaining uncovered sections. An example of a shell-and-tube heat exchanger is a condenser. The outer flow is steam that condenses and leaves as water following heat exchange with the inner tubes carrying cold water.

The main purpose of a heat exchanger is to transfer heat between fluids at different temperatures or phases. In the case of heat exchange between

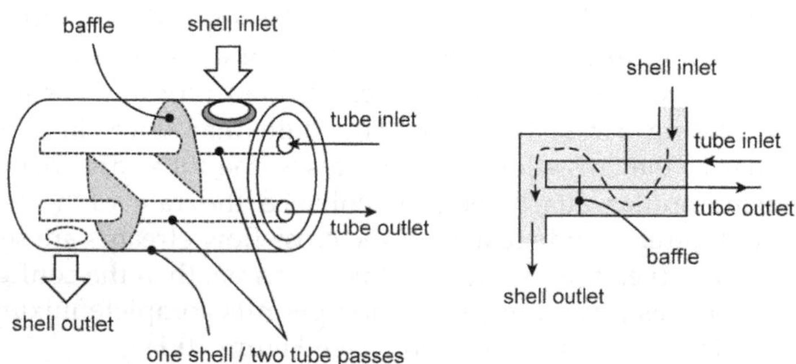

FIGURE 10.2
Shell-and-tube heat exchanger.

fluids of the same phase, but at different temperatures, the heat gained by the colder fluid stream, q_c, and heat lost by the hotter fluid stream, q_h, are given by

$$q_c = \dot{m}_c c_c (T_{c,o} - T_{c,i}) \tag{10.1}$$

$$q_h = \dot{m}_h c_h (T_{h,i} - T_{h,o}) \tag{10.2}$$

where the subscripts c, h, o, and i refer to cold, hot, outlet, and inlet, respectively (note: inlet and outlet denoted by sections 1 and 2 in Figure 10.3). Assuming that heat losses to the surroundings are negligible, then $q_h = q_c = q$. Equations (10.1) and (10.2) represent heat balances performed for the cold and hot fluid streams, respectively, over the entire heat exchanger (from the inlet to the outlet).

The packing of tubes within heat exchangers varies with different applications. This packing involves a surface area density spectrum of surfaces, i.e., a number and diameter of tubes within the heat exchanger. It is based on the hydraulic diameter of tubes, represented by $D_h = 4\ A_s/P$, where A_s and P refer to the surface area and perimeter, respectively. For example, a typical range is $0.8 < D_h < 5$ cm for shell-and-tube heat exchangers, $0.2 < D_h < 0.5$ cm for automobile radiators, and $0.05 < D_h < 0.1$ cm for gas turbine regenerators. Heat exchange also occurs widely in

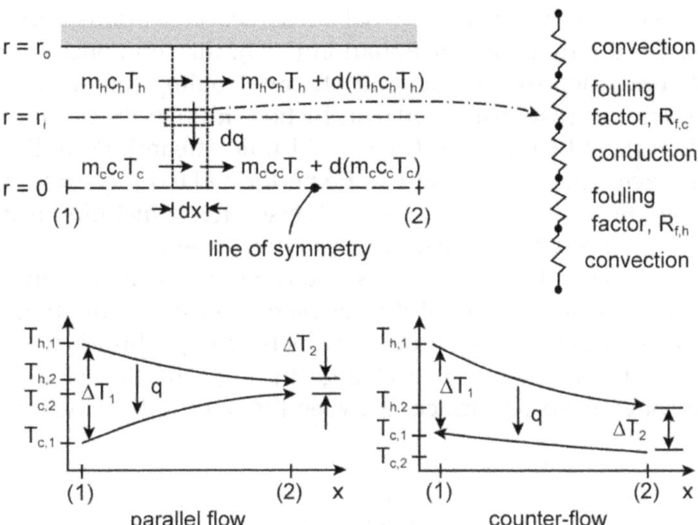

FIGURE 10.3
Thermal resistances in a concentric tube heat exchanger.

biological systems. For instance, heat exchange occurs in the range $0.01 < D_h < 0.02$ cm in human lungs.

10.2 Tubular Heat Exchangers

Heat transfer between two fluid streams in a concentric tube heat exchanger can be expressed in terms of temperature differences between the fluid streams at the inlet (section 1) and outlet (section 2). In particular, the following analysis will compute this heat transfer in the radial direction based on the temperature differences, ΔT_1 (inlet) and ΔT_2 (outlet). The quantities r_o, r_i, dx, and U will be defined as the outer tube radius, inner tube radius, width of the differential control volume in the axial (x) direction, and the total heat transfer coefficient, respectively (see Figure 10.3).

Based on all thermal resistances in series in the radial direction, the overall heat transfer coefficient may be written as

$$\frac{1}{UA} = \frac{1}{h_h A} + \frac{R_{f,h}}{A} + \frac{ln(r_o/r_i)}{2\pi kL} + \frac{R_{f,c}}{A} + \frac{1}{h_c A} \tag{10.3}$$

where A refers to the cross-sectional area (i.e., area normal to the flow direction). On the right side of Equation (10.3), the component resistances represent convection (outer hot fluid stream), fouling (outer, due to fluid impurities such as rust formation), conduction through the inner pipe wall, fouling (inner), and convection (inner cold fluid stream). Periodic cleaning of the heat exchanger surfaces should be performed to reduce and minimize the adverse effects of fouling. These adverse effects include an increased pressure drop and reduced heat transfer effectiveness.

Finned surfaces in heat exchangers could be represented through multiplication of the denominators of the convection terms in Equation (10.3) by the overall surface efficiency of the finned surface, η_o. This efficiency would include the efficiency for heat exchange through the fin, as well as heat transfer through the base surface (between the fins, as described in Chapter 2), i.e.,

$$\eta_o = 1 - \frac{A_f}{A}(1 - \eta_f) \tag{10.4}$$

where η_f, A_f, and A refer to fin efficiency, total fin area, and total surface area of the finned surface, respectively. For example, the fin efficiency for a uniform fin with an insulated tip was derived in Chapter 2 as follows:

$$\eta_f = \frac{tanh(mL)}{mL} \tag{10.5}$$

where $m = \sqrt{2h/kt}$ and t refers to the fin thickness. These results will be used in the present analysis when finned surfaces are included within the heat exchanger.

Within a heat exchanger, the differential heat transfer, dq, from the outer tube to the inner tube section of width dx can be computed by a steady-state heat balance as follows (see Figure 10.3):

$$\dot{m}_h c_h T_h = dq + [\dot{m}_h c_h T_h + d(\dot{m}_h c_h T_h)] \tag{10.6}$$

Performing a similar heat balance within the inner tube, combining with Equation (10.6), and using the overall heat transfer coefficient from Equation (10.3),

$$-\frac{dq}{\dot{m}_h c_h} - \frac{dq}{\dot{m}_c c_c} = dT_h - dT_c = d(\Delta T) = -\frac{U\Delta T dA}{\dot{m}_h c_h} - \frac{U\Delta T dA}{\dot{m}_c c_c} \tag{10.7}$$

Alternatively,

$$\frac{d(\Delta T)}{\Delta T} = -U\left(\frac{1}{\dot{m}_h c_h} + \frac{1}{\dot{m}_c c_c}\right) dA \tag{10.8}$$

Integrating Equation (10.8) from the inlet (section 1) to the outlet (section 2),

$$ln\left(\frac{\Delta T_1}{\Delta T_2}\right) = -UA\left(\frac{T_{h,1} - T_{h,2}}{q} + \frac{T_{c,2} - T_{c,1}}{q}\right) \tag{10.9}$$

which can be rearranged to yield

$$q = UA\left[\frac{(T_{h,2} - T_{c,2}) - (T_{h,1} - T_{c,1})}{ln(\Delta T_2/\Delta T_1)}\right] \equiv UA\left[\frac{\Delta T_2 - \Delta T_1}{ln(\Delta T_2/\Delta T_1)}\right] \tag{10.10}$$

As a result,

$$q = UA\Delta T_{lm} \tag{10.11}$$

where ΔT_{lm} refers to the *log mean temperature difference* as defined by the expression inside the square brackets in Equation (10.10).

The result in Equation (10.11) gives the total heat transfer from the hot fluid stream to the cold stream between the inlet (1) and outlet (2), based on given (or measured) temperature differences, ΔT_1 and ΔT_2. The previous analysis and results were derived for a parallel flow configuration. A similar analysis can be derived for a counterflow heat exchanger. For $\dot{m}_c$ in the opposite direction, the same result in Equation (10.11) may be used, except that the variables are replaced by $\Delta T_1 = T_{h,1} - T_{c,1} = T_h(inlet) - T_c(outlet)$ and $\Delta T_2 = T_{h,2} - T_{c,2} = T_h(outlet) - T_c(inlet)$.

For a parallel flow heat exchanger, the highest temperature difference is encountered between the two incoming fluid streams. Heat transfer from the hot stream to the cold stream reduces the temperature difference between the fluids in the streamwise direction (see Figure 10.3). On the other hand, the temperature difference increases in the flow direction for a counterflow arrangement (as viewed by the cold fluid stream). The temperature of the incoming cold fluid stream increases due to heat transfer from the hot stream flowing in the opposite direction. If the same inlet and outlet temperatures are considered, then the log mean temperature difference of the counterflow arrangement exceeds the difference for a parallel flow heat exchanger. Thus, a counterflow heat exchanger is usually considered to be more effective, since a smaller surface area is required to achieve the same heat transfer (assuming equivalent heat transfer coefficients between the fluid streams).

10.3 Cross-Flow and Shell-and-Tube Heat Exchangers

For more complicated geometrical configurations, such as cross-flow and shell-and-tube heat exchangers, the previous results can be modified through appropriate correction factors, i.e.,

$$q = FUA\Delta T_{lm} \tag{10.12}$$

In Equation (10.12), the correction factor, F, is usually based on experimental data to account for baffles and other geometrical parameters. For a one-shell-pass, one-tube-pass heat exchanger, the correction factor is $F = 1$. The value of F depends on the type of heat exchanger.

The correction factor, F, can be graphically depicted based on the following parameters (see Figure 10.4):

$$R = \frac{T_i - T_o}{t_o - t_i} \tag{10.13}$$

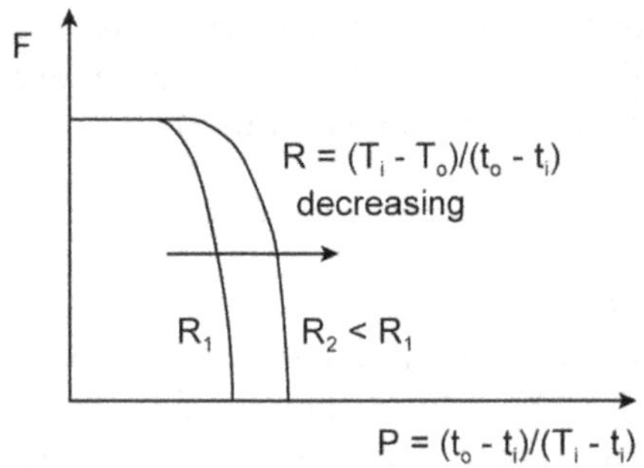

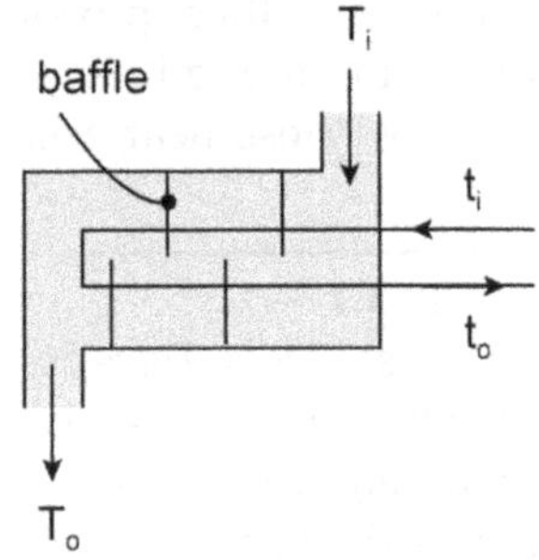

FIGURE 10.4
Correction factor for shell-and-tube heat exchanger.

$$P = \frac{t_o - t_i}{T_i - t_i} \tag{10.14}$$

where lowercase and uppercase values of temperature refer to inner flow (i.e., tube) and outer flow (i.e., shell), respectively. The subscripts o and i refer to outlet and inlet, respectively. For example, t_i refers to the inlet temperature of the tube flow, and T_o denotes the outlet temperature of the shell flow. As $(t_o - t_i)$ increases (at a fixed value of P), R decreases and the heat exchanger charts show that F increases (i.e., q increases). Results of correction factors for a variety of heat exchanger configurations have been presented and graphically illustrated by Bowman et al. (1940), Bejan (1993), Incropera and Dewitt (1990), and others. Also, the *Standards of the Tubular Exchange Manufacturers Association* (8th edition, New York, 1999) provides these results, which can be given as algebraic expressions or graphical representations.

For example, the correction factor for a one-pass shell side, with any multiple of two tube passes (Bowman et al., 1940), can be expressed by

$$F = \frac{[\sqrt{R^2 + 1}/(R - 1)] \cdot log[(1 - P)/(1 - PR)]}{log[(a + \sqrt{R^2 + 1})/(a - \sqrt{R^2 + 1})]} \tag{10.15}$$

For two shell passes and any multiple of four tube passes,

$$F = \frac{[\sqrt{R^2 + 1}/(2R - 2)] \cdot log[(1 - P)/(1 - PR)]}{log[(a + b + \sqrt{R^2 + 1})/(a + b - \sqrt{R^2 + 1})]} \tag{10.16}$$

where $a = (2/P) - 1 - R$ and $b = (2/P)[(1 - P)(1 - PR)]^{1/2}$. The trends of correction factor indicate that a higher temperature drop within the tube flow occurs with enhanced heat transfer to the external flow (outside the tube). Various other physical trends can be observed based on the correlations involving F, R, and P. The following problem gives an example showing how these heat exchanger factors can be computed.

Example 10.3.1

Two-Shell-Pass, Eight-Tube-Pass Heat Exchanger.

The overall heat transfer coefficient for a shell-and-tube heat exchanger (two shells, eight tube passes) is $U = 1300$ W/m^{2}K. Hot fluid ($c_p = 4.95$ kJ/kgK) enters the heat exchanger at 340°C with a mass flow rate of 2 kg/sec. The cold fluid stream ($c_p = 4.195$ kJ/kgK) enters the inlet tube at 20°C at a rate of 3 kg/sec. If the total length of tubing within the heat exchanger is 60 m, find the required tube diameter to cool the hot stream to 180°C at the outlet.

Steady-state conditions and negligible heat losses to the surroundings will be assumed in this problem. Constant thermophysical properties are evaluated at $\overline{T}_c = 80$°C and $\overline{T}_h = 260$°C. Then, based on an energy balance for the hot fluid through the entire heat exchanger,

$$q = -\dot{m}_h c_h (T_{h,o} - T_{h,i}) = -2(4950)(340 - 180) = -1.58 \times 10^6 \text{ W} \tag{10.17}$$

Similarly, based on an energy balance for the cold fluid stream,

$$q = \dot{m}_c c_c (T_{c,o} - T_{c,i}) \tag{10.18}$$

which yields an outlet temperature of

$$T_{c,o} = 20 + \frac{1.58 \times 10^6}{3(4195)} = 145.5°\text{C} \tag{10.19}$$

The outlet temperature can now be used to compute the heat exchange factors as follows:

$$P = \frac{145.5 - 20}{340 - 20} = 0.39 \tag{10.20}$$

$$R = \frac{340 - 180}{145.5 - 20} = 1.27 \tag{10.21}$$

which together yield a correction factor of $F \approx 0.98$.

Thus, the total heat transfer is given by

$$q = FUA\Delta T_{lm} \tag{10.22}$$

which yields a required surface area of

$$A = \frac{1.58 \times 10^6 ln[(340 - 145.5)/(180 - 20)]}{0.98 \times 1300(194.5 - 160)} = 7.02 \text{ m}^2 \quad (10.23)$$

Since the total surface area of tubes is $A = 2\pi DL$ and $L = 60$ m, the required tube diameter is $7.02/(2\pi 60) = 1.9$ cm, which corresponds to a tube diameter of approximately 3/4 in.

The previous example, as well as other heat exchanger problems involving the log mean temperature difference, is based on radial heat transfer between the hot and cold fluid streams. Axial heat conduction was neglected in the derivation of the heat transfer rate based on ΔT_{lm}. However, axial temperature gradients within the wall of a tube lead to some axial heat conduction and a reduced mean temperature difference between the hot and cold fluid streams. As a result, axial heat conduction reduces the actual heat transfer between the fluid streams. In some cases, this reduction can be up to approximately $\pm 5\%$, in comparison to computations with neglected axial conduction. The actual difference between computations with or without axial conduction is dependent on the magnitude of the radial conductance relative to the wall temperature gradients in the axial direction.

In heat exchanger construction, the competing influences of pressure drop and heat exchange are important considerations in the final system. In particular, increased heat transfer can usually be achieved by increasing the packing of tubes within the heat exchanger or using baffles or other heat enhancement devices. However, this enhanced heat transfer often arises at the expense of an increased pressure drop, which is disadvantageous in view of the additional pumping power required to move the fluid through the system at a specified mass flow rate. Conversely, fewer heat exchange tubes can lead to a smaller pressure drop, but often at the expense of lower heat transfer, in comparison to a design with a high surface area density. An optimal solution is desirable to provide an effective balance between the heat exchange and pressure losses. This optimal condition may be viewed in terms of the Second Law of Thermodynamics, whereby the entropy production rate is minimized in terms of a specific design parameter, such as packing density or diameter of tubes within the heat exchanger.

In view of the importance of pressure losses, additional correlations are needed to predict the pressure drop in heat exchangers. In particular, for a cross-flow perpendicular to a group of uniformly spaced tubes in a heat exchanger,

$$\Delta P = \frac{1}{2}\rho^2 G^2 v_1 \left[(1 + \sigma^2)\left(\frac{v_2}{v_1} - 1 \right) + f \frac{v_m}{v_1} \frac{L}{r_h} \right] \quad (10.24)$$

where G is the maximum mass velocity (density times velocity); v_1, v_2, and v_m are the specific volumes at sections 1 and 2 and the mean value (average of v_1 and v_2); σ is the ratio of the free-flow area, A_{ff}, of the finned passages to the frontal area of the heat exchanger, A_{fr}; f is the friction factor; L is the flow length in the axial direction; and r_h is the flow passage hydraulic radius (total heat exchanger volume divided by the total heat transfer surface area).

Friction factors and Colburn j factors (see Equation (3.87) in Chapter 3) are documented extensively by Kays and London (1984) for a variety of heat exchangers, including finned and various tubular configurations.

10.4 Effectiveness–NTU Method of Analysis

The *effectiveness–number of transfer units (NTU) method* is a useful technique for the analysis of heat exchangers. The effectiveness of a heat exchanger, ϵ, may be defined as the ratio of the actual heat transfer to the maximum possible heat transfer, i.e.,

$$\epsilon = \frac{q}{q_{max}} \tag{10.25}$$

As outlined near the end of Section 10.2, the maximum heat transfer could be achieved in a counterflow arrangement, when one of the fluids experience the maximum possible temperature difference, so that

$$q_{max} = (\dot{m}c)_{min}(T_{h,i} - T_{c,i}) \tag{10.26}$$

and c refers to the fluid specific heat. The minimum product of $\dot{m}c$ is required in Equation (10.26). Conservation of energy in balancing Equations (10.1) and (10.2) suggests that the fluid with $(\dot{m}c)_{max}$ does not have a larger temperature change.

For convenience, the heat capacity rates are defined as follows:

$$C_c = (\dot{m}c)_c \tag{10.27}$$

$$C_h = (\dot{m}c)_h \tag{10.28}$$

where the subscripts c and h refer to the cold and hot fluid streams, respectively. Then, combining Equations (10.1), (10.2), and (10.26),

$$\epsilon = \frac{C_h(T_{h,i} - T_{h,o})}{C_{min}(T_{h,i} - T_{c,i})} = \frac{C_c(T_{c,o} - T_{c,i})}{C_{min}(T_{h,i} - T_{c,i})} \tag{10.29}$$

As a result, for $C_h > C_c$,

$$\epsilon_c = \frac{T_{c,o} - T_{c,i}}{T_{h,i} - T_{c,i}} \tag{10.30}$$

whereas for the case of $C_h < C_c$,

$$\epsilon_h = \frac{T_{h,i} - T_{h,o}}{T_{h,i} - T_{c,i}} \tag{10.31}$$

The maximum heat capacity rate is $C_{max} = (\dot{m}c)_{max}$, whereas the minimum heat capacity rate is $C_{min} = (\dot{m}c)_{min}$. Furthermore,

$$NTU = \frac{UA}{C_{min}} \tag{10.32}$$

and U is the overall heat transfer coefficient. The effectiveness of a particular heat exchanger, ϵ, is often tabulated and graphically depicted as a function of C_{min}/C_{max} (denoted by C_r) and NTU. These concepts are further clarified in the following example.

Example 10.4.1

Oil Cooling in a Single-Shell-Pass, Four-Tube-Pass Heat Exchanger.

Oil is cooled from 55 to 35°C as it flows at a rate of 0.4 kg/sec through the tube within a single-shell-pass, four-tube-pass heat exchanger. On the shell side of the heat exchanger, water enters at 10°C at a rate of 1 kg/sec. Under modified operating conditions, the oil flow rate is reduced to 0.3 kg/sec. Up to what maximum temperature of oil can be introduced at the inlet of the tube under the same water operating conditions, while not exceeding an oil outlet temperature of 25°C? Constant thermophysical properties (i.e., c_p = 2100 J/kgK for oil) may be adopted in this problem.

Based on an overall heat balance for the water (subscript c, cold) and oil (subscript h, hot),

$$q = \dot{m}_c c_c(T_{c,o} - T_{c,i}) = \dot{m}_h c_h(T_{h,i} - T_{h,o}) = 0.4(2100)(55 - 35)$$
$$= 16800 \text{ W} \tag{10.33}$$

which yields the following water outlet temperature:

$$T_{c,o} = 10 + \frac{16800}{1(4181)} = 14°C \tag{10.34}$$

Thus, the heat exchanger factors can be computed as follows:

$$P = \frac{55 - 35}{55 - 10} = 0.44 \tag{10.35}$$

$$R = \frac{14 - 10}{55 - 35} = 0.2 \tag{10.36}$$

which yields a correction factor of $F = 0.99$.

The log mean temperature difference is given by

$$\Delta T_{lm} = \frac{(T_{h,i} - T_{c,o}) - (T_{h,o} - T_{c,i})}{ln((T_{h,i} - T_{c,o})/(T_{h,o} - T_{c,i}))} = \frac{(55 - 14) - (35 - 10)}{ln(41/25)}$$
$$= 32.3°C \tag{10.37}$$

Then, based on the definition of the heat transfer rate in terms of the log mean temperature difference,

$$UA = \frac{q}{F\Delta T_{lm}} = \frac{16800}{0.99 \times 32.3} = 525.4 \text{ W/°C} \tag{10.38}$$

Under the modified operating conditions,

$$C_c = \dot{m}_c c_c = 1 \times 4181 = 4181 \text{ W/°C} = C_{max} \tag{10.39}$$

$$C_h = \dot{m}_h c_h = 0.3 \times 2100 = 630 \text{ W/°C} = C_{min} \tag{10.40}$$

and so $C_r = C_{min}/C_{max} = 0.15$. In addition,

$$NTU = \frac{UA}{C_{min}} = \frac{525.9}{630} = 0.83 \tag{10.41}$$

Based on the parameters of C_r and NTU, the effectiveness charts (Figure 10.5) yield $\epsilon = 0.55$.

Furthermore, this effectiveness can be used to compute the oil inlet temperature, i.e.,

$$\epsilon = \epsilon_h = \frac{T_{h,i} - T_{h,o}}{T_{h,i} - T_{c,i}} \tag{10.42}$$

or alternatively,

$$T_{h,i} = \frac{T_{h,o} - \epsilon_h T_{c,i}}{1 - \epsilon_h} = \frac{25 - 0.55(10)}{1 - 0.55} = 43.3°\text{C} \tag{10.43}$$

Temperatures above this maximum value will likely yield an outlet oil temperature above the specified 25°C under the modified operating conditions.

Based on the definition of *NTU* in Equation (10.32), the number of transfer units may be interpreted as the ratio of the total heat conductance (or reciprocal of the total thermal resistance, R) to the minimum heat capacity rate. Also, a time scale can be used to interpret the meaning of *NTU*. In particular, if a specified amount of mass, m, resides in the heat exchanger at some stage of time, and the rate of mass flow through the heat exchanger (minimum fluid) is $\dot{m}$, then the ratio of $m/\dot{m}$ can be used to represent a mass (or fluid) residence time, t_r, in the heat exchanger. Based on this mass residence time, Equation (10.32) can be written as

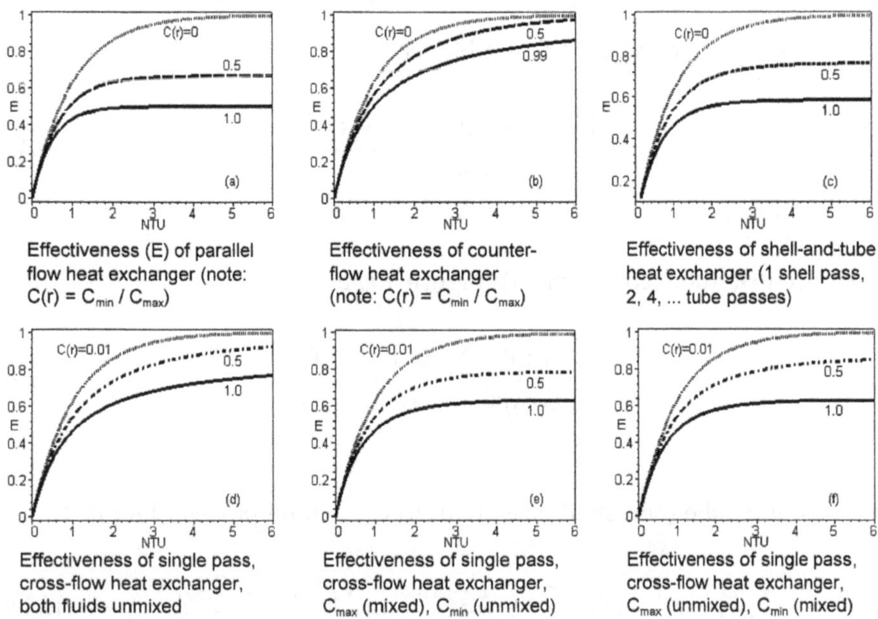

FIGURE 10.5
Effectiveness curves of various heat exchangers (note: E denotes ϵ).

$$NTU = \frac{UA}{C_{min}} = \frac{UA}{(\dot{m}c)_{min}} = \frac{1}{(1/UA)(mc/t_r)_{min}} = \frac{t_r}{(Rmc)_{min}} \tag{10.44}$$

The units of the denominator are units of time. As a result, NTU can be interpreted as the ratio of the fluid resistance time to a representative time constant of the C_{min} fluid. This time constant indicates a representative time required for the fluid to experience a fixed change in temperature when a unit amount (i.e., 1 J) of heat is transferred to the fluid. For example, values of $0.45 < NTU < 1$ are typically representative of automotive radiators, and other ranges of NTU can be used to identify the range of other heat exchanger applications.

The effectiveness of selected heat exchangers is illustrated in Figure 10.5. Algebraic expressions involving ϵ, NTU, and C_r have been developed and summarized for a variety of heat exchangers by Kays and London (1984). These results may be represented graphically, such as the forms shown in Figure 10.5.

The algebraic expressions corresponding to Figures 10.5a–f are listed below (Kays and London, 1984), where ϵ refers to the heat exchanger effectiveness (i.e., E along the vertical axis). The heat capacity ratio is denoted by $C_r = C_{min}/C_{max}$ (shown as $C(r)$ in Figure 10.5).

- Concentric tube (parallel flow): Figure 10.5a

$$\epsilon = \frac{1 - exp(-NTU(1 + C_r))}{1 + C_r} \tag{10.45}$$

- Concentric tube (counterflow): Figure 10.5b

$$\epsilon = \frac{1 - exp(-NTU(1 - C_r))}{1 - C_r exp(-NTU(1 - C_r))} \tag{10.46}$$

- Shell-and-tube (one shell pass; two, four, ... tube passes): Figure 10.5c

$$\epsilon = 2\left[1 + C_r + \sqrt{1 + C_r^2} \times \frac{1 + exp(-NTU\sqrt{1 + C_r^2})}{1 - exp(-NTU\sqrt{1 + C_r^2})}\right]^{-1} \tag{10.47}$$

- Cross-flow (single pass), both fluids unmixed: Figure 10.5d

$$\epsilon = 1 - exp\left[\frac{NTU^{0.22}}{C_r}(exp(-C_r(NTU)^{0.78}) - 1)\right] \quad (10.48)$$

- Cross-flow (C_{max} mixed, C_{min} unmixed): Figure 10.5e

$$\epsilon = \left(\frac{1}{C_r}\right)(1 - exp[-C_r(1 - exp(-NTU))]) \quad (10.49)$$

- Cross-flow (C_{max} unmixed, C_{min} mixed): Figure 10.5f

$$\epsilon = 1 - exp\left\{-\left(\frac{1}{C_r}\right)[1 - exp(-C_r NTU)]\right\} \quad (10.50)$$

As discussed in earlier chapters, fins can often substantially improve the heat transfer from surfaces. The following example demonstrates how previous results involving fins, such as Equation (10.4), can be incorporated into the analysis of heat exchangers.

Example 10.4.2

Finned Tubes in a Cross-Flow Heat Exchanger.

A set of 40 finned tubes (with 94% fin efficiency) is arranged uniformly in rows and perpendicular to an incoming airflow at 15°C and 1 kg/sec into a cross-flow heat exchanger. The following parameters are specified in the heat exchanger design: $A_c = 0.05$ m^2 (free-flow area), $l = 0.2$ m (flow length), $D_h = 1$ cm, and $A = 1.6$ m^2 (total heat transfer area). The Colburn and friction factors are $j = 0.02$ and $f = 0.04$, respectively. Hot water at 90°C enters the tubes (each of inner diameter of 2 cm) with a mean velocity of 1 m/sec. If the estimated thermal resistance due to fouling is 10^{-6} °C/W, find the outlet temperature of the air.

In this problem, the subscripts a, w, i, and o refer to air, water, inlet, and outlet, respectively. For heat transfer on the air side, the following thermophysical properties at 288 K are adopted: $\rho = 1.22$ kg/m^3, $\mu = 1.79 \times 10^{-5}$ kg/msec, $c_p = 1007$ J/kgK, and $Pr = 0.7$. Also,

$$G = \frac{\dot{m}_a}{A_c} = \frac{1}{0.05} = 20 \text{ kg/sec m}^2 \quad (10.51)$$

$$Re = \frac{D_h G}{\mu} = \frac{0.01(20)}{1.79 \times 10^{-5}} = 11,173 \tag{10.52}$$

Then, based on the specified Colburn factor and Equation (3.87)

$$h = \frac{jGc_p}{Pr^{2/3}} = \frac{0.02(20)1,007}{0.7^{2/3}} = 511 \ \ \mathrm{W/m^2K} \tag{10.53}$$

Furthermore, since the fin efficiency is 0.94, the total thermal resistance on the air side of the heat exchanger is given by

$$R_o = \frac{1}{\eta_o hA} = \frac{1}{0.94(511)1.6} = 0.0013 \tag{10.54}$$

For water flow within the tubes, the following thermophyical properties are evaluated at 90°C ≈ 365 K: c_p = 4209 J/kgK, ρ = 963 kg/m^3, k = 0.677 W/mK, Pr = 1.91, and = 3.1 × 10^{-4} kg/msec. Based on these values,

$$Re = \frac{\rho VD}{\mu} = \frac{963(1)0.02}{3.1 \times 10^{-4}} = 6.2 \times 10^4 \tag{10.55}$$

which suggests the following correlation for internal flow within a pipe:

$$\overline{Nu}_D = 0.023 Re_D^{0.8} Pr^{0.33} \tag{10.56}$$

$$\bar{h} = \frac{k}{D} \ Nu = \frac{0.677}{0.02} \ 0.023(6.2 \times 10^4)^{0.8}(1.91)^{0.33} = 6,577 \ \ \mathrm{W/m^2K} \tag{10.57}$$

Thus, the thermal resistance based on water flow within the pipes is given by

$$R_w = \frac{1}{h_w A} = \frac{1}{6577\pi(0.02)(0.2)40} = 0.0003 \ \mathrm{K/W} \tag{10.58}$$

The water mass flow rate is given by

$$\dot{m}_w = \rho VAN = 963(1)\frac{\pi}{4} \ 0.02^2(40) = 12.1 \ \ \mathrm{kg/sec} \tag{10.59}$$

As a result, the overall heat transfer coefficient, including fouling, as well as the air and water side convection, may now be estimated, i.e.,

$$U_{tot} = \frac{1}{AR_{tot}} = \frac{1}{1.6(0.0013 + 0.0003 + 10^{-6})} = 390.4 \ \text{W/m}^2\text{K} \tag{10.60}$$

Also, the heat capacity rates can be computed as follows:

$$\dot{m}_a c_a = 1(1007) = 1007 \ \text{W/K} = C_{min} \tag{10.61}$$

$$\dot{m}_w c_w = 12.1(42.09) = 50929 \ \text{W/K} = C_{max} \tag{10.62}$$

which yields a ratio of $C_{min}/C_{max} = 0.02$. Then the number of thermal units becomes

$$NTU = \frac{U_{tot}A}{C_{min}} = \frac{390.4(1.6)}{1007} = 0.62 \tag{10.63}$$

Based on these C_{min}/C_{max} and NTU values, the effectiveness charts yield $\epsilon \approx 0.5$, i.e.,

$$\epsilon = \frac{T_{ao} - T_{ai}}{T_{wi} - T_{ai}} \tag{10.64}$$

Thus, the air outlet temperature becomes

$$T_{ao} = 15 + 0.5(90 - 15) = 52.5°\text{C} \tag{10.65}$$

The resulting heat gained by the air can be equated with the heat removed from the water within the tubes, such that the outlet water temperature can be computed.

In the previous example, a finned tube cross-flow configuration was used for heat exchange between the liquid and gas flows. However, plate–fin heat exchangers are more commonly used in applications involving heat exchange between gas–gas streams. For example, air–air heat exchangers are typically based on plate–fin arrangements. These fins are classified into various types, such as plain fins, strip fins, pin fins, perforated fins, and others. Detailed design data involving these types of heat exchangers are outlined in Kays and London (1984).

10.5 Condensers and Evaporators

Condensers and evaporators are two-phase heat exchangers used in various engineering systems, such as power generation and refrigeration systems. The analysis of evaporators and condensers involves a combination of results derived earlier in this chapter for heat exchangers, such as effectiveness–NTU techniques, as well as correlations derived in Chapter 6 for condensation and boiling heat transfer. In particular, correlations for vertical and horizontal tubes and finned surfaces would be required.

If a fluid stream within a heat exchanger experiences a change of phase (i.e., it evaporates or condenses), then it is usually more useful to evaluate enthalpy (rather than temperature) in the energy balances. Temperature is assumed to remain nearly constant during the phase change, even though heat is transferred between the fluid streams. The relevant energy balances include Equation (10.6), leading to the derivation of the mean log temperature (or enthalpy) difference in Equation (10.10). Based on the enthalpy difference, the analysis would include both the latent and sensible heat portions of the energy transfer between different fluid streams in the heat exchanger.

Correction factors, together with the R and P factors in Equations (10.12) to (10.14), were utilized in previous sections for the analysis of cross-flow and shell-and-tube heat exchangers. Based on the definitions of the R and P factors, it can be observed that $P \to 0$ when the fluid stream in the tube flow experiences no change of temperature. This situation arises in condensers and evaporators when the temperature remains approximately constant (assuming a sufficiently small pressure drop through the tubes) due to the change of phase. If the other fluid is the condensing or evaporating stream, then $R \to 0$ instead. Based on these scenarios, the correction factor is $F = 1$. Since the temperature of the fluid undergoing phase change does not appreciably change along the flow path, it may interpreted in view of a single phase stream with a capacity rate, C, approaching an infinite value. For example, a single phase stream would approach isothermal flow when its flow rate becomes large.

A main difficulty in the analysis of condensers and evaporators, unlike previously discussed heat exchangers involving single phase flows, is the range of phase change regimes experienced by the fluid stream. A numerical study of a condenser in a power plant is described by Zhang et al. (1993). Since the heat transfer coefficient depends on the local phase fraction, which varies throughout the flow path, this heat transfer coefficient becomes position dependent. However, the phase distribution is typically unknown until the flow field solution is obtained. This suggests the importance of a systematic procedure for analyzing the heat transfer processes in condensers and evaporators.

A systematic procedure will be outlined based on relevant heat balances in a manner similar to that in Equation (10.6). These energy balances will involve enthalpy (rather than temperature) in order to include the latent heat of phase change, as well as sensible heat portions of the heat exchange. In particular, heat transfer is assumed to occur between a fluid stream undergoing phase change at a temperature of T_h and the other single phase fluid at T_c. Across a control volume of thickness dx in a tubular heat exchanger, heat transfer from the hot fluid stream (subscript h), dq, balances the energy gained by the other fluid stream in the form of an enthalpy rise, i.e.,

$$dq = \dot{m}_c dH = U_i(T_h - T_c)dA_i \tag{10.66}$$

where H and $\dot{m}_c$ will be used to refer to the fluid enthalpy and mass flow rate, respectively. Also, A_i refers to the inner area of the tube, i.e., based on d_i (inner diameter), rather than d_o (outer diameter).

The overall heat transfer coefficient, U, can be expressed in terms of the heat transfer coefficients based on the inner and outer tube surfaces, U_i and U_o, respectively, as follows:

$$\frac{1}{UA} = \frac{1}{U_i A_i} = \frac{1}{U_o A_o} \tag{10.67}$$

As a result, the heat transfer coefficient on the inner side can be determined by

$$U_i(\pi d_i) = \left[\frac{1}{h_c(\pi d_i)} + \frac{1}{h_h(\pi d_o)}\right]^{-1} \tag{10.68}$$

which can be rewritten to yield the following result:

$$U_i = \left[\frac{1}{h_c} + \frac{d_i}{h_h d_o}\right]^{-1} \tag{10.69}$$

where h_c and h_h refer to convection coefficients for the hot and cold streams, respectively, along the surface, dx, separating both fluid streams.

Since the fluid possibly undergoes various phase change regimes, the tube length is subdivided into N discrete elements and the previous energy balances are applied individually over each element. In particular, for the jth element, Equation (10.66) may be discretized as

$$\dot{m}_c(H_{j+1} - H_j) = U_{i,j}(T_{h,j} - T_{c,j})\pi d_i \Delta x \tag{10.70}$$

where $j = 1, 2, \ldots N$. This discrete energy balance is based on steady-state conditions and a uniform subdivision of the tube length into N elements of width Δx. Rearranging Equation (10.70),

$$H_{j+1} = \frac{U_{i,j}\pi d_i \Delta x}{\dot{m}_c}(T_{h,j} - T_{c,j}) + H_j \tag{10.71}$$

In this way, the enthalpy of each element can be based on previous (upstream) values of enthalpy and temperatures of the cold and hot fluid streams.

Once the enthalpy in a particular element is computed, its value may exceed the saturated vapor value of enthalpy at the flow pressure. In this case, the enthalpy can be used to determine the temperature of the superheated vapor based on thermodynamic property tables or computer-generated tabulation of the superheated property values. Alternatively, the change of enthalpy between elements can be used to find the corresponding temperature rise, i.e.,

$$c_p(T_{j+1} - T_j) \approx H_{j+1} - H_j \tag{10.72}$$

which yields T_{j+1} in terms of T_j (computed in the previous upstream element) and the computed enthalpy change from Equation (10.71). Since Equation (10.72) represents a discrete approximation of the definition of specific heat in a particular phase, it assumes a locally constant value of c_p in the linearization. It requires a sufficiently small temperature change between elements (i.e., sufficiently small Δx) so that significant changes of c_p with temperature are not realized. This approach can be used when the fluid exists entirely as a superheated vapor. In this case, the vapor specific heat at the mean temperature (between T_j and T_{j+1}) can be used.

On the other hand, if the enthalpy from Equation (10.71) is computed to be less than the saturated vapor enthalpy, then the quality (mass fraction of vapor), x, in element $j+1$ becomes

$$x_{v,j+1} = \frac{H_{j+1} - h_f}{h_g - h_f} \tag{10.73}$$

where h_g and h_f refer to the enthalpy of saturated vapor and liquid, respectively. Based on this phase fraction, an updated estimate of the convection coefficient can be calculated based on the flow regime associated with this vapor fraction. In particular, the transition factor (Chapter 6) is computed and used to identify the flow regime and appropriate correlation for heat transfer. Then an updated overall heat transfer coefficient is computed based on Equation (10.69), i.e.,

$$U_{i,j+1} = \left[\frac{1}{h_{c,j+1}} + \frac{d_i}{h_{h,j+1} d_o} \right]^{-1} \tag{10.74}$$

During the change of phase, the wall temperature between the two fluid streams is assumed to remain constant within the discrete element.

In this approach, a marching type of procedure is used whereby the problem variables are computed in the first element, $j = 1$, and then variables in successive elements are computed based on the previous element's values. For example, once $U_{i,1}$ is calculated from Equation (10.74), then the enthalpy, temperature, and quality for the following element, $j = 2$, can be calculated from Equations (10.71) to (10.73). Then the values are computed in the following element, and so on, over the entire length of the tube, for $j = 3, 4, \ldots N$. The entire procedure can be summarized as follows.

1. A boundary condition is applied within the initial element ($j = 1$).
2. A suitable forced convection correlation is used up to the element where phase change is first realized. A phase transition is identified when the mass fraction of the other phase (i.e., other than the initial inlet flow) becomes nonzero. The phase fraction can be detected from the temperature (with respect to the phase change temperature) or enthalpy (with respect to an equation of state).
3. In the first element experiencing phase change, the phase fraction is computed and the appropriate flow region is identified. Then the appropriate heat transfer correlation for that flow regime can be adopted based on correlations outlined in Chapter 6.
4. Near the saturation points (i.e., $0 \leq x < 0.001$ or $0.99 < x \leq 1$), a suitable convection correlation may be adopted with property values evaluated along the saturated liquid and vapor lines.
5. Repeat this sequence of steps for each element in the domain.

A similar procedure can be adopted for both types of two-phase flows: condensation or boiling. In boiling problems, a two-phase flow map would be typically utilized to identify the flow regime based on the computed phase fraction. This two-phase flow mapping would distinguish between flow regimes, such as the wavy, annular, and slug flow regimes. For example, when saturated liquid enters a heated horizontal pipe, boiling occurs initially by bubbles forming and growing along the heating surface. As the vapor fraction increases, transition occurs to plug flow, followed by slug flow. Transition to annular flow occurs at approximately a vapor fraction of $x = 0.04$, and then wavy flow at $x = 0.94$. Once the mixture becomes saturated vapor, subsequent heat addition is transferred to the vapor in a superheated state. In the assembly of elements, the average heat transfer coefficient can be computed as the arithmetic average of coefficients obtained over each element. In addition, based on Equation (10.66) over the entire tube, the total heat exchange between the fluids can be computed by the mass flow rate multiplied by the enthalpy difference between the inlet and outlet.

In addition to these heat exchange calculations, other design features and aspects of maintenance are important in terms of the effective performance of condensers and evaporators. For example, tubes should be readily cleanable on a regular basis, either through removable water heads or other means. Higher liquid velocities within the heat exchanger can reduce fouling (i.e., buildup of scale and dirt on the walls), reduce service, and extend the life of the heat exchanger. Also, higher operating efficiencies are often achieved by placement of tubes in stacks with metal-to-metal contact between fins to permit better drainage of the condensate. A lower thermal resistance is realized when less liquid accumulates on the fins, thereby improving the thermal efficiency of heat exchange. Another desirable feature is a light and compact design, which requires less space and reduces difficulties in installation and moving, while often reducing costs associated with maintenance. Similar considerations are applicable to the design and operation of evaporators.

Safety is another important factor in proper operation of condensers and evaporators. In systems operating at pressures different than the surrounding ambient (i.e., atmospheric) pressure, leakage can occur. A solution of soap or detergent can be brushed onto the surfaces where leakage is suspected, and if bubbles form therein, the specific points of leakage can be detected and repaired. Alternatively, pressurizing the system further and observing changes in pressure over time can indicate the tightness of the system (but not necessarily the location of leakage). Some chemical leaks can be detected individually. For example, sulfur dioxide can be detected by the white smoke forming when ammonia is brought into close contact with the leakage point.

The materials must be properly selected in conjunction with the working fluids. Although most refrigerants can be used well under normal conditions with most metals (such as steel, aluminum, and iron), some materials and liquids should never be used together. For example, methyl chloride fluid with aluminum shell and tubes can produce flammable gas by-products. For other heat exchanger materials, such as elastomers, the tensile strength, hardness, and other properties of the exposed elastomer must be fully considered under all operating conditions. In some cases, such as plastics formed by a variety of components, the resulting effects on a refrigerant liquid can often be difficult to predict, particularly due to the rapid rise in the number and types of polymer materials. As a result, an effective overall design of condensers and evaporators requires a thorough investigation of both quantitative and multidisciplinary design issues. For further detailed information regarding heat exchange design, the reader is referred to books by Fraas (1989), Hewitt (1998) and Kabac et al. (1981).

References

A. Bejan. 1993. *Heat Transfer*, New York: John Wiley & Sons.

R.A. Bowman, A.C. Mueller, and W.M. Nagle. Mean temperature difference in design, *Trans. ASME*, 62, 1940, pp. 283–294.

A.P. Fraas. 1989. *Heat Exchanger Design*, 2nd ed., New York: Wiley.

Heat Exchanger Design Handbook, G.F. Hewitt, ed., New York, Begell House, 1998.

F.P. Incropera and D.P. DeWitt. 1990. *Fundamentals of Heat and Mass Transfer*, 3rd ed., New York: John Wiley & Sons.

Heat Exchangers: Thermal-Hydraulic Fundamentals and Design, S. Kakac, A.E. Bergles, and F. Mayinger, eds., Washington DC: Hemisphere 1981.

W.M. Kays and A.L. London. 1984. *Compact Heat Exchangers*, 3rd ed., New York: McGraw-Hill.

G.F. Naterer, J. Boros, and X. Wang. 1996. "A turbulence integral model for the evaluation of the thermal effectiveness of a static damper in gas-fired water heaters," *CSME Trans.* 20: 437–452.

Standards of the Tubular Exchange Manufacturers Association, 8th ed., Tubular Exchange Manufacturers Association, Tarrytown, NY, 1999.

C. Zhang, A.C.M. Sousa, and J.E.S. Venart. 1993. "The numerical and experimental study of a power plant condenser," *ASME J. Heat Transfer* 115: 435–445.

Problems

1. In a cross-flow heat exchanger, water flows over a copper pipe (1.9-cm inner diameter; 2.4-cm outer diameter) with a convection coefficient of 210 W/m^{2}K. Oil flows through the pipe. Find the convection coefficient of the oil flow that is required to provide an overall heat transfer coefficient of 140 W/m^{2}K per unit length of pipe (based on the inner tube area).
2. In a counterflow heat exchanger, water enters at 20°C at 3 kg/sec and cools oil ($c_p = 2.2$ kJ/kgK) flowing at 2 kg/sec with an inlet

temperature of 160°C. The heat exchanger area is 12 m^2. What overall heat transfer coefficient is required when the water outlet temperature is 50°C?

3. Water flows through a copper pipe (3.2-cm inner diameter; 3.5-cm outer diameter) in a cross-flow heat exchanger. Air flows across the pipe. The convective heat transfer coefficients for the air and water sides of the pipe are 120 and 2400 W/m^2K, respectively. Up to what additional fouling resistance can be tolerated if its effect must not reduce the overall heat transfer coefficient (based on the outside area without fouling) by more than 5%?
4. A tubular heat exchanger operates in a counterflow arrangement. A design modification requires a higher mass flow rate of fluid in the inner tube. Under this modification, what change of tube length is required to maintain the same inlet and outlet temperatures of both fluids? Express your answer in terms of the tube diameter, thermophysical properties, fluid velocity, and convection coefficient of the external flow in the outer annulus. It may be assumed that this external convection coefficient remains approximately identical under both flow conditions.
5. A concentric tube heat exchanger is used to condense steam in the annulus between the inner tube (copper; 1.6-cm inner diameter, 0.2-cm wall thickness) and the surface of the outer tube (4-cm outer diameter). The water flows at 6 m/sec with an average temperature of 25°C through the inner pipe. The thermal resistances due to steam convection and fouling (outer surface of inner tube) are 2×10^{-5} and 10^{-6} K/W, respectively. If the total length of the inner pipe is 80 m, then find the total heat transfer to the water and the average inner wall temperature of the inner pipe.
6. A set of finned tubes is arranged uniformly in rows and placed normal to an incoming airflow at 15°C with a flow rate of 300 m^3/h into a cross-flow heat exchanger. Hot water at 85°C enters the tubes (each of inner and outer diameters of 10 and 12 mm, respectively) with a mean velocity of 0.22 m/sec. The heat exchanger volume is 0.006 m^3, and the Colburn and friction factors are $j = 0.014$ and $f = 0.049$, respectively. If the estimated thermal resistance due to fouling is 10^{-5} °C/W, find the number of finned tubes required to produce an air outlet temperature of 64°C. The following additional parameters are specified in the design of the heat exchanger: $A_{fr} = 0.1\ m^2$ (frontal area), $\sigma = 0.3$, $D_h = 5$ mm, copper fin diameter and thickness of 18 and 0.2 mm, respectively, $A_f/A = 0.9$, $a = 350\ m^2/m^3$ (ratio of the total area to the volume of the heat exchanger), and $l = 40$ cm (flow length).
7. A cross-flow heat exchanger in a power plant is used for intercooling between two compressor stages. Air enters the heat exchanger at 400°C

with a flow rate of 20 kg/sec and flows past finned tubes with a fin efficiency of 90%. The Colburn and friction factors are 0.008 and 0.03, respectively. If cooling water enters the heat exchanger at 10°C with a flow rate of 40 kg/sec, find the outlet temperature and pressure drop of the air. The following additional parameters are specified in the heat exchanger design: $A_{c,w} = 0.4$ m^2, $D_{h,a} = 2$ cm $= D_{h,w}$, $A_{c,a} = 2$ m^2, $A_f/A = 0.8$, $A_w/A_a = 0.2$, $\alpha = 350$ m^2/m^3, and $\sigma = 0.8$. Also, the dimensions of the heat exchanger are 2 m (length) × 0.75 m (height) × 0.8 m (width), and the thermal resistance due to tube fouling is 10^{-5} °C/W.

8. Derive Equation (10.45) for the effectiveness of a concentric tube (parallel flow) heat exchanger.
9. In a cross-flow heat exchanger, air is heated from 10 to 30°C (unmixed stream) by another airstream entering at 90°C (mixed stream). The mass flow rates of the cold and hot airstreams are 1 and 2 kg/sec, respectively. What percentage increase of the overall heat transfer coefficient is required if the cold stream must be heated to 35°C at the same flow rates of air? Assume that the heat transfer area is maintained equally in both cases (although its configuration may be altered to achieve higher heat transfer coefficients).
10. In a shell-and-tube heat exchanger, the inner and outer diameters of each copper tube are 1.5 and 1.9 cm, respectively. The convection coefficients for fluid flow inside and over the tubes are 4800 and 4600 W/m^{2}K, respectively. What change in the number of tubes will increase the overall heat transfer coefficient (based on the outside area, per unit length) by a factor of 30%? Assume that this change does not affect the shell-side fluid velocity and the flow rate in the tubes remains constant for both cases.
11. Oil is cooled by water flowing through tubes in a shell-and-tube heat exchanger with a single shell and two tube passes. The pipe's outer diameter, wall thickness, thermal conductivity, and length are 1.9 cm, 2 mm, 26 W/mK, and 4 m, respectively. Water flows at 0.3 kg/sec on the shell side with a convection coefficient of 4950 W/m^{2}K and an inlet temperature of 280 K. Oil flows at 0.2 kg/sec in the tubes with a convection coefficient of 4800 W/m^{2}K (note: $c_p = 2.2$ kJ/kgK for oil). If the oil inlet temperature is 360 K, determine its outlet temperature.
12. Water enters a shell-and-tube heat exchanger (single-shell, four-tube pass) at 30°C. It is heated by an oil flow ($c_p = 2.3$ kJ/kgK) that enters the tube at 240°C and leaves at 140°C. The mass flow rate (tube side) and heat exchanger area are 2 kg/sec and 18 m^2, respectively. What overall heat transfer coefficient is required to yield a water outlet temperature of 70°C on the shell side?
13. Air is heated by water flowing through tubes at 2 kg/sec in a cross-flow heat exchanger (water unmixed, air mixed). The overall heat transfer

coefficient is 500 W/m^{2}K, and the length and diameter of each tube are 3 m and 1.9 cm, respectively. The water inlet temperature is 80°C. Air enters the heat exchanger at 3 kg/sec and 10°C. How many tubes are required to produce an air outlet temperature of 60°C?

14. A shell-and-tube heat exchanger contains one shell pass and two tube passes. The overall heat transfer coefficient is 480 W/m^{2}K, and the total surface area of the tubes is 16 m^2. Water enters the shell side at 40°C and leaves at 80°C. Find the required mass flow rate of water to cool the fluid in the tubes from 260°C (at the tube inlet) to 120°C (at the tube outlet). The specific heat of the fluid in the tube flow is 2.4 kJ/kgK.
15. Water is heated as it flows through a single-pass shell-and-tube heat exchanger with N tubes internally. The outside surface of each tube (diameter D) is heated by steam condensing at a temperature of T_{sat}. Find the number of tubes that are required to condense a specified mass flow rate of steam, $\dot{m}_s$. Express your answer in terms of the mass flow rate of water, $\dot{m}_w$, water inlet and outlet temperatures, and thermophysical properties of water.
16. Discuss the main design features of importance when selecting a shell-and-tube heat exchanger for condensing refrigerant fluid.
17. What techniques can be used for heat transfer enhancement in heat exchangers?
18. Perform a review of the technical literature to determine the degree of compactness of various types of heat exchangers, such as automotive radiators and plate heat exchangers. The compactness of a heat exchanger can be calculated from the heat transfer area density, i.e., square feet of surface area per cubic feet of heat exchanger volume.
19. A square-shaped array of 1-cm-diameter tubes is located within a condenser. Water at 20°C enters at a velocity of 1 m/sec into the tubes. The heat exchanger is required to condense 4 kg/sec of refrigerant 12 (with $h_{fg} \approx 130$ kJ/kg) at 45°C outside the tubes.
    (a) How many tubes are required to keep the temperature rise of the water below 3°C?
    (b) What total surface area is required to produce the required condensate mass flow rate of refrigerant fluid?
20. A condenser consists of a heat exchanger with a single shell pass and internal copper tubes (diameter of 1.9 cm, length of 1 m). Steam condenses at atmospheric pressure on the outer surface of the tubes (to be arranged in a square array). The outside wall temperature of the tube is 92°C.
    (a) How many tubes (per row) are required to produce a rate of steam condensation of 60 g/sec?
    (b) What temperature rise of the cooling water is encountered for this tube configuration? The flow rate is 0.1 kg/sec per tube.

11

Computational Heat Transfer

Computational methods are often required for the analysis of engineering problems in complicated geometrical configurations. Several types of methods are available for the numerical solution of the heat transfer equations. These types of methods include: (i) integral formulations, (ii) direct methods (such as finite difference methods), and (iii) methods based on conservation principles (such as finite volume methods). Integral formulations may be further subdivided into weighted residual and variational methods. In the former case, an error type residual is minimized over the problem domain, whereas variational methods seek to minimize a certain function to find undetermined coefficients in the numerical solution. In this chapter, these numerical methods will be considered.

11.1 Finite Difference Methods

In a numerical formulation, the problem domain is subdivided into an assembly of subregions, called a grid or mesh, over which the governing equations are discretized and solved. For example, considering two-dimensional heat conduction,

$$\rho c_p \frac{\partial T}{\partial t} = k \frac{\partial^2 T}{\partial x^2} + k \frac{\partial^2 T}{\partial y^2} + \dot{q} \tag{11.1}$$

This equation is discretized over each subregion, and a resulting set of algebraic equations is obtained. More accurate results can be obtained with further mesh refinements, but more equations would then be required (i.e., more computational time and memory). In this section, the discretized equations will be derived based on finite difference approximations of the governing equations.

11.1.1 Steady-State Solutions for Heat Conduction

11.1.1.1 Taylor Series Expansions

A possible approach for deriving the finite difference equations is based on Taylor series methods. Performing a Taylor series expansion of temperatures about a nodal point (i, j), where i and j refer to column and row numbers of the discrete mesh respectively,

$$T_{i+1,j} = T_{i,j} + \left.\frac{\partial T}{\partial x}\right|_i \Delta x + \left.\frac{\partial^2 T}{\partial x^2}\right|_i \frac{\Delta x^2}{2} + \left.\frac{\partial^3 T}{\partial x^3}\right|_i \frac{\Delta x^3}{6} + \ldots \tag{11.2}$$

$$T_{i-1,j} = T_{i,j} - \left.\frac{\partial T}{\partial x}\right|_i \Delta x + \left.\frac{\partial^2 T}{\partial x^2}\right|_i \frac{\Delta x^2}{2} - \left.\frac{\partial^3 T}{\partial x^3}\right|_i \frac{\Delta x^3}{6} + \ldots \tag{11.3}$$

where Δx refers to the grid spacing in the x direction. A uniform grid spacing will be assumed in the current analysis.

Adding Equations (11.2) and (11.3) together and rearranging,

$$\frac{\partial^2 T}{\partial x^2} = \frac{T_{i+1,j} - 2T_{i,j} + T_{i-1,j}}{\Delta x^2} + \mathcal{O}(\Delta x^2) \tag{11.4}$$

where $\mathcal{O}(\Delta x^2)$ refers to an order of magnitude proportional to Δx^2. Writing a similar expression for the y direction derivative and combining with Equation (11.4),

$$\frac{\partial^2 T}{\partial x^2} + \frac{\partial^2 T}{\partial y^2} \approx \frac{T_{i+1,j} - 2T_{i,j} + T_{i-1,j}}{\Delta x^2} + \frac{T_{i,j+1} - 2T_{i,j} + T_{i,j-1}}{\Delta y^2} \approx 0 \tag{11.5}$$

which can be considered to be *second-order accurate*, since the truncation errors correspond to terms truncated above the second order from the original Taylor series expansions. For a uniform grid with $\Delta x = \Delta y$, Equation (11.5) becomes

$$T_{i,j} = \frac{T_{i+1,j} + T_{i-1,j} + T_{i,j+1} + T_{i,j-1}}{4} \tag{11.6}$$

which correctly indicates that the steady-state diffusion equation yields temperatures that are arithmetic averages of their surrounding nodal values.

11.1.1.2 Energy Balances

Another approach for deriving the finite difference equations is based on energy balances. In this case, the grid is subdivided into finite volumes, where the nodal point is located at the center of the volume. Then discrete energy balances are applied within each control volume. For example, considering a control volume at point p, or (i, j),

$$Q_n + Q_w + Q_s + Q_e = 0 \tag{11.7}$$

where the subscripts n, w, s, and e refer to the north, west, south, and east edges, respectively, of the control volume about point p. These edges are located halfway between the appropriate nodal points. For example, point n is located halfway between (i, j) and $(i, j+1)$.

Using a grid spacing of Δx and Δy in the x and y directions, respectively, together with Fourier's law and a linear interpolation of temperature between nodal points, Equation (11.7) yields

$$-k\Delta x\left(\frac{T_{i,j} - T_{i,j+1}}{\Delta y}\right) - k\Delta y\left(\frac{T_{i,j} - T_{i-1,j}}{\Delta x}\right) - k\Delta x\left(\frac{T_{i,j} - T_{i,j-1}}{\Delta y}\right)$$

$$-k\Delta y\left(\frac{T_{i,j} - T_{i+1,j}}{\Delta x}\right) = 0 \tag{11.8}$$

per unit depth. For a uniform grid ($\Delta x = \Delta y$), it can be readily verified that Equation (11.8) is reduced to Equation (11.6). In other words, the same finite difference result is obtained for both the Taylor series method and energy balances (as expected). Once the finite difference equations are obtained, the resulting set of algebraic equations must be solved to yield the final temperature values at all nodes within the domain.

Example 11.1.1.2.1

Mixed Conduction and Convective Cooling of a Brick Column.

Consider a uniform nine-noded discretization of a brick column cooled by convection on its left side by a fluid at 40°C with a convection coefficient of 60 W/m^{2}K. The square cross-section of the brick column has a side width of 20 cm, and the thermal conductivity of the clay brick is 1 W/mK. The temperatures of the top and bottom boundaries are 240 and 120°C, respectively, and the right boundary is well insulated. Find the temperature values at nodal points along the horizontal midplane of the column.

Using a node numbering scheme starting from the top left corner and proceeding by row and column to the lower right boundary, the boundary conditions are

$$T_1 = T_2 = T_3 = 240 \quad (11.9)$$

$$T_7 = T_8 = T_9 = 120 \quad (11.10)$$

and the unknown temperatures are T_4, T_5, and T_6.

Applying an energy balance to the internal node 5,

$$Q_{2-5} + Q_{4-5} + Q_{6-5} + Q_{8-5} = 0 \quad (11.11)$$

$$-k(0.1)\left(\frac{T_5 - T_2}{0.1}\right) - k(0.1)\left(\frac{T_5 - T_4}{0.1}\right) - k(0.1)\left(\frac{T_5 - T_6}{0.1}\right) - k(0.1) \times \left(\frac{T_5 - T_8}{0.1}\right) = 0 \quad (11.12)$$

Simplifying and rearranging,

$$T_4 - 4T_5 + T_6 = -360 \quad (11.13)$$

At the boundary node 4, the convective cooling condition is applied through Newton's law of cooling at the edge of the control volume, i.e.,

$$Q_b + Q_{1-4} + Q_{7-4} + Q_{5-4} = 0 \quad (11.14)$$

$$60(0.1)(40 - T_4) - k(0.05)\left(\frac{T_4 - T_1}{0.1}\right) - k(0.05)\left(\frac{T_4 - T_7}{0.1}\right) - k(0.1) \times \left(\frac{T_4 - T_5}{0.1}\right) = 0 \quad (11.15)$$

This equation may be simplified to the following form:

$$-8T_4 + T_5 = -420 \quad (11.16)$$

In the case of the energy balance for node 6, the boundary contribution to heat flow is $Q_b = 0$ since the right boundary is insulated, thereby yielding after simplification,

$$T_5 - 2T_6 = -180 \quad (11.17)$$

Equations (11.13), (11.16), and (11.17) can be solved by standard algebraic solvers, such as Gaussian elimination, iterative, or matrix inversion techniques (Cheney and Kincaid, 1985). Following this solution, the following results are obtained: $T_4 = 71.1°C$, $T_5 = 148.9°C$, and $T_6 = 164.4°C$. An overall heat balance may be used to validate the accuracy of these results. In particular, the total heat flow across the top boundary of the domain, joined by nodes 1 (top left corner), 2, and 3 (top right corner), is given by $Q_{123} = 813.3$ W/m. Similarly, the heat flows across the bottom and left boundaries are computed to be $Q_{789} = 213.3$ W/m and $Q_{147} = -1026.6$ W/m, respectively. Based on these results,

$$Q_{123} + Q_{789} + Q_{147} = 813.3 + 213.3 - 1026.6 = 0 \tag{11.18}$$

which indicates that the finite difference results yield individual and overall energy conservation under steady-state conditions.

If the nodes are located along a curved boundary, special consideration is required to obtain the appropriate finite difference equations. The algorithm for establishing the discretized equations becomes dependent on the boundary configuration. As a result, the generality of the algorithm can become compromised for complex grids. In an upcoming section, finite element methods will be described for complete geometric flexibility in the discretization of the governing equations and problem domain.

11.1.2 Transient Solutions

For transient problems, the temperature variations with spatial coordinates, i.e., x, y, and z, as well as time, t, are required. For example, considering one-dimensional heat conduction with zero source terms,

$$\rho c_p \frac{\partial T}{\partial t} = k \frac{\partial^2 T}{\partial x^2} \tag{11.19}$$

In this section, two types of transient formulations (explicit and implicit) will be briefly presented using the energy balance method (described for interior nodes).

11.1.2.1 Explicit Formulation

Using a backward difference approximation in time, together with a one-dimensional energy balance associated with Equation (11.19),

$$(A\Delta x)\rho c_p \left(\frac{T_i^{n+1} - T_i^n}{\Delta t} \right) = kA \left(\frac{T_{i+1}^n - T_i^n}{\Delta x} \right) - kA \left(\frac{T_i^n - T_{i-1}^n}{\Delta x} \right) \tag{11.20}$$

where the superscripts $n+1$ and n refer to current and previous time levels, respectively. Also, Δt is the discrete time step. In an explicit formulation, the heat terms of Q_e and Q_w on the right side of Equation (11.20) are evaluated at the previous time level.

Although an explicit approach typically reduces the computational time required over a time step, since the matrix of diffusion coefficients becomes diagonal, a time step restriction is usually imposed to maintain numerical stability. In particular, rearranging Equation (11.20),

$$T_i^{n+1} = (1 - 2Fo)T_i^n + Fo(T_{i+1}^n + T_{i-1}^n) \tag{11.21}$$

where $Fo = \alpha\Delta t/\Delta x^2$ is the Fourier modulus (nondimensional time) and $\alpha = k/(\rho c_p)$ is the thermal diffusivity. Based on the first term on the right side of Equation (11.20), the following stability criterion is obtained for interior nodes:

$$Fo \leq \frac{1}{2} \tag{11.22}$$

This result is required to maintain a positive leading coefficient of T_i^n in Equation (11.21). Otherwise, an increase of T_i^n could lead to a decrease of T_i^{n+1} and potentially numerical oscillations with a lack of solution convergence. In the following section, it will be shown that unconditional stability can be achieved by evaluating the diffusive terms in Equation (11.20) at the current time step $(n+1)$, rather than the previous time level (n).

11.1.2.2 Implicit Formulation

In a manner similar to that in the derivation of Equation (11.20), while evaluating the diffusive flux terms at the current time level $(n+1)$,

$$(A\Delta x)\rho c_p\left(\frac{T_i^{n+1} - T_i^n}{\Delta t}\right) = kA\left(\frac{T_{i+1}^{n+1} - T_i^{n+1}}{\Delta x}\right) - kA\left(\frac{T_i^{n+1} - T_{i-1}^{n+1}}{\Delta x}\right) \tag{11.23}$$

This formulation becomes implicit since the transient and diffusive flux terms involve temperatures that are currently sought (rather than known from a previous time level). Rearranging Equation (11.23),

$$(1 + 2Fo)T_i^{n+1} - Fo(T_{i-1}^{n+1} + T_{i+1}^{n+1}) = T_i^n \tag{11.24}$$

It can be observed that there is not a stability limit involving Δt. However, in this case, a set of simultaneous equations needs to be solved at each time step. Thus, there is a trade-off between more time steps (explicit formulation) and fewer time steps (implicit formulation), with more computational effort required at each time step. In many practical problems, implicit solutions usually offer the most economical and reliable method of numerical heat transfer analysis.

11.2 Weighted Residual Methods

In the previous section, structured grids involving a regular pattern of rows and columns in a mesh were used. However, if the number of grid points per grid line is not constant, or if no grid structure is apparent, then methods of dealing with unstructured grids are required. In this section, a general method for providing these unstructured grid capabilities (called the method of weighted residuals) will be discussed.

The governing equation for the transport of a general conserved quantity, B, can be written as

$$\mathscr{L}(B) = \nabla \cdot (\rho \mathbf{v} B) + \nabla \cdot (\Gamma \nabla B) + S = 0 \tag{11.25}$$

In this equation, $\mathscr{L}$ and Γ refer to an operator (acting on its argument, such as B) and diffusion coefficient, respectively. Also, S refers to source type terms, such as thermal buoyancy, phase change, or others. The numerical solution seeks to find an approximate solution of Equation (11.25), namely $\tilde{B}$. In this approach, $\mathscr{L}(\tilde{B}) = R(\mathbf{x})$ (called the solution *residual*) gives a measure of how well the governing partial differential equation is satisfied.

The residual function $R(\mathbf{x})$ gives the difference obtained in substituting the exact solution, B, into the governing equation, in comparison to substitution of the approximate solution, $\tilde{B}$. In particular, we will use a linear combination of appropriate basis functions, $B_n(\mathbf{x})$, to represent the approximate solution, i.e.,

$$\tilde{B} = a_1 B_1(\mathbf{x}) + \ldots + a_n B_n(\mathbf{x}) = \sum_{i=1}^{n} a_i B_i(\mathbf{x}) \tag{11.26}$$

The function in Equation (11.26) contains unknown coefficients ($a_1, a_2, \ldots, a_n$) that need to be determined from the numerical solution.

In general, the residual will not equal zero, since it is unlikely that our initial approximate solution will precisely match the exact solution. As a result, the coefficients in Equation (11.26) are determined by minimizing the residual over the domain or setting the integrated and weighted value of the residual to zero. In particular, we will pose n constraints on the coefficients $a_1, \ldots, a_n$ by formulating the following weighted residual model:

$$\int_D W_i \mathscr{L}(\tilde{B}) dD = \int_D W_i R dD = 0 \tag{11.27}$$

where D and W_i refer to the domain and weight functions, respectively. In this weighted residual method, we choose n weight functions, $W_1, \ldots, W_n$, and obtain n equations for the unknown coefficients $a_1, \ldots, a_n$. In practice, the unknown coefficients would represent problem variables such as temperature (energy equation) or velocity (momentum equation). We will obtain n integral equations for these n coefficients, thereby yielding an $n \times n$ matrix for the resulting set of linear equations.

We may interpret Equation (11.27) as carefully distributing the error distribution throughout the solution domain. If the weight function becomes large, then the residual approaches zero and the governing equation is satisfied exactly at the given point where the large weight function is applied. Alternatively, if the weight function is nonzero, then the residual is nonzero and errors are distributed away from the grid points. If the weight functions are selected to be equal to the basis functions (shape functions) of the approximate solution, then this method is called a Galerkin weighted residual method. In the following section, a Galerkin-based finite element method will be described.

11.3 Finite Element Method

11.3.1 Fundamental Concepts

In the finite element method, the problem domain is discretized and represented by an assembly of finite elements. Unlike the previously described finite difference methods, the finite element method yields discretized equations that are entirely local to an element. As a result, the discrete equations are developed in isolation and independent of the mesh configuration. In this way, finite elements can readily accommodate unstructured and complex grids, unlike finite difference methods requiring structured ($i =$ row number, $j =$ column number) grid formats. Finite element methods provide complete geometric flexibility. After the local element equations are formed, assembly rules are required to reconstruct the entire domain from its parts (elements).

The overall steps in the finite element method can be summarized as follows:

1. Discretize the domain by specifying the number of elements, shape of elements, and their distribution.
2. Select a type of interpolation for the dependent variable, such as linear or quadratic interpolation.

3. Determine the *element property equations* (or *stiffness equations*). For example, find the temperature equations for a heat transfer analysis.
4. Assemble the elements by following specified assembly rules.
5. Apply the boundary conditions.
6. Solve the discrete equation set and postprocess the results.

The following example outlines these fundamental concepts through an application to one-dimensional heat transfer in a nuclear fuel element.

Example 11.3.1.1

Heat Conduction in a Nuclear Fuel Element.

The purpose of this example is to apply the finite element method to a problem involving one-dimensional heat conduction within a nuclear fuel rod (see Figure 11.1). Specified temperature conditions are given at both edges of the rod and the governing equation is given by

$$\mathscr{L}(T) = \frac{d^2T}{dx^2} + \frac{\dot{P}}{k} = 0 \tag{11.28}$$

The one-dimensional domain is subdivided into four nodes with three linear elements located uniformly within the domain of length L. Use the finite element method to find the temperatures at each node within this domain.

1. The domain is discretized into three equal elements. Further grid refinements may be required to find the final converged solution.

2. Interpolation within each element is performed based on a locally linear approximate solution. Using dimensionless coordinates, given by

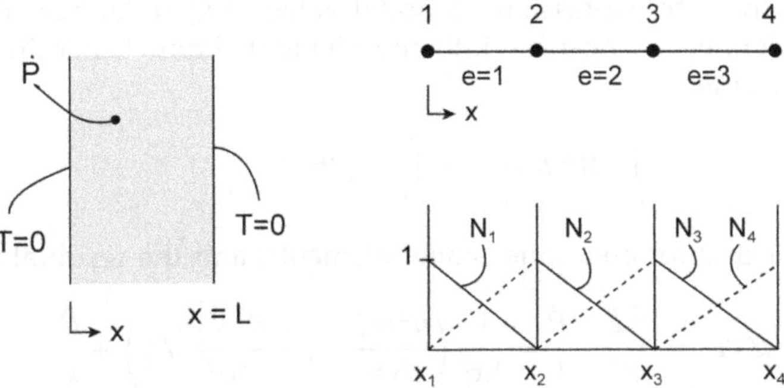

FIGURE 11.1
Conduction with internal heat generation (schematic and sample grid).

$x^* = x/L$, within each element,

$$x^* = \frac{x - x_1^e}{x_2^e - x_1^e} \tag{11.29}$$

where the subscript refers to the local node number (i.e., two local nodes for linear one-dimensional element), while the superscript e indicates local evaluation within the element. Also, based on a locally linear temperature profile within each element,

$$\tilde{T}^e = a_1^e + a_2^e x^* \tag{11.30}$$

where the $\tilde{T}$ notation (denoting approximate solution) will be dropped in subsequent equations for brevity.

If the temperatures at the local nodes are T_1^e (at $x^* = 0$) and T_2^e (at $x^* = 1$), then the coefficients in Equation (11.30) can be readily evaluated to yield

$$T^e = T_1^e + (T_2^e - T_1^e)x^* \tag{11.31}$$

Alternatively, Equation (11.31) can be written as

$$T^e = N_1^e(x^*)T_1^e + N_2^e(x^*)T_2^e \tag{11.32}$$

where the shape functions (or interpolation functions) are given by

$$N_1^e(x^*) = 1 - x^* \tag{11.33}$$

$$N_2^e(x^*) = x^* \tag{11.34}$$

These expressions, particularly Equation (11.32), indicate that there are two nodal degrees of freedom in the interpolation with linear elements.

3. The third step involves the derivation of the element property equations. In these expressions, a nodal value of Q (total heat flow, units of W) is defined as positive into the element. From Galerkin's method within an element,

$$\int_{V^e} W_i^e R^e dV^e = \int_{V^e} N_i^e R^e dV^e \tag{11.35}$$

where $i = 1$ or 2 for one-dimensional elements, and the residual is

$$R^e = \mathcal{L}(\tilde{T}) = \frac{d^2\tilde{T}}{dx^2} + \frac{\dot{P}}{k} = \frac{1}{\Delta x^2}\left(\frac{d^2N_1^e}{dx^{*2}}T_1^e + \frac{d^2N_2^e}{dx^{*2}}T_2^e\right) + \frac{\dot{P}}{k} \tag{11.36}$$

Then, based on Equation (11.35) for both weight functions within the element, the local finite element equations (indicated by the superscript e)

can be written in the following matrix form:

$$[c]\{T^e\} = \{r^e\} \tag{11.37}$$

where the 2 × 2 stiffness matrix, c_{mn}, is given by

$$c_{mn} = \frac{1}{\Delta x^2}\int_{V^e} N_m^e \frac{d^2 N_n^e}{dx^{*2}}\, dV^e \tag{11.38}$$

and the right-side vector, r_m, is

$$r_m = WR_m^e - \int_{V^e} N_m^e \frac{\dot{P}}{k}\, dV^e \tag{11.39}$$

After the assembly of all elements, the sum of residual terms becomes zero (based on the weighted residual method). As a result, the first term on the right side of Equation (11.39) can be effectively dismissed from further consideration.

From Equation (11.38), two problems can be observed in the formulation: (i) singular matrix due to zero second-order derivatives arising from linear shape functions, and (ii) no direct mechanism to readily invoke boundary conditions other than temperature-specified conditions. A possible approach to these difficulties is to maintain more interelement continuity. The value of T and all of its derivatives, up to and including one less than the highest order derivative appearing in the element property equations, should be continuous across element boundaries. For example, piecewise cubic interpolation functions could be used in our problem. Alternatively, we can use integration by parts to reduce the order of the highest derivative in the element equations. This method prevents a singular matrix and readily permits other types of boundary conditions, such as heat flux-specified conditions.

Using Galerkin's formulation with Equations (11.35) and (11.36), together with integration by parts and the local coordinates,

$$\int_{V^e} N_i^e \left(\frac{d^2 T}{dx^2} + \frac{\dot{P}}{k}\right) dV^e$$

$$= \frac{1}{\Delta x} N_i^e \frac{dT}{dx^*}\bigg|_0^1 - \frac{1}{\Delta x^2}\int_0^1 \frac{dN_i^e}{dx^*}\frac{dT}{dx^*}(\Delta x dx^*) + \int_0^1 N_i^e \frac{\dot{P}}{k}(\Delta x dx^*) = WR_i^e \tag{11.40}$$

The result in Equation (11.40) involves essentially three main terms between the equality signs. In the first term, the shape function, N_i^e, may be evaluated at both local nodes, yielding

$$\frac{1}{\Delta x} N_i^e \frac{dT}{dx^*}\bigg|_1 - \frac{1}{\Delta x} N_i^e \frac{dT}{dx^*}\bigg|_0 = \frac{Q_2^e}{kA} + \frac{Q_1^e}{kA} \tag{11.41}$$

where A refers to the cross-sectional area of the one-dimensional element in the direction of heat flow. From Equation (11.41), using local node 1 as an example, it can be observed that a negative temperature gradient yields a positive Q_1^e, i.e., heat flow into the element. Conversely, a positive temperature gradient at node 1 yields a negative heat flow (heat flow out of the element). In both cases, our original sign convention has been retained since both Q values are positive into the element.

For the second term in Equation (11.40),

$$\frac{1}{\Delta x}\int_0^1 \frac{dN_i^e}{dx^*}\frac{dT}{dx^*}\,dx^* = \frac{1}{\Delta x}\int_0^1 \frac{dN_i^e}{dx^*}\left(\frac{dN_1^e}{dx^*} T_1^e + \frac{dN_2^e}{dx^*} T_2^e\right)dx^* \tag{11.42}$$

Substituting $i = 1$, the right side of Equation (11.42) becomes $(T_1^e - T_2^e)/\Delta x$, whereas $i = 2$ yields the same result, except with a leading negative sign. Furthermore, for the third term in Equation (11.40),

$$\Delta x \int_0^1 N_i^e \frac{\dot{P}}{k}\,dx^* = \frac{1}{2}\,\Delta x \frac{\dot{P}}{k} \tag{11.43}$$

where $i = 1, 2$.

The finite element equations can now be obtained by adding all terms from Equations (11.41) to (11.43) into Equation (11.40). The stiffness matrix in Equation (11.38) is multiplied by kA (conductivity × area). Then this matrix becomes $kA/\Delta x$ along the diagonal and $-kA/\Delta x$ along the off-diagonal entries. Also, computing the right side in Equation (11.39),

$$\hat{r} \equiv (kA)r_m = Q_m^e + \frac{1}{2}\,\Delta x A\dot{P} - kAWR_i^e \tag{11.44}$$

Analogous results can be derived for certain stress analysis problems, but temperature is replaced by displacement, heat flow is replaced by force, and thermal conductivity is replaced by the modulus of elasticity. However, in stress analysis problems, special attention must be given to the sign conventions for directed quantities, since F (force) is a vector quantity.

4. In the fourth step of the finite element method, all elements are assembled within the mesh. The assembled WR_i^e terms (summed over all elements) become zero, while the sum of Q_i^e over all elements becomes Q_i (heat flow at node i). In the assembly process, the temperature at a specific node is unique, and so a suitable nodal mapping between local and global nodes is required. For example, the temperature at global node 2, T_2, is

equivalent to T_2^1, as well as T_1^2 (see Figure 11.1), where the subscript refers to the local node and the superscript refers to the element number. The following values indicate the first few components of the nodal mapping array, $ie(e, j)$, where e, j, and $ie(e, j)$ refer to the element, local node, and corresponding global node: $ie(1, 1) = 1$, $ie(1, 2) = 2$, $ie(2, 1) = 2$, $ie(2, 2) = 3$, and so on. These values are typically stored in a mesh data file.

Also, since all nodal Q values are positive into the element, their sum must be supplied at the node (externally). Thus, the sum of elemental Q contributions from adjacent elements must coincide with the resulting nodal Q value, i.e., $Q_2 = Q_2^1 + Q_1^2$ for node 2. In the current problem, no other heat sources (other than heat generation represented by $\dot{P}$) are experienced. As a result, assembling the element equations and using the nodal mapping,

$$\frac{kA}{\Delta x}\begin{bmatrix} 1 & -1 & 0 & 0 \\ -1 & 2 & -1 & 0 \\ 0 & -1 & 2 & -1 \\ 0 & 0 & -1 & 1 \end{bmatrix}\begin{Bmatrix} T_1 \\ T_2 \\ T_3 \\ T_4 \end{Bmatrix} = \begin{Bmatrix} \Delta x A\dot{P}/2 \\ \Delta x A\dot{P} \\ \Delta x A\dot{P} \\ \Delta x A\dot{P}/2 \end{Bmatrix} \tag{11.45}$$

The coefficients in the left side of the matrix constitute the *global stiffness matrix*, whereas the right side is called the *global right-side vector*.

5. In the fifth step, boundary conditions are applied. In this problem, $T_1 = 0 = T_4$ at the left and right boundaries, respectively. These boundary conditions can replace the first and fourth rows of the global stiffness matrix and right-side vector.

6. Finally, the discrete equation set is solved with a method such as Gaussian elimination (Cheney and Kincaid, 1985). The interior temperatures are obtained as $T_2 = \Delta x^2 \dot{P}/k = T_3$. The analytic solution can be readily obtained for comparison purposes by solving Equation (11.28) subject to the boundary conditions. It is expected that the accuracy of the predictions can be improved by using more grid points in the numerical discretization.

Certain important properties of the stiffness matrix can be observed from the Galerkin formulation. First, for steady-state diffusion-type problems, the stiffness matrix is symmetric (i.e., $c_{mn} = c_{nm}$). Second, for our convention of positive Q values into the element, the diagonal coefficients are positive (i.e., $c_{mm} > 0$). Third, for steady-state and linear problems, entries along a row or column of the stiffness matrix are summed to be zero before boundary conditions are applied. For example, in steady-state heat conduction with zero source terms, if T_n represents the solution of nodal temperature values, then T_n plus any constant, C, is also a solution since it is a linear problem. As a result,

$$\sum_{n=1}^{nnp} c_{mn}(T_n + C) = \sum_{n=1}^{nnp} c_{mn} T_n + C \sum_{n=1}^{nnp} c_{mn} = 0 \tag{11.46}$$

where *nnp* refers to the number of nodal points. The summed term involving T_n in Equation (11.46) becomes zero since T_n is a solution itself, thereby requiring that $\Sigma c_{mn} = 0$ (along a row or column). This suggests that the entries along a column or row must sum to zero for linear problems.

The matrix system in Equation (11.37) has represented the element stiffness equations. The global stiffness equations are obtained after the assembly process is completed (i.e., step 4 in the previous example). Since a finite element mesh typically involves many elements, an automated procedure to describe step 4 is required. A sample FORTRAN subroutine to perform this assembly task is shown below.

```
    do 30 e = 1, nel

       call stiff ( ) (obtain element stiffness matrix and
right side)

       do 20 iw = 1, nnpe

          i = ie(e, iw)

          r(i) = r(i) + rq(iw)

          do 10 jw = 1, nnpe

             j = ie(e, jw) -ie(e, iw) + isemi

             a(i, j) = a(i, j) + aq(iw, jw)

10        continue

20     continue

30  continue
```

The latter two expressions in the calculation of j are used for banded matrix storage (i.e., use only j = ie(e, jw) for square matrix storage). In the previous program listing, the variables refer to the number of elements (nel), number of nodal points per element (nnpe), global right-side vector (r), local elemental right side (rq), global stiffness matrix (a), local elemental stiffness matrix (aq), and ie (nodal map array).

The variable isemi refers to the semibandwidth of the matrix. After the assembly process, the global stiffness matrix is banded since there is a limited domain of influence of each node. For quadrilateral finite elements, an interior node is surrounded by eight other nodes whose diffusion coefficients influence the resulting solution at that node. The row of matrix entries contains a region of nonzero entries as wide as the largest node difference between these nodes. The following program listing gives a sample FORTRAN program to calculate isemi and iband (full bandwidth):

```
    maxdif = 0
    do 30 i = 1, nel
      do 20 j = 1, nnpe
        do 10 k = 1, nnpe
          l = abs(ie(i, j) -ie(i, k))
          if(l .gt. maxdif) maxdif = l
10      continue
20    continue
30  continue
    isemi = maxdif +1
    iband = 2*isemi -1
```

In order to reduce and minimize the bandwidth of a banded matrix, the node numbers in the mesh should be kept as close together as possible.

The previous example outlined the main steps in the finite element method for one-dimensional elements, but in many (most) cases, practical problems of interest require a multidimensional analysis. The geometric preliminaries for two-dimensional finite element analysis will be described in the following section.

11.3.2 Two-Dimensional Elements

Commonly encountered elements in two-dimensional configurations include a linear triangle, bilinear rectangle, and bilinear quadrilateral (see Figure 11.2). Triangular elements are well suited to irregular boundaries, and spatial and scalar interpolation can be accommodated in terms of a linear polynomial. In this case, bilinear rectangular elements have sides that remain parallel to the $x-y$ axes, and thus they cannot be arbitrarily oriented. More geometric flexibility is achieved with quadrilateral elements, whereby interpolation within each element can be expressed in terms of local coordinates. Although better suited to irregular boundaries, more geometrical computations are required. As a result, selection of element shapes for discretization of the spatial domain is based on trade-offs involving flexibility and computational effort.

In the first previously described case, a linear triangle refers to linear interpolation along a side or within the element. Higher order elements, such as a cubic triangle, require a higher order polynomial approximation (i.e., third order) of the interpolating function. For interpolation involving a scalar, ϕ, within a linear triangle,

$$\phi = a_1 + a_1 x + a_3 y \tag{11.47}$$

where the unknown coefficients, a_1, a_2, and a_3, can be determined based on the substitution of nodal values. For example, at node 1, the position is (x_1, y_1) and the scalar value is ϕ_1.

Using similar conditions at local nodes 2 and 3 and inverting the resulting 3 × 3 system from Equation (11.47) in terms of the unknown coefficients,

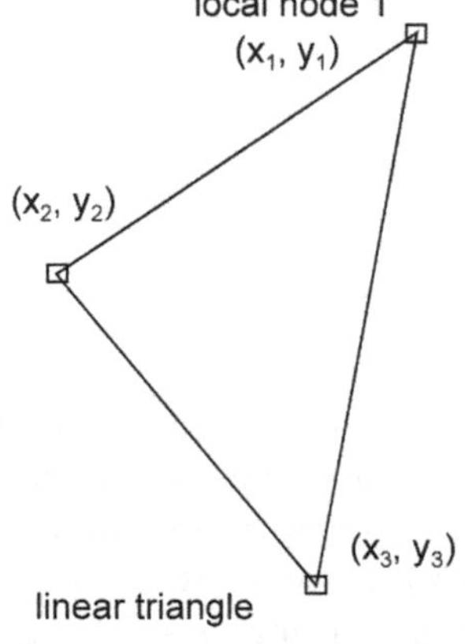

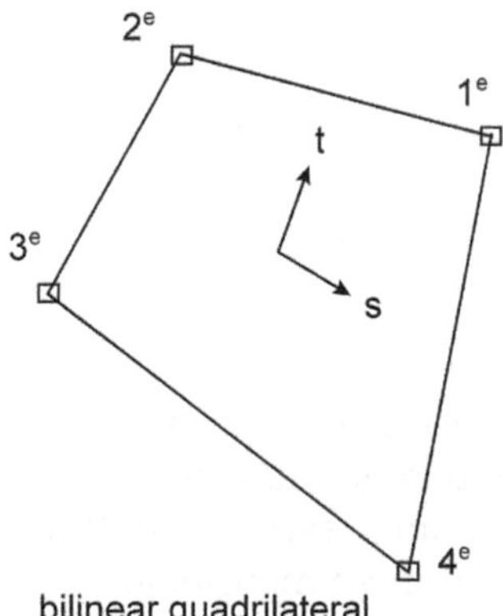

FIGURE 11.2
Different types of finite elements.

$$a_1 = \frac{1}{2A}[(x_2y_3 - y_2x_3)\phi_1 + (x_3y_1 - x_1y_3)\phi_2 + (x_1y_2 - x_2y_1)\phi_3] \quad (11.48)$$

$$a_2 = \frac{1}{2A}[(y_2 - y_3)\phi_1 + (y_3 - y_1)\phi_2 + (y_1 - y_2)\phi_3] \quad (11.49)$$

$$a_3 = \frac{1}{2A}[(x_3 - x_2)\phi_1 + (y_3 - y_1)\phi_2 + (x_2 - x_1)\phi_3] \quad (11.50)$$

where A is the area of the triangle, given by

$$A = \frac{1}{2}(x_2y_3 - x_3y_2 - x_1y_3 + x_1y_2 + y_1x_3 - y_1x_2) \quad (11.51)$$

Rearranging terms in Equations (11.48) to (11.51), the interpolation in Equation (11.47) can be written in terms of the shape functions, N_1, N_2, and N_3, as follows:

$$\phi = N_1\phi_1 + N_2\phi_2 + N_3\phi_3 \quad (11.52)$$

where

$$N_1 = \frac{1}{2A}(a_1 + b_1x + c_1y) \quad (11.53)$$

$$N_2 = \frac{1}{2A}(a_2 + b_2x + c_2y) \quad (11.54)$$

$$N_3 = \frac{1}{2A}(a_3 + b_3x + c_3y) \quad (11.55)$$

and

$$a_1 = x_2y_3 - x_3y_2; \quad a_2 = x_3y_1 - x_1y_3; \quad a_3 = x_1y_2 - x_2y_1 \quad (11.56)$$

$$b_1 = y_2 - y_3; \quad b_2 = y_3 - y_1; \quad b_3 = y_1 - y_2 \quad (11.57)$$

$$c_1 = x_3 - x_2; \quad c_2 = x_1 - x_3; \quad c_3 = x_2 - x_1 \quad (11.58)$$

Based on these definitions, it can be observed from Equations (11.51) and (11.56) that $2A = a_1 + a_2 + a_3$.

Also, the spatial derivatives of ϕ can be readily determined. From Equation (11.52),

$$\frac{\partial \phi}{\partial x} = \frac{\partial N_1}{\partial x}\phi_1 + \frac{\partial N_2}{\partial x}\phi_2 + \frac{\partial N_3}{\partial x}\phi_3 \tag{11.59}$$

$$\frac{\partial \phi}{\partial y} = \frac{\partial N_1}{\partial y}\phi_1 + \frac{\partial N_2}{\partial y}\phi_2 + \frac{\partial N_3}{\partial y}\phi_3 \tag{11.60}$$

where $\partial N_j/\partial x = b_j/2A$ and $\partial N_j/\partial y = c_j/2A$ $(j = 1, 2, 3)$. The following example uses the shape functions for interpolation of a scalar variable and derivatives inside a triangular element.

Example 11.3.2.1

Interpolation in a Triangular Element.

A triangular element with nodal points at (0.14, 0.01), (0.22, 0.05), and (0.14, 0.15) yields scalar values of $\phi = 160$, 140, and 180 at nodal points 1, 2, and 3, within the element, respectively. The purpose of this problem is to calculate the value of ϕ and its derivatives within the element, particularly at the point (0.2, 0.05). At what point along the bottom side (1–2) of the element does the $\phi = 150$ contour intersect that side?

Based on the nodal coordinates of the triangular element and Equations (11.56) to (11.58),

$$a_1 = x_2 y_3 - x_3 y_2 = 0.22(0.15) - 0.14(0.05) = 0.026 \tag{11.61}$$

and similarly, $a_2 = -0.0196$, $a_3 = 0.0048$, $b_1 = -0.08$, $b_2 = 0.14$, $b_3 = -0.04$, $c_1 = -0.08$, $c_2 = 0$, and $c_3 = 0.08$. Also, the area of the triangle is determined based on

$$2A = a_1 + a_2 + a_3 = 0.0112 \tag{11.62}$$

Using Equations (11.53) to (11.55),

$$N_1 = \frac{1}{0.0112}(0.026 - 0.1x - 0.08y) = 2.321 - 8.929x - 7.143y \tag{11.63}$$

$$N_2 = -1.75 + 12.5x \tag{11.64}$$

$$N_3 = 0.429 - 3.571x + 7.143y \tag{11.65}$$

Multiplying the shape function values by the nodal values of ϕ as outlined in Equation (11.52) at the point (0.2, 0.05),

$$\phi(0.2, 0.05) = N_1(0.2, 0.05)\phi_1 + N_2(0.2, 0.05)\phi_2 + N_3(0.2, 0.05)\phi_3$$

$$= 0.178(160) + 0.75(140) + 0.072(180) = 146.4 \quad (11.66)$$

Furthermore, the derivatives may be computed from Equation (11.59),

$$\frac{\partial \phi}{\partial x} = \frac{1}{2A}(b_1\phi_1 + b_2\phi_2 + b_3\phi_3) = -321.4 \quad (11.67)$$

and similarly, using Equation (11.60), $\partial\phi/\partial y = 142.0$ at the point (0.2, 0.05).

In terms of the $\phi = 150$ contour intersecting the edges of the element, linear interpolation along side 1–2 requires that

$$\frac{0.14 - 0.22}{160 - 140} = \frac{0.14 - x}{160 - 150} \quad (11.68)$$

which yields an intersection point of $x = 0.18$ on that edge. Performing similar calculations for y, we obtain $y = 0.03$ along side 1–2. Thus, the $\phi =$ 150 contour intersects the bottom edge of the element at the point (0.18, 0.03).

11.3.3 Coordinate Systems

It has been observed that dimensionless coordinates for a finite element allow the local geometry and governing equation calculations to be performed locally (with an element) without requiring a global dependence. Also, numerical integrations are often simplified in local coordinate systems since the limits of integration can be reduced to a simpler range (i.e., $0 \leq x^* \leq 1$), thereby providing a standard format for quadrature routines such as the Gauss–Legendre method. In this section, local coordinates for triangular and quadrilateral elements will be described.

11.3.3.1 Linear Triangular Elements

Within a triangular element, global coordinates (x and y) can be expressed in terms of local coordinates (ζ_1 and ζ_2) in the following fashion:

$$x = x_3 + \zeta_1(x_1 - x_3) + \zeta_3(x_2 - x_3) \quad (11.69)$$

$$y = y_3 + \zeta_1(y_1 - y_3) + \zeta_2(y_2 - y_3) \quad (11.70)$$

These equations can be inverted or isolated in terms of the local coordinates, yielding

$$\zeta_1 = \frac{a_1 + b_1 x + c_1 y}{a_1 + a_2 + a_3} = \frac{1}{2A}(a_1 + b_1 x + c_1 y) \equiv N_1 \quad (11.71)$$

and similarly,

$$\zeta_2 = \frac{1}{2A}(a_2 + b_2 x + c_2 y) \equiv N_2 \tag{11.72}$$

Since there are three nodal points in the linear triangular element, Equation (11.52) suggests that three shape functions are required. Consider a point within the element and lines connecting this point to the nodes of the element. Then the local coordinates may be interpreted as ratios of each subarea to the total area of the element. In particular, $\zeta_1 = A_1/A$, where A_1 refers to the subarea consisting of side 2–3 joined with the interior specified point. Similar expressions are obtained for the other subareas and local coordinates, including ζ_3, which is given by the ratio of A_3/A. Since the sum of individual subareas is A, dividing this summed equation by A yields

$$\zeta_3 = 1 - \zeta_1 - \zeta_2 \equiv N_3 \tag{11.73}$$

which suggests that there are only two independent local coordinates (ζ_1 and ζ_2) for linear triangular elements.

The local coordinates can be used for interpolation of the dependent scalar, ϕ, as well as geometry interpolation. Thus, in addition to interpolation indicated by Equation (11.52),

$$x = N_1(\zeta_1, \zeta_2)x_1 + N_2(\zeta_1, \zeta_2)x_2 + N_3(\zeta_1, \zeta_2)x_3 \tag{11.74}$$

Based on the type of interpolation, several classifications of elements can be defined. The following list indicates three types of elements:

- Subparametric element: geometry has lower interpolation than ϕ
- Isoparametric element: geometry has the same interpolation as ϕ (most common case)
- Superparametric element: geometry has higher interpolation than ϕ

For example, a superparametric quadratic triangular element consists of three nodes along each side of a triangle (midpoint and endpoints) to permit quadratic (second order) interpolation along each edge. Then both geometrical (x and y) and scalar interpolation are based on six shape functions, i.e.,

$$\phi = \sum_{i=1}^{6} N_i(\zeta_1, \zeta_2, \zeta_3)\Phi_i \tag{11.75}$$

where

$$N_j = (2\zeta_j - 1)\zeta_j \tag{11.76}$$

for $j = 1$, 2, and 3, and

$$N_4 = 4\zeta_1\zeta_2; \quad N_5 = 4\zeta_2\zeta_3; \quad N_6 = 4\zeta_3\zeta_1 \tag{11.77}$$

Using higher order interpolation may increase the accuracy of interpolation within an element, but often at the expense of increased computational time and storage.

Local and global coordinates must be related to each another for differentiation of the scalar variable involving triangular elements. Using the chain rule of calculus,

$$\frac{\partial \phi}{\partial \zeta_1} = \frac{\partial \phi}{\partial x}\frac{\partial x}{\partial \zeta_1} + \frac{\partial \phi}{\partial y}\frac{\partial y}{\partial \zeta_1} \tag{11.78}$$

$$\frac{\partial \phi}{\partial \zeta_2} = \frac{\partial \phi}{\partial x}\frac{\partial x}{\partial \zeta_2} + \frac{\partial \phi}{\partial y}\frac{\partial y}{\partial \zeta_2} \tag{11.79}$$

Rearranging and inverting Equations (11.78) and (11.79) by writing the derivatives in global coordinates in terms of the local derivatives,

$$\begin{Bmatrix} \dfrac{\partial \phi}{\partial x} \\ \dfrac{\partial \phi}{\partial y} \end{Bmatrix} = \frac{1}{|J|}\begin{pmatrix} \dfrac{\partial y}{\partial \zeta_2} & -\dfrac{\partial y}{\partial \zeta_1} \\ -\dfrac{\partial x}{\partial \zeta_2} & \dfrac{\partial x}{\partial \zeta_1} \end{pmatrix}\begin{Bmatrix} \dfrac{\partial \phi}{\partial \zeta_1} \\ \dfrac{\partial \phi}{\partial \zeta_2} \end{Bmatrix} \tag{11.80}$$

where

$$|J| = \frac{\partial x}{\partial \zeta_1}\frac{\partial y}{\partial \zeta_2} - \frac{\partial y}{\partial \zeta_1}\frac{\partial x}{\partial \zeta_2} \tag{11.81}$$

is the Jacobian determinant (also denoted by $det(J)$).

In addition, integration of various expressions within a triangular element is required in the finite element formulation. From calculus, a differential area element, dA, may be written as follows:

$$d\mathbf{A} = \mathbf{dx} \times \mathbf{dy} \tag{11.82}$$

where

$$\mathbf{dx} = \frac{\partial x}{\partial \zeta_1}\, d\zeta_1\hat{\zeta}_1 + \frac{\partial x}{\partial \zeta_2}\, d\zeta_2\hat{\zeta}_2 \tag{11.83}$$

$$\mathbf{dy} = \frac{\partial y}{\partial \zeta_1} d\zeta_1 \hat{\zeta}_1 + \frac{\partial y}{\partial \zeta_2} d\zeta_2 \hat{\zeta}_2 \quad (11.84)$$

and $\hat{\zeta}_1$ and $\hat{\zeta}_2$ are unit vectors in the ζ_1 and ζ_2 directions, respectively.

From Equations (11.82) to (11.84),

$$\mathbf{dA} = \left(\frac{\partial x}{\partial \zeta_1} \frac{\partial y}{\partial \zeta_2} - \frac{\partial x}{\partial \zeta_2} \frac{\partial y}{\partial \zeta_1} \right) d\zeta_1 d\zeta_2 \hat{\zeta}_3 \quad (11.85)$$

where ζ is the unit vector normal to the ζ_1 and ζ_2 planes. It can be observed that the magnitude of Equation (11.85) may be written as

$$|\mathbf{dA}| = |det(J)| d\zeta_1 d\zeta_2 \quad (11.86)$$

where $det(J)$ is the same Jacobian determinant as given in Equation (11.81).

Functional expressions can now be integrated in local coordinates for triangular elements, i.e.,

$$\iint_A f(\zeta_1, \zeta_2, \zeta_3) dA = \int_0^1 \int_{\zeta_2=0}^{1-\zeta_1} f(\zeta_1, \zeta_2, \zeta_3) |det(J)| d\zeta_2 d\zeta_1 \quad (11.87)$$

For example, if an exponential heat source appears in the energy equation, then a spatial integration of this heat generation within the element can be carried out based on Equation (11.87).

Some integrals can be calculated in a closed form in Equation (11.87), whereas for more complicated integrands, a closed-form analytic integration may not be possible. In that case, numerical integration is typically required. A commonly used numerical integration is Gauss quadrature, which can be written as follows for triangular elements:

$$\int_0^1 \int_0^{1-\zeta_1} f(\zeta_1, \zeta_2, \zeta_3) d\zeta_2 d\zeta_1 = \sum_{i=1}^{n} W_i f(\zeta_1^i, \ \zeta_2^i, \ \zeta_3^i) \quad (11.88)$$

where the weights, W_i, and abscissae, ζ_j^i, are shown for sample elements in Table 11.1. The subgrid quadrature can have a significant impact on the scheme's overall accuracy, particularly in transient flow problems (Naterer, 1997).

The order of accuracy indicates the magnitude of integration error in terms of a characteristic length (i.e., side length) of the triangular element. For example, for a linear triangle, the integration error is approximately reduced by a factor of 4 when the element size is reduced in half, whereas a quadratic element yields a reduction by a factor of 8 (i.e., proportional to h^3 rather than h^2).

11.3.3.2 Linear Quadrilateral Elements

The shape functions, N_i, for linear quadrilateral elements can be inferred by considering a sample four-noded quadrilateral element and bilinear interpolation. In this element, it is required that $N_i = 1$ at local node i and 0 at the other nodes. Using a local orthogonal coordinate system, denoted by coordinates s and t (see Figure 11.1; $-1 \leq s \leq 1$, $-1 \leq t \leq 1$), the shape functions can be constructed individually and combined in an appropriate manner. For example, along side 1–4, where $s = 1$, the shape function N_4 must satisfy the following relation:

$$N_4(s = 1) = \frac{1}{2}(1 - t) \tag{11.89}$$

since the interpolation is linear in t along that edge. Similarly, along side 3–4, where $t = -1$,

$$N_4(t = -1) = \frac{1}{2}(1 + s) \tag{11.90}$$

Since N_4 must satisfy both Equations (11.89) and (11.90), the product describes the appropriate bilinear interpolation in terms of both local coordinates,

$$N_4(s,\ t) = \frac{1}{4}(1 + s)(1 - t) \tag{11.91}$$

The remaining shape functions become

$$N_1(s,\ t) = \frac{1}{4}(1 + s)(1 + t) \tag{11.92}$$

TABLE 11.1

Gauss Quadrature for Triangular Elements

n	i	ζ_1^i	ζ_2^i	ζ_3^i	$\mathscr{W}_i$	**Order of Accuracy**
1 (linear)	1	1/3	1/3	1/3	1/2	$\mathcal{O}(h^2)$
3 (quadratic)	1	1/2	1/2	0	1/6	$\mathcal{O}(h^3)$
	2	0	1/2	1/2	1/6	
	3	1/2	0	1/2	1/6	

$$N_2(s,\ t) = \frac{1}{4}(1 - s)(1 + t) \tag{11.93}$$

$$N_3(s,\ t) = \frac{1}{4}(1 - s)(1 - t) \tag{11.94}$$

For linear quadrilateral elements, spatial derivatives of ϕ may also be obtained based on the shape functions in Equations (11.91) to (11.94), i.e.,

$$\left.\frac{\partial \phi}{\partial x}\right|_{(s,t)} = \sum_{i=1}^{4} \left.\frac{\partial N_i}{\partial x}\right|_{(s,t)} \Phi_i \tag{11.95}$$

$$\left.\frac{\partial \phi}{\partial y}\right|_{(s,t)} = \sum_{i=1}^{4} \left.\frac{\partial N_i}{\partial y}\right|_{(s,t)} \Phi_i \tag{11.96}$$

where the shape function derivatives are computed based on the chain rule of calculus as follows:

$$\begin{Bmatrix} \dfrac{\partial N_i}{\partial x} \\ \dfrac{\partial N_i}{\partial y} \end{Bmatrix} = \frac{1}{|J|} \begin{pmatrix} \dfrac{\partial y}{\partial t} & -\dfrac{\partial y}{\partial s} \\ -\dfrac{\partial x}{\partial t} & \dfrac{\partial x}{\partial s} \end{pmatrix} \begin{Bmatrix} \dfrac{\partial N_i}{\partial s} \\ \dfrac{\partial N_i}{\partial t} \end{Bmatrix} \tag{11.97}$$

and

$$|J| = \frac{\partial x}{\partial s}\frac{\partial y}{\partial t} - \frac{\partial y}{\partial s}\frac{\partial x}{\partial t} \tag{11.98}$$

is the Jacobian determinant.

Also,

$$\frac{\partial x}{\partial s} = \sum_{i=1}^{4} \frac{\partial N_i}{\partial s} x_i; \qquad \frac{\partial x}{\partial t} = \sum_{i=1}^{4} \frac{\partial N_i}{\partial t} x_i \tag{11.99}$$

$$\frac{\partial y}{\partial s} = \sum_{i=1}^{4} \frac{\partial N_i}{\partial s} y_i; \qquad \frac{\partial y}{\partial t} = \sum_{i=1}^{4} \frac{\partial N_i}{\partial t} y_i \tag{11.100}$$

The shape function derivatives with respect to the local coordinates can be readily evaluated from the expressions for $N_i\ (s,\ t)$ in Equations (11.91) to (11.94).

Also, functional expressions can be integrated based on the following area transformation:

$$\iint_A f(x, y)dxdy = \int_{-1}^{1}\int_{-1}^{1} f(s, t)|det(J)|dsdt \tag{11.101}$$

This integration can be numerically approximated by the Gauss–Legendre quadrature, i.e.,

$$\int_{-1}^{1}\int_{-1}^{1} f(s, t)|det(J)|dsdt \approx \sum_{j=1}^{ngp}\sum_{i=1}^{ngp} f(s_i, t_j)|det(J)|\mathcal{W}_i\mathcal{W}_j \tag{11.102}$$

where (s_i, t_i), $\mathcal{W}_i$, and $\mathcal{W}_j$ correspond to the Gauss points and weights, respectively (see Table 11.2), and ngp refers to the number of Gauss points.

If analytic expressions for integrated quantities over the finite element are difficult or impractical to evaluate, then the previous quadrature rules can be effectively used to find accurate approximations of the integrations.

11.3.4 Element Stiffness Equations

A variety of methods can be used to derive the finite element equations. In this section, the Galerkin weighted residual method will be adopted, whereas a hybrid method involving finite elements and volumes will be described in an upcoming section. Galerkin's method provides the correct number of basis functions. Also, it provides the same resulting equations as methods based on other (more complicated) types of formulations.

Galerkin's method selects the weight functions equal to the basis functions (shape functions) of the approximate solution. A motivation of this selection is based on Fourier analysis used in solutions of partial differential equations. In Fourier analysis (see Equation (2.77) in Chapter 2), recall that the temperature distribution was written as a product of functions of x and y, respectively, for certain steady-state conduction

TABLE 11.2

Gauss–Legendre Quadrature for Quadrilateral Elements

n	Number of Gauss Points	Gauss Points	Weights	Error
1	$1\times 1=1$	$s, t=0$	4 (at center)	$O(h^2)$
2	$2\times 2=4$	$s, t=\pm 1/\sqrt{3}$	1 (at each point)	$O(h^4)$
3	$3\times 3=9$	$s, t=\pm\sqrt{3}/\sqrt{5}$	25/81 (at outer corners) 40/81 (along midplanes) 64/81 (at midpoint)	$O(h^6)$

problems. In that case, the boundary conditions were used to eliminate the constants of integration arising from the separated ordinary differential equations and, consequently, find the eigenvalues and eigenfunctions. In an way analogous to the Galerkin weighted residual procedure, the same functions are multiplied together (forming so-called *orthogonal functions*) and integrated across the domain. In this way, the resulting integrated summation yields zero in all terms except a term with matching integer indices in the Fourier analysis. Analogous steps appear throughout the Galerkin procedure. Multiplying by the orthogonal function in the Fourier analysis yields desirable properties analogous to multiplying by the same basis functions in the Galerkin analysis. The following example demonstrates the overall steps in the Galerkin method for a two-dimensional problem.

Example 11.3.4.1

Heat Conduction in a Uniform Wall.

A section of wall material is subjected to a uniform heating rate of 100 W along its left boundary, and a temperature of 0°C is maintained along the right boundary (see Figure 11.3). The horizontal boundaries are adiabatic, and the thermal conductivity of the 0.2-m square section is $k = 1$ W/mK. The purpose of this problem is to use the Galerkin-based finite element method to find the steady-state nodal temperature values. Use eight triangular elements for the spatial discretization of the square domain.

1. The first step in the finite element solution involves discretization of the domain into eight uniformly distributed triangular elements (see Figure 11.3). It should be noted that there are different possible methods of

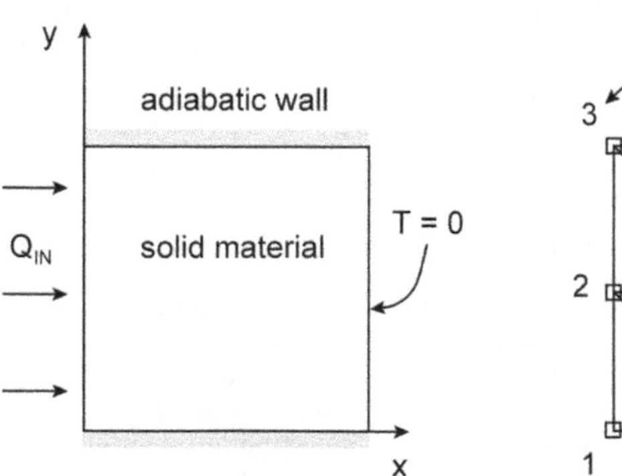

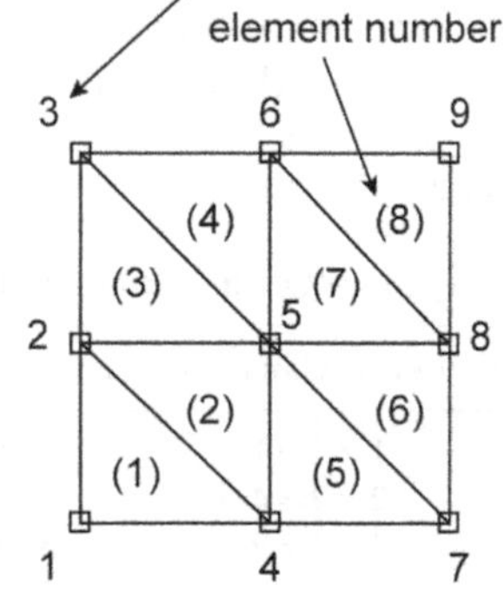

FIGURE 11.3
Physical domain and finite element discretization.

subdividing the domain into finite elements. For example, the discretization could involve four subsquares, each of which contains the same triangular alignment, or eight triangles all joined by a common point at the center of the domain. Although both methods would yield the same final solution, the former approach is adopted since the latter method (common center point) would contain a larger final matrix bandwidth since the center point would contain contributions from all nodes.

2. Interpolation is handled with linear, triangular, and isoparametric elements, i.e.,

$$T = N_1 T_1^e + N_2 T_2^e + N_3 T_3^e \tag{11.103}$$

where the shape functions are defined by Equations (11.53) to (11.58). Although the shape function coefficients are generally computed separately for each element, these coefficients are identical in the current problem due to the particular orientation of elements within the domain. As a result, $a_1 = 0.01$, $a_2 = 0 = a_3 = b_3 = c_2$, $b_1 = -0.1 = c_1$, and $b_2 = 0.1 = c_3$. Also, $2A = 0.01$, and so Equation (11.104) yields

$$T^e = (1 - 10x - 10y)T_1^e + (10x)T_2^e + (10y)T_3^e \tag{11.104}$$

in terms of the global coordinates.

3. Using Galerkin's weighted residual method for the steady-state heat conduction equation,

$$\iint_{A^e} N_i \left[\frac{\partial}{\partial x}\left(k\frac{\partial T}{\partial x}\right) + \frac{\partial}{\partial y}\left(k\frac{\partial T}{\partial y}\right)\right] dx^e dy^e = WR_i^e \tag{11.105}$$

where WR_i^e refers to the weighted residual and the superscript e refers to elemental (i.e., local within the element). Using integration by parts for the first integral in Equation (11.105),

$$\iint_{A^e} N_i \left[\frac{\partial}{\partial x}\left(k\frac{\partial T}{\partial x}\right)\right] dx^e dy^e$$

$$= \int_{y(x+)} \left(N_i k \frac{\partial T}{\partial x}\right)_{y(x+)} dy^e - \int_{y(x-)} \left(N_i k \frac{\partial T}{\partial x}\right)_{y(x-)} dy^e - \int_{y^e}\int_{x^e} \frac{\partial N_i}{\partial x} k \frac{\partial T}{\partial x} dx^e dy^e \tag{11.106}$$

The first and second terms on the right side of Equation (11.106) can be visualized by reference to a sample triangular element subdivided into differential segments of thickness dy with endpoints given by $y(x^-)$ (left side) and $y(x^+)$ (right side). Within a specified segment, dS and θ are used to represent the segment length along the triangle edge and the angle between the end edge and the y (vertical) axis, respectively. Then, based on these definitions,

$$dy = |\mathbf{dS}|\,cos(\theta) \tag{11.107}$$

which applies to either the left or right sides of the segment. In terms of the direction cosine with respect to the x axis, denoted by l_1,

$$dy = |\mathbf{dS}|\,cos(\theta_1) = l_1|\mathbf{dS}| \tag{11.108}$$

which applies to the right side at $y(x^+)$, where $\theta_1 = \theta$. From geometrical considerations of the component of the heat flux vector normal to the element's edge, it can be shown that

$$k\frac{\partial T}{\partial x} = k\frac{\partial T}{\partial n}l_1 \tag{11.109}$$

Then Equation (11.106) becomes

$$\int\int_{A^e} N_i\left[\frac{\partial}{\partial x}\left(k\frac{\partial T}{\partial x}\right)\right]dx^e dy^e$$
$$= \int_{bndy} N_i k\frac{\partial T}{\partial n}\bigg|_{bndy} l_1^2|\mathbf{dS}|\hat{\mathbf{i}} - \int_{y^e}\int_{x^e}\frac{\partial N_i}{\partial x}k\frac{\partial T}{\partial x}dx^e dy^e \tag{11.110}$$

where $\hat{i}$ is the unit x direction vector and *bndy* refers to the boundary. A similar result is obtained for the second term in Equation (11.105), except that l_1 is replaced by l_2 (direction cosine with respect to the y axis). Then, combining Equation (11.110) with the analogous y direction result into Equation (11.105),

$$\int_{bndy} N_i k\frac{\partial T}{\partial n}dS^e - \int_{y^e}\int_{x^e}\left(\frac{\partial N_i}{\partial x}k\frac{\partial T}{\partial x} + \frac{\partial N_i}{\partial y}k\frac{\partial T}{\partial y}\right)dx^e dy^e = WR_i^e \tag{11.111}$$

where the direction cosines have been summed to 1, thereby not appearing individually in Equation (11.111).

Various boundary conditions can be incorporated through the boundary integral term in Equation (11.111). Consider the following general form of the boundary equation:

$$k\frac{\partial T}{\partial n} = -h^e T + C^e \tag{11.112}$$

where h^e and C^e refer to the convection coefficient and a suitable constant describing the boundary condition, respectively. For example, the following list gives sample values of h^e and C^e for various commonly encountered boundary conditions:

- Dirichlet: $h \rightarrow$ *large value* and $C \rightarrow hT_s$ (where T_s refers to the specified surface temperature)
- Neumann: $h \rightarrow 0$ and $C \rightarrow q_s$ (where q_s is the specified heat flux)
- Adiabatic (insulated boundary): $h \rightarrow 0$ and $C \rightarrow 0$
- Robin (convection): $h \rightarrow h$ (unchanged convection coefficient) and $C \rightarrow hT_\infty$

If a boundary condition is not specified, then a default zero-flux (adiabatic) condition is assumed within the numerical formulation.

Substituting Equation (11.112) into Equation (11.111),

$$\int_{y^e}\int_{x^e}\left(\frac{\partial N_i}{\partial x}k\frac{\partial T}{\partial x}+\frac{\partial N_i}{\partial y}k\frac{\partial T}{\partial y}\right)dx^e dy^e+\int_{S^e}N_i h^e T dS^e$$
$$=\int_{S^e}N_i C^e dS^e - WR_i^e \tag{11.113}$$

The first term represents contributions from all elements in the domain, whereas the second and third terms (surface integral terms) represent only the boundary element contributions, since all internal adjacent elements have common sides where the surface integrals mutually self-cancel.

The temperature is approximated based on interpolation through the shape functions,

$$T^e=\sum_{j=1}^{3}N_j T_j^e \tag{11.114}$$

Based on Equations (11.113) and (11.114), the elemental stiffness equations become

$$[k^e]\{T^e\}=\{R^e\} \tag{11.115}$$

where

$$k_{ij}^e=[k_d^e]+[k_g^e]$$
$$=\int_{y^e}\int_{x^e}\left(\frac{\partial N_i}{\partial x}k\frac{\partial N_j}{\partial x}+\frac{\partial N_i}{\partial y}k\frac{\partial N_j}{\partial y}\right)dx^e dy^e+\int_{S^e}N_i h^e N_j dS^e \tag{11.116}$$

$$r_i^e=\{r_g^e\}+\{r_r^e\}=\int_{S^e}N_i C^e dS^e - WR_i^e \tag{11.117}$$

In these equations, the subscripts d, g, and r refer to interior elements, boundary elements, and residual terms, respectively.

The values of the coefficients in the stiffness matrix are now evaluated for triangular elements. Based on Equations (11.53) to (11.60), the interior coefficients in Equation (11.116) are obtained as follows:

$$k_{ij,d}^{e} = \int_{y^e}\int_{x^e}\left(\frac{b_i}{2A}k\frac{b_j}{2A} + \frac{c_i}{2A}k\frac{c_j}{2A}\right)dx^e dy^e = k\frac{b_i b_j}{4A} + k\frac{c_i c_j}{4A} \tag{11.118}$$

For the specific domain discretization in this example problem (shown in Figure 11.3), the coefficients in Equation (11.3) can be calculated so that the stiffness matrix becomes

$$[k_d^e] = \frac{1}{2}\begin{bmatrix} 2 & -1 & -1 \\ -1 & 1 & 0 \\ -1 & 0 & 1 \end{bmatrix} \tag{11.119}$$

It can be observed that the stiffness matrix is symmetric and the entries along a column or row are summed to zero (as expected). Also, r_r^e in Equation (11.118) is not analyzed since it becomes zero after summation over all elements (in accordance with the weighted residual method).

4. In this step of the finite element method, all elements within the domain are assembled together to construct the entire mesh. An element definition array, called $ie(e, j)$ (where e is the element number and j denotes the local node number), is used as a mapping between local nodes and global nodes in this assembly process. Table 11.3 outlines this mapping for the mesh discretization adopted in Figure 11.3. The values within the table represent the $ie(e, j)$ array, where the top row refers to the node numbers (j).

For example, the mapping array specifies that local node 2 in element 3 corresponds to global node 5 in the finite element mesh. The mapping array,

TABLE 11.3

Mapping between Local and Global Nodes

e	$j = 1$	$j = 2$	$j = 3$
1	1	4	2
2	5	2	4
3	2	5	3
4	6	3	5
5	4	7	5
6	8	5	7
7	5	8	6
8	9	6	8

$ie(e, j)$, identifies the row and column in the global stiffness matrix that corresponds to the entry in the local stiffness matrix. For example, performing a loop over all rows, i, and columns, j, in the computer algorithm,

$$k_{ij} = k_{ij}^o + k_{iw,jw}^e \tag{11.120}$$

where $i = ie(e, iw)$ and $j = ie(e, jw)$ are the row and column numbers in the global matrix, respectively, and $iw = 1, 2, 3$ and $jw = 1, 2, 3$ are the row and column numbers of the local stiffness matrix, respectively. The updates in the loop involving Equation (11.120) mean that the global stiffness matrix entry is based on the previous global matrix value plus the current value in the local stiffness matrix. In the sample program listings, Equation (11.120) is represented by

$$a(i,\ j) = a(i,\ j) + aq(iw,\ jw) \tag{11.121}$$

and similarly for the right side,

$$r(i) = r(i) + rq(iw) \tag{11.122}$$

Thus, the entries from the local stiffness matrix are placed in the row and column for which the global node corresponds to the local node from the given element. For example, in the local stiffness matrix for element 1, the second row is placed in the fourth row of the global matrix since $ie(1, 2) = 4$. This proper placement is obtained when the $ie(1, 2)$ entry is used to identify the index i in the calculation of the row number for $a(i, j)$ in Equation (11.121).

After all local stiffness matrices and right-side vectors are assembled into the global matrix, the following result is obtained for the current problem:

$$\frac{1}{2}\begin{bmatrix} 2 & -1 & 0 & -1 & 0 & 0 & 0 & 0 & 0 \\ -1 & 4 & -1 & 0 & -2 & 0 & 0 & 0 & 0 \\ 0 & -1 & 2 & 0 & 0 & -1 & 0 & 0 & 0 \\ -1 & 0 & 0 & 4 & -2 & 0 & -1 & 0 & 0 \\ 0 & -2 & 0 & -2 & 8 & -2 & 0 & -2 & 0 \\ 0 & 0 & -1 & 0 & -2 & 4 & 0 & 0 & -1 \\ 0 & 0 & 0 & -1 & 0 & 0 & 2 & -1 & 0 \\ 0 & 0 & 0 & 0 & -2 & 0 & -1 & 4 & -1 \\ 0 & 0 & 0 & 0 & 0 & -1 & 0 & -1 & 2 \end{bmatrix} \begin{Bmatrix} T_1 \\ T_2 \\ T_3 \\ T_4 \\ T_5 \\ T_6 \\ T_7 \\ T_8 \\ T_9 \end{Bmatrix} = \begin{Bmatrix} 0 \\ 0 \\ 0 \\ 0 \\ 0 \\ 0 \\ 0 \\ 0 \\ 0 \end{Bmatrix} \tag{11.123}$$

It can be observed that entries along a row or column are summed to zero, and the global stiffness matrix is symmetrical (as expected).

5. In this step, the boundary conditions for the physical problem are applied. The boundary integral terms in Equation (11.111) must be added

into the appropriate places in the local and global stiffness equations. In this procedure, each elemental boundary region is approximated by a linear section between the two local nodes on the boundary. For example, considering a counterclockwise ordering convention for an element with local nodes 1 and 2 on the domain boundary, the following boundary integral is added to k_{ij}^e, as outlined in Equation (11.116):

$$k_{ij,g}^e = \int_{S^e} N_i h^e N_j dS^e \tag{11.124}$$

where $dS^e = \Delta S^e d\zeta_1$ along side 1–2. Also, $N_1 = \zeta_1$, $N_2 = 1 - \zeta_1$, and $N_3 = 0$ on S_e along side 1–2 of the element.

As a result, the integrations in Equation (11.124) yield

$$[k_g^e] = \frac{h^e \Delta S^e}{6} \begin{bmatrix} 2 & 1 & 0 \\ 1 & 2 & 0 \\ 0 & 0 & 0 \end{bmatrix} \tag{11.125}$$

and similarly,

$$\{r_g^e\} = \frac{C^e \Delta S^e}{2} \begin{Bmatrix} 1 \\ 1 \\ 0 \end{Bmatrix} \tag{11.126}$$

which is added to the right-side vector in the global system on the domain boundaries.

Since the results in Equations (11.125) and (11.126) were obtained for side 1–2, the results must be generalized for cases where the boundary lies on a side other than side 1–2, i.e.,

$$k_{ij,g}^e = \chi \frac{h^e \Delta S^e}{6} \tag{11.127}$$

where $\chi = 2$ if $i = j$ and the boundary is side $i-j$, $\chi = 1$ if $i \neq j$ and the boundary is side $i-j$, and $\chi = 0$ if the boundary is not on side $i-j$. Similarly, for the right side,

$$r_{i,g}^e = \frac{C^e \Delta S^e}{2} \tag{11.128}$$

when local node i lies on the current boundary, and $r_{i,g}^e = 0$ otherwise (i.e., node i does not lie on the current boundary).

The implementation of boundary conditions also requires the mapping between local and global nodes to place the boundary stiffness matrix components and right-side entries correctly in the global system. For the current problem, Table 11.4 shows the global nodes, N1 and N2, together with boundary condition coefficients, h and C, for each surface, s, along the domain boundaries.

Using this boundary node mapping to add the boundary stiffness matrix, Equation (11.127), and right side, Equation (11.128), to the global system, the following final global system of algebraic equations is obtained:

$$\frac{1}{2}\begin{bmatrix} 2 & -1 & 0 & -1 & 0 & 0 & 0 & 0 & 0 \\ -1 & 4 & -1 & 0 & -2 & 0 & 0 & 0 & 0 \\ 0 & -1 & 2 & 0 & 0 & -1 & 0 & 0 & 0 \\ -1 & 0 & 0 & 4 & -2 & 0 & -1 & 0 & 0 \\ 0 & -2 & 0 & -2 & 8 & -2 & 0 & -2 & 0 \\ 0 & 0 & -1 & 0 & -2 & 4 & 0 & 0 & -1 \\ 0 & 0 & 0 & -1 & 0 & 0 & 3335.3 & 4999 & 0 \\ 0 & 0 & 0 & 0 & -2 & 0 & 1665.7 & 3337.3 & 1665.7 \\ 0 & 0 & 0 & 0 & 0 & -1 & 0 & 1665.7 & 3335.3 \end{bmatrix}\begin{Bmatrix} T_1 \\ T_2 \\ T_3 \\ T_4 \\ T_5 \\ T_6 \\ T_7 \\ T_8 \\ T_9 \end{Bmatrix} = \begin{Bmatrix} 5 \\ 10 \\ 5 \\ 0 \\ 0 \\ 0 \\ 0 \\ 0 \\ 0 \end{Bmatrix} \tag{11.129}$$

6. In this last step, the previous global equation set is solved by a method such as Gaussian elimination (Cheney and Kincaid, 1985). The final solution yields $T_1 = 20°\text{C} = T_2 = T_3$ along the left boundary, $T_4 = 10°\text{C} = T_5 = T_6$ along the vertical midplane, and $T_7 = 0 = T_8 = T_9$ along the right boundary. It can be readily verified by Fourier's law that these temperature results imply that the heat flow through the solid is 100 W, which correctly balances the specified heat inflow from the left boundary under steady-state conditions. Furthermore, using the shape functions and nodal temperature

TABLE 11.4

Mapping between Boundary Surfaces and Nodes

s	N1	N2	h	C
1	2	1	0	100
2	3	2	0	100
3	6	3	0	0
4	9	6	0	0
5	1	4	0	0
6	4	7	0	0
7	7	8	10^5	0
8	8	9	10^5	0

values in elements along the right boundary, it can also be verified that the heat flow there is also 100 W (as expected).

The results obtained by the finite element solution should be grid independent. In other words, the results should converge to certain final values as finer and finer grids are used in the domain discretization. A convergence study involves close observation of an important problem variable as the mesh is refined. This problem variable should be important in terms of its influence on the overall design of the system of interest, and it should display sensitivity to the numerical error. For example, the midpoint temperature in the two-dimensional domain could be used as the main variable in a convergence study of the previous problem. Some general guidelines can be formulated for effective convergence studies. For example, although it cannot always be practically enforced, all grids within a convergence study should be contained within each subsequent refined mesh. Also, all material within the domain should be fully contained by each mesh used in the convergence studies. Furthermore, the type of interpolation of problem variables should not change from one mesh to another mesh. This requirement includes completeness in terms of the degrees of freedom adopted in the interpolated variable profiles.

11.3.5 Time-Dependent Problems

For time-dependent problems with source terms (denoted by Q) and heat conduction, the governing equation is

$$\mathscr{L}(\phi) = D\frac{\partial^2 \phi}{\partial x^2} + D\frac{\partial^2 \phi}{\partial y^2} + Q - \lambda\frac{\partial \phi}{\partial t} = 0 \tag{11.130}$$

where D and λ refer to the diffusion coefficient and capacitance, respectively. In particular, $D = k$ (thermal conductivity) and $\lambda = \rho c_p$ for heat transfer problems. In the numerical solution of transient problems, two different Galerkin-based approaches can be used. First, new shape functions can be defined, such as N_i (x, y, t), where a time dependence is included. This approach can lead to considerable complexity, particularly in terms of the resulting mathematical expressions obtained in the weighted residual integrations.

Second, the transient derivative of ϕ (given by $\partial\phi/\partial t$) can be treated as a new temporary variable, called $\dot{\phi}$. Then it can be approximated in the following manner:

$$\dot{\phi} = \sum_{j=1}^{nnpe} N_j^t \dot{\phi}_j \tag{11.131}$$

where the superscript t refers to transient and *nnpe* is the number of nodal points per element (three for triangular elements). The finite element formulation then becomes

$$[k_{ij}]\{\phi_j\} + [c_{ij}]\{\dot{\phi}\} = \{r_i\} \tag{11.132}$$

where c_{ij} is called the capacitance matrix. Also, in a manner similar to that in the derivation leading to Equations (11.116) and (11.117),

$$k_{ij} = \sum_{e=1}^{nel} \left\{ \int_{V^e} \left(\frac{\partial N_i}{\partial x} D \frac{\partial N_j}{\partial x} + \frac{\partial N_i}{\partial y} D \frac{\partial N_j}{\partial y} \right) dV^e + \int_{S^e} N_i h^e N_j dS^e \right\} \tag{11.133}$$

$$r_i = \sum_{e=1}^{nel} \left\{ \int_{V^e} N_i Q^e dV^e + \int_{S^e} N_i C^e dS^e \right\} \tag{11.134}$$

$$c_{ij} = \sum_{e=1}^{nel} \left\{ \int_{V^e} N_i \lambda N_j dV^e \right\} \tag{11.135}$$

where the latter terms in Equations (11.133) and (11.134) are evaluated along the external boundaries of the domain only.

In a *consistent formulation*, the transient term, $\dot{\phi}$, is assumed to vary spatially in the subelement about the node (consistent with linear interpolation used in the diffusion terms). Substituting linear shape functions and performing the resulting integrations in Equation (11.135), the entries in the capacitance matrix, c_{ij}, become $\lambda \Delta x/3$ along the diagonal $(i = j)$ and $\lambda \Delta x/6$ in the off-diagonal entries $(i \neq j)$ for one-dimensional elements (2 × 2 matrix). An analogous 3 × 3 matrix is obtained for triangular elements, but in that case, the entries are multiplied by a factor of one half of the values for the one-dimensional element.

On the other hand, in a *lumped formulation*, the transient integral is taken as constant at a time level across the subelement. In this case, a step function is required for the shape function in Equation (11.135), thereby yielding a value of zero in the left half-element (i.e., $x^* < 1/2$ in one-dimensional elements) and 1 in the right half-element. Using this approach and performing the resulting integrations in Equation (11.135), the capacitance matrix entries become $\lambda \Delta x/2$ along the diagonal $(i = j)$ and zero for off-diagonal entries $(i \neq j)$.

Then the previously described transient variable, $\dot{\phi}$, can be replaced by a backward difference in time to rewrite Equation (11.132) in terms of ϕ alone, i.e.,

$$\dot{\phi}_j = \frac{\phi_j^{t+\Delta t} - \phi_j^t}{\Delta t} \tag{11.136}$$

where $t + \Delta t$ is the time level following time t. At a point in time between t and $t + \Delta t$, the variable ϕ_j is extrapolated forward from time t and backward from $t + \Delta t$.

The other terms in Equation (11.132) can now be evaluated at any intermediate point in time, i.e., at $t + \beta\Delta t$,

$$[k_{ij}]\{\phi_j\}^{t+\beta\Delta t} + [c_{ij}]\left\{\frac{\phi_j^{t+\Delta t} - \phi_j^t}{\Delta t}\right\} = \{r_i\} \tag{11.137}$$

where $0 \le \beta \le 1$. Assuming linearity in time,

$$\phi_j^{t+\beta\Delta t} = (1-\beta)\phi_j^t + \beta\phi_j^{t+\Delta t} \tag{11.138}$$

As a result, the local finite element equation becomes

$$\left(\beta[k_{ij}] + \frac{1}{\Delta t}[c_{ij}]\right)\{\phi_j\}^{t+\Delta t}$$
$$= \left(-(1-\beta)[k_{ij}] + \frac{1}{\Delta t}\{c_{ij}\}\right)\{\phi_j\}^t + \{r_i\}^{t+\beta\Delta t} \tag{11.139}$$

It can be observed that the first term in brackets that premultiplies $\{\phi_j\}^{t+\beta\Delta t}$ is the new effective stiffness matrix for transient problems. The expressions following the equal sign constitute the new effective right-side vector.

Various types of transient methods, such as explicit and implicit methods, are obtained as special cases of Equation (11.139). For example, $\beta = 0$ yields an explicit formulation, where future ϕ_j values are based on computations of diffusion and other terms at the previous time level. Explicit time advance is usually subject to certain guidelines to prevent numerical instability. In the present case, positive eigenvalues in the stiffness matrix and restricted time step sizes are required to prevent numerical oscillations. For a consistent explicit model, substituting $\beta = 0$ into Equation (11.139) yields

$$\frac{[c_{ij}]}{\Delta t}\{\phi_j\}^{t+\Delta t} = \left(-[k_{ij}] + \frac{[c_{ij}]}{\Delta t}\right)\{\phi_j\}^t + \{r_i\}^t \tag{11.140}$$

Once values of c_{ij} from a consistent formulation are substituted into Equation (11.140), a full local stiffness matrix is obtained (i.e., all nonzero entries in the 2×2 matrix for one-dimensional elements). An anticipated benefit of consistent formulations (i.e., reduced computational effort due to less stringent time step restrictions) is often not realized in local matrix inversion. As a result, a consistent formulation is usually not used in explicit schemes.

Instead, a lumped explicit formulation yields diagonal c_{ij} and local stiffness matrices, which can largely reduce the computational effort required to solve the resulting algebraic equations. A related stability criterion can be observed from Equation (11.140), particularly the first term that premultiplies $\{\phi_j\}$ on the right side of that equation. For small time steps, if all conduction heat flows are directed into a control volume, the right side of Equation (11.140) indicates that ϕ in the control volume will increase (as expected) due to the net heat inflow. However, for large time steps, the right side of Equation (11.140) may become negative, thereby causing ϕ to decrease in a control volume even though there is a net heat inflow to the control volume. This situation leads us to recognize that a time step limitation is required to maintain numerical stability in explicit formulations. A positive diagonal matrix to ensure numerical stability requires positive right-side values in Equation (11.140). If the first term on the right side of Equation (11.140) is required to stay positive, then for one-dimensional problems,

$$\Delta t \leq \frac{\lambda \Delta x^2}{2D} \tag{11.141}$$

and similarly for two-dimensional problems,

$$\Delta t \leq \left. \frac{\lambda(\Delta x^2 + \Delta y^2)}{2D} \right|_{min} \tag{11.142}$$

where the subscript *min* refers to the minimum value in the domain. These stringent time step restrictions often make explicit methods undesirable, except under certain classes of problems. The limitations of explicit models are noted particularly under limiting values of $\Delta t \to 0$ when the grid is refined.

Alternatively, implicit methods utilize $\beta = 1$ in Equation (11.140) so that diffusion and source terms are evaluated at the new time level (commonly used approach). Substituting $\beta = 1$ in Equation (11.140),

$$\left(\{k_{ij}\} + \frac{[c_{ij}]}{\Delta t}\right)\{\phi_j\}^{t+\Delta t} = \frac{1}{\Delta t}[c_{ij}]\{\phi_j\}^t + \{r_i\}^{t+\Delta t} \tag{11.143}$$

Increased temporal accuracy can be achieved by setting $\beta = 1/2$ in Equation (11.140). This corresponds to a *Crank–Nicolson scheme*, whereby diffusion and source terms are evaluated at the intermediate time level $(t+\Delta t/2)$. Substituting $\beta = 1/2$ into Equation (11.140),

$$\left(\frac{1}{2}\{k_{ij}\} + \frac{\{c_{ij}\}}{\Delta t}\right)\{\phi_j\}^{t+\Delta t} = \left(-\frac{1}{2}\{k_{ij}\} + \frac{1}{\Delta t}\{c_{ij}\}\right)\{\phi_j\}^{t} + \{r_i\}^{t+\Delta t/2} \quad (11.144)$$

which can be shown to have second-order accuracy in time. Despite its increased accuracy, this method is often not feasible because of other disadvantages, namely, practical difficulties in storing values at $t+\Delta t/2$, while also storing values of $[k_{ij}]$ and $\{\phi_j\}$ at time t. As a result of these trade-offs, the implicit method is more commonly adopted.

11.3.6 Fluid Flow and Heat Transfer

In convection problems, the velocity distribution throughout the flow field is required for the solution of the energy equation. Thus, the fluid flow equations become an important component of the overall formulation. Once the velocity field is obtained, these velocities are substituted into the convective terms in the energy equation. If the energy and momentum equations are fully coupled, then interequation iterations are required. Otherwise, the energy equation can be solved directly once the velocity results are known. Since the fluid flow equations introduce a new aspect of numerical modeling, this section will examine some main features of computational fluid flow.

In the present analysis, the governing equations of fluid flow are represented by the Navier–Stokes equations (described in Chapter 2). Linearizing the steady-state form of these equations (note: superscript o denotes linearized value),

$$\mathcal{L}_1(u,\ v,\ p) = \rho u^o \frac{\partial u}{\partial x} + \rho v^o \frac{\partial u}{\partial y} + \frac{\partial p}{\partial x} - \frac{\partial}{\partial x}\left(\mu \frac{\partial u}{\partial x}\right) - \frac{\partial}{\partial y}\left(\mu \frac{\partial u}{\partial y}\right) = 0 \quad (11.145)$$

$$\mathcal{L}_2(u,\ v,\ p) = \rho u^o \frac{\partial v}{\partial x} + \rho v^o \frac{\partial v}{\partial y} + \frac{\partial p}{\partial y} - \frac{\partial}{\partial y}\left(\mu \frac{\partial v}{\partial x}\right) - \frac{\partial}{\partial y}\left(\mu \frac{\partial v}{\partial y}\right) = 0 \quad (11.146)$$

$$\mathcal{L}_3(u,\ v,\ p) = \frac{\partial u}{\partial x} + \frac{\partial v}{\partial y} = 0 \quad (11.147)$$

These equations outline a key difference with our earlier problems involving heat transfer alone. In the case of fluid flow, we have three operators due to the three coupled equations for mass, x momentum, and y momentum transport. The two velocity components (u, v) and pressure (p) are the three resulting variables for two-dimensional problems.

In Equations (11.145) and (11.146), the superscript o refers to linearized (or lagged) value. Since the convection terms are nonlinear, iteration is required in the numerical solution. In other words, prior to an evaluation of the convection terms, the velocity is required, but this velocity is unknown until the solution of the fluid flow equations is obtained. As a result, an initial estimate is used in the first solution, and then this tentative estimate is compared with the resulting solution. The solution is repeated (using the new solution as the next linearized estimate) until an acceptable convergence, or agreement between previous and current values, is achieved.

Using the weighted residual method with Equations (11.145) to (11.147),

$$\int_V W_i^u \left[\rho u^o \frac{\partial \tilde{u}}{\partial x} + \rho v^o \frac{\partial \tilde{u}}{\partial y} + \frac{\partial \tilde{p}}{\partial x} - \frac{\partial}{\partial x}\left(\mu \frac{\partial \tilde{u}}{\partial x}\right) - \frac{\partial}{\partial y}\left(\mu \frac{\partial \tilde{u}}{\partial y}\right) \right] dV = 0 \tag{11.148}$$

$$\int_V W_i^v \left[\rho u^o \frac{\partial \tilde{v}}{\partial x} + \rho v^o \frac{\partial \tilde{v}}{\partial y} + \frac{\partial \tilde{p}}{\partial y} - \frac{\partial}{\partial y}\left(\mu \frac{\partial \tilde{v}}{\partial x}\right) - \frac{\partial}{\partial y}\left(\mu \frac{\partial \tilde{v}}{\partial y}\right) \right] dV = 0 \tag{11.149}$$

$$\int_V W_i^p \left[\frac{\partial \tilde{u}}{\partial x} + \frac{\partial \tilde{v}}{\partial y} \right] dV = 0 \tag{11.150}$$

Also, define the approximate local variable as a linear combination of nodal values weighted by the shape functions at their corresponding nodes. In other words,

$$\tilde{u} = \sum_{i=1}^{nnpe} N_i^u u_i; \quad \tilde{v} = \sum_{i=1}^{nnpe} N_i^v v_i; \quad \tilde{p} = \sum_{i=1}^{nnpe} N_i^p p_i \tag{11.151}$$

where $nnpe$ refers to the number of nodal points per element (i.e., $nnpe = 2$ for linear elements and $nnpe = 4$ for quadrilateral elements). Furthermore, based on the Galerkin approach, we will select the weight function to equal the shape function so that $W_i^u = N_i^u$, $W_i^v = N_i^v$, and $W_i^p = N_i^p$.

Assembling all finite elements and invoking integration by parts in the x momentum equation analogously to the method carried out earlier for heat conduction, we obtain

$$\sum_{e=1}^{nel} \int_{V^e} \left[N_i^u \left(\rho u^o \frac{\partial \tilde{u}}{\partial x} + \rho v^o \frac{\partial \tilde{u}}{\partial y} + \frac{\partial \tilde{p}}{\partial x} + \frac{\partial N_i}{\partial x} \mu \frac{\partial \tilde{u}}{\partial x} + \frac{\partial N_i}{\partial y} \mu \frac{\partial \tilde{u}}{\partial y} \right) \right] dV^e$$

$$- \sum_{e=1}^{nel} \int_{S^e} N_i^u \mu \left. \frac{\partial \tilde{u}}{\partial n} \right|_{S^e} dS^e = 0 \tag{11.152}$$

where the second summation represents the equivalent nodal force for the element due to shear stresses in the x direction. It is analogous to the surface heat flux in the heat conduction equation.

Rewriting the equations in matrix form, we obtain

$$[k^{uu}]\{u_i\} + [k^{up}]\{p_i\} = \{r_i^u\} \tag{11.153}$$

$$[k^{vv}]\{v_i\} + [k^{vp}]\{p_i\} = \{r_i^v\} \tag{11.154}$$

$$[k^{pu}]\{u_i\} + [k^{pv}]\{v_i\} = \{r_i^p\} \tag{11.155}$$

where the individual stiffness matrices are given by

$$k_{ij}^{uu} = \sum_{e=1}^{nel} \int_{V^e} \left[N_i^u \left(\rho u^o \frac{\partial N_j^u}{\partial x} + \rho v^o \frac{\partial N_j^u}{\partial y} \right) + \left(\frac{\partial N_i^u}{\partial x} \mu \frac{\partial N_j^u}{\partial x} + \frac{\partial N_i^u}{\partial y} \mu \frac{\partial N_j^u}{\partial y} \right) \right] dV^e \tag{11.156}$$

$$k_{ij}^{up} = \sum_{e=1}^{nel} \int_{V^e} N_i^u \frac{\partial N_j^p}{\partial x} \, dV^e \tag{11.157}$$

$$r_i^u = \sum_{e=1}^{nel} \int_{S^e} N_i^u \tau_u^e dS^e \tag{11.158}$$

$$k_{ij}^{vv} = \sum_{e=1}^{nel} \int_{V^e} \left[N_i^v \left(\rho u^o \frac{\partial N_j^v}{\partial x} + \rho v^o \frac{\partial N_j^v}{\partial y} \right) + \left(\frac{\partial N_i^v}{\partial x} \mu \frac{\partial N_j^v}{\partial x} + \frac{\partial N_i^v}{\partial y} \mu \frac{\partial N_j^v}{\partial y} \right) \right] dV^e \tag{11.159}$$

$$k_{ij}^{vp} = \sum_{e=1}^{nel} \int_{V^e} N_i^v \frac{\partial N_j^p}{\partial y} \, dV^e \tag{11.160}$$

$$r_i^v = \sum_{e=1}^{nel} \int_{S^e} N_i^v \tau_u^e dS^e \tag{11.161}$$

$$k_{ij}^{pu} = \sum_{e=1}^{nel} \int_{V^e} N_i^p \frac{\partial N_j^u}{\partial x} \, dV^e \tag{11.162}$$

$$k_{ij}^{pv} = \sum_{e=1}^{nel} \int_{V^e} N_i^p \frac{\partial N_j^v}{\partial x} dV^e \tag{11.163}$$

$$r_i^p = 0 \tag{11.164}$$

In the previous expressions, the surface integral terms, r_i^u and r_i^v, are evaluated only at external surfaces. Also, the previous expressions may be written in a local form (locally within an element) by removal of the summation over all elements. In this way, the local contribution of each element is assembled into the global system on an element-by-element basis.

Also, boundary conditions must be specified for closure of the problem definition. Consider an example involving heat transfer and fluid flow through a diffuser (i.e., expanding section of a duct). At the inlet, specified temperature and velocity profiles, based on experimental data or other correlations, can be given. If the problem is symmetric about some axis in the domain, then we have known relationships of zero spatial gradients of problem variables normal to this axis at the symmetry point. Also, if the outlet condition is unknown, then we can impose a condition at some outflow location where any errors in this specification will not propagate upstream (i.e., downstream of the recirculation cell in the diffuser). If the flow at this location is approaching a fully developed character, then temperature or velocity derivatives in the streamwise direction may be assumed to be approximately zero. A zero streamwise gradient condition is often mild enough to permit the computations to proceed. It often serves as an effective basis for subsequent refinements in the grid layout or boundary conditions. In overall terms, we can generally specify the boundary conditions in terms of Dirichlet, Neumann, or Robin types of conditions.

Unlike linear problems, the prediction of fluid flow involves nonlinear equations. As a result, an iterative procedure is required for the nonlinear convection terms. The list below summarizes the main steps in this iterative solution:

- Make initial estimates of the u^o and v^o values (i.e., start-up value or value from a previous iteration). These values should satisfy proper conservation of mass.
- Solve the finite element equations for the new (updated) u, v, and p values.
- Update the estimate in the first step as follows: $u^o \rightarrow u^o + \omega(u - u^o)$ and $v^o \rightarrow v^o + \omega(v - v^o)$. Overrelaxation $(1 \leq \omega \leq 2)$ is used if convergence is occurring and it is desirable to speed up the iterations. On the other hand, underrelaxation $(0 \leq \omega \leq 1)$ is used if the results

are diverging or oscillating and the iterations need to proceed more slowly.

- Check the convergence status by calculating whether *max* $|(u-u_o)/u_{ref}|$ (or some other residual) is less than a specified tolerance.
- Returning to the first step, repeat these steps until convergence is achieved.

Various computational issues affect the performance of the model for computational fluid dynamics (CFD). For example, the bandwidth of the coupled velocity–pressure equations can be excessively large if the stiffness matrices of individual variables (u, v, and p) are not carefully placed. Rather than ordering of variables in the usual fashion (i.e., u followed by v and p), rearrangement of the ordering sequence can largely reduce the overall bandwidth. For instance, for each node, the continuity and v momentum equations are interchanged, thereby ensuring nonzero entries on the diagonal and reducing the overall system bandwidth.

11.4 Hybrid Methods

Hybrid methods refer to numerical techniques that combine various features of previously mentioned methods. For example, the control volume-based finite element method (CVFEM) combines the advantages of geometrical flexibility of finite elements with the conservation-based features of finite volumes. In this section, the fundamentals of a particular type of CVFEM will be outlined.

Consider a CVFEM that subdivides the problem domain into linear quadrilateral elements and a local nonorthogonal (s, t) coordinate system within each element (see Figure 11.4). The discrete conservation equations can be obtained by integration of the conservation equations over finite control volumes and time intervals. Each control volume is defined by further subdivision of four internal or sub-control volumes (SCVs) within an element, each of which is associated with a control volume and its corresponding element node. The sub-control volume boundaries, or subsurfaces (SS), are coincident with the element exterior boundaries and the local coordinate surfaces defined by $s = 0$ and $t = 0$. An integration point (*ip*) is defined at the midpoint of each subsurface.

Finite element shape functions are required to relate global Cartesian coordinates and scalar values, ϕ, to local element values in a bilinear fashion, i.e.,

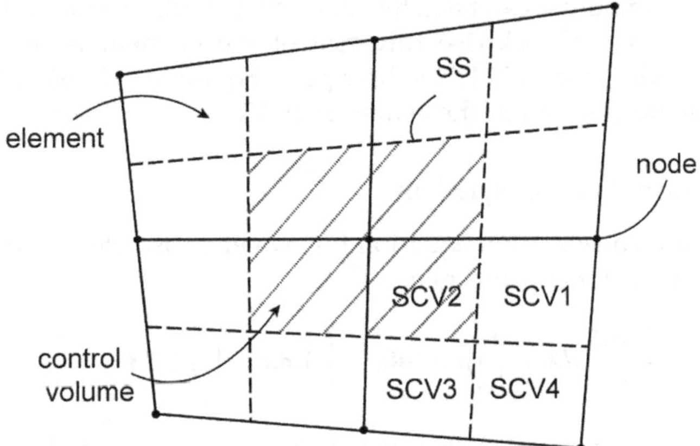

FIGURE 11.4
Schematic for a CVFEM.

$$\phi(s,\ t) = \sum_{i=1}^{4} N_i(s,\ t)\Phi_i \tag{11.165}$$

where the shape functions and their derivatives for quadrilateral elements have been defined in Equations (11.91) to (11.94). Also, the uppercase notation (Φ) refers to nodal variables, whereas the lowercase notation (ϕ) refers to integration point values.

The subelement area, A, defined by $0 \leq s \leq 1$ and $0 \leq t \leq 1$ for SCV1, and the surface outward normal vector, Δs, at an integration point, ip, between points a and b on the SS, may be computed by

$$A = \int_0^1 \int_0^1 |J|\, dsdt \approx |J|_{(1/2,1/2)} \tag{11.166}$$

$$\mathbf{\Delta s} = \Delta y \hat{\mathbf{i}} - \Delta x \hat{\mathbf{j}} \tag{11.167}$$

where

$$\Delta x = \left.\frac{\partial x}{\partial s}\right|_{ip} (s_b - s_a) + \left.\frac{\partial x}{\partial t}\right|_{ip} (t_b - t_a) \tag{11.168}$$

and

$$\Delta y = \left.\frac{\partial y}{\partial s}\right|_{ip} (s_b - s_a) + \left.\frac{\partial y}{\partial t}\right|_{ip} (t_b - t_a) \tag{11.169}$$

Other subelement areas can be determined in a similar fashion. It should be noted that a counterclockwise traversal of the element is used in these calculations. Also, the 1/2, 1/2 subscript in Equation (11.166) refers to the local coordinate position at the center of SCV1.

11.4.1 Control Volume Equations

Integrating the conservation equation for ϕ over a discrete volume (or area, A, in two-dimensional problems),

$$\int_A \frac{\partial(\rho\phi)}{\partial t}\, dA + \int_S (\rho\phi\mathbf{v})\cdot\mathbf{dn} + \int_S \mathbf{j}\cdot\mathbf{dn} = \int_A \rho\hat{S}dA \tag{11.170}$$

where $\mathbf{dn}$, $\mathbf{j}$, and $\hat{S}$ refer to the surface normal, diffusive flux (including pressure in momentum transport), and source terms, respectively. An implicit formulation will be considered, wherein convection, diffusion, and source terms are evaluated at the current time level, rather than a previous time level in an explicit approach.

With reference to SCV1 within an element (see Figure 11.4), a backward difference is often adopted for the approximation of the transient terms in Equation (11.170), i.e.,

$$\int_{SCV1} \frac{\partial(\rho\phi)}{\partial t}\, dA \approx J_1\left[\frac{(\rho\Phi)_1 - (\rho\Phi)_1^o}{\Delta t}\right] \tag{11.171}$$

where the subscript 1 and superscript o refer to local node 1 and previous time level, respectively. Here J_1 represents the Jacobian determinant (area of SCV1).

For the source (right side) term in Equation (11.170), a similar lumped approximation is adopted,

$$\int_{SCV1} \rho\hat{S}dA \approx J_1(\rho\hat{S})_{1/2,1/2} \tag{11.172}$$

where the 1/2, 1/2 subscript refers to the local coordinate position in the center of SCV1.

The diffusive term in Equation (11.170) must be evaluated at both SS1 and SS4 surfaces. In the case of the SS1 approximation,

$$\int_{SS1} \mathbf{j}_x\cdot\mathbf{dn} \approx \mathbf{j}_y|_{ip1}\Delta y_1 - j|_{ip1}\Delta x_1 \tag{11.173}$$

where a midpoint approximation has been used. The diffusive flux term, $\mathbf{j}$, can be related to the scalar variable, ϕ, through a suitable phenomenological law, such as Fourier's law for the energy equation. The following example shows how this diffusive term can be assembled for the case of heat conduction alone (without convection).

Example 11.4.1.1

CVFEM Stiffness Coefficients for Heat Conduction.

Find the stiffness matrix coefficients based on a CVFEM for the heat conduction equation. Use linear, quadrilateral elements and a lumped approximation for time-varying quantities.

In heat conduction problems, the energy balance for a control volume requires that the rate of energy change with time balances the net rate of heat inflow into the control volume plus the heat generated inside the control volume. With reference to node 1 of SCV1 (see Figure 11.4), this energy balance can be stated mathematically as follows:

$$Q_{2,1} + Q_{4,1} + Q_{e1,1} + Q_{e2,1} + \int_V \dot{q}dV = \frac{\partial}{\partial t}\int_V \rho c_p T dV \tag{11.174}$$

where the subscripts $e1,1$ and $e2,1$ refer to external elements that contribute their heat flows to the energy balance for SCV1. Equation (11.174) has been derived based on a discrete balance of heat flows. Alternatively, it can be derived by direct integration of the transient heat conduction equation over the discrete volume, SCV1.

The heat flows within the element, including $Q_{2,1}$ (from SCV2 into SCV1) and $Q_{4,1}$ (from SCV4 into SCV1) are computed by

$$Q = -\int_S \mathbf{q} \cdot \mathbf{dn} \tag{11.175}$$

where

$$\mathbf{q} = -k\frac{\partial T}{\partial x}\,\mathbf{i} - k\frac{\partial T}{\partial y}\,\mathbf{j} \tag{11.176}$$

is the two-dimensional Fourier heat flux. Also, **i** and **j** refer to the unit vectors in the x and y directions, respectively.

Performing the integration along the subsurface for $Q_{2,1}$,

$$Q_{2,1} = \int_{t=0}^{t=1} \left(-k\frac{\partial T}{\partial x}\,dy - k\frac{\partial T}{\partial y}\,dx \right)_{s=0} dt \tag{11.177}$$

Along the path of integration carried out along SS1, the differentials dx and dy may be written in terms of a single local coordinate, t. This occurs since the chain rule written for these differentials involves ds, which vanishes during the integration, since $s = 0$ along SS1. Substituting those expressions for the differentials and utilizing the shape functions for the evaluation of the temperature derivatives,

$$Q_{2,1} = -\sum_{i=1}^{4}\left\{\int_0^1 \left(k\frac{\partial N_i}{\partial x}\frac{\partial y}{\partial t} - k\frac{\partial N_i}{\partial N_y}\frac{\partial x}{\partial t}\right)_{s=0} dt\right\} T_i \tag{11.178}$$

Similar expressions are obtained for the other heat conduction terms in Equation (11.174). It should be noted that the remaining heat flows satisfy $Q_{i,j} = -Q_{j,i}$ (where $i \neq j$) since the heat entering a given SCV is equivalent to the flow leaving the adjacent SCV.

Furthermore, the heat generation and energy storage terms for SCV1 may be approximated in the following manner:

$$\dot{q}_1 = \int_0^1\int_0^1 \dot{q}\,|J|\,ds dt \tag{11.179}$$

$$\rho c_p \frac{\partial T}{\partial t} \approx \rho c_p |J_1| \frac{\partial T_1^{n+1} - T_1^n}{\Delta t} \tag{11.180}$$

where the superscripts $n+1$ and n refer to current and previous time levels, respectively. Also, the subscript 1 and $|J_1|$ refer to local node 1 and the Jacobian determinant of SCV1, respectively. The approximation in Equation (11.180) is a *lumped approximation* since the transient term is lumped at the nodal value, without interpolation throughout the element.

Based on Equations (11.174) and (11.178) to (11.180), the algebraic equation for SCV1 can now be written in the following form (Schneider and Zedan, 1982):

$$\sum_{j=1}^{4} k_{1,j}^e T_j^e = r_1^e \tag{11.181}$$

where $k_{1,j}^e$ refers to the row of local (elemental) stiffness matrix coefficients for SCV1, i.e.,

$$k_{1,j}^e = \int_0^1 \left(k\frac{\partial N_j}{\partial x}\frac{\partial y}{\partial t} - k\frac{\partial N_j}{\partial y}\frac{\partial x}{\partial t}\right)_{s=0} dt - \int_0^1 \left(k\frac{\partial N_j}{\partial x}\frac{\partial y}{\partial s} - k\frac{\partial N_j}{\partial y}\frac{\partial x}{\partial s}\right)_{t=0} ds$$
$$+\frac{1}{\Delta t}\rho c_p |J|_1 \tag{11.182}$$

Also, r_1^e represents the right-side vector for SCV1, where

$$r_1^e = \int_0^1\int_0^1 \dot{q}\,|J|\,dsdt + \frac{1}{\Delta t}\,\rho c_p det(J) T_1^n \tag{11.183}$$

Similar terms are constructed for the other SCVs such that the entire 4 × 4 stiffness coefficient matrix, $k_{i,j}^e$, can be completed. The previous equations have been constructed locally within the element so that the subscripts refer to local (not global) nodes. For example, T_1 refers to the temperature at local node 1 within the current element.

In a computer program, the local stiffness coefficient matrix can be efficiently computed by a double loop over $i = 1, 2, 3, 4$ and $j = 1, 2, 3, 4$. In particular, the first loop is indicated by the range of subscript j in Equation (11.182). The second loop would cover the four SCVs within the entire element. Furthermore, it can be observed that the heat conduction terms from external elements, i.e., $Q_{e1,1}$ and $Q_{e2,1}$ in Equation (11.174), are not required in the stiffness coefficient in Equation (11.182), since these terms mutually self-cancel each other once all elements in the mesh are assembled. In other words, the heat flow entering SCV1 from $e1$ balances the heat flow leaving the adjacent SCV from the external element, and thus both terms self-cancel each other. An exception is elements located along the external boundaries of the domain, where external flows (or boundary temperatures) must be specified through appropriate boundary conditions.

Once the elemental stiffness matrix is completed, a standard assembly procedure is performed for all finite elements in the mesh. This assembly implies that the energy conservation is applied over an *effective control volume* rather than individual SCVs within an element. The full control volume consists of all SCVs associated with a particular node in the mesh. In this way, the geometric flexibility of the finite element method is retained when the governing equations are formed locally within an element (independent of the mesh configuration and layout).

In the previous example, the diffusive flux referred to heat conduction, whereas diffusion involves shear stresses in the momentum equations of fluid flow. Thus, in comparing with the previous analysis, k is replaced with μ, while T is replaced by the appropriate velocity component. In addition to the diffusive terms, the convective term in Equation (11.170) requires evaluation at both SS1 and SS4 subsurfaces since both surfaces contribute to the convective transport of ϕ for SCV1. Considering the SS1 evaluation,

$$\int_{SS1} (\rho\phi\mathbf{v})\cdot\mathbf{dn} \approx (\rho\phi u)_{ip1}\Delta y_1 - (\rho\phi v)_{ip1}\Delta x_1 \tag{11.184}$$

where $ip1$ refers to integration point 1 (i.e., midpoint of SS1 in Figure 11.4). A similar expression is obtained for the SS4 integration. In the momentum equations, ϕ represents a velocity component, so that nonlinear convection terms are obtained. Thus, interequation iterations are typically required to obtain the final converged solution. In the momentum equation, the convective term involves a "convected" velocity (a linearized or lagged velocity) and a "convecting" velocity (active mass-conserving velocity). Successive iterative updates are required following each solution of the fluid flow equations until suitable convergence and agreement are achieved between both velocities.

Based on the previous approximations for transient, diffusion, and convection terms, the control volume equations can now be completed once all elements are assembled. The sub-control volume contributions to a conservation equation for a particular global node are completed after all elements are considered. But it remains that integration point values, such as integration point velocities u_{ip1} and v_{ip1} in Equation (11.184), must still be related to nodal variables in order to provide a well-posed algebraic system of equations. Conventional schemes for this convection modeling include the upwind differencing scheme (UDS) and central differencing scheme (CDS) (Minkowycz et al., 1988). A generalization of these schemes that includes the proper physical limiting trends for high and low flow rates is called physical influence scheme (PINS). In this approach, a discrete approximation involving transport processes at the integration point is constructed (Schneider and Raw, 1987; Naterer and Schneider, 1995). This approach will be described hereafter.

11.4.2 Integration Point Equations

It has been observed that the integration point values, such as u_{ip1} and v_{ip1}, are required for closure of the control volume equations since they are required in the convection terms. In this section, these integration point values will be determined from a local balance of transport processes, including convection and pressure. More specifically, the integration point velocities will be approximated from the discretized form of the momentum equations. In the x and y directions, respectively,

$$\rho\frac{\partial u}{\partial t} + \rho u\frac{\partial u}{\partial x} + \rho v\frac{\partial u}{\partial y} = -\frac{\partial p}{\partial x} + \mu\left(\frac{\partial^2 u}{\partial x^2} + \frac{\partial^2 u}{\partial y^2}\right) \tag{11.185}$$

$$\rho\frac{\partial v}{\partial t} + \rho u\frac{\partial v}{\partial x} + \rho v\frac{\partial v}{\partial y} = -\frac{\partial p}{\partial y} + \mu\left(\frac{\partial^2 v}{\partial x^2} + \frac{\partial^2 v}{\partial y^2}\right) - \rho g \tag{11.186}$$

Equations (11.185) and (11.186) involve five distinct operators: transient (first term), convection (second and third terms), pressure (fourth term), diffusion (fifth and sixth terms), and source (seventh term) terms. Once discrete approximations of these operators are assembled together in Equations (11.185) and (11.186), the integration point velocities are obtained in terms of nodal variables. Then these variables are substituted back into the control volume equations, yielding a set of algebraic equations in terms of nodal variables alone.

Numerical approximations of each integration point operator are required. Firstly, the transient term at the integration point is approximated by a backward difference time, i.e., at integration point 1,

$$\left.\frac{\partial \phi}{\partial t}\right|_{ip1} \approx \frac{\phi_{ip1} - \phi_{ip1}^{o}}{\Delta t} \tag{11.187}$$

where the superscript o refers to the previous time level.

Secondly, an upstream difference approximation is used to represent the convection operator. In terms of the arbitrary scalar ϕ at $ip1$,

$$\rho u \frac{\partial \phi}{\partial x} + \rho v \frac{\partial \phi}{\partial y} \approx \rho V \left(\frac{\phi_{ip1} - \phi_{u}}{L_c} \right) \tag{11.188}$$

where $V = \sqrt{u^2 + v^2}$ represents the fluid velocity magnitude. In Equation (11.188), L_c is the convection length scale in the streamwise direction and ϕ_u represents the upwind value of ϕ. The direction of the line segment between ϕ_{ip1} and ϕ_u is based on skewed upwinding. The upstream value, ϕ_u, is calculated by an interpolation upstream of the subvolume edge where the local streamline through the integration point intersects that edge. For example, if the line constructed in the upwind direction intersects the quadrant edge between local nodes 2 and 3,

$$\phi_u \approx \frac{a}{b} \Phi_2 + \left(1 - \frac{a}{b}\right) \Phi_3 \tag{11.189}$$

where Φ denotes nodal values. In Equation (11.189), a and b refer to coefficients corresponding to linear interpolation for ϕ_u in terms of Φ_2 and Φ_3 along the intersected edge. This skewed upwinding strives to retain both the directional and strength influences of convection at the integration point, thereby reducing differencing errors in the convection modeling.

Thirdly, shape functions are used in the approximation of pressure gradient, i.e., at $ip1$,

$$\left.\frac{\partial p}{\partial x}\right|_{ip1} = \sum_{i=1}^{4} \frac{\partial N_i}{\partial x} P_i \tag{11.190}$$

where P_i refers to the pressure at local node i. In a similar fashion, the pressure gradient in the y direction is approximated in terms of bilinear interpolation using the shape functions.

Fourth, the diffusion (Laplacian) operator is approximated by a central difference, so that

$$\left.\frac{\partial^2 u}{\partial x^2} + \frac{\partial^2 u}{\partial y^2}\right|_{ip1} \approx \frac{1}{L_d^2}\left(\sum_{j=1}^{4} N_j \Phi_j - \phi_{ip1}\right) \tag{11.191}$$

where L_d is a diffusion length scale,

$$L_d^2 = \left(\frac{2}{\Delta x^2} + \frac{8}{3\Delta y^2}\right)^{-1} \tag{11.192}$$

It can be verified that these expressions yield the correct scaling properties of the Laplacian operator.

Finally, local source terms, such as thermal buoyancy (based on the Boussinesq approximation) in the y momentum equation, can be evaluated by either direct substitution of corresponding integration point values or interpolation of nodal values using the shape functions. After all integration point operators are assembled back into Equations (11.185) and (11.186), a local matrix inversion is required to express the integration point values in terms of nodal values alone. These inverted matrices are called influence coefficient matrices since individual coefficients express the relative contributions of convection, diffusion, pressure, and source terms. The matrices provide the required coupling between integration point and nodal point values. After they are substituted into the convection terms of the control volume equations, the global system of equations can be solved.

11.5 Numerical Methods for Other Applications

11.5.1 Phase Change Heat Transfer

Additional latent heat terms appear in the governing equations of phase change heat transfer. For example, in one-dimensional heat conduction problems with solidification or melting,

$$\rho c_p \frac{\partial T}{\partial t} = k \frac{\partial^2 T}{\partial x^2} + \rho L \frac{\partial f_s}{\partial t} \tag{11.193}$$

where f_s represents the solid fraction and L is the latent heat of fusion. The latent heat released or absorbed at the phase interface represents a nonlinear term requiring iterations in the governing equation to establish its converged value. Once the phase distribution is determined following this iterative procedure, the thermophysical properties from the appropriate phase can be substituted into the previous equation for its solution. In the numerical formulation, the latent heat of phase transformation can be handled by source-based or apparent heat capacity methods.

11.5.1.1 Source-Based Method

In Equation (11.193), the latent heat term can be modeled like a heat source (or sink) that is released (or absorbed) in control volumes containing the phase interface. As mentioned earlier, an iterative approach is required, whereby a tentative phase distribution is selected, and then the equations are discretized and solved separately in each phase according to this distribution. In each phase, this procedure follows analogously to previous numerical analysis of single phase problems. Following this solution, iterations are continued until the predicted and tentative phase distributions agree with one another. Alternatively, convergence can be established when temperature differences between successive iterations are sufficiently small.

11.5.1.2 Apparent Heat Capacity Method

Alternatively, Equation (11.193) can be written in terms of an *apparent* (or *effective*) *specific heat*. Using the chain rule of calculus for the change of solid fraction for pure materials in the last term,

$$\frac{\partial f_s}{\partial t} = \frac{\partial f_s}{\partial T} \frac{\partial T}{\partial t} \tag{11.194}$$

Substituting Equation (11.194) into Equation (11.193), rearranging, and using implicit finite difference approximations,

$$\rho \left(c_p - L \frac{\partial f_s}{\partial T} \right)_i^{n+1} \left(\frac{T_i^{n+1} - T_i^n}{\Delta t} \right) = k \left(\frac{T_{i+1}^{n+1} - 2T_i^{n+1} - T_{i-1}^{n+1}}{\Delta x^2} \right) \tag{11.195}$$

where the superscripts n and $n+1$ refer to the previous and current time levels, respectively. The leading term in parentheses on the left side of Equation (11.195) represents an apparent specific heat since it includes both sensible and latent heat components. Once a suitable correlation is available for the solid fraction in terms of temperature (as discussed in Chapters 5 and 8), Equation (11.195) can be solved throughout the solid and liquid phases simultaneously.

Unlike linear problems such as steady-state heat conduction, phase change heat transfer is nonlinear due to the latent heat term in the apparent specific heat. Thus, iterations are required in the numerical procedure. For example, in single phase regions, only the regular specific heat is adopted, but in the two-phase region, both sensible and latent heat parts are retained in the effective specific heat. In this way, the latent heat is released (or absorbed) at nodes containing the phase interface. Iterations are required in order to determine whether the discrete nodal location contains the phase interface. If the phase interface is inside the control volume, then the latent heat of phase transformation is included. Otherwise, only the sensible heat portion is retained in the specific heat. The nodal location containing the phase interface is identified by the location where the temperature equals the phase change temperature.

An orderly sequence of steps can be followed for an efficient iteration procedure within the energy equation for phase change problems (Naterer and Schneider, 1995). For example, a control volume is required to pass through a two-phase region during the phase transition. This rule ensures that the latent heat is released or absorbed during the phase transition. Also, in the absence of other heat sources or sinks in the problem domain, phase change in one control volume cannot occur without a neighboring control volume changing phase first. These guidelines can ensure that the iterations are performed systematically without randomly cycling through phases within a time step.

Extending Equation (11.195) to consider two-dimensional problems including convection,

$$\rho c_p \frac{\partial T}{\partial t} + \rho c_l \frac{\partial (f_l u T)}{\partial x} + \rho c_l \frac{\partial (f_l v T)}{\partial y} = k\left(\frac{\partial^2 T}{\partial x^2} + \frac{\partial^2 T}{\partial y^2}\right) + \rho L\left(\frac{\partial f_s}{\partial t}\right) \tag{11.196}$$

where the solid phase has been assumed to be motionless. Using the standard finite volume procedure, Equation (11.196) can be integrated on a term-by-term basis over the control volume to yield the relevant discrete equations.

11.5.2 Computational Modeling of Turbulence

Designers of fluid engineering systems are often mainly interested in the broad characteristics of a turbulent flow, such as the level of mixing between various fluid streams or dispersed phases within the flow. As a result, in many cases, the time-averaged turbulence equations with a low-order closure (such as a $k-\epsilon$ model) are suitable for the computational predictions. However, it is unlikely that any single model can accurately represent all types of turbulent flows. Thus, we should usually consider these models and numerical procedures to be engineering approximations, rather than exact laws governing all detailed aspects of the turbulence.

A variety of techniques is available for the numerical modeling of turbulent flows. For example, a *direct numerical simulation* (DNS) solves the Navier–Stokes equations directly to the smallest scales in the flow (called the Kolmogorov scales). In this approach, the main problem is the wide range of eddy scales. Computational costs and memory requirements for realistic industrial flow simulations quickly become prohibitive in the DNS approach, even with the most recent computers.

A second approach involves *large eddy simulations*. These techniques apply models of the small scales in the flow to resolve the large eddies. *Reynolds stress models* represent a third possible approach for turbulence modeling. In this case, separate stress transport equations are solved for individual turbulent stress (or heat flux) terms. A variety of higher order turbulence quantities, such as pressure strain (i.e., interaction of fluctuating pressure with fluctuating strain rates), is introduced, and the resulting complexity again becomes prohibitive in many applications. A fourth possibility involves the numerical solution of the $k-\epsilon$ equations. This approach represents one of the most popular techniques for predicting turbulent flows in industrial applications.

The eddy viscosity hypothesis, together with the $k-\epsilon$ model, can provide reasonably good results for many flows. This procedure initially solves the mean flow equations (i.e., mass, momentum, energy equations). Then the equations for the transport of turbulent kinetic energy (k) and dissipation rate (ϵ) are solved, based on the mean flow results. The eddy viscosity is determined based on the k and ϵ fields. Then this eddy viscosity is combined with the molecular viscosity in the mean momentum equations, yielding the effective stresses due to molecular diffusion and turbulence transport. In a similar way, the molecular diffusivity is replaced by the sum of molecular and turbulent heat diffusivities in the energy equation.

In this approach, two additional partial differential equations (involving k and ϵ) must be solved in the formulation. A difficulty arises because the time scales associated with turbulent mixing are much smaller than corresponding scales associated with the mean flow. As a result, special attention is required for the iterative procedure involving the coupled mean flow and turbulence equations. A common approach is first calculating the mean flow quantities based on the eddy viscosity and diffusivity from a previous iteration or time step. Then the $k-\epsilon$ equations are solved based on the computed mean flow quantities. Nonlinear terms in the $k-\epsilon$ equations must be linearized in a suitable fashion. They can be linearized similarly to the procedure adopted in the mean flow equations. Iterations between the mean flow and turbulence equations are performed until suitable convergence between those equations is obtained.

In turbulent flow simulations, convergence may be difficult to obtain since numerical instability can arise from a large time step leading to negative (nonphysical) values of k or ϵ. As a result, underrelaxation is often adopted during iterations in these equations. Instead of using the updated velocity in the next iteration, a combination of this value and the previous estimate are adopted. Also, the model must be analytically and computationally realizable. This means that positive definite quantities, such as turbulent kinetic energy, approach zero in an asymptotic manner. In the following example, we will consider the constraint of realizability in three alternative turbulence discretizations.

Example 11.5.2.1

Realizability of Transient Turbulence Predictions.

Consider the decay of turbulence governed by the following reduced $k-\epsilon$ equations:

$$\frac{\partial k}{\partial t} = -\epsilon \tag{11.197}$$

$$\frac{\partial \epsilon}{\partial t} = -c_{\epsilon 2}\frac{\epsilon^2}{k} \tag{11.198}$$

Outline a numerical procedure that ensures that the solutions of these equations remain positive.

In an explicit formulation, all terms on the right sides of the previous equations are evaluated at the previous time level (superscript n), thereby yielding

$$k^{n+1} = k^n - \epsilon^n \Delta t \tag{11.199}$$

$$\epsilon^{n+1} = \epsilon^n - c_{\epsilon 2}\frac{\epsilon^{2,n}}{k^n}\Delta t \tag{11.200}$$

Since this k equation can possibly lead to negative values, it is not a suitable formulation. The turbulent kinetic energy must not produce a negative value.

Instead, if we consider an implicit formulation, the right-side terms are evaluated at the new time level (superscript $n+1$), so that

$$k^{n+1} = k^n - \epsilon^{n+1}\Delta t \tag{11.201}$$

$$\epsilon^{n+1} = \epsilon^n - c_{\epsilon 2}\left(\frac{\epsilon^n}{k^n}\right)\epsilon^{n+1}\Delta t \tag{11.202}$$

We find that the k equation can again yield negative values of k; therefore, this formulation is still not suitable.

Consider a semi-implicit, uncoupled formulation whereby k and ϵ are rearranged from the previous implicit formulations as follows:

$$k^{n+1} = k^n - \left(\frac{\epsilon^n}{k^n}\right)k^{n+1}\Delta t \tag{11.203}$$

$$\epsilon^{n+1} = \frac{\epsilon^n}{1 + c_{\epsilon 2}\epsilon^n \Delta t / k^n} \tag{11.204}$$

Equation (11.203) can be rearranged to give

$$k^{n+1} = \frac{k^n}{1 + \epsilon^n \Delta t / k^n} \tag{11.205}$$

which represents a suitable and realizable model since both k and ϵ remain as positive-definite quantities.

Although a finer grid should be used for the turbulence quantities (compared to the mean flow) due to smaller turbulence length scales, this approach is not usually implemented. However, an alternative upwinding approach in the convection terms for k and ϵ is considered. For example, local blending of a central difference scheme with a low-order upwind discretization can be used. This approach often leads to certain discretization errors for k and ϵ, but it may be necessary for computations to be performed on a single grid. Furthermore, boundary equations are needed for the turbulence equations. A frequently adopted approach involves using experimentally validated near-wall velocity profiles (such as the *law of the wall*) to apply boundary conditions, rather than refining the grid near the wall.

For example, consider the application of boundary conditions in a finite volume scheme for turbulent channel flows. We will construct the mesh so that finite volumes along the boundary are placed within the overlap region of the boundary layer (i.e., $30 < yu_\tau/\nu < 100$, where $u_\tau = \sqrt{\tau_w/\rho}$). The finite volume representation of the discrete x direction momentum equation can be written as

$$a_p u_p = \sum a_{nb} u_{nb} - \rho u_\tau^2 \Delta x - V \left. \frac{\partial p}{\partial x} \right|_p + \hat{S} \tag{11.206}$$

where V and $\hat{S}$ refer to the volume and source terms, respectively. Also, the subscripts p and nb refer to nodes p (center of the control volume) and neighboring nodes, respectively. For example, the summation in Equation (11.206) is performed over eight neighboring nodes for a two-dimensional structured grid at an internal node.

Also, the second term on the right side of Equation (11.206) represents the wall shear stress. In the near-wall region, this term is well represented by an empirical, experimentally validated velocity profile (such as the log law of the wall). Thus, this profile can be directly substituted into the shear stress term in Equation (11.206) and combined with the left side of that equation. The updated velocity, u_p^{n+1}, is factored out of the shear stress expression so that the resulting coefficient can be combined with a_p on the left side of Equation (11.206). In particular, the wall stress expression is linearized by $u_p^2 = u_p^n u_p^{n+1}$ in order to maintain numerical stability with positive coefficients preceding u_p^{n+1}. Thus, the mean flow equation (involving u_p) can be solved in the control volume near the wall, rather than refining the mesh therein.

Furthermore, boundary conditions are needed for the $k - \epsilon$ equations. For nodal points along a wall boundary, we can set $k = 0$. However, the dissipation is not zero at the wall. Instead, a commonly adopted approach is to assume that the derivative of ϵ normal to the wall is zero. At a high Reynolds number, the viscous sublayer of the boundary layer is so thin that it is often difficult to use enough grid points to resolve the flow behavior in this region. As discussed earlier, wall functions that describe the log law can be used. At an inflow boundary, k and ϵ are often unknown. If these values are available from measurements, they should be used. Otherwise, if k is not known, a sufficiently small value, such as $10^{-4}\overline{u}^2$ (where $\overline{u}$ is the mean velocity), should be adopted. The value of ϵ is often selected so that the length scale associated with its definition (i.e., $\epsilon = k^{3/2}/l$) is approximately an order of magnitude smaller than the width of the shear layer in the flow.

This discussion has involved certain overall aspects of numerical turbulence modeling. Additional detailed formulations of turbulence are described by Minkowycz et al. (1988). In the following section, numerical analysis of multiphase flows with droplets is examined.

11.5.3 Multiphase Flows with Droplets and Phase Change

Multiphase flows with droplets (or particles) can be modeled through Lagrangian or Eulerian techniques. Eulerian modeling of multiphase flows is based on spatial averaging of the governing equations over fixed control volumes including both carrier and dispersed phases (as discussed in Chapter 9). For example, in reacting flows, the carrier phase is the incoming airflow, while the dispersed phase refers to the droplets or particles, such as injected fuel droplets or pulverized coal particles. In this section, numerical modeling of multiphase flows with droplets and phase change will be considered.

Various techniques can be used for the numerical analysis of multiphase flows. These types of problems arise in various engineering and scientific applications, e.g., droplet combustion in fuel injectors, icing of aircraft and power lines (Draganoiu et al., 1996), and chemically reacting flows. In certain applications, the main issues in the numerical modeling deal with interactions between the droplet flow and surrounding flow field. For example, freestream turbulence largely affects the mixing and reaction rates of droplets in an internal combustion engine. In this case, computational models require detailed droplet tracking algorithms to predict the subgrid heat transfer rates due to droplet evaporation and combustion. However, in other applications, such as freezing of incoming supercooled droplets on aircraft surfaces, less emphasis may be required in the droplet–gas interaction in the freestream. In that case, knowledge of the impinging droplet flux on the surface is more important.

From Chapter 9, the governing equations for multiphase flows with droplets and phase change were derived based on spatial averaging procedures. In particular, spatial averaging was performed over a multiphase control volume containing more than one phase (see Figure 9.1). This spatial averaging provided the governing differential equations for each phase (Banerjee and Chan, 1980). Integrating Equation (9.1) over a discrete control volume for purposes of the current numerical analysis,

$$\int_A \frac{\partial(\rho_k \phi_k)}{\partial t}\, dA + \int_S (\rho_k \phi_k \mathbf{v}_k)\cdot \mathbf{dn} + \int_S (\mathbf{j}_k)\cdot \mathbf{dn} = \int_A \hat{S}_k dA \tag{11.207}$$

In an implicit formulation, convection, diffusion, and source terms are evaluated at the current time level, whereas an explicit formulation would evaluate these terms at the previous time level.

In reference to a CVFEM, the individual terms of Equation (11.207) could be evaluated in a manner analogous to that discussed in Section 11.4. For example, a backward difference in time is used for the approximation of the transient term. Both this term and the source (right side) term can be modeled through a lumped approximation. The source term on the right side of Equation (9.1) includes the cross-phase interactions in the multiphase flow, i.e., resistance between the droplets and airstream, evaporation, etc. The convective term in Equation (11.207) can be modeled through upwinding techniques discussed earlier, such as direct upwinding or pressure-weighted skewed upwinding. Diffusive terms (i.e., molecular and turbulent diffusion of momentum and heat) could utilize bilinear interpolation with the shape functions in the context of a CVFEM.

In the numerical upwinding for the convective term in Equation (11.207), the integration point values must be related to the appropriate nodal variables. As discussed in Section 11.4, this procedure can be carried out through a local balance of transport processes at the integration point. For example, the integration point velocity components (air and droplets separately) can be determined from the transport forms of Equation (9.1). Once discrete representations of these terms are assembled together, a local matrix inversion would be performed to express the four integration point variables (per element) in terms of the appropriate nodal quantities. In this way, the resulting global system of equations can be solved in terms of nodal quantities alone.

These procedures are based on techniques discussed earlier in this chapter (i.e., Section 11.4) for the carrier (air) phase. However, additional equations must be solved for the dispersed (droplet) phase. In particular, the droplet velocities are required for the mass influx of droplets into control volumes located along a wall or moving phase interface. In this case, the subscript k in Equation (11.207) denotes the liquid phase, rather than the air (continuous) phase. Then the source term includes cross-phase interactions arising from the spatial averaging procedures. Furthermore, the interpretation of certain terms requires special treatment in the dispersed phase. For example, although pressure is well defined in the continuous phase, the discrete droplets are dispersed within the continuous phase. The pressure gradient in the continuous phase is largely responsible for fluid acceleration of the multiphase flow, rather than internal pressure variations within individual droplets. Also, the resistance law between the dispersed and continuous phase requires detailed modeling of the droplet parameters (Tsuboi and Kimura, 1998).

In some applications, such as spray deposition and solidification of impinging droplets on a substrate in manufacturing processes, additional steps are required to predict the impinging droplets with phase change. Another example is impinging supercooled droplets and ice accretion on an aircraft surface. In these cases, the surface's shape is altered due to droplet freezing and ice accretion on the solid boundary. This type of moving boundary problem involves a discrete change of the phase interface position at each time step. Grid transformations to allow this moving phase boundary generally entail difficulties such as interpolation of scalar values from the previous grid to new nodal locations on the updated grid. Also, grid or coordinate transformations are time-consuming and difficult to implement into conventional CFD codes using a fixed domain. These difficulties require innovative solutions in numerical modeling.

Equation (11.207) can be adapted to the scalar transport of the droplet phase fraction throughout the physical domain. The variable ϕ can be interpreted similarly to species concentration, but for the volume fraction of droplets instead. Along solid walls, including boundaries covered with solidified droplets or ice, the droplet outflow must be altered due to the presence of the surface. Also, additional freezing of incoming supercooled droplets increases the liquid fraction of the volume (potentially consisting of air, liquid, and solid ice, simultaneously). Once the mass influx yields a filled control volume (i.e., filled with water or ice), the multiphase flow can no longer pass through this occupied volume and the edge of the solid interface must move somehow into a neighboring control volume. In this approach, this surface location and the liquid fraction within a volume are both used to identify the location of the moving phase interface.

Along the solid–air interface (receiving incoming droplets), the solid and liquid (droplet) phase fractions can accumulate until the control volume at the surface becomes filled. Beyond this point, the mass fraction of water (liquid or solid) becomes unity and the advective term in the mass and momentum equations must be reduced to zero since the volume is filled. The phase interface position is then identified within the adjacent volume containing a liquid fraction between zero and 1. During a time step when a control volume is changing from a partially filled stage to a completely filled stage, the amount of droplet influx into the partially filled volume may exceed the volume available. In this case, the excess amount must be transferred to the adjacent control volume for proper mass conservation. In other words, the phase fraction is a bounded quantity (between zero and 1), and thus a droplet mass influx into a nearly filled volume must not surpass this bound.

In this transition stage, as the interface moves between adjacent control volumes, the excess mass (i.e., "overfilled" mass) from the former control volume is implicitly added to the adjacent volume once the phase change is completed in the following iteration or time step. This procedure ensures that mass is conserved. However, additional difficulties may be encountered if the ice (solid–air) interface moves more than two control volumes (or elements) away from its present position, since a direct connection between nodes is no longer apparent. Although it appears that a time step limitation is required for numerical stability, this limitation would generally be much less stringent than conventional time step limitations (Anderson et al., 1984).

11.5.4 Multiphase Flows with Free Surfaces and Phase Change

Some technological problems involve free surfaces between the gas and liquid–solid phases, as well as the presence of phase transition (such as solidification or melting). For example, a free surface is established between the liquid and gas phases during volumetric shrinkage in casting solidification problems (Xu, Li, 1991). After a free surface is established in this example, a numerical method is required to adjust and predict the movement of the free surface in conjunction with the phase transition. In this section, a widely used technique (called the volume of fluid (VOF) method) will be described for the analysis of free surface flows.

The VOF method (Hirt and Nichols, 1981) defines a fluid fraction, F, within each discrete cell (i.e., finite volume or element) of the mesh. Values of $F = 1$, $F = 0$, and $0 < F < 1$ represent a full cell, empty cell, and cell that contains a free surface (see Figure 11.5), respectively. In the previously cited example of solidification shrinkage, the free surface would be contained within a cell containing liquid and gas phases. Marker particles are defined as cell intersection points that define where the fluid is located in the cell. The normal direction to the free boundary lies in the direction of the gradient of F. When both the normal direction and the value of F in a boundary cell are known, a line through the cell can be constructed to approximate the interface. Boundary conditions, including surface tension forces, can then be applied at the free surface.

In the VOF method, the following variables are used: A, acceptor cell (gaining fluid volume); D, donor cell (losing fluid volume); u, normal velocity at the face of the cell; Δx, cell spacing; V, fluid and void volume;

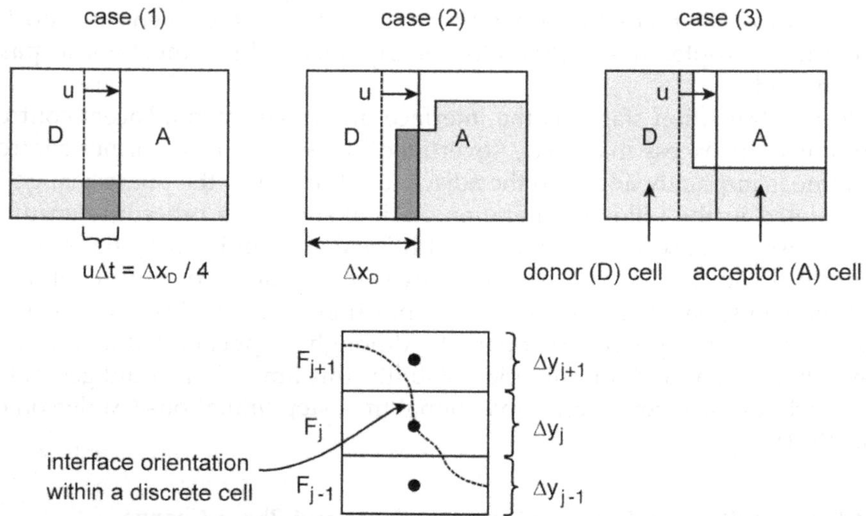

FIGURE 11.5
Schematic of the VOF method.

and F, fluid fraction in the cell. Based on these definitions, the quantity of F advected across the edge of the cell in a time step of Δt is given by ΔF, where

$$\Delta F = Min\{F_{AD}|V| + CF,\ F_D\Delta x_D\} \tag{11.208}$$

where

$$CF = Max\{(1 - F_{AD})|V| - (1 - F_D)\Delta x_D,\ 0\} \tag{11.209}$$

The following example illustrates how this technique can be used to find the movement of a free surface between donor and acceptor cells.

Example 11.5.4.1
Volume of Fluid Method for Free Surface Movement.
Calculate the quantity of F advected across the cell face between donor and acceptor cells over a time step of Δt for the following three cases:

1. $F_D = \frac{1}{4}$, $F_{AD} = F_D = \frac{1}{4}$, and $\Delta t = \Delta x_D/(4u)$
2. $F_D = \frac{3}{25}$, $F_{AD} = F_A = \frac{19}{25}$, and $\Delta t = 3\Delta x_D/(10u)$
3. $F_D = \frac{22}{25}$, $F_{AD} = F_A = \frac{10}{25}$, and $\Delta t = 3\Delta x_D/(10u)$

It may be assumed that the cell spacing and normal velocity at the face can be denoted by Δx and u, respectively. These three examples are illustrated in Figure 11.5.

1. In the first case, the cells have $F_D = F_{AD} = 1/4$ and $u\Delta t = \Delta x_D/4$. In this case, we would expect that $\Delta F = \Delta x_D/16 = u\Delta t/4$, which is less than the maximum permissible amount that could be convected into A. Equations (11.208) and (11.209) verify these trends since

$$CF = Max\left\{\left(1 - \frac{1}{4}\right)u\Delta t - \left(1 - \frac{1}{4}\right)\Delta x_D,\ 0\right\} = 0 \tag{11.210}$$

since $\Delta x > u\Delta t$. Then

$$\Delta F = Min\left\{\frac{1}{4}\ u\Delta t + 0,\ \frac{1}{4}\ \Delta x_D\right\} = \frac{1}{4}\ u\Delta t \tag{11.211}$$

If the donor cell is almost empty, then the *Min* condition in Equation (11.208) prevents the advection of more fluid from the donor cell than it has to give. Similarly, the *Max* feature of Equation (11.209) adds an additional flux, *CF*, which is available through advection in the time step Δt.

2. In this case, we would expect that

$$\Delta F = F_D \Delta x_D = \frac{3}{25}\ \Delta x_D \tag{11.212}$$

Using the VOF method, the amount advected into the acceptor cell is computed as follows:

$$CF = Max\left\{\left(1-\frac{19}{25}\right)u\Delta t - \left(1-\frac{3}{25}\right)\Delta x_D,\ 0\right\} = 0 \tag{11.213}$$

since $u\Delta t = 3\Delta x_D/10$. As a result,

$$\Delta F = Min\left\{\frac{19}{25}\ u\Delta t + 0,\ \frac{3}{25}\ \Delta x_D\right\} = \frac{3}{25}\ u\Delta t \tag{11.214}$$

It can be observed that cell D advects all of its fluid into cell A.

3. In the third case, we would expect that

$$\Delta F = F_D \Delta x_D = \frac{22}{25}\ \Delta x_D \tag{11.215}$$

which represents more fluid than can be accepted by cell A. Also,

$$CF = Max\left\{\left(1-\frac{10}{25}\right)u\Delta t - \left(1-\frac{22}{25}\right)\Delta x_D,\ 0\right\} = \frac{3}{50}\ \Delta x_D \tag{11.216}$$

since $u\Delta t = 3\Delta x_D/10$ (see Figure 11.4). Then

$$\Delta F = Min\left\{\frac{10}{25}\left(\frac{3}{10}\ x_D\right) + \frac{3}{25}\ \Delta x_D,\ \frac{22}{25}\ \Delta x_D\right\} = \frac{6}{25}\ \Delta x_D \tag{11.217}$$

In a way similar to that in the first case, the *Max* feature of Equation (11.209) adds the additional flux, *CF*, which is available due to advection over the time step of Δt.

Another important consideration is the interface orientation inside a cell that contains the free surface. The VOF method assumes that this boundary can be approximated by a straight line cutting through the cell. By determining the line's shape, it can be moved across the cell to a position such that it intersects the cell's edges to provide the known amount of fluid volume in the cell. The free surface can be represented by either $Y(x)$ or $X(y)$ (single valued) by taking the weighted average (in terms of $F_{i,j}$) of three adjacent cell boundary positions. If the magnitude of dY/dx is greater than the magnitude of dX/dy, the surface is more horizontal than vertical. If $dX/dy < 0$, fluid lies below the surface. With the slope and its sign, a line is constructed in the cell with a correct amount of fluid beneath it (see Figure 11.5). Then local surface tension forces that affect the dynamics of the free surface movement can be estimated based on the $Y(x)$ and $X(y)$ profiles.

Finally, the motion of F (such as void movement) can be predicted based on the solution of an appropriate governing equation, such as the following two-dimensional advection equation:

$$\frac{\partial F}{\partial t} + u\frac{\partial F}{\partial x} + v\frac{\partial F}{\partial y} = 0 \tag{11.218}$$

This governing equation states that the transient accumulation of F in a control volume balances its net inflow across the cell surface. For example, the net liquid fraction inflow into a control volume accounts for the transient change of liquid fraction or the local wave growth. Equation (11.218) may be applicable to free surface flows without phase change, but likely not to phase change problems. For example, solid–liquid phase change in enclosures may behave differently since voids are typically transmitted to boundary regions instantaneously, and not convected with the flow as indicated by Equation (11.218).

Appropriate boundary conditions must be applied along a free surface, such as matching conditions of shear stress, temperature, and velocity. Also, surface tension forces act on the fluid–gas interface. In free surface problems, the position of the moving interface is often unknown. As a result, iterative or matching procedures from both sides of the free surface are required to track its movement.

11.6 Accuracy and Efficiency Improvements

In order to improve the solution accuracy of numerical simulations, refinement of the grid spacing or time step is often required. Also, improvements in solution accuracy are closely linked with how well the problem physics is modeled. Although the accuracy may be improved, a trade-off between higher accuracy and additional computation time or storage is often realized. The grid spacing and time step refinements are performed until solution convergence is achieved. The values of important parameters in the problem, such as the location of the boundary layer detachment, should become sufficiently independent of the grid spacing and time step size. In some cases, these refinements may become cost or time prohibitive. As a result, we can often assess suitable convergence behavior by reduction of the solution residual below some specified tolerance.

Solution convergence is an important characteristic of the numerical model. The convergence rate of conventional iterative techniques is often fast during the first few time steps or iterations; however, the more slowly varying components of error are more difficult to eliminate. We may consider discretization errors as composed of both small-wavelength and long-wavelength components. For example, heat transfer in the vicinity of a phase interface may be considered in terms of a short wavelength due to the proximity of the error to the phase interface. However, a large recirculation cell in the flow field may be composed of larger wavelength errors, based on a larger distance associated with propagation of disturbances throughout the flow field.

A useful method of improving the efficiency of a numerical algorithm is the *multigrid method*. In this method, a sequence of grids is constructed with a subdivision of grid spacings by factors of 2 in each finer mesh. If the convergence of the solution on the fine mesh becomes too slow, the solution is transferred to a coarser grid, where longer wavelength components of the solution error are more effectively handled. The solution on the finer grid is then corrected in an appropriate fashion to reflect the reduction of long-wavelength error components.

We refer to full multigrid procedures whenever the iterative process begins with the coarsest grid. Conversely, we can refer to a cycling type of procedure whenever the process begins with the finest grid. In linear problems, such as steady-state heat conduction without phase change, we can work with only corrections to the solution from another grid (called *correction storage*). However, in nonlinear problems, such as phase change heat transfer, we need to work with the full solution from both grids (called *full approximation storage*). Also, we refer to the *restriction operator* as the discrete operator transferring the error from the fine grid to the coarse grid. The *prolongation operator* is used for the transfer of the correction from the coarse grid to the fine grid. These operators are typically achieved by suitable weighting of nodal values between various grids.

Another promising approach for accuracy and efficiency improvements is based on the Second Law of Thermodynamics. Numerical solutions are initially obtained from the conservation equations. These results are then considered in view of the second law. For example, the second law can identify the appropriate time step constraints for stable time advance (Camberos, 2000). If the second law is violated locally within the domain (i.e., due to discretization errors), then a corrective step is applied based on the magnitude of negative entropy production within the control volume

(Naterer and Schneider, 1994). This corrective step is applied to the numerical solution prior to advancement to the next time step, so that better solution accuracy can be achieved. In closing, the second law has important implications in regard to establishing error bounds, numerical stability, robustness, system optimization, and other important features of a numerical simulation (Naterer and Camberos, 2001, Naterer, 1999).

References

D.A. Anderson, J.C. Tannehill, and R.H. Pletcher. 1984. *Computational Fluid Mechanics and Heat Transfer*, Washington, DC: Hemisphere Publishing Corp.

S. Banerjee and A. Chan. 1980. "Separated flow models: I. Analysis of the time averaged and local instantaneous formulations," *Int. J. Multiphase Flow* 6: 1–24.

J.A. Camberos. 2000. "Non-linear time-step constraints for simulation of fluid flow based on the Second Law of Thermodynamics," *AIAA J. Thermophys. Heat Transfer* 14: 177–185.

W. Cheney and D. Kincaid. 1985. *Numerical Mathematics and Computing*, 2nd ed., Monterey, CA: Brooks/Cole Publishing Company.

G. Draganoiu, L. Lamarche, and P. McComber. 1996. "A computer model of glaze accretion on wires," *ASME J. Offshore Mech. Arctic Eng.* 118: 148–157.

C.W. Hirt and B.D. Nichols. 1981. "Volume of fluid (VOF) method for the dynamics of free boundaries," *J. Comput. Phys.* 39: 201–225.

W.J. Minkowycz et al. 1988. *Handbook of Numerical Heat Transfer*, New York: John Wiley & Sons, Inc.

G.F. Naterer. 1997. "Sub-grid volumetric quadrature accuracy for transient compressible flow predictions," *Int. J. Numerical Methods Fluids* 25: 143–149.

G.F. Naterer. 1999. "Constructing an entropy-stable upwind scheme for compressible fluid flow computations," *AIAA J.* 37: 303–312.

G.F. Naterer and J.A. Camberos. The Role of Entropy and the Second Law in Computational Thermofluids, AIAA Paper 2001-2758, paper presented at AIAA 35th Thermophysics Conference, Anaheim, CA, June 11–14, 2001.

G.F. Naterer and G.E. Schneider. 1994. "Use of the second law for artificial dissipation in compressible flow discrete analysis," *AIAA J. Thermophys. Heat Transfer* 8: 500–506.

G.F. Naterer and G.E. Schneider. 1995. "PHASES model of binary constituent solid-liquid phase change: part 1: numerical method," *Numerical Heat Transfer B* 28: 111–126.

G.E. Schneider and M.J. Raw. 1987. "Control-volume finite-element method for heat transfer and fluid flow using co-located variables: 1. Computational procedure," *Numerical Heat Transfer* 11: 363–390.

G.E. Schenider and M. Zedan. Control Volume Based Finite Element Formulation of the Heat Conduction Equation, AIAA Paper 82-0909, paper presented at AIAA/ASME 3rd Joint Thermophysics, Fluids, Plasma and Heat Transfer Conference, St. Louis, MO, June 7–11, 1982.

K. Tsuboi and S. Kimura. Numerical Study of the Effect of Droplet Distribution in Incompressible Droplet Flows, AIAA Paper 98-2561, paper presented at AIAA 29th Fluid Dynamics Conference, Albuquerque, NM, 1998.

D. Xu and Q. Li. 1991. "Gravity and solidification induced shrinkage in a horizontally solidified ingot," *Numerical Heat Transfer* 20: 203–221.

Problems

1. Annular tapered fins with a thickness of c/r_2 (where c is a constant) are designed to enhance the rate of heat transfer from air-cooled tubes of a combustion cylinder. One-dimensional, steady-state heat transfer occurs from the base ($r = a$) to the tip ($r = b$) of each fin in the radial (r) direction. The ambient air flows past the fins with a temperature and convection coefficient of T_∞ and h, respectively, and the temperature at the base of the fin is T_b.
   (a) Derive the finite difference equations for the interior and boundary nodes of the domain. It may be assumed that a constant thermal conductivity, k, within the fin is specified.
   (b) Explain how the numerical solution procedure in part (a) would need to be modified if variations of thermal conductivity with temperature become significant.
2. A plane wall with a width of 4 cm is cooled along its right boundary by a fluid at 20°C and a convection coefficient of $h = 800\ \text{W/m}^2\text{K}$. The left boundary is maintained at 160°C, and the initial wall temperature is 60°C. The thermal conductivity and diffusivity are 50 W/mK and $10^{-5}\ \text{m}^2/\text{sec}$, respectively. Using a five-node discretization (one-dimensional domain) and an explicit finite difference formulation, calculate the steady-state temperatures at the midplane and right boundary. Explain how you would expect the trends in your results to change if $2Fo\,(1 + Bi) > 1$, where Fo and Bi refer to the Fourier and Biot numbers, respectively.

3. The top edge of a silicon chip in an electronic assembly is convectively cooled by a fluid at 20°C with a convection coefficient of 600 W/m^{2}K. A two-dimensional explicit finite difference method ($\Delta x = 6$ mm, $\Delta y = 5$ mm) is used to predict the transient temperature variation within the chip ($\rho c_p = 1659$ kJ/m^{3}K, $k = 84$ W/mK). The rate of internal heat generation within the chip is 10^8 W/m^3.
   (a) What minimum time step is required to ensure stable time advance for a node on the top boundary? Determine the stability criterion by deriving the relevant finite difference equation on the top boundary.
   (b) Would your answer be more restrictive for a fixed heat flux boundary condition? Explain your response.
4. A weighted residual method uses $W_i = P$ at nodal points (where P represents a very large number), but $W_i = 0$ away from the nodal points. Describe how this selection influences the residual and error distributions throughout the problem domain.
5. State three important advantages of the finite element method over other conventional numerical methods developed for structured grids.
6. Are the shape functions linear along the edge of a four-node quadrilateral element? Explain your answer.
7. Describe the differences between a residual solution error and the order of accuracy in the context of finite element analysis.
8. Can the elemental boundary integral ($\int_S N_i h N_j ds$) be a full (entirely nonzero) 3×3 matrix for triangular elements in linear and steady diffusion problems? Explain your response.
9. Explain how overrelaxation and underrelaxation terms play a role in finite element analysis of fluid dynamics problems.
10. Explain how unstructured grid capabilities provide a key advantage of finite element methods in comparison to other conventional numerical techniques using structured grids.
11. Under what circumstances would you recommend that Gauss quadrature with quadrilateral elements be adopted in the weighted residual method?
12. A specified velocity profile is provided at the inlet of a diffuser, but the outflow conditions are unknown. What boundary location and conditions at the outlet would be suitable for a finite element simulation of this internal fluid flow?
13. Determine the requirement that the shape functions must satisfy in order to model the condition that a scalar, ϕ, equals a constant within the finite element.
14. Consider the following differential equation subject to two boundary conditions:

$$\frac{d^2u}{dx^2}+4=0, -1 \le x \le 1 \text{ subject to } u(-1)=0,\ u(1)=0 \qquad (11.219)$$

(a) Use the trial function $\hat{u} = a_1 cos(\pi x/2)$ with a single element and find a Galerkin solution of the differential equation (note: $sin^2x + cos^2x = 1$).

(b) Compare the approximate solution in part (a) with the exact solution. How can closer agreement be achieved in this problem?

(c) Consider this same problem, but select a trial function of $\hat{u} = a_1(1-x^2)+a_2(1-x^2)^2+a_3(1-x^2)^3$ instead. What values of a_1, a_2, and a_3 would the weighted residual method give in this case? Does your answer depend on the type of weighted residual method?

15. A weighted residual formulation based on the method of moments is used to solve the steady-state heat conduction equation. The approximate temperature function is written in terms of three exponential basis functions in the following manner:

$$\tilde{T} = a_0e^{ix} + a_1e^{2ix} + a_2e^{3ix} \qquad (11.220)$$

where i refers to the complex imaginary number. Find the discrete equations for this one-dimensional heat conduction problem using the weight functions of $W_n = x_n$, where $n = 0$, 1, and 2.

16. A composite wall consists of the following three different layers: 9-cm-thick section 1 ($k_1 = 5$ W/mK), 6-cm-thick section 2 ($k_2 = 0.8$ W/mK), and 7-cm-thick section 3 ($k_3 = 16$ W/mK). The left and right boundaries are maintained at $T = 400$°C and $T = 100$°C, respectively. Use a one-dimensional finite element procedure (per unit wall area) to determine the interface temperature values, as well as the heat flux through the third (right) section.

17. Heat transfer in a circular fin is governed by the following fin equation:

$$kA\frac{d^2T}{dx^2} - hP(T - T_f) = 0$$

where $h = 1$ W/cm^{2}K for a fluid flowing past the fin at a temperature of $T_f = 12$°C. The thermal conductivity of the fin is 3 W/cmK, and the fin diameter and length are 1 and 4 cm, respectively. The base temperature of the fin is maintained at 60°C. Use a Galerkin finite element method with four equal one-dimensional elements to find the nodal temperature distribution within the fin.

18. A nuclear fuel element of thickness $2L$ is covered with a steel cladding of thickness a. Heat is generated within the nuclear fuel at a rate of $\dot{q}$, and it is removed from the left surface by a fluid at $T_\infty = 90°\text{C}$ with a convection coefficient of $h = 6{,}000\ \text{W/m}^2\text{K}$. The right surface is well insulated, and the thermal conductivities of the fuel and steel are $k_f = 60\ \text{W/mK}$ and $k_s = 64\ \text{W/mK}$, respectively. Use the finite element method with one-dimensional elements to find the temperatures at the external boundaries, steel–fuel interfaces, and fuel element centerline.
19. The cross section of a metal component is triangular. The left and lower boundaries are insulated, and the other surface is exposed to a convection condition with $h = 5\ \text{W/m}^2\text{K}$. A uniform internal heat source provides $\dot{q} = 2\ \text{kW/m}^3$ within the element. Use a single triangular element to estimate the nodal temperature distribution in terms of the positions of the nodal points.
20. Consider two adjacent linear quadrilateral elements. A local coordinate system (s, t) is defined within each element in order to interpolate appropriate values for the problem variable ϕ. Evaluate ϕ along the common edge from both elements and determine whether there is a unique (single) distribution for both cases.
21. Consider two adjacent linear triangular elements along the external boundary of a problem domain. The purpose of this problem is to outline the proper implementation of an insulation (zero gradient) boundary condition along the upper boundary, i.e., $\partial T/\partial y = 0$ along a horizontal upper boundary.
    (a) Use the isoparametric shape functions for triangular elements to find the boundary temperature at node 3 that satisfies the insulated boundary requirement. The temperatures at nodes 1, 2, and 4 are 170, 120, and 210°C, respectively.
    (b) Determine the $x-y$ coordinates where the contour line for this temperature (node 3, part (a)) intersects the element boundary along side 1–2 in element 1.
    (c) Use the linear shape functions to check that this intersection point lies on the line describing side 1–2 in element 1.
    (d) Use the linear shape functions to check that the temperature at each point along the contour line is constant and equal to the value obtained at node 3.
22. A layer of viscous liquid of constant thickness flows under gravity down an inclined surface in a chemical purification process. There is no fluid velocity perpendicular to the plate. Under steady-state conditions, the film motion is governed by a balance between frictional and gravitational forces, i.e.,

$$\mu \frac{d^2u}{dy^2} + \rho g sin(\alpha) = 0 \tag{11.221}$$

It may be assumed that the air resistance is negligible, such that the shear stress at the free surface is zero (i.e., $\mu(du/dy) = 0$ on the free surface).

(a) Obtain a finite element solution for the film velocity, $u = u(y)$. Use three linear (one-dimensional) elements.

(b) Find the exact solution and compare it with the above computed solution from part (a).

(c) How can we obtain a closer agreement between the computed and exact solutions?

23. Consider a thin, wide plate with an externally applied heat flux at the top boundary and a well-insulated bottom surface. The initial plate temperature is 0°C. Using an explicit lumped formulation with a time step size of 1 sec, assess whether the resulting formulation is numerically stable. In particular, explain whether numerical oscillations will occur in the computations.

24. The purpose of this question is to write four computer subroutines (STIFF, ASSMBL, SHAPE, and BNDRY) to solve the following heat equation with the Galerkin weighted residual method and triangular finite elements.

$$k\frac{\partial^2 T}{\partial x^2} + k\frac{\partial^2 T}{\partial y^2} = 0 \tag{11.222}$$

The thermal conductivity, k, is assumed to be constant and independent of temperature in this analysis. The following STIFF and ASSMBL subroutines find the element properties for the two-dimensional field equation (stiffness matrix and right side) and perform the assembly of all elements, respectively:

- Subroutine STIFF (e, nelm, nnpe, x, y, na, nb, nc, iside, cond, h, tinf, rq, aq).
- Subroutine ASSMBL (ne, nelm, nnpm, nnpe, x, y, ie, iside, h, tinf, cond, na, nb, nc, isemi, aq, rq, a, r).

For each element, e, the ASSMBL subroutine should call the stiffness generator, STIFF, for the actual matrix entries. Your program should import the mesh data from an external file. Also, shape function information (i.e., coefficients a, b, c, and the element area) should be obtained from a separate subroutine (called SHAPE).

- Subroutine SHAPE (e, nelm, nnpe, x, y, na, nb, nc, ar4)

The shape functions possess various important features, including (i) $N_i = 1$ at node i and $N_i = 0$ at nodes j, k, and (ii) $N_1 + N_2 + N_3 = 1$. Use a sample element to verify these features of the shape functions.

Consider boundary conditions that include (i) temperature-specified, (ii) adiabatic, and (iii) convection-specified conditions. Write a subroutine (BCAPPL) to treat the first boundary type, and modify your stiffness generator (STIFF) to handle the second and third boundary types. A banded solver can be used to solve the final set of discrete algebraic equations.

As a validation problem, consider heat conduction in a rectangular-shaped material subjected to convective cooling ($h = 20$ Btu/h ft^2 °F and $T_\infty = 70$°C) on the right boundary and a uniform temperature of 100°F on the left boundary. The insulated horizontal boundaries are 2 ft apart, and the width of the domain is 4 ft. The conductivity of the material is 25 Btu/h ft °F. Find the steady-state temperature distribution using the previously described finite element procedure and eight triangular elements.

Note: e = element number, ne = total number of elements, nelm = maximum total elements, nnpm = maximum total nodes, nnpe = number of nodes per element, x(nelm, nnpe) = nodal x values, y(nelm, nnnpe) = nodal y values, ie(nelm, nnpe) = local–global node bookkeeping array, aq(nnpe, nnpe) = element stiffness matrix, a(nnpm, nnpm) = global stiffness matrix, rq(nnpe) = element right-hand-side vector, r(nnpm) = global right-hand-side vector, na/nb/nb = shape function coefficients, ar4 = four times element area, isemi = semiband-width, cond = conductivity, h = convection coefficient, tinf = reference temperature.

25. The purpose of this question is to write three computer subroutines (SHAPE, STIFF, and BNDRY) to solve the two-dimensional field equation with a CVFEM and quadrilateral finite elements.

$$k\frac{\partial^2 \phi}{\partial x^2} + k\frac{\partial^2 \phi}{\partial y^2} - G\phi + Q = 0 \quad (2-D \;\; Field \;\; Equation)$$

Write a subroutine, SHAPE, to evaluate the shape functions and their derivatives for a linear, quadrilateral, isoparametric, and finite element.

- Subroutine SHAPE (nnp, nnpe, nel, ie, e, x, y, s, t, dn, dx, dy, jac, dqdx, dqdy)

The shape functions possess several features, such as (i) $N_i = 1$ and $N_j = 0$ $(j \neq i)$ at local node i, and (ii) $N_i = 0$ on sides opposite to node i. Use a numerical example of a sample element to verify these features in your program. Mesh data should be imported from a file, such as the example file summarized below:

- 9, 8, 8 (number of nodal points, elements, and boundary surfaces)
- 5, 2, 1, 4 (counterclockwise listing of nodes in the first element, stored within the *ie* array)
- ... (Other elements)
- 0, 0 (x and y global coordinates of the first node, stored in the x, y arrays)
- ... (Other nodes)

Write two subroutines (called STIFF and ASSMBL) that find the element properties for the two-dimensional field equation (stiffness matrix and right side), and perform the assembly functions of each.

- Subroutine STIFF (e, nel, nnp, nnpe, x, y, lc, s, t, o, ie, dn, cond, flux, aq, rq)

For each element, e, an ASSMBL subroutine should call STIFF to obtain the entries of the stiffness matrix. Shape function information should be obtained from the SHAPE subroutine.

In particular, consider diffusion transport $(G = 0 = Q)$ and boundary conditions, including (i) temperature-specified and (ii) flux-specified conditions. Write a subroutine (BNDRY) to specify the boundary conditions by giving the appropriate coefficients, A, B, and C, at the boundary, where

$$A\frac{\partial \phi}{\partial n} + B\phi = C$$

- Subroutine BNDRY (ie, o, nel, nnpe, nnp, nsrf, bgn, bel, x, y, bc, c, r)

  The specified boundary condition data should be imported from a file, such as the example file outlined below:

- 2, 1 (pair of global nodes on the side of the first boundary element, stored in the bgn array)
- ... (Other boundary nodes and elements)
- 1 (element number corresponding to the first boundary element, stored in the bel array)
- ... (Other boundary elements)
- $0, 1, 10$ (A, B, C values for the first node of the first element, stored in the bc array)
- ... (Other boundary nodes and elements)

As a validation problem, consider the following steady-state heat conduction problem (note: $\phi = T$ herein). The top and bottom boundaries are insulated, and the left and right boundaries are subjected to specified heat flux and temperature conditions, respectively. Find the temperature distribution and present your results for the nodal temperature values. A banded solver can be used to solve the set of algebraic equations. A grid refinement should be performed to show the grid independence of your results. Problem parameters, such as conductivity, should be imported from a separate project data file (i.e., sample.prj).

In addition, select another engineering heat transfer problem of practical or industrial relevance and prepare a numerical solution of it with your finite element program. Your presentation should include a brief description of the finite element model and a discussion of the results. Submit your source program; input files and numerical results for both validation and application problems. For the application problem, provide additional background information, problem parameters, boundary conditions, and a discussion of solution errors.

Note: e = element number; nel = number of elements; nnp = number of nodal points; nnpe = number of nodes per element; x, y(nnp) = nodal x and y values; ie(nel, nnpe) = local to global node bookkeeping array; aq(nnpe, nnpe) = element stiffness matrix; c(nnp, nnp) = global stiffness matrix; rq(nnpe) = element right side; r(nnp) = global right side; dn = shape functions and derivatives; isemi = semibandwidth; cond = conductivity; s, t = local coordinates; dx, dy = delta x and y; jac = Jacobian; dqdx, dqdy = scalar derivatives; lc = integration point local coordinates; o = orientation array for ip and SVCV numbers; flux = diffusion coefficients; nsrf = number of boundary surface elements; bgn = boundary global nodes; bel = boundary elements; bc = boundary condition coefficients (A, B, C).

26. In this question, numerical modeling of the scalar conservation equation, including diffusion and convection, is considered. A scalar quantity (called ϕ), such as energy or concentration of a pollutant in an airstream, is transported throughout a flow field by diffusion and convection. The purpose of this project is to write two computer subroutines (IPOINT, STIFFPHI), based on a CVFEM, to solve the following two-dimensional scalar conservation equation:

$$\frac{\partial(\rho\phi)}{\partial t} + \nabla\cdot(\rho\mathbf{v}\phi) + \nabla\cdot(\Gamma\nabla\phi) = \hat{S} \tag{11.223}$$

Integrating Equation (11.223) over a discrete volume (or two-dimensional area, A, encompassed by a surface, S) and using the Gauss theorem,

$$\int_A \frac{\partial(\rho\phi)}{\partial t}\,dA + \int_S (\rho\mathbf{v}\phi)\cdot d\mathbf{n} + \int_S (\mathbf{j})\cdot d\mathbf{n} = \int_A \hat{S}\,dA \tag{11.224}$$

where $\mathbf{v}$ and $\mathbf{n}$ refer to the velocity and unit normal vector at the surface, respectively.

In Equation (11.224), it can be observed that integration point values of ϕ (i.e., values at the midpoint of the subsurface) must be related to nodal quantities since the final algebraic equations involve nodal Φ variables only. As an example, consider a reduced case of one-dimensional convection–diffusion modeling at an integration point, $i+1/2$, between nodal points i and $i+1$. The following list shows conventional schemes for obtaining $\phi_{i+1/2}$ in terms of nodal Φ_i and Φ_{i+1} :

- CDS: $\phi_{i+1/2} = 0.5\Phi_i + 0.5\Phi_{i+1}$
- UDS: $\phi_{i+1/2} = \Phi_i$ (flow from left to right)
- Exponential differencing scheme (EDS): $\phi_{i+1/2} = ((1+\alpha)/2)\Phi_i + ((1-\alpha)/2)\Phi_{i+1}$, where $\alpha \approx Pe^2/(5+Pe^2)$ and $Pe = \rho u_i \Delta x_i/\Gamma$ (called the Peclet number)
- PINS: $\phi_{i+1/2}$ dependence on Φ_i and Φ_{i+1} obtained by a local approximation of the governing equation at the integration point $(i+1/2)$

The relative influences of the upstream and downstream nodes are closely related to the Peclet number, Pe, and they are identified by the influence coefficients that premultiply each nodal value. Write a computer subroutine (IPOINT) to specify two-dimensional values of the influence coefficient array, called ic, and velocities corresponding to the selected problem of interest at the integration points. Verify that these influence coefficients approach CDS and UDS for low- and high-mass flow rates, respectively, and that the proper influences in the x and y directions are achieved. Using the previously discussed influence coefficients, modify your STIFF subroutine to accommodate convection, transient, and source terms. See portions of the following sample code for an overall structure of the program:

```
      subroutine stiffphi(e,nel,nnp,nnpe,x,y,lc,s,
       t,ic, ...)
      integer ie(nel,nnpe), o(11)
      real x(nnp), y(nnp), s(3), t(3), ds(nnpe,4),
       dn(nnpe,4) ...
      real ic(2,2,nel,nnpe,2*nnpe), vi(4,nel,nnpe),
       cnn(nnp) ...
      do 900 i = 1, nnpe
       ...
        mflow = vi(1,e,i)*ds(i,2) -vi(2,e,i)*ds(i,1)
        call shape(nnp,nnpe,nel,ie,e,x,y,s,t,dn,
         dx,dy ...)
        do 800 j = 1, nnpe
          dfsc = ... diffusion coefficients
          advc = ... -ic(1,1,e,i,4+j)*mflow
          flux(1,i,j) = advc +dfsc
800     continue
900   continue
      do 2200 i = 1, nnpe
        n = ie(e,i)
        do 1800 j = 1, nnpe
          if(i .eq. j) aq(1,i,j) = ... ds(i,3)/dt-
           flux(1,i,j) + flux(1,o(i),j)
          if(i .ne. j) aq(1,i,j) = ... -flux(1,i,j)+
           flux(1,o(i),j)
1800  continue
      rq(i) = ... cnn(n)*ds(i,3)/dt+source terms ...
2200  continue
```

Select a validation problem to assess the accuracy of your formulation, and present your numerical results for this problem. An example of a useful test case is combined convection–diffusion for flow over a sharp step. The velocity components may be specified in the IPOINT subroutine as uniform throughout the flow field for this test case. In practice, a separate CFD solver is required to find the $\mathbf{v}$, p, and *ic* values. For a transient problem with a specified ϕ boundary condition at the inlet (above the step), verify that this upstream boundary condition is propagated over time and maintained through the flow. In particular, for high *Pe* values, verify that convection influences are dominant in your results. Conversely, for low *Pe* values (including the limit of zero flow), check that diffusion effects are smearing the ϕ distribution across the flow.

Also, select an engineering heat transfer problem of practical or industrial relevance involving scalar transport with combined convection–diffusion. For example, consider the transport equations for k (turbulent kinetic energy) and ϵ (dissipation rate). Then find the numerical solution of the problem with your finite element program.

Your presentation should include a brief description of the numerical model and a discussion of the results. Submit your source program; input files and numerical results for both validation and application problems. For the application problem, provide appropriate background information, problem parameters, boundary conditions, and a discussion of solution errors.

Note: e = element number; nel = number of elements; nnp = number of nodal points; nnpe = number of nodes per element; x, y(nnp) = nodal x and y values; ie(nel, nnpe) = local to global node bookkeeping array; aq(nnpe, nnpe) = element stiffness matrix; rq(nnpe) = element right side; dn = shape functions and derivatives; diff = diffusivity; s, t = local coordinates; lc = integration point local coordinates; o = orientation array for ip and SVCV numbers; mflow = SS mass flow rate; cnn(n) = scalar value at node n (previous time step); df = time step size; ic = influence coefficients.

27. Consider a one-dimensional discretization of the validation problem described in the previous question. Describe the sources of error that arise due to conventional upwinding (such as UDS) for *transient convection* of a step function. Suggest alternatives to overcome these sources of errors. Explain your responses through references to a specific example, such as a five-node discretization with a step function of $\phi = 1$ convected into the domain over time.

Appendices

TABLE A1

Conversion of Units and Constants

Conversion Factors

Acceleration	1 m/sec^2	= 4.252 × 10^7 ft/hr^2	= 3.2808 ft/sec^2
Area	1 m^2	= 1550.0 in.^{2}	= 10.764 ft^2
Density	1 kg/m^3	= 0.06243 lbm/ft	
Dynamic viscosity	1 kg/m·sec	= 1 N·sec/m^2	= 2419.1 lbm/ft·h
Energy	1 kJ	= 0.9478 Btu	= 737.56 ft·lbf
Force	1 N	= 1 kg·m/sec^2	= 0.22481 lbf
Heat flux	1 W/m^2	= 1 kg/sec^3	= 0.3171 Btu/h·ft^2
Heat transfer coefficient	1 W/m^{2}K	= 0.1761 Btu/ft^2 h °F	
Heat transfer rate	1 W	= 3.4123 Btu/h	= 1.341 × 10^{-3} hp
Kinematic viscosity	1 m^2/sec	= 10.7636 ft^2/sec	
Latent heat	1 kJ/kg	= 0.4299 Btu/lbm	
Length	1 m	= 39.37 in.	= 3.2808 ft
Mass	1 kg	= 2.2046 lbm	= 1.1023 × 10^{-3} U.S. tons
Mass diffusivity	1 m^2/sec	= 10.7636 ft^2/sec	= 3.875 × 10^4 ft^2/h
Mass flow rate	1 kg/sec	= 7936.6 lbm/h	
Mass transfer coefficient	1 m/sec	= 1.181 × 10^4 ft/h	
Pressure, stress	1 Pa	= 1 N/m^2	= 1.4504 × 10^{-4} lbf/in.^{2}
Specific heat	1 kJ/kgK	= 0.2388 Btu/lbm·°F	= 0.2389 cal/g·°C
Temperature	K	= °C + 273.15	= (5/9)(°F + 459.67)
	R	= °F + 459.67	= (9/5)(°K)
Temperature difference	1 K	= 1 °C	= (9/5)°F
Thermal conductivity	1 W/mK	= 0.57782 Btu/h·ft·°F	
Thermal diffusivity	1 m^2/sec	= 10.7636 ft^2/sec	= 3.875 × 10^4 ft^2/h
Thermal resistance	1 K/W	= 0.5275°F/h·Btu	
Velocity	1 m/sec	= 3.2808 ft/sec	= 3.6 km/h
Volume	1 m^3	= 264.17 gal (U.S.)	= 1000 L
Volume flow rate	1 m^3/sec	= 1.585 × 10^4 gal/min	= 2118.9 ft^3/min

SI Unit Conversions

Prefix (Symbol)	Multiplier
Tera (T)	10^{12}
Giga (G)	10^9
Mega (M)	10^6
Kilo (k)	10^3

Milli (m)	10^{-3}
Micro (μ)	10^{-6}
Nano (n)	10^{-9}
Pico (p)	10^{-12}
Femto (f)	10^{-15}

Constants

Atmospheric pressure (P_{atm})	= 101,325 N/m^2	= 14.69 lbf/in.^{2}
e	= 2.7182818	
Gravitational acceleration (g)	= 9.807 m/sec^2	
1 mole	= 6.022×10^{23} molecules	= 10^{-3} kmol
π	= 3.1415927	
Speed of light in a vacuum (c)	= 2.998×10^8 m/sec	
Stefan– Boltzmann constant (σ)	= 5.67×10^{-8} W/m^{2}K^4	= 0.1714×10^{-8} Btu/h ft^2 R^4
Universal gas constant (R)	= 8.315 kJ/kmol·K	= 1.9872 Btu/lb mol·R

TABLE A2

Properties of Metals at STP

Metal	Melting Point (°C)	Boiling Point (°C)	Thermal Conductivity (W/mK)	Specific Heat, c_p (kJ/kgK)	Coefficient of Expansion ($\times 10^6$/K)	Density, ρ (kg/m^3)	Heat of Fusion (kJ/kg)
Aluminum	660	2441	237.0	0.900	25	2700	397.8
Antimony	630	1440	18.5	0.209	9		
Beryllium	1285	2475	218	1.825	12		
Bismuth	271.4	1660	8.4	0.126	13		
Cadmium	321	767	93	0.230	30		
Chromium	1860	2670	91	0.460	6	7150	330.8
Cobalt	1495	2925	69	0.419	12	8860	276.4
Copper	1084	2575	398	0.385	16.6	8960	205.2
Gold	1063	2800	315	0.130	14.2	19,300	62.8
Iridium	2450	4390	147	0.130	6		
Iron	1536	2870	80.3	0.452	12	7870	272.2
Lead	327.5	1750	34.6	0.130	29	11,300	23.0
Magnesium	650	1090	159	1.017	25		
Manganese	1244	2060	7.8	0.477	22		
Mercury	−38.86	356.55	8.39	0.138			
Molybdenum	2620	4651	140	0.251	5	10,200	288.9
Nickel	1453	2800	89.9	0.444	13	8900	297.3
Niobium	2470	4740	52	0.268	7		
Osmium	3025	4225	61	0.130	5		
Platinum	1770	3825	73	0.134	9		
Plutonium	640	3230	8	0.134	54		
Potassium	63.3	760	99	0.753	83		
Rhodium	1965	3700	150	0.243	8		
Selenium	217	700	0.5	0.322	37		
Silicon	1411	3280	83.5	0.712	3		
Silver	961	2212	427	0.239	19	10,500	111.0
Sodium	97.83	884	134	1.226	70		
Tantalum	2980	5365	54	0.142	6.5		
Thorium	1750	4800	41	0.126	12		
Tin	232	2600	64	0.226	20	7280	59.0
Titanium	1670	3290	20	0.523	8.5	4500	418.8
Tungsten	3400	5550	178	0.134	4.5		
Uranium	1132	4140	25	0.117	13.4		
Vanadium	1900	3400	60	0.486	8		
Zinc	419.5	910	115	0.389	35		

Source: Data reprinted with permission from Hewitt, G.F., et al., eds., *International Encyclopedia of Heat and Mass Transfer*, CRC Press, Boca Raton, FL, 1997; Weast, R.C., ed., *CRC Handbook of Tables for Applied Engineering Science*, CRC Press, 1970.

TABLE A3

Properties of Nonmetals and Phase Change Materials
Nonmetals[a]

Material	Density, ρ (kg/m^3)	Thermal Conductivity, k (W/mK)	Specific Heat, c_p (kJ/kgK)
Asbestos millboard	1400	0.14	0.837
Asphalt	1100		1.67
Brick, common	1750	0.71	0.920
Brick, hard	2000	1.3	1.00
Chalk	2000	0.84	0.900
Charcoal, wood	400	0.088	1.00
Coal, anthracite	1500	0.26	1.26
Concrete, light	1400	0.42	0.962
Concrete, stone	2200	1.7	0.753
Corkboard	200	0.04	1.88
Earth, dry	1400	1.5	1.26
Fiberboard, light	240	0.058	2.51
Fiberboard, hard	1100	0.2	2.09
Firebrick	2100	1.4	1.05
Glass, window	2500	0.96	0.837
Gypsum board	800	0.17	1.09
Ice (0°C)	900	2.2	2.09
Leather, dry	900	0.2	1.51
Limestone	2500	1.9	0.908
Marble	2600	2.6	0.879
Mica	2700	0.71	0.502
Mineral wool blanket	100	0.04	0.837
Paper	900	0.1	1.38
Paraffin wax	900	0.2	2.89
Plaster, light	700	0.2	1.00
Plaster, sand	1800	0.71	0.920
Plastics, foamed	200	0.03	1.26
Plastics, solid	1200	0.19	1.67
Porcelain	2500	1.5	0.920
Sandstone	2300	1.7	0.920
Sawdust	150	0.08	0.879
Silica aerogel	110	0.02	0.837
Vermiculite	130	0.058	0.837
Wood, balsa	160	0.050	2.93
Wood, oak	700	0.17	2.09
Wood, white pine	500	0.12	2.51

Phase Change Materials

Material	Melting Point (°C)	Latent Heat of Fusion (kJ/kg)
Calcium chloride hexahydrate	29.4	193
Sodium carbonate decahydrate	33	251
Disodium phosphate dodecahydrate	36	280
Sodium sulfate decahydrate	32.4	253

Material	Melting Point (°C)	Latent Heat of Fusion (kJ/kg)
Sodium thiosulfate pentahydrate	49	200
n-Octadecane	28.0	243
n-Eicosane	36.7	247
Polyethylene glycol 600	20– 25	146
Stearic acid	69.4	199
Water	0.0	333
Tristearin	56	191
Paraffin wax, Sunoco 116	47	209

[a]*Source:* Data reprinted with permission from Lane, G.A., *Solar Heat Storage: Latent Heat Materials*, Vols. 1 and 2, CRC Press, Boca Raton, FL, 1996.

[b]*Source:* Data reprinted with permission from Weast, R.C., ed., *CRC Handbook of Tables for Applied Engineering Science*, CRC Press, Boca Raton, FL, 1970.

TABLE A4

Properties of Air at Atmospheric Pressure

Temperature, T (K)	Density, ρ (kg/m^3)	Specific Heat, c_p (kJ/kgK)	Viscosity, μ (kg/msec)	Thermal Conductivity, k (W/mK)	Pr
150	2.367	1.010	10.28×10^{-6}	0.014	0.758
200	1.769	1.006	13.28×10^{-6}	0.018	0.739
250	1.413	1.005	15.99×10^{-6}	0.022	0.722
260	1.359	1.005	16.50×10^{-6}	0.023	0.719
270	1.308	1.006	17.00×10^{-6}	0.024	0.716
275	1.285	1.006	17.26×10^{-6}	0.024	0.715
280	1.261	1.006	17.50×10^{-6}	0.025	0.713
290	1.218	1.006	17.98×10^{-6}	0.025	0.710
300	1.177	1.006	18.46×10^{-6}	0.026	0.708
310	1.139	1.007	18.93×10^{-6}	0.027	0.705
320	1.103	1.007	19.39×10^{-6}	0.028	0.703
330	1.070	1.008	19.85×10^{-6}	0.029	0.701
340	1.038	1.008	20.30×10^{-6}	0.029	0.699
350	1.008	1.009	20.75×10^{-6}	0.030	0.697
400	0.882	1.014	22.86×10^{-6}	0.034	0.689
450	0.784	1.021	24.85×10^{-6}	0.037	0.684
500	0.706	1.030	26.70×10^{-6}	0.040	0.680
550	0.642	1.040	28.48×10^{-6}	0.044	0.680
600	0.588	1.051	30.17×10^{-6}	0.047	0.680
700	0.504	1.075	33.32×10^{-6}	0.052	0.684
800	0.441	1.099	36.24×10^{-6}	0.058	0.689
900	0.392	1.121	38.97×10^{-6}	0.063	0.696
1000	0.353	1.142	41.53×10^{-6}	0.068	0.702

Source: Data reprinted with permission from Hewitt, G.F., et al., eds., *International Encyclopedia of Heat and Mass Transfer*, CRC Press, Boca Raton, FL, 1997; Weast, R.C., ed., *CRC Handbook of Tables for Applied Engineering Science*, CRC Press, 1970.

TABLE A5

Properties of Other Gases (1 atm, 298 K)

Gas	Density, ρ (kg/m^3)	Specific Heat, c_p (kJ/kgK)	Gas Constant (J/kg°C)	Thermal Conductivity, k (W/mK)	Dynamic Viscosity, μ (kg/msec)
Acetylene, C_2H_2	1.075	1.674	319	0.024	1.0×10^{-5}
Ammonia, NH_3	0.699	2.175	488	0.026	1.0×10^{-5}
Argon, Ar	1.608	0.523	208	0.0172	2.0×10^{-5}
n-Butane, C_4H_{10}	2.469	1.675	143	0.017	0.7×10^{-5}
Carbon dioxide, CO_2	1.818	0.876	189	0.017	1.4×10^{-5}
Carbon monoxide, CO	1.144	1.046	297	0.024	1.8×10^{-5}
Chlorine, Cl_2	2.907	0.477	117	0.0087	1.4×10^{-5}
Ethane, C_2H_6	1.227	1.715	276	0.017	9.5×10^{-5}
Ethylene, C_2H_4	0.072	1.548	296	0.017	1.0×10^{-5}
Fluorine, F_2	0.097	0.828	219	0.028	2.4×10^{-5}
Helium, He	0.164	5.188	2077	0.149	2.0×10^{-5}
Hydrogen, H_2	0.083	14.310	4126	0.0182	0.9×10^{-5}
Hydrogen sulfide, H_2S	10.753	0.962	244	0.014	1.3×10^{-5}
Methane, CH_4	0.662	2.260	518	0.035	1.1×10^{-5}
Methyl chloride, CH_3Cl	2.165	0.837	165	0.010	1.1×10^{-5}
Nitric oxide, NO	1.229	0.983	277	0.026	1.9×10^{-5}
Nitrogen, N_2	1.147	1.040	297	0.026	1.8×10^{-5}
Nitrous oxide, N_2O	1.802	0.879	189	0.017	1.5×10^{-5}
Oxygen, O_2	1.309	0.920	260	0.026	2.0×10^{-5}
Ozone, O_3	1.965	0.820	173	0.033	1.3×10^{-5}
Propane, C_3H_8	1.812	1.630	188	0.017	8.0×10^{-5}
Propylene, C_3H_6	1.724	1.506	197	0.017	8.5×10^{-5}
Sulfur dioxide, SO_2	2.622	0.460	130	0.010	1.3×10^{-5}
Xenon, Xe	5.375	0.481	63.5	0.0052	2.3×10^{-5}

Source: Data reprinted with permission from Weast, R.C., ed., *CRC Handbook of Tables for Applied Engineering Science*, CRC Press, 1970.

Properties of Other Gases (1 atm, 298 K)

Gas	Boiling Point (°C)	Latent Heat of Evaporation (kJ/kg)	Melting Point (°C)	Latent Heat of Fusion (kJ/kg)	Heat of Combustion (kJ/kg)
Acetylene, C_2H_2	−75	614.0	−82.2	53.5	50,200
Ammonia, NH_3	−33.3	1,373.0	−77.7	332.3	—
Argon, Ar	−186	163.0			—
n-Butane, C_4H_{10}	−0.4	386.0	−138	44.7	49,700
Carbon dioxide, CO_2	−78.5	572.0			—
Carbon monoxide, CO	−191.5	216.0	−205		10,100
Chlorine, Cl_2	−34.0	288.0	−101	95.4	—
Ethane, C_2H_6	−88.3	488.0	−172.2	95.3	51,800
Ethylene, C_2H_4	−103.8	484.0	−169	120.0	47,800

Gas	Boiling Point (°C)	Latent Heat of Evaporation (kJ/kg)	Melting Point (°C)	Latent Heat of Fusion (kJ/kg)	Heat of Combustion (kJ/kg)
Fluorine, F_2	– 188.0	172.0	– 220	25.6	—
Helium, He	4.22 K	23.3			—
Hydrogen, H_2	20.4 K	447.0	– 259.1	58.0	144,000
Hydrogen sulfide, H_2S	– 60	544.0	– 84	70.2	18,600
Methane, CH_4		510.0	– 182.6	32.6	5,327
Methyl chloride, CH_3Cl	– 23.7	428.0	– 97.8	130.0	
Nitric oxide, NO	– 151.5		– 161	76.5	—
Nitrogen, N_2	– 195.8	199.0	– 210	25.8	—
Nitrous oxide, N_2O	– 88.5	376.0	– 90.8	149.0	—
Oxygen, O_2	– 182.97	213.0	– 218.4	13.7	—
Ozone, O_3	– 112.0		– 193	226.0	—
Propane, C_3H_8	– 42.2	428.0	– 189.9	44.4	50,340
Propylene, C_3H_6	– 48.3	438.0	– 185		50,000
Sulfur dioxide, SO_2	– 10.0	362.0	– 75.5	135.0	—
Xenon, Xe	108.0	96.0	– 140	23.3	—

Source: Data reprinted with permission from Weast, R.C., ed., *CRC Handbook of Tables for Applied Engineering Science*, CRC Press, 1970.

Properties of Other Gases (Effects of Temperature)

Gas	Temperature, T (°C)	Density, ρ (kg/m³)	Specific Heat, c_p (kJ/kgK)	Thermal Conductivity, k (W/mK)	Dynamic Viscosity, μ (kg/msec)
Ammonia, NH_3	0	0.956	2.176	0.022	9.18×10^{-6}
	20	0.894	2.176	0.024	9.82×10^{-6}
	50	0.811	2.176	0.027	1.09×10^{-5}
Argon, Ar	– 13	1.87	0.523	0.016	2.04×10^{-5}
	– 3	1.81	0.523	0.016	2.11×10^{-5}
	7	1.74	0.523	0.017	2.17×10^{-5}
	27	1.62	0.523	0.018	2.30×10^{-5}
	77	1.39	0.519	0.020	2.59×10^{-5}
	227	0.974	0.519	0.026	3.37×10^{-5}
	727	0.487	0.519	0.043	5.42×10^{-5}
	1227	0.325	0.519	0.055	7.08×10^{-5}
Butane, C_4H_{10}	0	2.59	1.591	0.013	6.84×10^{-6}
	100	1.90	2.026	0.023	9.26×10^{-6}
	200	1.50	2.454	0.036	1.17×10^{-5}
	300	1.24	2.812	0.052	1.40×10^{-5}
	400	1.05	3.127	0.069	1.64×10^{-5}
	500	0.916	3.402	0.090	1.87×10^{-5}
	600	0.812	3.642	0.113	2.11×10^{-5}

Gas	Temperature, T (°C)	Density, ρ (kg/m^3)	Specific Heat, c_p (kJ/kgK)	Thermal Conductivity, k (W/mK)	Dynamic Viscosity, μ (kg/msec)
Carbon dioxide, CO_2	−13	2.08	0.813	0.014	1.31×10^{-5}
	−3	2.00	0.823	0.014	1.36×10^{-5}
	7	1.93	0.832	0.015	1.40×10^{-5}
	17	1.86	0.842	0.016	1.45×10^{-5}
	27	1.80	0.851	0.017	1.49×10^{-5}
	77	1.54	0.898	0.020	1.72×10^{-5}
	227	1.07	1.014	0.034	2.32×10^{-5}
Carbon monoxide, CO	−13	1.31	1.041	0.022	1.59×10^{-5}
	−3	1.27	1.041	0.023	1.64×10^{-5}
	7	1.22	1.041	0.024	1.69×10^{-5}
	17	1.18	1.041	0.025	1.74×10^{-5}
	27	1.14	1.041	0.025	1.79×10^{-5}
	77	0.975	1.043	0.029	2.01×10^{-5}
	227	0.682	1.064	0.039	2.61×10^{-5}
Ethane, C_2H_6	0	1.342	1.646	0.019	8.60×10^{-6}
	100	0.983	2.066	0.032	1.14×10^{-5}
	200	0.776	2.488	0.047	1.41×10^{-5}
	300	0.640	2.868	0.065	1.68×10^{-5}
	400	0.545	3.212	0.085	1.93×10^{-5}
	500	0.474	3.517	0.108	2.20×10^{-5}
	600	0.420	3.784	0.132	2.45×10^{-5}
Ethanol, C_2H_5OH	100	1.49	1.686	0.023	1.08×10^{-5}
	200	1.18	2.008	0.035	1.37×10^{-5}
	300	0.974	2.318	0.050	1.67×10^{-5}
	400	0.828	2.611	0.067	1.97×10^{-5}
	500	0.720	2.891	0.086	2.26×10^{-5}
Helium, He	0	0.368	5.146	0.142	1.86×10^{-5}
	20	0.167	5.188	0.149	1.94×10^{-5}
	40	0.156	5.188	0.155	2.03×10^{-5}
Hydrogen, H_2	−13	0.0944	14.133	0.162	8.14×10^{-6}
	−3	0.0910	14.175	0.167	8.35×10^{-6}
	7	0.0877	14.226	0.172	8.55×10^{-6}
	27	0.0847	14.267	0.177	8.76×10^{-6}
	77	0.0819	14.301	0.182	8.96×10^{-6}
	727	0.04912	14.506	0.272	1.26×10^{-5}
Methane, CH_4	0	0.716	2.164	0.031	1.04×10^{-5}
	100	0.525	2.447	0.046	1.32×10^{-5}
	200	0.414	2.805	0.064	1.59×10^{-5}
	300	0.342	3.173	0.082	1.83×10^{-5}
	400	0.291	3.527	0.102	2.07×10^{-5}
	500	0.253	3.853	0.122	2.29×10^{-5}
	600	0.224	4.150	0.144	2.52×10^{-5}

Gas	Temperature, T (°C)	Density, ρ (kg/m^3)	Specific Heat, c_p (kJ/kgK)	Thermal Conductivity, k (W/mK)	Dynamic Viscosity, μ (kg/msec)
Nitrogen, N_2	77	1.14	1.041	0.026	1.79×10^{-5}
	227	9.75	1.042	0.030	2.00×10^{-5}
	727	6.82	1.056	0.040	2.57×10^{-5}
Oxygen, O_2	−13	1.50	0.915	0.023	1.85×10^{-5}
	−3	1.45	0.916	0.024	1.90×10^{-5}
	7	1.39	0.918	0.025	1.96×10^{-5}
	27	1.35	0.918	0.026	2.01×10^{-5}
	77	1.30	0.920	0.027	2.06×10^{-5}
	227	1.11	0.929	0.031	2.32×10^{-5}
	727	7.80	0.972	0.042	2.99×10^{-5}
Propane, C_3H_8	0	1.97	1.548	0.015	7.50×10^{-6}
	100	1.44	2.015	0.026	1.00×10^{-5}
	200	1.14	2.456	0.040	1.25×10^{-5}
	300	0.939	2.833	0.056	1.40×10^{-5}
	400	0.799	3.159	0.074	1.72×10^{-5}
	500	0.694	3.446	0.095	1.94×10^{-5}
	600	0.616	3.695	0.118	2.18×10^{-5}

Source: Data reprinted with permission from Weast, R.C., ed., *CRC Handbook of Tables for Applied Engineering Science*, CRC Press, 1970.

TABLE A6(a)

Properties of Liquids (300 K, 1 atm, except [a] at 297 K)

Liquid	Density, ρ (kg/m^3)	Specific Heat, c_p (kJ/kgK)	Dynamic Viscosity, μ (kg/msec)	Thermal Conductivity, k (W/mK)	Freezing Point (K)
Acetic acid	1049	2.18	0.001155	0.171	290
Acetone	784.6	2.15	0.000316	0.161	179.0
Alcohol, ethyl	785.1	2.44	0.001095	0.171	158.6
Alcohol, methyl	786.5	2.54	0.00056	0.202	175.5
Alcohol, propyl	800.0	2.37	0.00192	0.161	146
Ammonia	823.5	4.38		0.353	
Benzene	873.8	1.73	0.000601	0.144	278.68
Bromine		0.473	0.00095		245.84
Carbon disulfide	1261	0.992	0.00036	0.161	161.2
Carbon tetrachloride	1584	0.866	0.00091	0.104	250.35
Castor oil	956.1	1.97	0.650	0.180	263.2
Chloroform	1465	1.05	0.00053	0.118	209.6
Decane	726.3	2.21	0.000859	0.147	243.5
Dodecane	754.6	2.21	0.001374	0.140	247.18
Ether	713.5	2.21	0.000223	0.130	157
Ethylene glycol	1097	2.36	0.0162	0.258	260.2
Fluorine, R-11	1476	0.870[a]	0.00042	0.093[a]	162
Fluorine, R-12	1311	0.971[a]		0.071[a]	115
Fluorine, R-22	1194	1.26[a]		0.086[a]	113
Glycerine	1259	2.62	0.950	0.287	264.8
Heptane	679.5	2.24	0.000376	0.128	182.54
Hexane	654.8	2.26	0.000297	0.124	178.0
Iodine		2.15			386.6
Kerosene	820.1	2.09	0.00164	0.145	
Linseed oil	929.1	1.84	0.0331		253
Mercury		0.139	0.00153		234.3
Octane	698.6	2.15	0.00051	0.131	216.4
Phenol	1072	1.43	0.0080	0.190	316.2
Propane	493.5	2.41[a]	0.00011		85.5
Propylene	514.4	2.85	0.00009		87.9
Propylene glycol	965.3	2.50	0.042		213
Sea water	1025	3.76–4.10			270.6
Toluene	862.3	1.72	0.000550	0.133	178
Turpentine	868.2	1.78	0.001375	0.121	214
Water	997.1	4.18	0.00089	0.609	273

Source: Data reprinted with permission from Weast, R.C., ed., *CRC Handbook of Tables for Applied Engineering Science*, CRC Press, 1970.

TABLE A6(b)

Properties of Liquids (300 K, 1 atm, except [a] at 297 K)

Liquid	Latent Heat of Fusion (kJ/kg)	Boiling Point (°C)	Latent Heat of Evaporation (kJ/kg)	Coefficient of Expansion (1/K)
Acetic acid	181	391	402	0.0011
Acetone	98.3	329	518	0.0015
Alcohol, ethyl	108	351.46	846	0.0011
Alcohol, methyl	98.8	337.8	1100	0.0014
Alcohol, propyl	86.5	371	779	
Benzene	126	353.3	390	0.0013
Bromine	66.7	331.6	193	0.0012
Carbon disulfide	57.6	319.40	351	0.0013
Carbon tetrachloride	174	349.6	194	0.0013
Chloroform	77.0	334.4	247	0.0013
Decane	201	447.2	263	
Dodecane	216	489.4	256	
Ether	96.2	307.7	372	0.0016
Ethylene glycol	181	470	800	
Fluorine, R-11		297.0	180.0	
Fluorine, R-12	34.4	243.4	165	
Fluorine, R-22	183	232.4	232	
Glycerine	200	563.4	974	0.00054
Heptane	140	371.5	318	
Hexane	152	341.84	365	
Iodine	62.2	457.5	164	
Kerosene			251	
Linseed oil		560		
Mercury	11.6	630	295	0.00018
Octane	181	398	298	0.00072
Phenol	121	455		0.00090
Propane	79.9	231.08	428	
Propylene	71.4	225.45	342	
Propylene glycol		460	914	
Toluene	71.8	383.6	363	
Turpentine		433	293	0.00099
Water	333	373	2260	0.00020

Source: Data reprinted with permission from Weast, R.C., ed., *CRC Handbook of Tables for Applied Engineering Science*, CRC Press, 1970.

TABLE A6(c)

Properties of Saturated Water

T (°C)	P (kPa)	ρ_f (kg/m^3)	ρ_v (kg/m^3)	h_{fg} (kJ/kg)	$c_{p,f}$ (kJ/kgK)	$\mu_f \cdot 10^6$ (kg/msec)	k_f (W/mK)	Pr_f	σ_f (N/m)
0.01	0.612	999.8	0.005	2501	4229	1791	0.561	13.50	0.0757
10	1.228	999.7	0.009	2477	4188	1308	0.580	9.444	0.0742
20	2.339	998.2	0.017	2453	4182	1003	0.598	7.010	0.0727
30	4.246	995.6	0.030	2430	4182	798	0.615	5.423	0.0712
40	7.381	992.2	0.051	2406	4183	653	0.631	4.332	0.0696
50	12.34	988.0	0.083	2382	4181	547.1	0.644	3.555	0.0680
60	19.93	983.2	0.130	2358	4183	466.8	0.654	2.984	0.0662
70	31.18	977.8	0.198	2333	4187	404.5	0.663	2.554	0.0645
80	47.37	971.8	0.293	2308	4196	355.0	0.670	2.223	0.0627
90	70.12	965.3	0.423	2283	4205	315.1	0.675	1.962	0.0608
100	101.3	958.4	0.597	2257	4217	282.3	0.679	1.753	0.0589
110	143.2	951.0	0.826	2230	4233	255.1	0.682	1.584	0.0570
120	198.5	943.2	1.121	2202	4249	232.2	0.683	1.444	0.0550
130	270.0	934.9	1.495	2174	4267	212.8	0.684	1.328	0.0529
140	361.2	926.2	1.965	2145	4288	196.3	0.683	1.232	0.0509
150	475.7	917.1	2.545	2114	4314	182.0	0.682	1.151	0.0488
160	617.7	907.5	3.256	2082	4338	169.6	0.680	1.082	0.0466
170	791.5	897.5	4.118	2049	4368	158.9	0.677	1.025	0.0444
180	1001.9	887.1	5.154	2015	4404	149.4	0.673	0.977	0.0422
190	1254.2	876.2	6.390	1978	4444	141.0	0.669	0.937	0.0400
200	1553.7	864.7	7.854	1940	4489	133.6	0.663	0.904	0.0377
220	2317.8	840.3	11.61	1858	4602	121.0	0.650	0.857	0.0331
240	3344.7	813.5	16.74	1766	4759	110.5	0.632	0.832	0.0284
260	4689.5	783.8	23.70	1662	4971	101.5	0.609	0.828	0.0237
280	6413.2	750.5	33.15	1543	5279	93.4	0.581	0.848	0.0190
300	8583.8	712.4	46.15	1405	5751	85.8	0.548	0.901	0.0144
320	11279	667.4	64.6	1239	6536	78.4	0.509	1.006	0.0099
340	14594	610.8	92.7	1028	8241	70.3	0.469	1.236	0.0056
360	18655	528.1	143.7	721	14686	60.2	0.428	2.068	0.0019
373	21799	402.4	242.7	276	21828	46.7	0.545	18.69	0.0001

Source: Data reprinted with permission from Hewitt, G.F., et al., eds., *International Encyclopedia of Heat and Mass Transfer*, CRC Press, Boca Raton, FL, 1997.

TABLE A7

Properties of Liquid Metals (1 atm)

Metal (Melting Point, °C)	T (°F)	T (°C)	Density, ρ (kg/m^3)	Specific Heat, c_p (kJ/kgK)	Thermal Conductivity, k (W/mK)	Dynamic Viscosity, μ (kg/msec)
Aluminum (1220)	1300	704	2370	1.084	104.2	2.8×10^{-3}
	1350	732	2360	1.084	109.7	2.4×10^{-3}
	1400	760	2350	1.084	111.3	2.0×10^{-3}
	1450	788	2340	1.084	121.0	1.6×10^{-3}
Bismuth (520)	600	316	10000	0.144	16.4	1.62×10^{-3}
	800	427	9870	0.149	15.6	1.34×10^{-3}
	1000	538	9740	0.154	15.6	1.10×10^{-3}
	1200	649	9610	0.159	15.6	0.923×10^{-3}
Lead (621)	700	371	10500	0.159	16.1	2.39×10^{-3}
	850	454	10400	0.155	15.6	2.05×10^{-3}
	1000	538	10400	0.155	15.4	1.74×10^{-3}
	1150	621	10200	0.155	15.1	1.52×10^{-3}
Lithium (355)	400	204	506	4.184	41.5	0.595×10^{-3}
	600	316	497	4.184	39.8	0.506×10^{-3}
	800	427	489	4.184	38.1	0.551×10^{-3}
Mercury (−38)	50	10	13600	0.138	8.3	1.59×10^{-3}
	200	93	13400	0.138	10.4	1.25×10^{-3}
	300	149	13200	0.138	11.6	1.10×10^{-3}
	400	204	13100	0.134	12.5	0.997×10^{-3}
	600	316	12800	0.134	14.0	0.863×10^{-3}
Tin (449)	500	260	6940	0.243	32.9	1.82×10^{-3}
	700	371	6860	0.251	33.6	1.46×10^{-3}
	850	454	6810	0.259	32.9	1.26×10^{-3}
	1000	538	6740	0.268	32.9	1.13×10^{-3}
	1200	649	6680	0.276	32.9	0.997×10^{-3}
Zinc (787)	850	454	6900	0.498	58.3	3.12×10^{-3}
	1000	538	6860	0.485	57.5	2.56×10^{-3}
	1200	649	6760	0.473	56.8	2.07×10^{-3}
	1500	816	6740	0.448	56.4	1.46×10^{-3}

Source: Data reprinted with permission from Weast, R.C., ed., *CRC Handbook of Tables for Applied Engineering Science*, CRC Press, 1970.

TABLE A8

Diffusion of Gases into Air (1 atm)

Substance	Diffusion Coefficient, D (m^2/sec) at 0°C	Diffusion Coefficient, D (m^2/sec) at 25°C	Schmidt Number, Sc (ν/D) at 0°C	Schmidt Number, Sc (ν/D) at 25°C
Hydrogen	6.11×10^{-5}	7.12×10^{-5}	0.217	0.216
Ammonia	1.98×10^{-5}	2.29×10^{-5}	0.669	0.673
Nitrogen	1.78×10^{-5}		0.744	
Oxygen	1.78×10^{-5}	2.06×10^{-5}	0.744	0.748
Carbon dioxide	1.42×10^{-5}	1.64×10^{-5}	0.933	0.940
Methyl alcohol	1.32×10^{-5}	1.59×10^{-5}	1.00	0.969
Formic acid	1.31×10^{-5}	1.59×10^{-5}	1.01	0.969
Acetic acid	1.06×10^{-5}	1.33×10^{-5}	1.25	1.16
Ethyl alcohol	1.02×10^{-5}	1.19×10^{-5}	1.30	1.29
Chloroform	9.1×10^{-6}		1.46	
Diethylamine	8.84×10^{-6}	1.05×10^{-5}	1.50	1.47
n-Propyl alcohol	8.5×10^{-6}	1.00×10^{-5}	1.56	1.54
Propionic acid	8.46×10^{-6}	9.9×10^{-6}	1.57	1.56
Methyl acetate	8.40×10^{-6}	1.00×10^{-5}	1.58	1.54
Butylamine	8.21×10^{-6}	1.01×10^{-5}	1.61	1.53
Ethyl Ether	7.86×10^{-6}	9.3×10^{-6}	1.69	1.66
Benzene	7.51×10^{-6}	8.8×10^{-6}	1.76	1.75
Ethyl acetate	7.15×10^{-6}	8.5×10^{-6}	1.85	1.81
Toluene	7.09×10^{-6}	8.4×10^{-6}	1.87	1.83
n-Butyl alcohol	7.03×10^{-6}	9.0×10^{-6}	1.88	1.71
i-Butyric acid	6.79×10^{-6}	8.1×10^{-6}	1.95	1.90
Chlorobenzene		7.3×10^{-6}		2.11
Aniline	6.10×10^{-6}	7.2×10^{-6}	2.17	2.14
Xylene	5.9×10^{-6}	7.1×10^{-6}	2.25	2.17
Amyl alcohol	5.89×10^{-6}	7.0×10^{-6}	2.25	2.20
n-octane	5.05×10^{-6}	6.0×10^{-6}	2.62	2.57
Naphthalene	5.13×10^{-6}	5.2×10^{-6}	2.58	2.96

Source: Data reprinted with permission from Hewitt et al., Eds., *International Encyclopedia of Heat and Mass Transfer*, CRC Press, Boca Raton, FL, 1997; Weast, R.C., ed., *CRC Handbook of Tables for Applied Engineering Science*, CRC Press, 1970.

TABLE A9(a)

Radiative Properties of Selected Materials

Total Radiation Emissivities

Material	0–38°C 32–100°F	260–538°C 500–1000°F
Metallic Materials (Clean, Dry)		
Polished aluminum, silver, gold, brass, tin	0.02–0.024	0.03–0.10
Polished brass, copper, steel, nickel	0.03–0.08	0.06–0.2
Polished chromium, platinum, mercury	0.03–0.08	0.06–0.2
Dull, smooth, clean aluminum and alloys	0.08–0.20	0.15–0.45
Dull, smooth, clean copper, brass, nickel, iron	0.08–0.20	0.15–0.45
Dull, smooth, clean stainless steel, lead, zinc	0.08–0.20	0.15–0.45
Rough-ground, smooth-machined castings	0.15–0.25	0.3–0.65
Steel mill products, sprayed metal, molten metal	0.15–0.25	0.3–0.65
Smooth, slightly oxidized aluminum, copper	0.2–0.4	0.3–0.7
Smooth, slightly oxidized brass, lead, zinc	0.2–0.4	0.3–0.7
Bright aluminum, gilt, bronze paints	0.3–0.55	0.4–0.7
Heavily oxidized and rough iron, steel	0.6–0.85	0.7–0.9
Heavily oxidized and rough copper, aluminum	0.6–0.85	0.7–0.9
Nonmetallic Materials		
White or light-colored paint, plaster, brick, porcelain	0.80–0.95	0.6–0.85
White or light-colored tile, paper, plastics, asbestos	0.80–0.95	0.6–0.85
Medium red, brown, green, and other colors of paint, brick, tile, inks, clays, stone, concrete, wood, water	0.85–0.95	0.70–0.85
Glass and translucent plastics, oil, varnish, ice	0.85–0.95	0.75–0.95
Carbon black, tar, asphalt, matte-black paints	0.90–0.97	0.90–0.97

Source: Data reprinted with permission from Weast, R.C., ed., *CRC Handbook of Tables for Applied Engineering Science*, CRC Press, 1970.

TABLE A9(b)

Normal Emissivities of Glass

Type of Glass	50°C	253°C	451°C	693°C
Thickness of 1/4 in. (6.35 mm)				
Borosilicate, low-expansion	0.89	0.90	0.88	0.78
96% silica	0.87	0.81	0.72	0.56
Soda-lime plate	0.91	0.91	0.88	0.71
Thickness of 1/2 in. (12.7 mm)				
Borosilicate, low expansion	0.89	0.90	0.89	0.81
96% silica	0.87	0.83	0.76	0.62
Soda-lime plate	0.91	0.92	0.90	0.83

Source: Data reprinted with permission from Weast, R.C., ed., *CRC Handbook of Tables for Applied Engineering Science*, CRC Press, 1970.

TABLE A9(c)

Total Solar Absorptivities

Surface Material	0.3–2.5 μm
White surfaces: paint, paper, plaster, plastics, fresh snow	0.1 – 0.3
Light-colored surfaces: paint, paper, textiles, stone, dry grass	0.25 – 0.5
Light-colored surfaces: concrete, wood, sand, bricks, plastics	0.25 – 0.5
Darker colors: paint, inks, brick, tile, slate, soil, rusted iron	0.4 – 0.8
Black asphalt, tar, slate, carbon, rubber, water	0.85 – 0.95
Clean, dark metals: iron and steel, lead, zinc; metallic paints	0.2 – 0.5
Polished, bright metals: aluminum, silver, magnesium	0.07 – 0.3
Polished, bright metals: tin, copper, chromium, nickel	0.07 – 0.3

Source: Data reprinted with permission from Weast, R.C., ed., *CRC Handbook of Tables for Applied Engineering Science*, CRC Press, 1970.

TABLE A9(d)

Ratios of Emissivity to Absorptivity

Surface Material	Emissivity (298 K)	Solar Absorptivity
Highly polished (white) metals, gold, yellow brass	0.02–0.08	0.1–0.4
Clean (dark) metals	0.1–0.35	0.3–0.6
Metallic-pigment paints	0.35–0.55	0.4–0.6
White, nonmetal surfaces	0.7–0.9	0.1–0.35
Dark-colored nonmetals	0.7–0.9	0.45–0.8
Black paint, asphalt, carbon, water	0.85–0.95	0.7–0.9

Source: Data reprinted with permission from Weast, R.C., ed., *CRC Handbook of Tables for Applied Engineering Science*, CRC Press, 1970.

TABLE A9(e)

Reflectivities of Various Surfaces

Reflector Surface (Zero Transmissivity)	**Reflectivity**
Polished silver, clean	0.95
Aluminized glass, front surface	0.92
Silvered mirror, back surface	0.88
Polished aluminum, specular	0.83
White porcelain or plastic, enamel	0.78
Smooth aluminum, diffuse	0.76
White paint, gloss	0.75
Chrome plate, specular	0.65
Stainless steel, specular	0.60
Bright aluminum paint	0.60

Source: Data reprinted with permission from Weast, R.C., ed., *CRC Handbook of Tables for Applied Engineering Science*, CRC Press, 1970.

TABLE A9(f)

Transmissivities of Various Surfaces

Diffuser or Enclosure	Transmissivity
Thin quartz or silica	0.90
Clear glass or plastic (1/8 in.)	0.90
Ground or frosted glass	0.75
Opal-white glass	0.50
Heat-absorbing plate glass (1/4 in.)	0.60

Source: Data reprinted with permission from Weast, R.C., ed., *CRC Handbook of Tables for Applied Engineering Science*, CRC Press, 1970.

TABLE A9(g)

Total Emissivity of Pure Carbon Dioxide and Water Vapor

$P \times L$ (ft × atm) (total pressure 1 atm)	500 R 4°C	1000 R 282°C	1500 R 560°C	2000 R 838°C	2500 R 1116°C	3000 R 1393°C
CO_2						
0.01	0.03	0.03	0.03	0.03	0.02	0.01
0.02	0.04	0.04	0.04	0.04	0.03	0.02
0.05	0.06	0.06	0.06	0.06	0.05	0.04
0.10	0.08	0.07	0.07	0.07	0.06	0.05
0.20	0.10	0.09	0.09	0.09	0.08	0.07
0.50	0.13	0.12	0.12	0.12	0.11	0.09
1.0	0.15	0.14	0.14	0.14	0.13	0.12
2.0	0.17	0.16	0.16	0.16	0.16	0.15
5.0	0.20	0.19	0.19	0.19	0.19	0.18
H_2O						
0.02	0.06	0.04	0.02	0.02	0.01	—
0.05	0.10	0.07	0.06	0.04	0.03	0.02
0.10	0.15	0.11	0.08	0.06	0.05	0.04
0.20	0.20	0.16	0.12	0.10	0.08	0.06
0.50	0.30	0.24	0.20	0.16	0.14	0.12
1.0	0.37	0.32	0.27	0.23	0.19	0.16
2.0	0.44	0.39	0.35	0.30	0.26	0.22
5.0	0.55	0.50	0.47	0.41	0.36	0.31

Source: Data reprinted with permission from Weast, R.C., ed., *CRC Handbook of Tables for Applied Engineering Science*, CRC Press, 1970.

TABLE A9(h)

Total Emissivities of Gases (Dependence on Partial Pressure)

Gas	*T* (R)	*T* (K)	*P* × *L* (0.01 atm × ft)	*P* × *L* (0.05 atm × ft)	*P* × *L* (0.1 atm × ft)	*P* × *L* (0.5 atm × ft)	*P* × *L* (1.0 atm × ft)
H_2O	1000	538	0.02	0.07	0.11	0.24	0.32
	1500	816	0.01	0.06	0.08	0.20	0.27
	2000	1093	—	0.04	0.06	0.16	0.23
	2500	1371	—	0.03	0.05	0.14	0.19
CO_2	1000	538	0.03	0.06	0.07	0.12	0.14
	1500	816	0.03	0.06	0.07	0.12	0.14
	2000	1093	0.03	0.06	0.07	0.12	0.14
	2500	1371	0.02	0.05	0.06	0.11	0.13
CH_4	1000	538	0.02	0.04	0.06	0.12	0.17
	1500	816	0.02	0.05	0.07	0.15	0.19
	2000	1093	0.02	0.05	0.07	0.15	0.19
	2500	1371	0.02	0.05	0.06	0.14	0.18
NH_3	1000	538	0.05	0.14	0.20	0.50	0.60
	1500	816	0.02	0.08	0.13	0.34	0.47
	2000	1093	0.01	0.04	0.07	0.20	0.30
CO	1000	538	0.01	0.02	0.03	0.05	0.06
	1500	816	0.02	0.04	0.05	0.08	0.10
	2000	1093	0.02	0.04	0.05	0.05	0.09
SO_2	1000	538	0.02	0.08	0.13	0.28	0.35
	1500	816	0.01	0.06	0.10	0.24	0.32
	2000	1093	0.01	0.04	0.07	0.20	0.28
	2500	1093	0.01	0.03	0.05	0.15	0.23

Source: Data reprinted with permission from Weast, R.C., ed., *CRC Handbook of Tables for Applied Engineering Science*, CRC Press, 1970.

TABLE A10

Atomic Weights of Elements

Element	kg/kmol	Element	kg/kmol	Element	kg/kmol
Hydrogen, H	1.01	Bromine, Br	79.90	Thulium, Tm	168.93
Helium, He	4.00	Krypton, Kr	83.80	Ytterbium, Yb	173.04
Lithium, Li	6.94	Rubidium, Rb	85.47	Lutetium, Lu	174.97
Beryllium, Be	9.01	Strontium, Sr	87.62	Hafnium, Hf	178.49
Boron, B	10.81	Yttrium, Y	88.91	Tantalum, Ta	180.95
Carbon, C	12.01	Zirconium, Zr	91.22	Tungsten, W	183.84
Nitrogen, N	14.01	Niobium, Nb	92.91	Rhenium, Re	186.21
Oxygen, O	16.00	Molybdenum, Mo	95.94	Osmium, Os	190.23
Fluorine, F	19.00	Technetium, Tc	97.91	Iridium, Ir	192.22
Neon, Ne	20.18	Ruthenium, Ru	101.07	Platinum, Pt	195.08
Sodium, Na	22.99	Rhodium, Rh	102.91	Gold, Au	196.97
Magnesium, Mg	24.31	Palladium, Pd	106.42	Mercury, Hg	200.59
Aluminum, Al	26.98	Silver, Ag	107.87	Thellium, Tl	204.38
Silicon, Si	28.09	Cadmium, Cd	112.41	Lead, Pb	207.2
Phosphorus, P	30.97	Indium, In	114.82	Bismuth, Bi	208.98
Sulfur, S	32.07	Tin, Sn	118.71	Polonium, Po	208.98
Chlorine, Cl	35.45	Antimony, Sb	121.76	Astatine, At	209.99
Argon, Ar	39.95	Tellerium, Te	127.60	Radon, Rn	222.02
Potassium, K	39.10	Iodine, I	126.90	Francium, Fr	223.02
Calcium, Ca	40.08	Xenon, Xe	131.29	Radium, Ra	226.03
Scandium, Sc	44.96	Caesium, Cs	132.91	Actinium, Ac	227.03
Titanium, Ti	47.87	Barium, Ba	137.33	Thorium, Th	232.04
Vanadium, V	50.94	Lanthanum, La	138.91	Protactinium, Pa	231.04
Chromium, Cr	52.00	Cerium, Ce	140.12	Uranium, U	238.03
Manganese, Mn	54.94	Praseodymium, Pr	140.91	Neptunium, Np	237.05
Iron, Fe	55.85	Neodymium, Nd	144.24	Plutonium, Pu	239.05
Cobalt, Co	58.93	Promethium, Pm	144.91	Americium, Am	243.06
Nickel, Ni	58.69	Samarium, Sm	150.36	Curium, Cm	247.07
Copper, Cu	63.55	Europium, Eu	151.97	Berkelium, Bk	249.07
Zinc, Zn	65.39	Gadolinium, Gd	157.25	Californium, Cf	251.08
Gallium, Ga	69.72	Terbium, Tb	158.93	Einsteinium, Es	252.08
Germanium, Ge	72.61	Dysprosium, Dy	162.50	Fermium, Fm	257.10
Arsenic, As	74.92	Holmium, Ho	164.93	Mendelevium, Md	258.10
Selenium, Se	78.96	Erbium, Er	167.26	Nobelium, No	259.10

Source: Data reprinted with permission from Hewitt, G.F., et al., eds., *International Encyclopedia of Heat and Mass Transfer*, CRC Press, Boca Raton, FL, 1997.

References

G.F. Hewitt, G.L. Shires, and Y.V. Polezhaev, eds., *International Encyclopedia of Heat and Mass Transfer*, Boca Raton, FL: CRC Press, 1997.

G.A. Lane. *Solar Heat Storage: Latent Heat Materials*, vols. 1 and 2, Boca Raton, FL: CRC Press, 1996.

R.C. Weast, ed., *CRC Handbook of Tables for Applied Engineering Science*, Boca Raton, FL: CRC Press, 1970.

Index

C

D

E

F

G

H

I

M

N

O

P

Q

R

S

T

U

V

W

Z

Milton Keynes UK
Ingram Content Group UK Ltd.
UKHW021933071024
449327UK00022B/1786